Passiflora

PASSIFLORA

Passionflowers of the World

Torsten Ulmer
&
John M. MacDougal

with drawings by
Bettina Ulmer

TIMBER PRESS
Portland · Cambridge

Frontispiece: *Passiflora jilekii*.

Translated (in part) from the German by Klaus D. Balser.

Published in 2004 by

Timber Press, Inc.
The Haseltine Building
133 S.W. Second Avenue, Suite 450
Portland, Oregon 97204-3527, U.S.A.

Timber Press
2 Station Road
Swavesey
Cambridge CB4 5QJ, U.K.

www.timberpress.com

Designed by Susan Applegate
Printed in China

Library of Congress Cataloging-in-Publication Data

Ulmer, Torsten.
Passiflora : Passionflowers of the world / Torsten Ulmer and John M. MacDougal; with drawings by Bettina Ulmer.
p. cm.
Includes bibliographical references (p.).
ISBN 0-88192-648-5 (hardcover)
1. Passiflora. I. MacDougal, John Mochrie, 1954– II. Title.
SB413.P3U58 2004 635.9'33456—dc22 2003023792

A catalog record for this book is also available from the British Library.

Contents

Foreword

In his 1938 monograph of the American Passifloraceae, Ellsworth P. Killip lamented that "a study of *Passiflora* should be profusely illustrated." Alas, his wasn't! Not only are the flowers of this taxon amazingly complex and difficult to interpret, but the vegetative morphology is fantastically varied as well. As Killip noted, "The leaves are always alternate, but probably in no other group of plants is their shape more striking or the variation in outline of a single plant more extreme than in *Passiflora*." Yet Killip gives us only text and a very few line drawings from which to imagine unimaginable diversity across more than 525 species.

For many years a ragged photocopy of Killip's work was the main reference available to those of us attempting to identify unknown *Passiflora*. Things have started changing, however, with the newsletter of the Passiflora Society International providing illustrations of many species, and John Vanderplank's *Passion Flowers* illustrating many cultivated species, in addition to Torsten and Bettina Ulmer's German-language editions of *Passionsblumen*. With the publication of this new book, Torsten Ulmer and John M. MacDougal have made available outstanding images of the majority of passionflower species, many seen only a few times in the wild, and have organized them according to the most up-to-date phylogenetic hypothesis. The latter, along with a classification table, is provided in the chapter on systematics by Christian Feuillet and John MacDougal. This chapter's previously unpublished information is based on two decades of work by these authors and greatly increases the value of the book.

Specialty chapters by other authors add diversity and spice. The history of passionflowers is covered by Emil Kugler and Leslie King; Klaus Balser discusses pests and diseases; Jan Meerman, Michael Schwerdtfeger, and Reto Gamma contribute chapters on the *Passiflora* species of Belize, the Andes, and Brazil, respectively; and Roland Fischer contributes a chapter on interspecific hybrids. In his chapter, Ron Boender discusses *Heliconius* butterflies. These insects are parasites of *Passiflora* in that their larvae consume the leaves of the plants, yet they are a glorious reflection of the presence of *Passiflora* species in Neotropical forests and in butterfly houses worldwide. Many *Passiflora* enthusiasts, including Ron Boender, discovered these plants through an initial interest in *Heliconius* and other passionvine butterflies. I believe that Boender's chapter may encourage *Passiflora* specialists to develop an interest in the relationships these plants have with insects—beyond killing caterpillars!

Naturalists active in the late 20th and early 21st centuries have witnessed unprecedented losses of the earth's wild habitats and species. On the other hand, we have enjoyed unprecedented opportunities to travel quickly and cheaply to all parts of the earth to observe or collect, and sophisticated photographic technologies have allowed us to document previously unknown or little-known species and their environments. More than 140 species of *Passiflora* have been described since Killip's great work. These facts—plus the availability of Web-based global communications and exchanges of information—were exploited by the authors to assemble this wonderful collection of images, systematic knowledge, and descriptions of wild species and artificial hybrids of the genus *Passiflora*.

Many of us who work with these plants would have loved to have written a book such as this, yet we all are quite happy to make small contributions of information and photographs in order to enjoy a more extensive collective work. I congratulate my two colleagues for their important effort. This book succeeds and excels both scientifically and aesthetically. If you want to illustrate biodiversity to your friends, show them this book. If you want to encourage students to pursue botany or conservation biology, give them a copy.

Professor Lawrence E. Gilbert
Section of Integrative Biology
University of Texas at Austin

Preface

PASSIONFLOWERS have experienced a real renaissance during the last few years. Though restricted to a small circle of enthusiasts as recently as the early 1990s, these charming climbers have become popular plants in the meantime. The wide spectrum of cultivation requirements among passifloras means that anybody can enjoy these fascinating plants.

The inspiration for this book was the German-language *Passionsblumen* (Ulmer and Ulmer 1997), which met with such approval that an entirely new, much larger, English-language book, aimed at an international readership, recommended itself. This new book offers a comprehensive survey of the genus *Passiflora* and provides botanists, hobbyists, and passionflower enthusiasts alike with up-to-date knowledge. The emphasis is on the description of individual species, with their incredible diversity in flower form and color. Various recommendable hybrids are also presented.

In selecting the described plants, we took into account a large portion of the species cultivated in the United States and Europe, as well as botanical peculiarities known only from the plants' habitats. Individual species are not presented alphabetically in a single list but are assigned to the 4 subgenera included in separate chapters, in line with the new systematics by John MacDougal and Christian Feuillet (see chapter 2). Species are listed alphabetically within these chapters. Supersection *Tacsonia* is presented in a chapter of its own to pay tribute to the fascinating flowers of these high-mountain species from the Andes.

Every species description contains information about native range, classification in the systematics, cultivation requirements, and further interesting facts; a detailed leaf drawing illustrating leaf shape; a shorthand characterization stating the plant's essential features; and, in the color section, a photo of the flower. Measurements are provided length by width.

A book like this could only be produced with the help of many experts and enthusiasts from around the world. John MacDougal and I wish to offer our sincere thanks to the following people, who contributed chapters, provided color photos and valuable information, made available plants and seeds, and supported this project through a period of several years: Hilaire Annonay, Wilhelm Barthlott, Giselher Behnken, Paul E. Berry, Armando Carlos Cervi, Geo Coppens, Heinrich Drexler, Thomas Fechtler, Tassilo Feuerer, Christian Feuillet, Roland Fischer, Mindy Fitch, Harald Förther, Axel Frank, Reto Gamma, Helga Gelewsky, Günter Gerlach, Lawrence E. Gilbert, Klaus Gilmer, Gerrit Graat, Guillermo Guzmán, Katie Hansen, Christian Houel, Peter Møller Jørgensen, Alan Jouy, Karl-Heinz Kahlert, Uli Katz, Elma Kay, Leslie A. King,

Maria Klein, Sandra Knapp, Emil Kugler, Erwin Kühnen, Jürgen Lautner, Eldon Leiter, Gwilym Lewis, Annika Lindberg, Bob Magill, Neal Maillet, James Mallet, Luis A. Mazariegos, Jan Meerman, Brigitte Mußgay, Mauro Peixoto, Pierre Pomié, Irmi Reiber, Harri Reimann, Thomas Schneider, Anette Schumacher, Michael Schwerdtfeger, Sergio Segura, Jürgen Skrabal, Carmen Ulloa, João Vasconcellos-Neto, Roberto Vásquez Chavez, Maurizio Vecchia, Stefan Vogel, Maximilian Weigend, Henk Wouters, and our parents.

I would also like to thank Klaus D. Balser, without whose translation work this book would not have been possible. While working on the book, many difficulties had to be overcome, and I was always able to rely on his aid and his words of encouragement. He translated the text into English (except chapters 1, 2, 3, 9, 19, and parts of chapter 14, which were written in English by their authors) with meticulous care, and also contributed a chapter of his own. I am enormously grateful for his help.

I also wish to express my sincerest thanks to my coauthor, who brought his immense expert knowledge to bear in the project. Besides contributing outstanding chapters, he suggested numerous changes and additions, harmonized the text with the new systematics, and put the finishing touches to the manuscript. The improved quality of the contents is his merit.

I am very grateful to Patrick J. Worley as well, who put in a word for the book with Timber Press and thus paved the way for its publication.

My special thanks also go to Ron Boender, who supported the project from the very beginning. He put countless slides of passionflowers at my disposal and wrote the impressive chapter on *Heliconius* butterflies, another field of interest he is passionate about.

Finally, my thanks are due to my wife, Bettina, who supplied the drawings and took care of many daily duties, and to Annika, our daughter. Both often had to do without me in recent years.

Torsten Ulmer
Witten, Germany

Introduction

Passionflowers may well be the most fascinating plants of the tropics. Their unique, almost lavishly beautiful flowers have a mysterious look and convey an exotic ambiance. The genus, with its extraordinary diversity in forms and colors, arouses an enormous passion for collecting among enthusiasts. Many species are easily cultivated as container plants, and in summer these charming climbers give the greenhouse, conservatory, balcony, or terrace a tropical flair.

Most of the presently known species of genus *Passiflora* are perennial climbers, but some are trees, shrubs, herbaceous plants, or even annuals. Their main dispersal area extends over Central America and, above all, South America. Colombia accommodates more than 100 species and is rich in nearly all sections of the genus. Only a few species advance to North America. Another 25 species are indigenous to Southeast Asia, Australia, and Oceania.

Among all plants from the New World, passionflowers were met with the greatest admiration by Europeans during the 16th and 17th centuries. This admiration was only partly due to the beauty of their blossoms, however. The main reason was that passionflowers were imagined to symbolize the sufferings and death, or passion, of Christ.

With the beginning of the age of science, legends about flowers faded and interest in *Passiflora* diminished. Only when numerous greenhouses in the increasingly popular botanical gardens were built did passionflowers again become sought-after cultivated plants, a status which increased as further species were discovered and brought to Europe. During the 2nd peak of their popularity, at the end of the 19th and beginning of the 20th century, many hybrids were raised that are still popular today.

During and after World Wars I and II, high maintenance costs and fuel shortages prevented many greenhouses in Europe from being kept warm. The available space in the few heated greenhouses that existed was not sufficient for all tropical plants. Because of their considerable floor space requirements, passionflowers were often among the first plants to be parted with. Unfortunately, many a hybrid was thus lost forever, and even gorgeous popular species like *Passiflora insignis* (Plate 2) and *P. kermesina* were not brought back into cultivation until the beginning of the 21st century.

After World War II, several years passed before botanical gardens replenished their passionflower collections, and during these years there was very little talk about the plants. Only when an increasing percentage of the public could afford a greenhouse or conservatory did passionflowers regain their former esteem. The absolute breakthrough, however, came with the late realization that these plants are more resistant and easier to grow than their reputation suggests, and that many of the hundreds of species flourish as container plants, either around the home or close by on the balcony.

Coevolution With *Heliconius* Butterflies

No other genus or family of plants has as variable a leaf formation as passionflowers, with their dramatically unlobed or 2- to 9-lobed leaves, sometimes camouflaged with variegation (see Plate 3). Leaf size is also remarkably variable: while the smallest leaves of the genus (*Passiflora fruticosa*) are 0.5 cm long, the leaves of *P. macrophylla* can reach a length of 95 cm. This diversity in leaf shape and size can be regarded as a result of the coevolution of the genus with *Heliconius* butterflies, also called longwings or maracuja butterflies. These brightly colored insects are indigenous to Central and South America and feed only on passionflowers during the caterpillar stage. Lawrence E. Gilbert and many of his students and former students, above all John Smiley and James Mallet, have investigated the interrelations between *Heliconius* species and passionflowers and published numerous papers on this amazing and changing relationship.

Since heliconid females hunt with their eyes for plants suitable for oviposition, passionflowers often imitate the look and leaf form of plants not eaten by the butterflies. They are sometimes amazingly well camouflaged as a result.

Varying leaf forms can even occur within one and the same species. In the rain forest, for example, young passionflowers imitate the leaf forms of plants growing on the ground, while plants that are several years old imitate the typical leaves of the canopy. As with all nymphalid butterflies, *Heliconius* butterflies have a specially developed pair of forelegs with which they tap and taste the hostplant to test whether it is a camouflaged passionflower or a plant that is unsuitable for oviposition. As a general defense, the leaves of most passionflower species contain toxins—especially cyanogenic glycosides, which yield cyanide. This metabolite, although toxic for the plant, is neutralized by agglutination to sugar and stored mainly in the leaves. As the leaves grow older, the concentration of cyanogenic glycosides diminishes. These leaf toxins have an effect on numerous pests but none on the caterpillars of *Heliconius* butterflies. Over the course of evolution the butterflies developed enzymes capable of degrading the cyanogenic glycosides. They also use these same enzymes in the reverse direction, to construct their own defensive cyanogenic glycosides, which are chemically different from those of their hostplants. These defensive chemicals and the butterflies' brilliant warning colors keep birds from eating them.

Other defense mechanisms of passionflowers are effective against the butterflies. Outside the blossom, extrafloral nectar glands produce a sweet nectar on the petioles, leaves, stipules, and in other locations (see Plate 4). With the help of these sugary glandular secretions, the plants attract ants and wasps, which feed on the eggs and caterpillars of the butterflies.

Moreover, some passionflower species very successfully imitate the eggs and even the oviposition patterns of maracuja butterflies. This phenomenon, called egg-mimicry, results in what look basically like yellow dots, or butterfly eggs. Such imitations are found in most locations where passifloras grow. *Passiflora auriculata*, for example, bears 2 conspicuous petiole glands, and *P. retipetala* has yellow-orange swellings on the filiform extensions of the stipules. Fake *Heliconius* eggs simulate that a butterfly has already laid eggs on the plant. Since a single passionflower vine cannot always provide enough room for several nests of eggs, and because the caterpillars of some species are often cannibalistic, the female generally looks for another place to lay eggs.

Other species, such as *Passiflora adenopoda* and *P. lobata*, have additional mechanical protection. Young caterpillars that become trapped in the dense, hooked hairs of the leaves are rendered immobile and, while trying to free themselves, are injured and sometimes die from loss of body fluids. More information about butterflies can be found in chapter 19.

Floral Ecology

The unique blossoms of passionflowers, with their extremely complicated forms, are quite variable in shape, structure, and color. Colors are green or white to spectacular yellow, blue, violet, purple, pink, orange, or red. Blossom size spans from less than 1 cm in *Passiflora suberosa* to 17–18 cm in some clones of *P. insignis* or *P. vitifolia*. A diameter of even 22 cm, found in a collection of *P. speciosa* in Brazil, is reported by Reto Gamma. The short-lived flowers usually remain open only for a day. In some representatives of supersection *Tacsonia* and in the representatives from Australia (*P. aurantia*, *P. cinnabarina*, and *P. herbertiana*), flowers remain open for up to 3 days. They appear singly or in pairs on peduncles of their own that emerge from the leaf axils or are multiflorous racemes (*P. racemosa*). Of botanical interest are a few species in which the more primitive condition exists, where it appears that the flowers are borne on the tendril, such as species from subgenus *Deidamioides*—section *Deidamioides* (*P. deidamioides*), section *Tryphostemmatoides* (for example, *P. gracillima* and *P. arbelaezii*), and section *Polyanthea* (*Passiflora cirrhiflora*)—and a few species from *Decaloba*, supersection *Pterosperma*.

The blossoms of genus *Passiflora* are nearly always composed of 5 sepals and 5 petals. Only in subgenus *Decaloba* is a small lineage of species with petals missing, the *P. suberosa* group. A passionflower's structure, scent, and the amount and composition of its nectar correspond to the needs of its respective pollinators. Species with a large and conspicuous corona (for example, *P. alata*, *P. amethystina*, *P. caerulea*, and *P. edulis*) generally exude a sweetish odor and are organized to be pollinated by large bees, bumble bees, and other flying insects. The bees land on the visibly banded corona, the location of the scent production. From here they can reach down with their mouthparts to the nectar located under the inner corona. There they touch the down-turned anthers with their back and carry the pollen to the next flower.

The otherwise conspicuous corona is generally stunted to little knots of 1–2 mm in most species of supersection *Tacsonia*. Because of the long floral tube and the androgynophore, which protrudes widely out of the flower, small insects are unable to reach down to the nectar at the base of the blossom and do not pollinate the flower during their attempts. Only hummingbirds can reach the nectar with their long beak. In this way, the pollen sticks onto the heads of the birds and is carried to the next flower. Coronas are not needed as a landing platform or perfume generator in the case of birds. In most tacsonias the floral tube is about 10 cm long, and in *Passiflora mixta* var. *pilaloensis* it may be up to 15 cm. Only a single rare species of hummingbird, the sword-billed hummingbird (*Ensifera ensifera*), has a beak that is 10 cm long, as long as the bird itself (see Plate 6). This is the only bird that is able to pollinate long-tubed tacsonias. Consequently, the native range of this rare hummingbird is the same as that of supersection *Tacsonia*: the northern Andes mountains.

Passionflower species that are pollinated by hummingbirds often have reddish or pink, long, odorless flowers that do not attract bees. Such red flowers are found in supersections *Distephana* and *Coccinea* (see Plate 7). Nevertheless, *Passiflora vitifolia* is visited by a stingless species of bee, but instead of pollinating the plant the bee steals the nectar. The upright corona, partially inclined toward the androgynophore, usually hinders nectar robbers from direct access. To get to the nectar and past this obstacle, the bee drills a hole with its powerful mouthparts into the base of the floral tube from the outside and sucks the nectar without pollinating the plant. To protect itself from such robbers, *P. vitifolia* has developed extrafloral nectar glands on its floral bracts, in addition to the glands on the petioles. This way the plant lures ants, who like to eat the sugary glandular secretions, and in return the ants dispel the uninvited bees (see Plate 8).

A special ecological position is occupied by species such as *Passiflora mucronata*, *P. trisecta*, *P. galbana*, *P. ovalis*, and *P. penduliflora*, which are pollinated by bats, who visit the flowers (see Plate 9). The white or pale green fragrant blossoms open at dusk and flower throughout the night. The flower shapes are adapted to the length and shape of the bats' snouts and tongues, which are rather plump in comparison with those of other pollinators. The flowers have a whitish or pale green floral tube that is broadly open, and the nectar is not blocked by the corona. The androgynophore often lies against one side, to one edge of the floral tube. They also may have a zygomorphic androecium, in which all anthers are inclined toward the side to which the bats are flying, and the blossom is raised (or in the case of *P. penduliflora*, hung down) to an exposed position beyond the plant by correspondingly long and vigorous peduncles.

The remarkable movement of the stigmas of passionflowers can be easily observed and has been investigated throughout the centuries. This movement helps prevent self-pollination, thereby guaranteeing a genetic variability of the seeds. Immediately upon opening of the flower, the styles with the stigmas are usually standing upright. Then only the filaments, with the anthers on their ends, are bent down. Although the pollen will stick to the first visitor, it is prevented from coming into contact with its own stigmas. The pollinators fly to the next blossom and within a short time carry on their back a mixture of pollen from different plants. Independent of the activity of the respective pollinator and the amount of nectar in the blossom, the styles will bend down within just a few minutes to several hours after opening. With the help of this procedure the plant raises the probability of being pollinated with the pollen from another flower. Even if there is no pollinator, some species, such as *Passiflora gracilis*, *P. capsularis*, *P. foetida*, *P. herbertiana*, *P. morifolia*, and *P. suberosa*, can pollinate themselves with their own pollen by bending down their styles to the anthers. Hence these species will also bear fruit when they are grown individually.

Modern technology permits travel to passionflower habitats that is easier, quicker, and cheaper than ever before. In addition, more and more remote regions in Central and South America have become accessible due to the often ruthless road construction in these areas. Despite the negative consequences of such changes, the positive effect is that an increasing number of *Passiflora* species are being discovered and collected, and because of this, certain endangered species will be protected from becoming extinct.

Countless photos and herbarium specimens were screened for the purposes of this book. Several species that came under examination could not be assigned unquestionably, or seemed to be new, or at least very rare (for example, see Plates 10 and 11). The current state of affairs is such that some passionflowers remain unidentified, and much work remains to be done to cover the genus *Passiflora* exhaustively.

1
A Brief History of the Passionflower

Emil E. Kugler & Leslie A. King

The European world did not learn about passionflowers until after the Spanish conquest of America. The early period of their discovery covers exactly 200 years and dates from the time of Pedro de Cieza de León in 1553 to the publication by Carolus Linnaeus (also known as Carl von Linné) of *Species Plantarum* in 1753. By then 24 species had been described. The Indians of North and South America had known the plants by various names—*maracock*, for example. The first Spanish explorers called them *granadillas* (small pomegranates). From the very beginning, and particularly after Nicolás Monardes had published the 2nd edition of his herbal in 1574, descriptions of passionflowers became intertwined with a religious and mystical belief that the various parts of the flower depicted the crucifixion of Christ. This symbolism became so embedded in the spirit of the age that even Linnaeus felt obliged to use *Passiflora*, or *flos passionis* (*flos* meaning "flower," *passio* meaning "suffering"), as the definitive name for the genus.

The modern period dates from the time of Linnaeus to the present day. These years have seen not only the discovery of more than 500 species, including representatives from the Old World, but also the establishment of commercial fruit production and the creation of more than 400 artificial hybrids. A renewed interest in passionflowers in the past decade or so is shown by the appearance of numerous books, 2 journals, and 2 on-line discussion groups devoted to the genus.

This historical account is based entirely on published work. The text is based on an article by Emil Kugler (1995), with modifications and further sources of information.

The 16th Century

The first report of a passionflower came from Cieza de León (1553). Clements R. Markham (1864) provides a vivid description of Cieza de León's early days:

> Cieza's work is, in many respects, one of the most remarkable literary productions of the age of the Spanish conquest in America. Written by a man, who had passed his life in the camp from early boyhood, it is conceived on a plan that would have done credit to the most thoughtful scholar, and is executed with care, judgement and fidelity. Cieza de León is believed to have been born in the year 1519 in the City of Seville, where he spent the early years of his life. In 1532 at the extraordinarily early age of fourteen, young Pedro embarked at Seville and set out to seek his fortunes in the New World. Seville and Cadiz were crowded with adventurers, all eagerly seeking for a

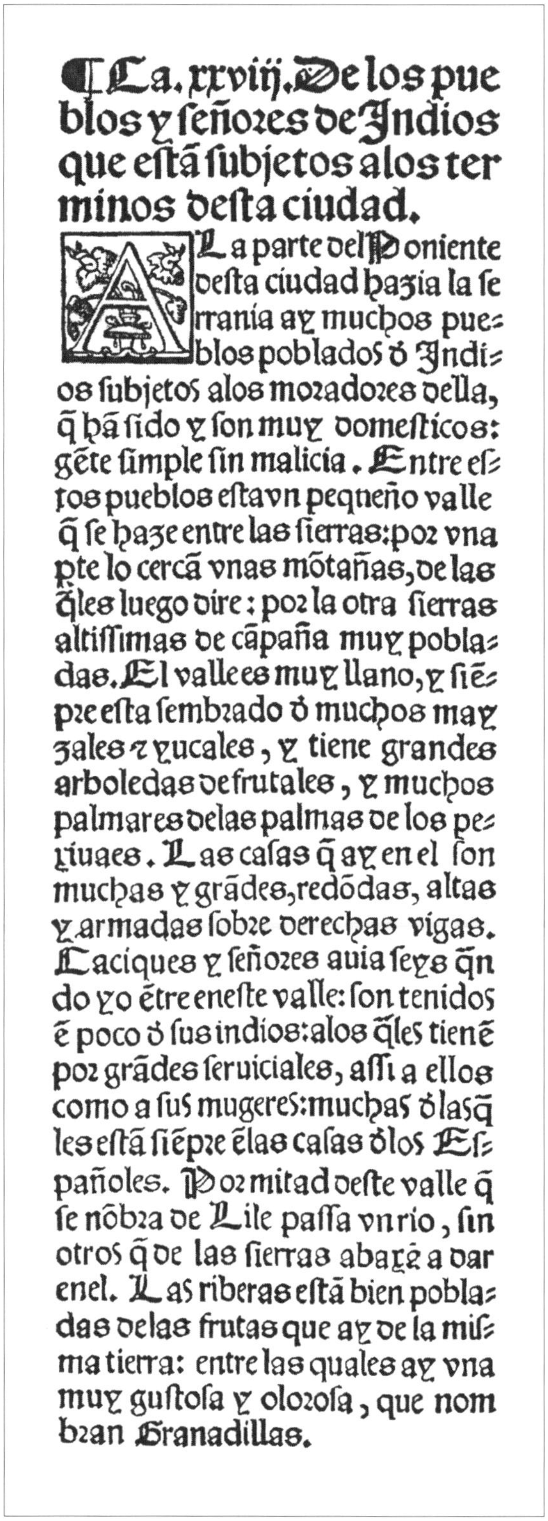

¶ Ca. xxviij. De los pueblos y señores de Indios que estã subjetos alos terminos desta ciudad.

A la parte del Poniente desta ciudad hazia la serranía ay muchos pueblos poblados d Indios subjetos alos moradores della, q̃ hã sido y son muy domesticos: gẽte simple sin malicia. Entre estos pueblos esta vn pequeño valle q̃ se haze entre las sierras: por vna pte lo cercã vnas mõtañas, de las q̃les luego dire: por la otra sierras altissimas de cãpaña muy pobladas. El valle es muy llano, y siẽpre esta sembrado d muchos mayzales y yucales, y tiene grandes arboledas de frutales, y muchos palmares delas palmas de los pexiuaes. Las casas q̃ ay en el son muchas y grãdes, redõdas, altas y armadas sobre derechas vigas. Caciques y señores auia seys q̃ndo yo ẽtre eneste valle: son tenidos ẽ poco d sus indios: alos q̃les tienẽ por grãdes seruiciales, assi a ellos como a sus mugeres: muchas d las q̃les estã siẽpre ẽlas casas d los Españoles. Por mitad deste valle q̃ se nõbra de Lile passa vn rio, sin otros q̃ de las sierras abaxẽ a dar enel. Las riberas estã bien pobladas delas frutas que ay de la misma tierra: entre las quales ay vna muy gustosa y olorosa, que nombran Granadillas.

Pedro de Cieza de León's text on the granadilla.

passage to that marvellous land beyond the setting sun. It was, indeed, a time of wild excitement. Every ship that returned from the Indies, and there were many, would bring reports of the discovery of new and previously undreamt of powerful empires. Cieza de León, at the time of his embarkation, was a mere boy. . . . His character was destined to be formed in a rough and savage school, and it is most remarkable that, so fine a fellow as our author really was, he should have been produced amidst the horrors of the Spanish American conquest. Humane, generous, full of noble sympathies, observant and methodical, he was bred amidst scenes of cruelty, pillage and wanton destruction, which were calculated to produce a far different character. It is not known in what fleet our boy soldier set out from Spain, but it is more than probable that he sailed from his native land in one of the ships of the expeditionary fleet of Don Pedro de Heredia. This expedition, which consisted of one galleon and 2 caravels, carrying in all about a hundred men, sailed from Cadiz in the end of 1532. On the 14th of January 1533, the expedition entered the bay of Carthagena in modern Colombia. In no part of Spanish America did the Indians more resolutely defend their homes than along the coast of Tierra Firme, as it was called, and young Pedro saw some very rough service. He began to write a journal in the year 1541, which formed the material for his future work, while serving under Robledo at Cartago in the Cauca Valley.

Until 1550 he stayed in Peru, a country that included what is now modern Colombia. As a civil servant of the Spanish administration he undertook a number of official journeys to learn about the structure of the country and the customs of the inhabitants. On his return home he wrote his history of Peru (Cieza de León 1553). In the first part of the book—the only part still in existence—he describes the provinces, the founding of new cities, the rites and customs of the Indians, and many other things. Chapter 28 concerns the villages and the Indians chiefs who lived within the jurisdiction of the town of Cali. Situated in the upper valley of the Río Cauca in the province Valle, about 260 km southwest of Bogotá, Cali is now the 3rd largest city in Colombia. In this section of Cieza de León's work, on sheet 33, is a paragraph describing the area in which the granadilla was found, which translates as follows:

> On the western side of this city, toward the mountains, there are many villages of Indians, who are very docile, a simple people void of malice. Amongst these villages there is a small valley closed in by mountains. The valley is level, and is always sown with maize and yuccas, besides having plantations of fruit trees, and of the palms called pixiuares. The houses in this valley are very large, round, lofty, and supported on straight poles. There were six chiefs when I entered this valley, who were held in small estimation by the Indians, many of whom are always in the houses of Spaniards. Through the centre of this valley, which is called Lile, a river flows, and is fed by many streams coming from the mountains. The banks of this river are well covered with fruit trees, amongst which there is one that is very delicious and fragrant called granadilla.

Unfortunately, Cieza de León does not say what name the Indians used to describe the granadilla. It is possible that the particular species in question is *Passiflora ligularis*, which is cultivated and found growing in the wild from Mexico to Bolivia, and which continues to be called the granadilla. The tasty juice in the arils is used mainly for preparing refreshing drinks. In 1787, German botanist Friedrich K. Medikus suggested that some species of *Passiflora* should be grouped as a new genus, named by him as *Cieca*, in honor of Cieza de León (or Cieca di Lione). Today *Cieca* ranks as a supersection in subgenus *Decaloba*.

In 1569, Nicolás Monardes, a physician in Spain, published his major work on the plants and herbal remedies used in the West Indies. In the 2nd edition of the book, translated into English in 1596, one monograph describes the granadilla:

> From the firme lande they brought mee certayne fruites which are hearbes which they call in the hils of the Peru where they growe Granadillas, and this name the Spaniards did give them, for the likenes and fashion that they have to our Granadas, which wee call Poungarnardes, for that they are well neere of the same greatnesse and colour when they are ripe, saving that they haue not a litle crowne, and now they are dry they seeme within y the seed is like to the graines of Peares somwhat greater, for they are all ful of little graines, verie faire, and thew very wel, they are white within, and without any savor. The plant that this fruite beareth, is like to Ivie, and so it runneth up and clingeth fast too anything that is neare to it wheresoever it be set: it is faire when it hath fruite, for the greatnesse it is a perticular hearbe, and onely in one place it is founde, it calfeth a flower like to a white Rose, and in the Leaves it hath figures which are signes of the Passion of our Lord, that it seemeth as though they were painted, wyth muche care, where the flower more particular than any other that has beene seene. The fruite is the little graines, which we haue spoken of, and when they are in season they be full of Licour, some what sharpe, and all full of seede, which are opened as one doeth open an Egge, and the Licour is to bee supped up with greate contentment of the Indians, and of the Spaniards. And when they haue suped it up, many do not feele paines in their stomacks, but rather they soften the bellie. They seeme to bee temperate, with some moysture.

The same passage, translated to modern English, reads something like this:

> From the continent they brought me the fruit of a plant, which grows wild in the mountains of Peru. The fruit is called granadilla. It is the name given by the Spaniards, because it is similar to our pomegranates, having the same size, and, when it is ripe, the same color, but it lacks the little crown. When dried, the seeds found inside are similar to those of the pear,

> but rather larger and pitted, making them appear quite unusual. The pulp is whitish and without taste. The plant, which produces this fruit, is like ivy; it climbs and creeps everywhere it can. It looks elegant when you see it loaded with fruits. The large flower is very similar to the flower of a white rose, and it appears to have been carefully made to show the representation of the Passion of Jesus Christ. When mature, it contains a slightly tart liquid and is filled with seeds. They are opened like eggs, and the liquor is drunk with great delight both by the Indians and Spaniards. The juice is used for treating stomach cramps and acts as a laxative. The fruit seems to be warm and rather moist.

In this way Monardes was the first to provide a short description of the granadilla and to refer to a certain similarity between the flower and the marks of the crucifixion of Christ. This initiated the symbolism and mysticism that found a use in the church for converting the Indians and that led eventually to the naming of the genus by Linnaeus (1742, 1753). The writings of Monardes are among the most remarkable in the history of herbal medicine. They are essentially objective and written in a scientific spirit, undistorted by legends and travel reports. They witness the enormous changes that occurred in the world and in the history of culture following the discovery of America. Monardes was also the first to describe tobacco and coca leaves in detail.

Monardes was born in Seville in 1493, and it was in Seville that he practiced medicine until his death in 1588. He never visited America. His knowledge of American remedies was gained from those who returned from the new continent and by a close study of the natural products that were imported. The idea that all material is composed of fire, water, air, and earth led to the concept that there are 4 qualities in a body, namely warmth, humidity, coldness, and dryness. This was an accepted truth right up to the 18th century. It was believed that any reduction in one of these qualities would cause an imbalance and lead to illness, and the cures for such illnesses were divided according to whether they had the characteristics of dryness, dampness, coldness, or warmth. The final sentence in Monardes's description of granadillas, "They seeme to bee temperate, with some moysture," now becomes understandable.

The Spanish Jesuit José de Acosta mentioned the flower of the granadilla in his *Historia Natural y Moral de las Indias* (1590), which was translated into English several years later:

> The flower of the Granadilla is held for an admirable thing, and they say it hath in it the markes of the Passion, and that therein they note the nailes, the pillar, the whips, and crowne of thornes, and the woundes, wherein they are not altogether without reason, and yet to finde out and observe these things, it requires some pitie to cause beleefe: but it is very exquisite and fair to the eye, although it have no smell. The fruite which they also call Granadilla is eaten and drunke; or to speake more properly, sucked for a refreshing; this fruite is sweete, and too sweete after some opinions.

The 17th Century

In Lisbon in 1602, Martin del Barco-Centenera published an epic poem, "Argentina y la conquista del Río de la Plata con otros acontecimientos de los regnos del Perú, Tucumán y estado del Brasil" (Argentina and the Conquest of the Rio de la Plata with Other Events of the Kingdoms of Peru, Tucuman, and the State of Brazil). The poem, actually a song, tells of the constituents of the earth, of reptiles, terrible vipers and snakes, sirens and carbuncles, butterflies that change into caterpillars and then into mice, and many other miracles. According to Barco-Centenera, he held the flower of a granadilla in his hand as he wrote about it, in the 2nd stanza of canto III:

> *La flor de la granada o granadilla*
> *De Indias, y misterios encerrados,*
> *A quien no causara gran marauilla,*
> *Figuranse los doze consagrados,*
> *De vna color verde, y amarilla,*
> *La corona y los clauos tres morados,*
> *Tan natural estan y casi al viuo,*
> *Que yo me admiro agora que lo escriuo.*

(The flower of the granada or granadilla
Of the Indians and concealed mysteries.
Who is not amazed?
Representing the twelve Apostles,
Of a green and yellow color,
The crown of thorns and three nails.
The flower is so natural and almost alive,
And now I admire it and so write this down.)

In 1608, Spanish Jesuits under Juan Romero presented Pope Paul V (Camollo Borghese, 1552–1621, pope since 1595) with dried plants and a color drawing of a passionflower. After 1609, many sheets with similar drawings were made available to a wider audience in Italy and Germany. The botanical features in these drawings were transformed into more understandable pictorial images. Changes were made in order to make the alleged agreement with the story appear as fact. These are the first representations of the passionflower to appear in Europe. With an accompanying text, they confirmed people's preconceptions about passionflowers, which endured well into the 18th century.

Simone Parlasca's representation of a passionflower in 1609.

Three figurative forms were developed between 1609 and 1610 (Stopp 2001). The first, in the 1609 pamphlet produced by the Dominican monk Simone Parlasca in Bologna, was probably based on original dried plants. However, careless conservation led to individual leaves overlapping each other, and as a consequence both 3- and 4-lobed leaf forms were depicted. The writing contains discussions and many poems in honor of the "Fiore della Granadiglia overo della passione di nostro Signore Giesu Christo" (flower of the granadilla or of the Passion of Our Lord Jesus Christ). In his 1996 book on passionflowers, John Vanderplank published 3 poems from Parlasca's work but mistakenly cited them as being from Giacomo Bosio, the church scholar and secretary of the Maltese Order. They were translated into English by Mauro Sericchio and the Amici Della Passiflora Society.

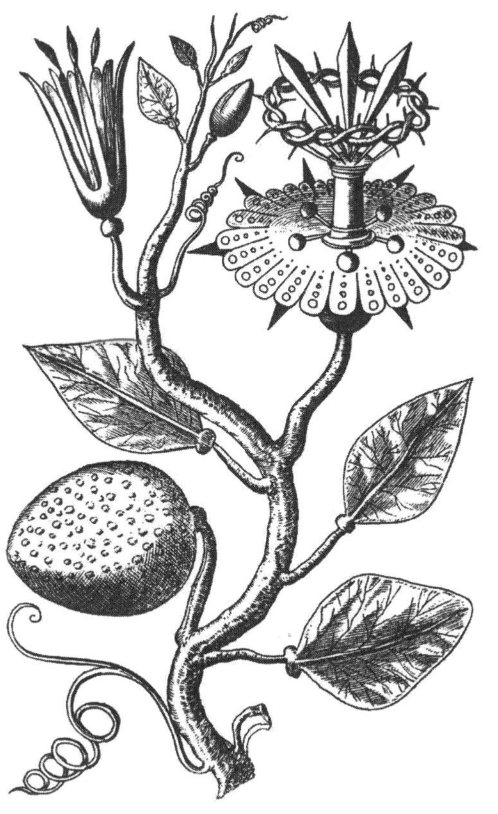

A passionflower depicted by Eugenio Petrelli in 1610.

An example of the 2nd representational form of the passionflower can be found in a pamphlet produced by Donato Rasciotti in Venice in 1609. In this drawing only the strongly hatched lance-shaped part of the lower right leaf was used, so that the leaves appear unlobed. This much-simplified form was widely used, such as in the pamphlets from

Giacomo Bosio's 1610 representation of a passionflower.

A drawing of *Passiflora incarnata* published by Jean Robin in 1614.

Augsburg, Germany (Mang 1609, Wörle 1609), and as a frontispiece in Antonio Possevino's book (Petrelli 1610).

The 3rd form, published in Rome in 1610, appears in the writings of Giacomo Bosio above a crucifix. The shape of the leaves described by Parlasca seemed so improbable to Bosio that he drew 2 separate unlobed leaves at each node. The drawing wrongly showed the petals surviving around the fruit and also showed the fruit forming between the bud and flowering stages. Bosio was much more critical than Parlasca when describing this miracle flower. But when the Augustinian monk Emmanuello de Villegas brought back a drawing from the New World, the authenticity of which he could guarantee, Bosio lost his skepticism. In his later work he described the granadilla in some detail.

Bosio reported that the Spaniards called the passionflower *La Flor de las cinco Llagas* (The Flower With the 5 Wounds). He also gave a detailed overview of the events of 1608 and 1609, described the flower in detail, and stressed the signs of the Passion as shown by parts of the plant. Stamens were scourges, the androgynophore was a flagellation column, styles were nails, 5 blood-red spots at the base of the flower were 5 wounds, 72 coronal filaments were 72 thorns in the crown, leaves were pointed lances, and 30 glands on the lower surface of the leaf were the 30 pieces of silver that Judas was paid to betray Christ. In later literature even more symbolic associations can be found.

But which species was this passionflower of 1608–1610? If one looks for clues amongst those plants that were in cultivation about 3 years later in Paris (the first in Europe), the answer should be *Passiflora incarnata*. However, Bosio distinguished between 2 plants: one from Peru whose sepals and petals were tawny, one from New Spain (Mexico) whose sepals and petals were white tinged with pink. Unfortunately, that does not help us, because *P. incarnata* is not indigenous to South America but to the United States. Furthermore, *P. incarnata* has 3-lobed leaves without glands on the lower leaf surface, and there are no marks on the fruit. Some have suggested it may be *P. ligularis*, which has unlobed leaves, the correct coloring in the petals and sepals, and the right geographical distribution. Its fruit is orange, densely covered with tiny light dots. On page 21 of *Passion Flowers and Passion Fruit* (1991), Vanderplank writes, "In Europe today the story is based on *P. caerulea*, most probably because it was the first passion flower to be introduced into the continent." However, it is certain that *P. incarnata* was the first species to be cultivated in Europe—in Paris in 1612, in Rome before 1619. By contrast, *P. caerulea* was probably growing in Paris in 1625 but did not appear in London until at least 1629.

The passionflower was soon in great demand as a miracle plant and was rapidly established in European gardens. Thus it flowered in August 1612 in the Paris garden of botanist Jean Robin, "Arboriste du Roy." His son, Vespasien, had brought it back from North America. To mark this special event, the flowers of this rare plant were depicted in a copperplate engraving (Anonymous 1614). This is an accurate representation and shows without doubt *Passiflora incarnata*.

Donato d'Eremita, Prefetto della Aromatoria di Sta. Catherina a Fornello di Napoli, employed 2 members of the Accademia dei Lincei (Lynx Academy) to work as illustrators. The Accademia dei Lincei had been given this name to reflect its main purpose: to resolve uncertainties in the depiction of nature by using "eyes as sharp as a lynx." The members were Giovan Fabri and Fabio Colonna. The latter had come to prominence in 1592 as the author of an important work called *Phytobasanos sive plantarum aliquot historia*. In 1619, Fabri made an accurate painting for Donato d'Eremita of a *Passiflora incarnata* growing in his garden. This was published as a pamphlet with a copperplate engraving in the same year (Donato d'Eremita 1619). In 1622, Colonna made a drawing of the same plant (Donato d'Eremita 1622). A further publication appeared in 1620 from Pietro Castelli (under the pseudonym Tobias Aldinus), who was at that time physician to cardinal Odoardo Farnese in Rome. Castelli was also responsible for the administration of the cardinal's garden. He was asked to put an end to the overwhelming religious superstition that had arisen from the symbolism of the passionflower. The pamphlet shows 2 drawings of *P. incarnata* in which it can be seen in its natural form. Similar drawings were published in Castelli's booklet (Aldinus 1625), in which he stated, "I see no cross, no spear, nor crown of 72 spines in this flower—it is pure fiction. By using my imagination, I could find the Passion in any number of flowers."

Shortly after 1600, explorers William Strachey and John Smith independently came across *Passiflora incarnata* in North America. Their reports were only available as unpublished manuscripts until the 19th century. Strachey (1849) described the region around the Kings River in southwestern Virginia from Point Comfort to Jamestown, at that time controlled by a chieftain (Strachey speaks of him as a "king") of the Powhatan Indians, including an account of passionflowers, known as maracocks:

> Yt is an ample and faire countrie indeed, an admirable porcion of land, comparatively high, wholsome, and fruictfull; the seat sometyme of a thowsand Indians and three hundred Indian howses and those Indians, as it may well appeare, better husbands [husbandmen] than in any parte ells that we have observed, which is the reason that so much ground is there cliered and opened, enough, with little labour, allreddy prepared, to receave corne, or make viniards of twoo or three thowsand acres; and where, beside, we find many fruict-trees, a kind of goosbery, cherries, and other plombes, the maricock apple, and many prettie copsies or boskes (as it weere) of mulberye trees . . . About their howses they have commonly square plotts of cleered grownd, which serve them for gardens, some one hundred, some two hundred foote square, wherein they sowe . . . They plant also the field apple, the maracock, a wyld fruit like a kind of pomegranett, which increaseth infinitlye, and ripens in August, contynuing untill the end of October, when all the other fruicts be gathered . . . [The maracock] groweth generally lowe, and creepeth in a manner amongst the corne (albeit I have seene yt, planted in a gardein within our fort, at James Towne, to spred and rise as high as the pale); yt is of the bignes of a queene apple,

> and hath manie azurine or blew karnells, like as a pomegranett, and yt bloometh a most sweet and delicate flower, and yt is a good sommer cooling fruict, and in every field where the Indians plant their corne be cartloades of them.

In 1608, Captain John Smith, president of Virginia and admiral of New England, the same man who achieved fame through his relationship with the Indian "princess" Pocahontas, wrote of passionflowers in his diary (1884): "The Indians plant also Maracocks, a wild fruit like a lemmon, which also increase infinitely: they begin to ripe[n] in September, and continue till the end of October."

Among the alternate names for "passionflower" that have been used in different parts of the world are the Algonkian (Virginia) *maracock* or *maricock*; the Powhatan (Virginia) *mahcawq*; *maypop*, used in the southeastern United States; the Caribbean (Guadeloupe) *merécuya*; the Guarani (South America) *mburucujâ*; and the Tupi (South America) *múrucuya*, derived from *muru* (meal) and *cuya* (container) to mean "meal in a dish" (Peckoldt 1909). Most of these words seem similar enough to suggest a linguistic connection. Seeds of *Passiflora incarnata* occur in many human archaeological sites in the United States from the Late Archaic (3500–800 BC) to the Historic era (AD 1550–1800) (McGuire 1999). Algonkin Indians in Virginia and Creek people in Florida ate its fruit, which they gathered from cultivated, semidomesticated, or wild plants.

In 1622, Daniel Rabel published an accurate picture of *Passiflora incarnata* under the name "Maraco Indica" or "flos passionis." A contribution by Federigo Cesi, prince of St. Angelus and St. Paulus and founder of the Accademia dei Lincei, now proves to be of special importance. The paper was released for publication in 1628 but did not appear until 1651. In it Cesi renames the "granadilla/flos passionis" as *Passiflora*. Thus it was that 2 members of the Accademia dei Lincei, Giovan Fabri and Fabio Colonna, tried to divest the "flos passionis" of its symbolism, but a 3rd member effectively set it in stone (Cesi 1651).

In his *Paradisi in sole paradisus terrestris* (1629), under "Maracoc siue Clematis Virginiana, The Virginian Climer" [sic: *Passiflora incarnata*], British pharmacist John Parkinson denied the symbolism: "I dare say God neuer willed his Priests to instruct his people with lyes: for they come from the diuell, the author of them."

In 1633, Jesuit and botanist Giovanni B. Ferrari summarized the history of the "granadiglia/maracot," described in detail the symbolic connection of the "miracolo dell'India occidentale" (miracle from the West Indies), and provided notes on its cultivation. In 1980, Swiss gardener Emil Iwert created a mystical passionflower legend that was woven around Ferrari and the events of that time. Iwert tried to make a precise connection between the individual parts of *Passiflora caerulea* and the marks of the crucifixion, which he then set out in drawings. In 1982, in Florida, S. Leigh Moore wrote a short book, *The Legend of the Passionflower*, in which he similarly brought together all earlier discussions and some of his own ideas. In the late 1990s, German biologist and artist Roland Spohn created a 4-part picture-cycle in which he linked the botanic and symbolic aspects of *P. caerulea* (Caesar and Spohn 1997).

Scientists Georg Marcgrav von Liebstad and Willem Piso reported in 1648 on the passionflowers they found during their journeys together in Brazil. They retained the Indian *murucujá* and *murucuyá* as collective terms for the species they described and pictured, including *Passiflora alata*, *P. filamentosa*, *P. laurifolia*, *P. edulis*, *P. cincinnata* (or possibly *P. serratodigitata*), and *P. suberosa*.

Francisco Hernandez, who was physician to Phillip II of Spain, produced drawings and descriptions of the herbal remedies of the Mexican Indians: Coanenepilli or Contrayerva (*Passiflora mexicana*) and Tzinacanatlapatli (*P. coriacea*). The former possesses an attractive pink or red corona and contrasting pale green sepals and petals. The woodcut leaves little doubt about its identity, particularly as 20 of them, complete with corona, can be seen thrown in a pot. Although the flower is not shown, *P. coriacea* is also unmistakable because of the characteristic elliptical shape of its leaves. Hernandez lived from 1514 to 1578, and this work was published posthumously in 1651.

In his catalog of the botanic garden at Padua in 1662, Georgius A. Turre describes the *Passiflora lutea* cultivated there as "granadilla pumila fl. paruo luteo" (dwarf granadilla with small yellow blooms). Here, for the first time, the name granadilla is used for a genus. Later a number of botanists before Linnaeus would place 18 *Passiflora* species in this group, namely *biflora, caerulea, capsularis, cupraea, foetida, heterophylla, holosericea, incarnata, laurifolia, maliformis, pedata, punctata, rotundifolia, serratifolia, serratodigitata, suberosa, tiliifolia,* and *vespertilio*. Of course, binary names were still not in use but were set out in a similar way to those used by Turre.

In 1691 the London physician and botanist Leonard Plukenet published drawings of *Passiflora foetida, P. laurifolia,* and *P. suberosa,* and in 1696 described *P. caerulea, P. coriacea, P. incarnata, P. mexicana,* and *P. multiflora*. In 1693 the French Franciscan Charles Plumier described and illustrated *P. foetida, P. laurifolia, P. maliformis, P. multiflora, P. murucuja, P. pedata, P. rubra, P. serratodigitata,* and *P. suberosa*. Catalogs of plants cultivated in European gardens from around 1660 to 1700 include the following species (together with others that cannot now be readily identified): *P. caerulea* Paris ca. 1625, England 1629, Leiden 1687, Naples 1697, Nuremberg 1700; *P. foetida* Altdorf 1662, Amsterdam 1689, Nuremberg 1700; *P. incarnata* Paris 1612, Rome 1619, London 1629, Messina 1640, Altdorf and Pisa 1662, Berlin 1663, Bologna 1666, Uppsala 1685, Leiden 1687, Naples 1697, Nuremberg 1700; *P. laurifolia* England 1690; *P. lutea* Padua 1662, Leiden 1687, Nuremberg 1700; and *P. suberosa* Leiden 1687, Amsterdam 1689.

More information about the symbolism of the

Francisco Hernandez's woodcuts of Coanenepilli (*Passiflora mexicana*), left, and Tzinacanatlapatli (*P. coriacea*), right.

passionflower can be found in Ulla-Britta Kuechen's work (1989). Emil Kugler (2002) has also described in some detail the symbolism of the later 17th century.

The 18th Century

In 1737, Linnaeus established the scientific name for the genus as *Passiflora*, a word filled with religious and mystical meaning, but he failed to mention its first use by Cesi. Many believed that this superstitious connection should have been rejected. *Maracotta* would have been a better-sounding name, as suggested by Abraham Cowley (1668) in his poem no. 59.

In 1745, Johann G. Hallman, a pupil of Linnaeus, wrote a thesis in Stockholm concerning the genus *Passiflora* in which he summarized the available botanical material. The thesis contains descriptions of 22 recognized species, with references to earlier synonyms and discussion of their history, names, classification, distribution, associated superstitions, and uses. The scientific names of the species are still shown in a polynomial form. For example, *P. incarnata* appears as *P. foliis trilobis serratis*, identifying it as a passionflower with 3-lobed, saw-edged leaves.

Linnaeus used Hallman's thesis, little changed, in 1753 as the basis for his final definition of the species in the genus. The modern era of taxonomy dates from the appearance of this *Species Plantarum*, in which Linnaeus used the binomial or binary nomenclature exclusively. It was not his own creation, having been used before, although somewhat inconsistently. From this point onward there was only one correct scientific name for each species, separating it from every other species in the same genus. This nomenclature was adopted by the scientific world without any particular problems, and the advantages were considerable. Before Linnaeus, it was often difficult for botanists working with polynomial names to know which species was being described. Some names were based on earlier authors, but others felt inclined to invent new names. Passionflowers had been assigned to several genera, a situation that did not help clarify their status. Linnaeus identified 24 species in *Passiflora*: *caerulea, capsularis, cupraea, foetida, hirsuta* [syn. *suberosa*], *holosericea, incarnata, laurifolia, lutea, maliformis, minima* [syn. *pallida*], *multiflora, murucuja, pallida, pedata, perfoliata, punctata, rotundifolia, rubra, serratifolia, serratodigitata, suberosa, tiliifolia*, and *vespertilio*. His species can still be found today as dried specimens in his herbarium in London.

Linnaeus established a taxonomic structure that endures to this day. After Linnaeus, progress took place in a well-defined framework and was more concerned with the introduction of ornamental and fruit plants into culture, the breeding of hybrids, and the discovery and scientific description of new species. Using morphological analysis, and later the chemical constituents, molecular biological, and other criteria, new taxonomic relationships were found that led to revisions in the organization of the genus, both below and above the species level. Such reorganizations should ultimately reflect the actual relationships that have arisen in the course of their evolution.

From the 19th Century to the Present Day

Today the production and processing of the edible fruit of certain passionflowers, particularly *Passiflora edulis*, is a major undertaking. The use of the dried leaves of *P. incarnata* and some other species is also significant in herbal medicine. Further developments after Linnaeus can only be briefly reported. The French biologist Jean Babtiste de Lamarck, founder of an evolutionary theory, was able to recognize 35 *Passiflora* species in 1789. Antonio J. Cavanilles, a Spanish cleric and professor of botany, devoted his 1790 work exclusively to *Passiflora*, describing 43 species and illustrating 32 of them. From then on many publications appeared with collections of color plates and accompanying text. The first of these, which remains in print, was *The Botanical Magazine*, founded in 1787 by William Curtis, author of *P. alata*. In the first volume of this series, Plate 28 illustrated *P. caerulea*. Among a number of short-lived series, the *Botanical Register*, in 17 volumes, founded by Sydenham T. Edwards in 1815, contained original descriptions of *Passiflora*. In French garden magazines, regular short contributions appeared with color drawings of passionflowers, which by this time were popular garden and glasshouse plants. In 1805 the French botanist Antoine Laurent de Jussieu suggested a tentative name for the family, which he designated "order" Passifloreae. This was firmly established in the following year using the modern spelling: Passifloraceae (Roussel 1806).

There next began the era of *Passiflora* hybrids. Different passionflower species can sometimes be crossed quite easily, and it is often possible to admire the hybrid flower in the following year. The first hybrids were *P.* ×*caeruleo-racemosa* (today correctly called *P.* ×*violacea*), described by Joseph Sabine in 1822; *P. alato-caerulea* (today known as *P.* ×*belotii*), described by John Lindley in 1824; and *P.* ×*colvillii* (from *P. incarnata* and *P. caerulea*), described by Robert Sweet in 1825 in his *British Flower Garden* (1823–1829). In all 3 hybrids, *P. caerulea* was the pollen parent. This process has continued to the present day, and there are now more than 400 hybrids.

Maxwell T. Masters, a British physician and botanist, was the recognized authority on the subject and wrote 69 relevant publications. Most appeared in London's *Gardeners Chronicle* (now *Horticulture Week*), for whom Masters was editor-in-chief for many years. In 1871 he wrote a monograph on the natural history of the Passifloraceae and in 1872 edited the section on this family in the *Flora Brasiliensis*. By 1877 he had produced a list of all known passionflowers growing in European gardens. Many of the original descriptions of *Passiflora* bear Masters's name.

German botanist Hermann A. Harms followed up the work of Masters, writing 36 publications on Passifloraceae. His thesis, written in 1893, concerned the evaluation of anatomical structure for the classification and organization of Passifloraceae. As with Masters, Harms named and described many new species and was regularly sent herbarium specimens from plant collectors and producers of local floras. His major work involved a reevaluation of the Passifloraceae for both editions of the *Natürlichen Pflanzenfamilien* (Natural Plant Families) in 1893 and 1925. The title refers to the progress that had been made in relation to the old Linnaean system, which had relied arbitrarily on the number of sexual organs on the bloom, and which had embodied a rather artificial organization, putting passionflowers and mallows together as sisters, for example.

In his 1938 monograph on the American Passifloraceae, North American botanist Ellsworth P. Killip based his ideas largely on the work of Harms. He described 355 species, 47 of which were new. Unfortunately, Killip ignored the species found in Africa, Asia, Australia, and Oceania, and his monograph contained few illustrations. A supplement appeared in 1960 (Killip and Cuatrecasas), in which 8 new species were described and illustrated.

In 1955 the Jesuit and botanist Lorenzo Uribe Uribe published an account of Passifloraceae in volume 27 of *Flora de la Real Expedición Botánica del Nuevo Reino de Granada* (Flora of the Royal Botanical Expedition to the New Kingdom of Granada [Colombia]), but this has been out of print for some time. Reproductions of paintings of passionflowers, which had been created under the leadership of the Spanish botanist José Celestino Mutis around 1790, are depicted on 42 large color plates.

In 1972, Willem J. J. O. de Wilde (Leiden) and Peter S. Green (Kew), working independently, dealt with the *Passiflora* of Australasia. However, they were not able to make any direct connection with the American species. To this day there is no single key with which to identify species from both the Old and the New World.

Danish botanist Lauritz B. Holm-Nielsen and his coworkers wrote a treatment for *Flora of Ecuador* in 1988 in which they described and illustrated 75 species of *Passiflora*. In the same year, the late American botanist Linda K. Escobar published her work on certain Passifloraceae for *Flora de Colombia*. It contains descriptions and illustrations of 32 species placed in the subgenera *Tacsonia*, *Rathea*, *Manicata*, and *Distephana*. John MacDougal published a monograph in 1994 based on his 1984 Ph.D. dissertation, *Revision of Passiflora Section Pseudodysosmia*. It dealt with 18 species, 4 of which were new. Floras for various countries continue to appear, but from a taxonomic viewpoint, few offer little that is new amongst the Passifloraceae. New species are being described, however, in single scientific articles in piecemeal fashion each year; there are now more than 525 recognized species, compared to 355 in Killip's time. So far, hundreds of theses and dissertations have been written that are concerned, at least in part, with passionflowers and cover fields of activity that extend from biology and botanical systematics to pharmacology and social sciences, to name but a few. A new complete taxonomic work to replace "Killip" and merge the Old World species with the New World species is long overdue.

Apart from the work of Uribe, and Florence Knock's *Passifloras for Your Garden* (1965), now out of print, there were for a time few specialist reference sources or books containing information for enthusiasts. One had to make do mainly with paragraphs in dictionaries and general publications on garden and houseplants. However, in the last few years there has been a surge of interest in this area. As a consequence of the increasing popularity of passionflowers, 2 dedicated journals have appeared regularly: the English-language *Passiflora*, the newsletter and journal of the Passiflora Society International (Coconut Creek, Florida); and the German-language *Passiflorunde*, the publication of the Interessengemeinschaft Passionsblumen (Vienna, Austria). In 2001, *Passiflorunde* produced a special issue containing a checklist of hybrids and cultivars of passionflowers in both German and English (Frank and Kugler 2001a, 2001b). An international registration authority sanctioned by the International Society for Horticultural Science was established in 2000 to keep track of and standardize cultivar and hybrid names; this is being operated through the Passiflora Society International. The first official register of cultivated passionflower names appeared in 2003 (Vanderplank). Two Internet discussion groups exist: one for English speakers at passiflora-l@nic.surfnet.nl, and one linked to the Interessengemeinschaft Passionsblumen at ig.passionsblumen@t-online.de.

A number of books are available, including John Vanderplank's *Passion Flowers and Passion Fruit* (1991), with further editions in 1996 and 2000, Peter Klock's *Das große Buch der Passionsblumen* (1996), and Bettina and Torsten Ulmer's *Passionsblumen: Eine faszinierende Gattung* (1997) and *Faszinierende Pflanzenwelt: Passionsblumen* (1999). It is hoped that this welcome development continues and that passionflower enthusiasts will enjoy interesting and new species and hybrids, informative literature, and beautiful color photographs.

2

Systematics

John M. MacDougal & Christian Feuillet

SYSTEMATICS is another word for taxonomy, or classification. It is humankind's effort to understand the world and its overwhelming diversity. In modern times, classification systems of living organisms have been hierarchical and based on an acceptance of the theory of evolution. Our classification is hierarchical because it is based on increasingly inclusive groupings called *ranks*. We say that an *individual* plant belongs to a local *population* of similar organisms, groups of similar populations make up a *species*, groups of related species make up a *series* or *section*, groups of related sections make up a *genus*, and so on, past *family* and *order* all the way to plant kingdom and beyond. Out in the forests and fields of the real world, however, only the categories of species and below are generally recognizable and "real." In nature, species do not breed with other species—that is often part of the definition. Categories above the rank of species, like section and genus, are basically arbitrary.

Evolution comes into our classification when we try to make it "natural," that is, when we try to make it accurately reflect the genealogy or blood lineage of the organisms. Without a time machine we cannot know the actual ancestors of a group, but we can guess. A reasonable assumption is that things which resemble one another are genetically related and have a common ancestor. Thus, the few species lacking petals in *Passiflora* may all be closely related and all descended from an ancestor with no petals. In fact, this approach works well in reconstructing the family tree of the passionflowers. The occasional misleading "parallel" and "convergent" evolution (for example, red flowers evolving more than once, in different lineages) is the fly in the jam, and each new system of classification for *Passiflora* has been able to sort it out better. Modern analysis of the DNA of the plants will lead to an even better understanding about which plant is related to which, but that is only now beginning and has not been used for the classification used here.

Passionflowers are the more than 525 species of genus *Passiflora*. What puts them in this genus? Basically, these species all have a corona of filaments, 5 stamens, and sex organs elevated on a conspicuous column, the androgynophore. About 12 related genera are tendril-bearing vines or lianas but differ in the placement and number of sex organs, including *Adenia*, *Basananthe*, and *Dilkea* (de Wilde 1974). *Passiflora* and these other genera, many of which are African, are lumped together into the passionflower tribe, Passifloreae. There is little controversy at this level, except for the genus *Tetrapathea* from New Zealand. Most botanists consider this to be an unusual *Passiflora* deserving its own subgenus (Green 1972, Vanderplank 1991, 1996a), but we keep it separate as a closely related genus and do not treat it here.

The passionflowers' closest cousins are probably the shrubs and trees of the African tribe Paropsieae (genus *Barteria*, *Paropsia*, and so on), and these are usually placed in the larger passionflower family. The shrubs and herbs of the families Turneraceae and Malesherbiaceae are also extremely closely related. Sometimes all these are placed in the passionflower family, Passifloraceae.

As discussed in chapter 1, Johann Hallman wrote a dissertation in 1745, at the dawn of modern botany, on everything known about passionflowers at that time. A few years later Linnaeus added 2 species to the list in his seminal *Species Plantarum*, bringing the total to 24 species known to science—95% of the species known today were yet undiscovered. Explorers sent back increasing numbers of specimens of additional species, and it soon became apparent that there were many kinds, shapes, colors, and sizes of passionflowers. To understand this variation, botanists started placing the species into groups and naming those groups at various ranks, or levels. Characteristics used to segregate passionflowers into groups included type of bracts, length of floral tube, corona, leaf shape, and so on.

The history of systematics and classification of passionflowers basically follows a pattern of alternate lumping and splitting, all in response to the discovery of new passionflowers. As more species and strange and diverse forms were discovered, more categories inside *Passiflora* were created by various authors to hold the diversity. Eventually there were too many categories for convenience, and as "missing links" were discovered, some categories had to be merged, or lumped. Thus came the part of the cycle when another author was able to take the big mess and simplify it. Later, as still more new species were discovered, additional categories were named again, or split. This basic cycle has been repeated for 250 years. The system used in this book (Feuillet and MacDougal 2004) lumps all species of *Passiflora* into only 4 main categories, or subgenera. This system was devised to simplify the system in previous use (Killip 1938, amended) with its 24 subgenera. There are still just as many small subcategories, or groups of closely related species, but kinship relations are more accurately represented among the old subgenera by grouping them together until there are only 4 groups.

The hierarchy of the ranks, or categories, is fixed by convention. We use here the following hierarchy, each rank inclusive of the ranks below it:

Family
Tribe
Genus
Subgenus
Supersection
Section
Series
Species
Subspecies
Variety
Forma

Sometimes one rank is skipped, which is acceptable as long as the order is correct. Cultivated selections or varieties of a species or hybrid are called cultivars and are always designated by a capitalized name in single quotes.

The latest system owes nearly all of its names and most of its sections to the many earlier authors who have loved and studied passionflowers. A number of the category names that are still used were coined by Augustin Pyrame de Candolle in 1822 and 1828 when he named sections like *Decaloba*, *Dysosmia*, *Distephana*, and *Astrophea*. Other names, like *Xerogona* and *Disemma*, came from botanists who, amazed at the differences between the berries and capsular fruit, and between flowers with giant coronas and long-tubed flowers with no corona, and so on, split the passionflowers into a number of genera (Labillardière 1825, Rafinesque 1838, Roemer 1846). At the end of the 19th century and into the early 20th century, 3 major technical monographs laid the modern foundation for *Passiflora* systematics, including works by Masters (1871, 1872), Harms (1925), and Killip (1938). These monographs should be a part of the library of every student of passionflowers. Killip's unillustrated book is the reigning bible of passionflowers and includes 355 species. A new bible is not yet available but is needed, since approximately 140 new species have been discovered since 1938. The approximately 20 Old World species were treated and illustrated by de Wilde (1972).

This book presents passionflower species alphabetically arranged in separate chapters for each of the 4 subgenera, and the tacsonias receive their own special chapter. Hybrids are also presented alphabetically in a separate chapter. The introductions to each subgenus chapter include pertinent notes on systematics. To make it easier to look up an unknown species, some guidelines are provided on how to recognize the 4 subgenera. These rough guidelines are very generalized and do not cover the full variation of characteristics within each subgenus. Nevertheless, they should be accurate for more than 90% of the species.

Subgenus *Astrophea*

57 Species; South and Central America

Woody lianas, sometimes small trees. All but 2 species from South America. Leaves unlobed, never variegated, venation pinnate; plant glabrous; stem terete; new growth erect, sometimes cernuous; stipules tiny; petiolar nectaries 2 near apex, sessile or scarlike; lamina without nectaries or these inconspicuous at or near margin; peduncles 1- to 5-flowered, rarely borne from base of tendril; bracts tiny; flowers pink to purple or red to orange and usually with yellow corona and obvious floral tube, or white with yellow corona; coronal filaments without red to purple bands; operculum spreading or erect, not plicate; styles often stiff and projecting upward; fruits usually yellow, or greenish with reddish marks; seeds mostly reticulate. Chromosome number $n = 12$? (2 counts).

Subgenus *Deidamioides*

13 Species; South and Central America

Rare vines or woody lianas, sometimes with forked tendrils or tendrils with apical disks. Leaves unlobed (or 3- to 9-compound in 2 species), never variegated, often with 2 lateral veins near base of leaf; plant glabrous; stem terete; new growth erect; stipules tiny; petiolar nectaries 2, sessile, scarlike, or only slightly raised; lamina without nectaries or these inconspicuous and at margin; peduncles 1- to 2-flowered or flowers borne on tendril; bracts tiny; flowers whitish or greenish white with whitish to yellow corona, or yellow with yellow and orange corona, without obvious floral tube; coronal filaments without red to purple bands except present in *Passiflora obovata*; operculum more or less plicate; styles not stiff and erect; fruits greenish, yellow, or orange; seeds mostly reticulate. Chromosome number $n = 12$? (one count).

Subgenus *Decaloba*

214 Species; North and South America, Southeast Asia, Australia

Vines to small climbers, small flowers, small fruits that are often purple, grooved seeds common. Leaves often 2-lobed or with the central lobe shortest, sometimes 3-lobed or unlobed, rarely 5-lobed,

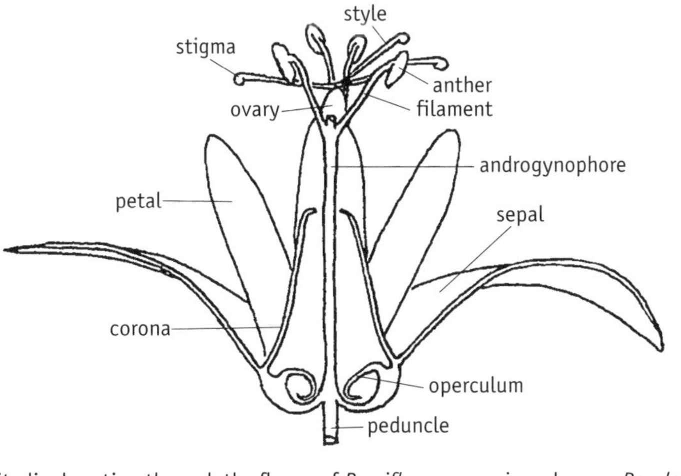

Longitudinal section through the flower of *Passiflora murucuja*, subgenus *Decaloba*.

sometimes variegated, with 3 main veins; plant glabrous to very pubescent; stem terete, striate, grooved, almost angulate, or 3- to 5-angled; new growth cernuous or erect; stipules usually bristle-like, sometimes leafy; petiolar nectaries absent or 2, rarely more than 4 and then the flower whitish, the nectaries (if present) raised up from the petiole; lamina with nectaries as dots between the main veins, or in a few species near the margin, or sometimes absent; peduncles 1- to 20-flowered, the flowers borne on tendrils only in 3 species of *Pterosperma*; bracts 3, tiny to bristlelike, or conspicuous and leafy, rarely 2 and leafy; flowers variable, generally small, yellow-green to white, or reddish, rarely with a floral tube; coronal filaments in 1–3 series, sometimes with red to purplish bands; operculum plicate; styles slender; fruits purplish or black to blue berries, or greenish, yellow, orange, or reddish, sometimes dehiscent; seeds reticulate or transversely sulcate. Chromosome number nearly all $n = 6$.

Subgenus *Passiflora*

236 Species; North and South America

Most abundant in South America. Vines to lianas, large colorful flowers often with complicated corona or sometimes with long tube, bracts usually conspicuous, fruits large. Leaves 3-lobed or unlobed, or up to 7-lobed, never variegated, with 3 (to 5) main veins or venation pinnate; plant glabrous to very pubescent; stem terete, striate, grooved, or almost angulate; new growth erect, or cernuous in some tacsonias; stipules inconspicuous or lanceolate to ovate, or ear-shaped and leafy, or deeply cleft; petiolar nectaries 2–6, rarely more than 6 or absent, usually raised up from the petiole; lamina with or without nectaries, if present then at margins, especially between the lobes of 3-lobed leaves; peduncles 1-flowered, the flowers never borne on tendrils; bracts 3, generally conspicuous and leafy, almost always wider than 0.25 cm; flowers quite variable, medium to large, purple, white, blue, violet, red, or mixtures of these, floral tube sometimes conspicuous; coronal filaments in (1–)3–20 series, if conspicuous then often with red or purplish bands; operculum not plicate, sometimes filamentous in upper part; styles often club-shaped; fruits greenish, yellow, orange, or reddish; seeds reticulate. Chromosome number mostly $n = 9$, also $n = 10$ or 11.

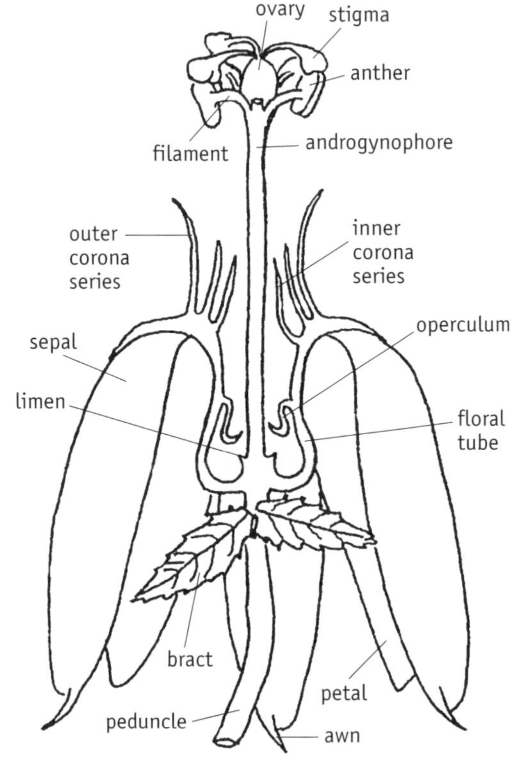

Longitudinal section through the flower of *Passiflora vitifolia*, subgenus *Passiflora*, supersection *Coccinea*.

Classification of Passionflowers

Passiflora **L.**	**520 species**
1. Subgenus *Astrophea* (DC.) Mast.	**57 species**
1.1. Supersection *Astrophea*	27 species
1.1.1. Section *Astrophea*	10 species
1.1.2. Section *Capreolata* MacDougal & Feuillet	15 species
1.1.3. Section *Leptopoda* Killip ex Feuillet & Cremers	02 species
1.2. Supersection *Pseudoastrophea* (Harms) Feuillet & MacDougal	30 species
1.2.1. Section *Pseudoastrophea* (Harms) Killip	17 species
1.2.2. Section *Botryastrophea* (Harms) Killip	13 species
1.2.2.1. Series *Botryastrophea* (Harms) MacDougal & Feuillet	06 species
1.2.2.2. Series *Carnae* Feuillet	07 species
2. Subgenus *Deidamioides* (Harms) Killip	**13 species**
2.1. Section *Polyanthea* DC.	01 species
2.2. Section *Deidamioides* (Harms) Feuillet & MacDougal	01 species
2.3. Section *Tetrastylis* (Barb. Rodr.) Harms	02 species
2.4. Section *Mayapathanthus* MacDougal & Feuillet	02 species
2.5. Section *Tryphostemmatoides* Harms	07 species
3. Subgenus *Decaloba* (DC.) Rchb.	**214 species**
3.1. Supersection *Pterosperma* Gilbert & MacDougal	04 species
3.2. Supersection *Hahniopathanthus* (Harms) MacDougal & Feuillet	05 species
3.3. Supersection *Disemma* (Labill.) MacDougal & Feuillet	21 species
3.3.1. Section *Octandranthus* Harms	17 species
3.3.2. Section *Disemma* (Labill.) MacDougal & Feuillet	03 species
3.3.3. Section *Hollrungiella* Harms	01 species
3.4. Supersection *Multiflora* (Small) MacDougal & Feuillet	019 species
3.5. Supersection *Auriculata* MacDougal & Feuillet	008 species
3.6. Supersection *Cieca* (Medic.) MacDougal & Feuillet	018 species
3.7. Supersection *Bryonioides* (Harms) MacDougal & Feuillet	020 species
3.8. Supersection *Decaloba* (DC.) MacDougal & Feuillet	119 species
3.8.1. Section *Decaloba* DC.	106 species
3.8.2. Section *Xerogona* (Raf.) Killip	013 species
4. Subgenus *Passiflora*	**236 species**
4.1. Supersection *Passiflora*	19 species
4.1.1. Series *Passiflora*	13 species
4.1.2. Series *Palmatisectae* Feuillet & MacDougal	01 species
4.1.3. Series *Pedatae* Killip ex Cervi	01 species
4.1.4. Series *Setaceae* Killip ex Cervi	04 species
4.2. Supersection *Stipulata* Feuillet & MacDougal	95 species
4.2.1. Section *Granadillastrum* Triana & Planch.	66 species
4.2.2. Section *Calopathanthus* Harms	01 species
4.2.3. Section *Tacsonioides* DC.	04 species
4.2.4. Section *Kermesinae* (Cervi) Feuillet & MacDougal	04 species
4.2.5. Section *Dysosmia* DC.	20 species
4.3. Supersection *Laurifolia* (Cervi) Feuillet & MacDougal	42 species
4.3.1. Series *Laurifoliae* Killip ex Cervi	21 species
4.3.2. Series *Quadrangulares* Feuillet & MacDougal	06 species
4.3.3. Series *Tiliifolia* Feuillet & MacDougal	14 species
4.3.4. Series *Marginatae* Killip ex Cervi	01 species
4.4. Supersection *Coccinea* Feuillet & MacDougal	14 species
4.5. Supersection *Distephana* (DC.) Feuillet & MacDougal	05 species
4.6. Supersection *Tacsonia* (Juss.) Feuillet & MacDougal	61 species
4.6.1. Section *Rathea* (Karst.) Harms	03 species
4.6.2. Section *Insignes* Feuillet & MacDougal	05 species
4.6.3. Section *Colombiana* Escobar	19 species
4.6.3.1. Series *Colombianae* Escobar	09 species
4.6.3.2. Series *Leptomischae* Escobar	08 species
4.6.3.3. Series *Quindiensae* Escobar	02 species
4.6.4. Section *Parritana* Escobar	02 species
4.6.5. Section *Fimbriatistipula* Escobar	02 species
4.6.6. Section *Tacsoniopsis* Triana & Planch.	02 species
4.6.7. Section *Elkea* Feuillet & MacDougal	15 species
4.6.8. Section *Tacsonia* (Juss.) Harms	05 species
4.6.9. Section *Boliviana* (Harms) Feuillet & MacDougal	02 species
4.6.10. Section *Trifoliata* Feuillet & MacDougal	01 species
4.6.11. Section *Manicata* (Harms) Feuillet & MacDougal	05 species

3
Morphology

THE MORPHOLOGY and anatomy of the family Passifloraceae and genus *Passiflora* are not well known, the most significant general works being more than a century old: Masters's monograph of 1871 and the works of Harms (1893a, 1898, 1925). Harms's publications, written in German, are brilliant but have been overlooked by many English-speaking students of the genus. Not counting the closely related Turneraceae and Malesherbiaceae, most modern scholars agree that the passionflower family is composed of 2 tribes: the tribe Paropsieae, a small group of African shrubs and trees, and the tribe Passifloreae, composed of about a dozen genera of mostly tendril-climbing vines. For those interested, brief descriptions of the morphology of 21 other genera of the family are presented in John Hutchinson's volume on the Passifloraceae (1967), the best modern synthesis of the family as a whole. Willem J. J. O. de Wilde (1971b) produced a key to the 16 genera of the Passifloraceae recognized at the time of publication (including tribe Paropsieae) and subsequently (1974) reexamined the "genus concept" and generic delineation in the tribe Passifloreae based on the morphology of the flowers. His article includes flower diagrams and line drawings of longitudinal sections of the flowers of "representative species" of the 9 Old World genera that he recognized. All have a corona of some sort. In a revision of supersection *Bryonioides*, John MacDougal (1994) reviewed the morphological literature for the genus in detail and listed many other references. More recently, Alex Bernhard (1999) examined the development of the flower in several genera, paying special attention to the corona, nectaries, and stamens. He suggested that the Old World genus *Adenia* was very primitive in the family. There are some good treatments of the morphology of certain organs and structures in the genus (for example: tendrils, Cusset 1968; pollen, Presting 1965, Spirlet 1965), but these are not as useful for general information on the genera.

Habit

Passionflowers range from large woody canopy lianas more than 35 m long to delicate climbing annuals that reach only knee-high. A few are shrubby or even small trees (subgenus *Astrophea*). This form is most unusual in *Passiflora* and holds a special fascination for almost all students of passionflowers. Some treelike forms are similar in habit to pawpaw trees: slender, understory, mostly single-stemmed. Others are more branched but still short, with a trunk only as big as one's arm. A few are truly trees, with at least 5 species known to reach 15 m tall and 20 cm diam. at breast height: *P. lindeniana*, *P. sphaerocarpa*, *P. arborea*, *P. tica*, and *P. emarginata*. The vast majority of passionflowers, however, are medium-sized vines that are woody (or woody at the base) or forest-edge climbers that are a few meters long. Generally, species of subgenus *Decaloba* are smaller plants than those of other subgenera. The few known annuals include *P. gracilis*, *P. tenella*, and forms of *P. foetida*. The rest of the passionflowers are perennial, although a few kinds from temperate or seasonal zones, such as *P. incarnata*, *P. lutea*, and many species in supersection *Bryonioides*, have annual stems that die back to the perennial roots during a season of dormancy each year. The usual or maximum size or habit is rarely recorded by collectors and is poorly known for most species. Vegetative reproduction by rhizomes or adventitious buds on roots is sometimes seen in a few species and is the rule in certain nontropical species, such as *P. lutea*, *P. incarnata* (Wehtje et al. 1985), and *P. caerulea*.

Root

The root is very rarely observed or collected in *Passiflora*. Masters (1871) went so far as to say that the root "offers few characters worthy of special comment . . . It is probably very uniformly fibrous throughout the [family]." Actually, roots of old plants commonly become quite woody and thick, at least the primary and secondary roots, and a few are known to have a woody taproot, such as *P. exsudans* and especially *P. karwinskii*. Roots remain

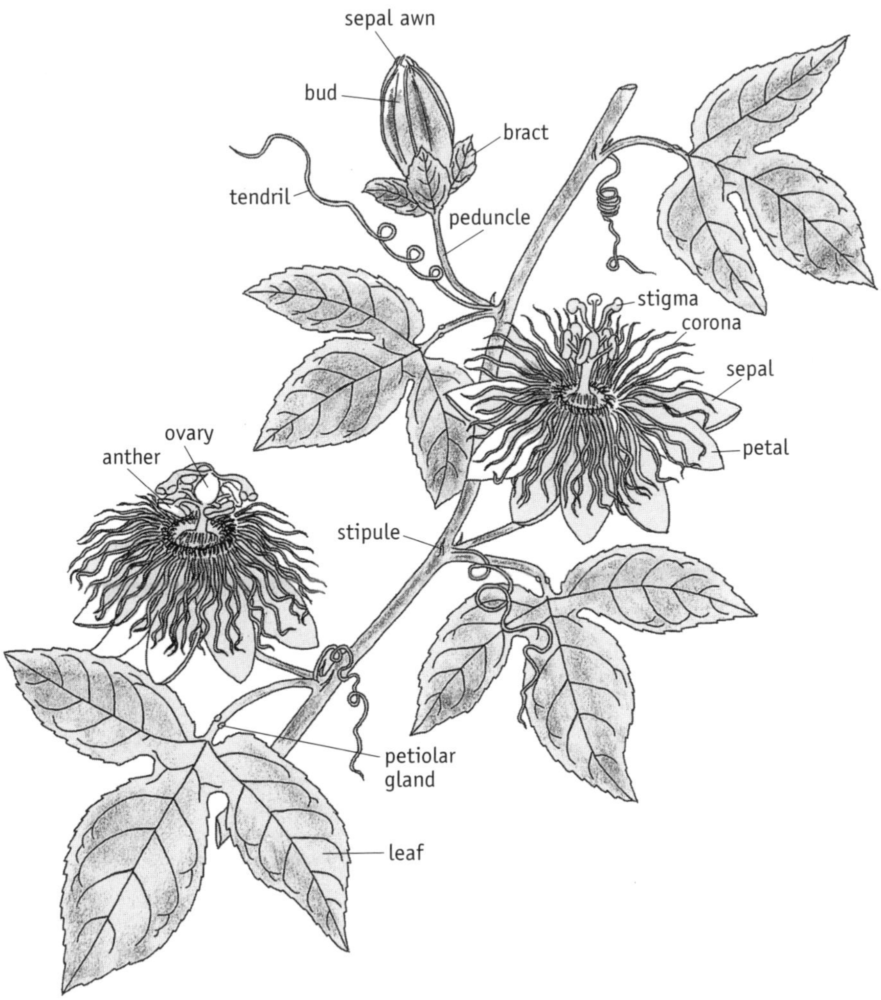

Part of a flowering branch of *Passiflora edulis*.

poorly studied, however, and those of most species are unknown.

Stems

The stems of the climbing passionflowers are slender, of course, and have minimal wood since they depend on the wood of other plants and trees to provide support in their quest for sunlight. The wood that is there is usually laid down in concentric rings, although there are no easily discernible annual growth rings that could be used to count the age of the plant. The few that are trees have normal wood. Anomalous secondary growth is found in a few species, such as *P. multiflora*, and in many species of subgenus *Astrophea* section *Botryastrophea* (Ayensu and Stern 1964).

The shape of the stems ranges from perfectly terete to roughly terete or obscurely angulate or sulcate, to definitely or sharply angled, or angled with sharp edges called wings, and a few stems are even flat and ribbonlike. Most stems are at least a little striate, and these become exaggerated when the stem is dried, as in a museum herbarium specimen. Sometimes stems are described as "subangulate and striate" when in fact the living plant has a terete stem; this may be because the description was made from a dried museum specimen. Remember that the stems of passionflower vines are basically straight; they climb by means of tendrils, never as twining vines where the stem wraps or spirals around the support.

In species of *Passiflora*, one important characteristic of the stem that has considerable taxonomic (and perhaps ecological) importance is the posture of the apex in relation to the main axis of the shoot (MacDougal 1994). The actively growing shoot tip may be more or less straight or somewhat zigzag, in any case, with the apex in general line with the rest of the stem (the very tip therefore facing away from the earth), or the tip may be cernuous and crook-shaped (the very tip facing the earth). In the species with the cernuous shoot growth, the very apex seems to be positively geotropic. The nodding shoot tips seem to be an advanced trait in the genus found in a relatively small number of groups of species and not in other genera of the family. The cernuous apex is presently known only in part of subgenus *Decaloba* (supersections *Pterosperma*, *Hahniopathanthus*, and *Decaloba*), part of subgenus *Astrophea* (at least section *Astrophea*), parts of supersection *Tacsonia* in subgenus *Passiflora* (section *Rathea*, part of section *Colombiana* including series *Leptomischae*, part of section *Elkea*), and occasionally in *P. setacea*. This recently discovered characteristic needs to be examined carefully across the genus, and home or garden observations are exceedingly valuable. The spotty occurrence in *Tacsonia* suggests a reexamination of the sectional taxonomy is needed.

Node

The node is the place where the leaf joins the stem. A close look at the stem of a *Passiflora* will reveal that a leaf, 2 stipules, and 2 axillary buds in a vertical series are present at each node. The lowermost bud, closest to the subtending leaf stalk, gives rise to a tendril or tendril-tipped inflorescence, whereas the upper, more distal bud (the "accessory bud") is tiny and vegetative and is located where the plant may branch. In the great majority of passionflower species, the stalk or peduncle of the inflorescence does not develop, and the flower or flowers appear to arise on either side of the tendril.

Nodes and their leaves are arranged in an alternating spiral along the stem in all species of Passifloraceae except the Southeast Asian *Passiflora moluccana* var. *teysmanniana* (syn. *P. horsfieldii*, *P. cochinchinensis*), in which the leaves are almost opposite or sometimes perfectly opposite (Harms 1925, de Wilde 1972a). The arrangement of leaves along a stem is called phyllotaxy and can be expressed by a simple fractional notation (Kirchoff 1984). The 2/5 phyllotactic arrangement is characteristic of the family as a whole. The stem is circled 2 times by leaf attachments before one leaf is directly over another, and it is the 5th leaf that is over the first leaf from where one may have started counting. There are a few exceptions in *Passiflora*. These are the apparent 1/3 arrangements in some species with triangular stems in section *Xerogona* and *P. trialata*, and the apparent 1/2 (distichous) arrangement in *P. quadrangularis* and its relatives, in

P. complanata and its relatives, and in variants of *P. misera*. Several other species of supersection *Decaloba* (formerly subgenus *Plectostemma*) have strongly flattened stems according to Killip (1938), and these, too, may diverge from the family norm.

Vestiture

The presence or absence of trichomes, and their nature and density, is called vestiture. Passionflowers range from glabrous to glabrescent, through a full range of degrees of hairiness, or pilosity, to densely pubescent. Various terms are used to describe the kinds of vestiture, such as *pilose* or *lanate*. The trichomes themselves are of various forms but are usually composed of one cell emerging from the epidermis of the plant. Different parts of the plant body may have or lack pubescence, though the upper surfaces of the leaves typically have the least, while the nodes of the stem and apices of the petioles have the most. In many species, only the ovary is pubescent. Sometimes the vestiture of the seedlings is very different from that of the adults.

Vestiture is variable across the genus *Passiflora*, and it is difficult to discern a simple pattern. In subgenus *Decaloba*, there are often 2 kinds of hairs on the plant body: nearly microscopic, short, bent-over, sausage-shaped hairs, and longer, more or less erect, straight hairs. Supersection *Bryonioides* is characterized by sharply hooked, or uncinate, trichomes that snag like miniature Velcro. On the other hand, species of supersection *Hahniopathanthus* are completely naked. Subgenus *Astrophea* has short pubescence of often yellowish or brownish curving hairs, though this is sometimes very sparse and seen only on the petioles. Subgenus *Deidamioides* appears to be glabrous. The nearly 250 species of subgenus *Passiflora* show a great range of pilosity, from completely glabrous (*P. subpeltata*, *P. quadrangularis*, *P. ligularis*) to hispid (*P. menispermifolia*), some covered in dense, short, brownish fuzz (*P. coactilis*), others in a thick coat of matted, lanate hairs (*P. lanata*). Species in section *Dysosmia* deserve special mention, for many species have glandular trichomes (trichomes that have a swollen tip with secretory function).

Prophylls

The uppermost, or distal, bud where each leaf joins the stem in *Passiflora* is a vegetative bud from which the plant may sprout a side branch. Of course, there will be leaves on that side branch, but the very first leaflike appendages, or prophylls, of that branch are the tiny bud scales already seen at the bud. These prophylls never expand when the side branch grows and are morphologically distinct from the other leaves of the branch. Most dicotyledonous plants have 2 prophylls (Esau 1965, Guédès 1979), but Harms (1925) noted that *Passiflora* may have 1 or 2. The presence of 2 prophylls is primitive in *Passiflora* and is seen in subgenus *Astrophea* and subgenus *Deidamioides*. Species of the subgenera *Decaloba* and *Passiflora* may have either 1 or 2, depending on the section and species. This microscopic characteristic can sometimes be used to identify species and sections. Unfortunately, the prophylls of the vegetative bud are hard to see without a magnifying glass and are extremely difficult to see on pubescent species. Those studying passionflowers in the future should document the variation in prophyll number and its usefulness in the taxonomy of passionflowers.

Stipules

The paired appendages on the stem at the base of the leaf stalk are called stipules. These are found in many plant families but have been lost in others. All species in the passionflower family have stipules. They range from microscopic or extremely small and deciduous, as with subgenus *Astrophea*, subgenus *Deidamioides*, and *Passiflora microstipula* and its relatives in subgenus *Decaloba*; to setaceous, threadlike, or falcate, as with supersection *Decaloba*; to lanceolate, as with many species in supersection *Laurifolia*; to slightly leafy, as with many species in subgenus *Passiflora*, especially supersection *Stipulata*; to fantastically large and leaflike, as with supersection *Hahniopathanthus* and *P. galbana*.

Some of the shapes of stipules can be hard to describe, and one common shape in particular has received many names in different books. This conspicuous stipule has a main vein and a leafy expansion mostly on one side of that vein. The stipule is joined to the stem slightly to the side of its base. The shape has been called semiovate, obliquely ovate, dimidiate-ovate, ovate-amplexicaule, reniform, semireniform, subreniform, auriculate, and ovate-auriculate. Of course, among the hundreds of species that have leafy stipules, there are many minor variations; for instance, some are pointed at the tips, while others are blunt. For the purposes of this book, this kind of stipule is described as semiovate or subreniform. Each of these terms may be more or less correct, but it is important to note that these are relatively similar shapes of leafy stipules, as opposed to the stipules that are setaceous or falcate.

In species with the larger stipules, the margins may be serrate, toothed, or glandular with extrafloral nectaries. In a few—*Passiflora pinnatistipula*, or section *Dysosmia*, for example—the stipule is deeply dissected into slender segments. Some species, such as *P. retipetala*, sport butterfly egg mimics on their stipules.

Leaves

A leaf has 2 basic parts: the petiole and the lamina. The petioles in passionflower vines are usually well developed and often slightly flattened, or may even have a little canal on their adaxial surface. Pubescence, if present, usually shows up here the strongest. There are often 2 or more nectary glands on the petiole, although the petiolar glands have been lost in a number of species and in all of supersection *Decaloba*.

The laminas are also well developed, ranging from thin and membranous to thick and coriaceous, and are between 0.5 cm and 1 m in length. If there is pubescence it is usually thicker on the abaxial leaf surface. The underside is also sometimes paler than the upper. In a few species that grow in deep shady rain forests, such as *Passiflora costaricensis*, the leaf is not flat but bullate. Laminas can be unlobed, 2-lobed, 3- to 5-lobed, or even 9-lobed. The 2-lobed leaves evolved from 3-lobed leaves. The venation of the leaves is basically palmate (3 or 5 main veins), but there are many unlobed species with pinnate venation (1 main central vein). Venation patterns of the leaves of many species of *Passiflora* were examined by Edward P. Klucking (1992), who published an excellent atlas of photographs of cleared leaves. Cordate leaves are a common feature of climbing plants in general, and lobed leaves in *Passiflora* are often cordate at their base. The base of the lamina can also be truncate, rounded, or cuneate. A very few species are compound, as is the case with *P. deidamioides*, *P. cirrhiflora*, *P. pedata*, and *P. trifoliata*. This is an advanced condition (Cusset 1965, 1970).

Independent of whether they are simple or compound, lobed or unlobed, the margins of the leaves of passifloras can be variously serrate, serrate-lobed, dentate, denticulate, or entire. The teeth are associated with a tiny vein ending, and sometimes the tooth has a small swollen glandular apex. The teeth are transformed into obvious marginal laminar nectary glands in some species. Cusset (1965, 1970), Roth (1974), and Harms (1893a) present a broad study of the morphology and ontogeny of the leaves of the family, including the homologies and derivation of extrafloral nectaries from marginal teeth.

The genus *Passiflora* itself exhibits the most diverse leaf shapes of any genus of plants, probably due in part to evolutionary pressures from passionvine-eating butterflies that identify the plants by sight (Gilbert 1975, 1982). Basically the different leaf shapes are a kind of camouflage for each species, meant to keep the pest butterflies from finding them. Many speculations have been made about certain passionflowers mimicking the leaves of other, nonpassiflora species with which they grow. Though probably true, this is hard to prove scientifically. The butterflies have also affected the shape and color of other parts of the passionvine (egg mimicry by stipules, petiolar nectaries, and so on) (Williams and Gilbert 1981; see also chapter 19). It is usual among passifloras for the leaves to change shape as the seedling, then young plant, grows into maturity. Often the young leaves are more deeply lobed; sometimes the adult leaves are more deeply lobed. This change in shapes along the stem is called heteroblasty. Some species have a more complicated heteroblastic series of leaves, with at least 3 main stages. In subgenus *Decaloba* the change is not just shape but also the presence or degree of pale stripes or variegation on the leaves. Variegation is always stronger and more extensive in juvenile plants, and species that have variegated leaves at flowering are rare. Scientists remain ignorant of these changes in most species because it is necessary to start with seeds and grow the plant through to maturity. This has never been done with the vast majority of *Passiflora* species, so even single observations of seedlings (photographs, notes, pressed specimens) are very valuable to science.

Extrafloral Nectaries

Glands functioning as extrafloral nectaries are common in the family and are thought to be homologous to and derived from glandular teeth of the margin of the lamina (Cusset 1965, 1970; Hickey and Wolfe 1975). This is obvious for marginal glands on the lamina itself, but the glands on the petiole, too, are thought to be modified and derived from these same teeth. In some species—for example, some of those in supersection *Tacsonia*—a complete transitional series of glandular teeth can be seen that start with the leaf margin and change to petiolar nectaries. Glands on the stipules and bracts are similar in that some of the same genes involved in making leaves are involved in making leafy stipules and bracts (themselves a modified leaf). Functionally, these glands attract and reward insects like ants and wasps, which then act as patrols and bodyguards, fending off animals that would eat the plant, such as *Heliconius* butterfly caterpillars (Gilbert 1982).

The presence or absence of these various nectaries, and their form, are extensively used in identifying passionflowers. One of the first characteristics to look for when identifying a species is the petiole. Does the plant have petiolar nectaries, and if so, how many? Only a few groups lack these, notably supersection *Decaloba*. The nectary glands are often paired, or more or less opposite, and in many species there are only 2. They often appear as raised bumps or resemble tiny barrels or cylinders; if narrowed where they attach, they may resemble trumpets or spoons. Sometimes they are not raised at all and look like a scar, and some are so far up the petiole as to actually be on the base of the main vein of the leaf itself.

Laminar nectaries occur in several forms, including as glandular marginal teeth, strictly marginal glands (as with *Passiflora platyloba*); submarginal glands associated with minor veins of the lower surface (as with many *Astrophea*, supersection *Coccinea*, supersection *Pterosperma*, and supersection *Cieca*); and as ocellae between the 3 major veins or scattered over the lower surface of the lamina (as with *P. auriculata*, section *Decaloba*). The lack of any laminar nectaries characterizes a few groups, such as section *Xerogona*. More details on extrafloral nectaries in passifloras can be found in works by Cusset (1965, 1970), Durkee (1977, 1982, 1983), and Roth (1974).

Tendrils

The evolutionary origin and morphological interpretation of the tendril in Passifloraceae is reviewed by Cusset (1968), who presents a literature review of the subject. The tendril is always perfectly axillary—that is, exactly in and above the crotch of the leaf. This fact is used in the field to easily distinguish passionflowers from Cucurbitaceae, vines of the cucumber family, which have tendrils coming off from the side of the node. The tendril is simple and unbranched in 95% of passionflowers, another very useful field characteristic, for many other families of vines usually or always have branched tendrils, such as Cucurbitaceae and Vitaceae. Three- to 5-parted tendrils are seen in section *Tryphostemmatoides*. A very few species regularly produce adhesive disks at the apices of the tendrils, allowing seedlings to climb straight up the smooth trunks of rain forest trees. This is a characteristic of the species of section *Tryphostemmatoides* and the juveniles of section *Mayapathanthus*. Only a few passionflowers never produce tendrils, the tree form species of section *Astrophea* being one example.

There has been much interest in how tendrils respond to touch, and Charles Darwin (1884) found the tendrils of *Passiflora gracilis* to be the most sensitive to touch of any tendril-bearing plant in his study. During their receptive stage, the tendrils of *Passiflora* move, sweeping their tips in circles or ellipses. Darwin reported that one passionflower vine in his study had the fastest-moving tendrils of all other climbing plants known to him: one tendril completed a revolution in about 43 minutes at 32°C. Once a receptive tendril makes contact and curls around an object, the tendril continues to grow apically 1 cm or more, further anchoring the tendril to the support.

Inflorescence

The ground plan of the inflorescence in *Passiflora* is closely related to the tendril. This is because primitive passionflowers had a stalked inflorescence at each node, each with groups of flowers. The common stalk is called the peduncle. The smaller stalks, each of which holds a flower, are called pedicels. The tendril originated as a modification of the central flower's pedicel. So the primitive condition is a stalked inflorescence with a tendril in the middle of several flowers. This is still seen in some passifloras, such as *P. cirrhiflora*, *P. deidamioides*, section *Tryphostemmatoides*, some *Astrophea*, supersection *Pterosperma*, and many other genera, like *Adenia*. Reduction and loss of that common peduncle of the inflorescence in nearly all *Passiflora* species has given rise to forms of a sessile inflorescence in which the pedicels arise side by side with the tendril in the axil of the leaf. In this derived arrangement it is easy to overlook that the tendril and 1 or 2 flower stalks are really part of the same unit. Now comes semantics: different authors use "peduncle" and "pedicel" in different ways. Since the peduncle common to the tendril and pedicels has been lost, most taxonomic works, including Killip's (1938), refer to the remaining pedicels as the "peduncles." Although this practice obscures the morphological relationships and homologies of the parts of the inflorescence between the genera in the family, most modern passionflower specialists also avoid the word "pedicel" and instead use "peduncle" in its loose sense in the descriptions, because it is easier for the nonspecialist to understand. It doesn't really matter what it's called, as long as it is clear what structure is being discussed. The evolution of inflorescences in *Passiflora* and *Adenia* is reviewed by Cusset (1968), and that of the Passifloraceae (especially *Adenia*) and the Flacourtiaceae is reviewed by de Wilde (1971a, 1971b).

The only confusion now lies with the modern use of the term "pedicel." The pedicels (or "peduncles," as we now call what is left) are jointed—that is, they have an abscission layer either just at the flower or at some distance below the flower. On the flower side of the joint, the tissue is similar in color to the outside of the flower, not the rest of the stalk. Sometimes a word is needed to refer to the part of the flower stalk that falls away with the flower or fruit. The stalk region that is distal to the articulation is often called the pedicel (Deginani 2001, Jørgensen et al. 1984, Sowerby 1794, Tillett 2003). De Wilde (1971a) started calling this the "stipe" of the flower, and this was adopted by MacDougal in his publications. In this book, we do not mention the articulation, except in unusual cases, and we do not use the word "pedicel" but instead call the whole flower stalk the "peduncle," acknowledging that it is jointed at some point along its length.

The peduncles in *Passiflora* are solitary or in pairs at each node. Sometimes they are branched, forming numerous flowers at each node, as with *P. complanata*, *P. sexflora*, and *P. holosericea*. A secondary aggregation of flowers in condensed terminal shoots or on short (rarely long) axillary branches is found in a number of passionflowers. Short leafless spurs off old wood, loaded with flowers, are not uncommon, as can be found in *P. amoena*, *P. pergrandis*, *P. involucrata*, and *P. rovirosae*. Longer spurs or pseudoracemes are also common, as with *P. racemosa*, *P. balbis*, *P. longiracemosa*, and *P. vitifolia*. In some species this inflorescence can resume vegetative growth. Leafless condensed terminal shoots are found in *P. coriacea*, *P. viridiflora*, and their relatives, and many species have shoots with reduced leaves and many flowers. Mass flowering, also called mast flowering or big-bang flowering, often produces boughs of flowers that function as a kind of leafy inflorescence, as occurs in *P. multiflora*.

Bracts

The presence on the "peduncle" of 3 bracts or bracteoles (prophylls) subtending each flower is a characteristic of most groups of *Passiflora*. This a rare condition in the plant world, since almost all flowering plants have 2 bracts beneath each flower. The bracts range from minute and setaceous to leafy and fused together so as to form a large cup around the bud or flower. The primitive condition in the genus is the presence of only 2 small bracts on each peduncle and/or pedicel—seen, for example, in some parts of subgenus *Deidamioides*. The presence of 3 bracts is considered more advanced, although by subsequent loss of a bract, a few species have reverted to a 2-bracteate condition (*P. hahnii*, for example, and forms of *P. suberosa*). The topology of the inflorescence and the homology of the bracts to prophylls of the branches of the inflorescence has been analyzed by Harms (1898) and Cusset (1968).

Bracts vary greatly in size and form and were among the first characteristics used to define subgroups in the genus *Passiflora* (de Candolle 1822). A good rule of thumb is that if the bracts are large (wider than 0.3 cm), the species belongs to subgenus *Passiflora*. They are often distinctive markers of sections: bracts in section *Dysosmia* are deeply dissected, with the segments often gland-tipped; there are no bracts in section *Xerogona*; bracts in subgenus *Astrophea* are uniformly very small; bracts in series *Tiliifolia* are very large and fused together at their bases; and bracts in section *Elkea* are medium and united into a cup.

Flowers

Those that grow passifloras often do so just to behold the exotic flowers, their fringed perfumed coronas or long floral tubes most unusual among flowering plants. With more than 525 species and many hybrids from which to choose, gardeners are confronted with myriad variations of color and form. We present here an overview of these variations, along with an analysis of the parts of the flower, which really are not so complicated as they first appear.

The flowers among various species range from tiny, as with those of *Passiflora multiflora* and *P. suberosa*, which are about 1 cm, to as large as a baby's head, as with *P. pergrandis* and *P. speciosa*, to long (almost 20 cm) and slender as a tube, as with some tacsonias. All of this variation reflects differences in the pollinators that feed at the flowers. Nearly all species have "one-day" flowers, flowers that remain open for less than a day or often for only 1/2 a day. Just a few species stay open 2–3 days, including *P. aurantia*, *P. jorullensis*, and a number of tacsonias from the cool Andes.

Floral Tube

The part of the flower where the sepals, petals, and corona attach, and which holds the nectary at its base, can be shaped like a cup, a flat saucer, or a short or long tube. If this structure is deep enough to be noticed, we call it the floral tube, following the usage by Masters (1871), Puri (1948), Brizicky (1961), and, in part, Tillett (1988, 2003). Stephen S.

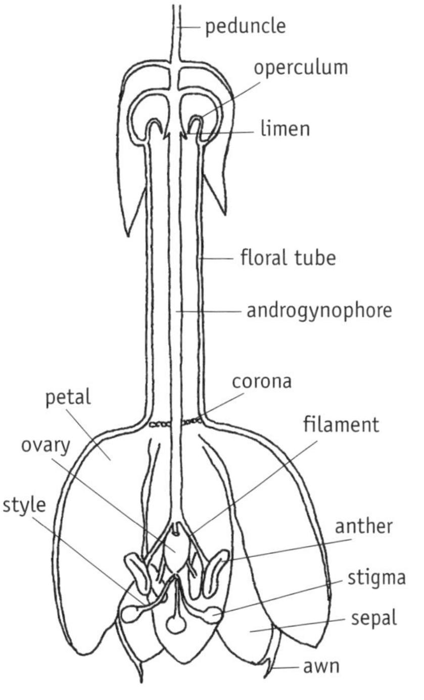

Longitudinal section through the flower of *Passiflora mixta*, subgenus *Passiflora*, supersection *Tacsonia*.

Tillett correctly notes that the bottom of the cup, the part that holds the nectary and androgynophore, is derived from stem and is thus of different developmental origin than the upper part of the flower from the operculum and corona upward to the sepals and petals. He calls only the lower part the floral cup and the upper part, derived from sepals and petals tissue, the floral tube. In this book, we lump it all together as the floral tube.

The structure of the floral tube has been given many names. It was called the receptacle by Harms (1925) and Bernhard (1999) and the calyx tube by Killip (1938) and Jørgensen and others (1984). It has been called the hypanthium by others (Deginani 2001; de Wilde 1971a, 1971b, 1972a, 1972b, 1974; Holm-Nielsen et al. 1988; MacDougal 1994). For our purposes, all these terms refer basically to the same thing.

The floral tube has been formed in slightly different ways in various species, depending which parts of the perianth and corona have fused. The best modern discussion of the parts and structures of the passionflower is Tillett's Spanish-language "Passionis Passifloris II" (1988). Bernhard (1999) provides the best modern analysis of floral development.

Sepals

Passifloras have 5 sepals that overlap their edges while in bud: 2 are wholly exterior, 2 are interior, and one has a margin covering an interior sepal and another margin covered by one of the exterior sepals. This appears to be true in all species in the family except in *Passiflora perfoliata* and the related *P. penduliflora*, in which the sepals and petals grow to maturity as free appendages. Because of the overlapping in bud, the sepals vary somewhat in shape and size in any one flower, being larger when they are outermost. This is most obvious in species with small flowers. Any edge of a sepal that is covered by another is thin and pale.

Some species have a notable ridge down the outside, and these are called carinate. A projection off the tip of the sepal, or just below the tip, is a common characteristic of certain groups, especially subgenus *Passiflora*. A few species in subgenus *Decaloba* also have it. This projection can be small and bristlelike, or like a small horn (a *cornus*), or longer (an *awn*). Sometimes it is associated with the keel on the sepal and can be quite long, laterally flattened, and even a little leafy. In a few species with exaggerated awns, the awns are hollow. Typically, the length and development of any corni present at the apices also vary with position of the sepal in bud, so that the outer 3 sepals have longer corni.

Petals

The 5 petals are more delicate and membranous than sepals and are generally equal in size and shape in any one flower. They may be the same size as the sepals, smaller (sometimes tiny in comparison, as with *Passiflora auriculata* and *P. gilbertiana*), or even absent (as with supersection *Cieca*, some clones of *P. apetala*, and some clones of *P. jatunsachensis*). In a very few cases the petals are slightly larger than the sepals, as with *P. pectinata*, in which the petals are longer, and *P. guatemalensis*, in which they are wider.

Corona

The corona is the hallmark of the genus *Passiflora* and probably the main source of the crucifixion symbolism. Although present to varying degrees in all genera in the family, it is most highly developed and differentiated in this genus. The other genera have one or a few rows of simple filaments or a membrane; illustrations can be found in works by de Wilde (1974) and Bernhard (1999). In *Passiflora* these rows show various degrees of specialization in size, color, fusion, and function. The corona is not a special sex organ, nor is it derived from the stamens. It is simply a bunch of outgrowths from the base of the petals and sepals, often at the top of the floral cup. The origin and nature of the corona have been investigated many times over the years, and it is generally agreed that the corona is simply derived from petals and sepals (Bernhard 1999).

The morphology of the passionflower, particularly of the corona in its various forms, has been analyzed and illustrated by a number of authors (de Wilde 1974, Harms 1925, Lindman 1906, Masters

1871, Puri 1948, Sowerby 1794, and Sprengel 1793, and especially Bernhard 1999, Jørgensen et al. 1984, and Tillett 1988). Various terminologies for the series of threads, membranes, and rings have been used, but since 1938, students of the family have mostly followed Killip. Tillett (1988), in an effort to stabilize these terms, has provided an excellent and complete survey of the floral morphology of passifloras, including a historical analysis of the changing uses of the terminology and suggestions for standard usage. Carl Axel Magnus Lindman, in his excellent work of 1906, introduced the terms *radii* and *pali* for the conspicuous outer and the shorter, less conspicuous inner coronal filaments, respectively, and these terms are used by many authors. The distinction between radii and pali is sometimes more or less arbitrary, depending on the degree of differentiation of the various rows of the filamentous corona. Although the terms are useful, we have here used simply *outer corona* and *inner corona* to refer to these rows of filaments.

Inside the outer and inner corona is an innermost coronal row that usually functions to cover the nectary chamber. In many passifloras this row is more or less fused into a membrane. Since it has such a specialized form and function, we refer to this not as the innermost coronal series but as the operculum (Sowerby 1794). The operculum has diverse forms in *Passiflora* and defines major taxonomic groups of species. In subgenus *Decaloba* it is a pleated membrane that is often slightly curved over toward the column. This was the source of the older subgeneric name, *Plectostemma*, which, meaning "pleated wreath," likened the operculum to a folded wreath around the base of column. In the *P. murucuja* group, the operculum is a conspicuous tubular membrane jutting far from the base of the flower, and is unusual enough in appearance to have spawned a separate genus or subgenus for that species group in the past. In subgenus *Astrophea*, the operculum is membranous at the base, often tubular and erect but sometimes spreading, and the distal part is fringed or cleft. There are many variations of this structure, and the large subgenus *Passiflora* features the most varieties. Here it can be an entire or denticulate nonfolded membrane, as with section *Dysosmia*, or a dependent membrane, hanging down toward the nectary with the very margin often upturned, as with supersection *Tacsonia* and supersection *Distephana*; in the rest it ranges from a single row of free filaments, or filaments joined at the base, to a membrane with filaments or small teeth at the apex, or can even be entire. Sometimes the flower must be cut open to see this structure. One species, *P. quadrangularis*, has 2 membranous opercula formed from its 2 innermost coronal rows. The operculum excludes small insects and ineffectual pollinators from the nectar, keeps out rain, and, in some species with pendent flowers, prevents nectar from dripping out of the flower. A few species, such as *P. penduliflora* and *P. andina*, have no operculum.

The outer corona is often reduced to 1 or 2 thick, colorful, specialized rows of filaments. It serves as a visual attractant for pollinators, as a landing platform in some species, as a high-surface-area perfume generator, and, through its physical interaction with parts of the androgynophore and the sex organs, as a way to screen or specify which pollinator may access the nectar reward. In subgenus *Astrophea* and subgenus *Decaloba* it is often yellow and swollen above the middle; in subgenus *Passiflora* it is variable, though often banded with purplish red; in supersection *Tacsonia* it is usually reduced or even nearly absent.

An additional 1 or 2 series of inner filaments are the norm, and in subgenus *Passiflora* (except supersection *Tacsonia*), there may be many rows of increasingly smaller capillary filaments or small teeth, sometimes growing larger again as the central column is approached. The inner coronal filaments are slender but often minutely swollen at their tips, or capitellate.

Nectary

Under the operculum, or between the column and the innermost coronal series, is the floral nectary. It usually appears as a circular trough, sometimes yellow, sometimes white. This produces sugary nectar for pollinating visitors. In many groups with larger flowers there is a low, raised annular ridge within the nectar trough that has been called the nectar ring (Harms 1925; Killip 1938; de Wilde 1972a, 1972b; Holm-Nielsen et al. 1988), annulus (Puri 1948; Tillett 1988, 2003), or nectary ring (Durkee 1981). This is not part of the corona but is merely an infolding of the inner surface of the nectary. In a few species, such as *Passiflora pyrrhantha*, the nectar chamber is divided into 5 containers of nectar.

Limen

Between the nectary and the base of the androgynophore, or near the base of the androgynophore, is tissue that is apparently derived from stamens or staminal tissue (Bernhard 1999, Puri 1948). Lindman (1906) called this the limen, a term used subsequently by those in the field. It often forms a ring or cuplike structure around the base of the androgynophore. In some species it is so flat or short that it is hard to see, and only its outer edge is free. In other species it is more conspicuous. The edge of the limen often interacts with the operculum and helps enclose and shelter the nectary. It is not part of the corona.

Androgynophore

The androgynophore (literally "male-female-bearer") is sometimes also called the column. It is composed of an ovary on a long stalk to which stamens have become attached. The staminal filaments form a connate tube, and this tube surrounds, or is adnate to, the ovary stalk. The stamens continue past the ovary, and their free parts hold the anthers. The androgynophore is straight in nearly all passifloras but is strongly curved in several, including *Passiflora mucronata*, *P. discophora*, *P. ovalis*, and *P. lobata*. In a very few it is reduced to almost nothing or is absent, in which case the ovary and stamens are not elevated, as with *P. multiflora*, *P. xiikzodz* subsp. *itzensis*, and *P. monadelpha*. In the nontubular species of subgenus *Passiflora* there is sometimes a large swelling near or at the base of the androgynophore. This is called the trochlea, and in some species it is related to the flower stalk, attaching deeply into an invagination in the base of the passionflower. Sometimes it has 2 swellings with a depression in the middle and is pulley-shaped. In one group of species in section *Pseudoastrophea*, a sometimes colorful trochlea is present well up the column. It is not known if this is a homologous structure.

Stamens and Pollen

Passionflowers have 5 stamens, basically radially symmetric. One or 2 rare species from the Old World may have up to 8 stamens, and certain taxa show more or less obvious deviations from a radial symmetry, especially in *Astrophea*, *Tacsonia*, and *Decaloba*. In species of *Passiflora* with a lateral orientation of the flowers, asymmetry of stamens is more common, presumably because pollinators visit the flower more from one direction. This is seen mainly in species pollinated by hummingbirds. In some of these species, the stamens become asymmetric only after the flower opens, as with the bat-pollinated *P. mucronata* (Sazima and Sazima 1978). The few species with curved androgynophores also have the stamens strongly zygomorphic. The anthers are all held to the lower side of the column.

The anthers of *Passiflora* are held one way in bud but flip over as the flowers open. In general, the long axis of the anther is maintained parallel to the long axis of its filament. By contrast, in many species of subgenus *Passiflora*, for example (especially in supersections *Passiflora*, *Stipulata*, and *Coccinea*), the anther is held crosswise. In such cases, after the flower opens and the anther flips over, it then turns 90° so that its long axis is perpendicular to the filament. The anther is lightly locked in that position by the interaction of 2 knobs of tissue at the distal end of the filament with a knob of tissue on the anther.

The pollen grains are subspheroidal to prolate and yellow or rarely white in *Passiflora*. They usually have 3 or 6 germination pores each associated with a groove, and they are fantastically ornamented when seen with an electron microscope. The rich variation in pollen morphology, especially in subgenus *Decaloba*, could give much information about the evolution of the genus, but this has not been approached systematically using modern techniques. Surveys of pollen morphology in the whole family were conducted by light microscopy by Dietrich Presting (1965) and Marie-Louise Spirlet (1965). In 2002, Maria T. Amela García and others presented a careful analysis of Argentinean species using transmission and electron microscopy, and provided a reference list to previous works on pollen.

Gynoecium

The gynoecium (literally "female house") includes the ovary, styles, and stigmas. The ovary in the family has one locule (a cavity in which ovules occur) and parietal placentation (in which ovules are attached to the outside walls). In *Passiflora* the gynoecium is composed of 3 fused identical portions, or carpels. A very few species have 4, including *P. ovalis*, *P. dioscoreifolia*, and *P. lobata*. Most species occasionally bear a flower with 4, but this is just a developmental "hiccup" and is of minor interest. The ovary is usually on a small stipe that extends 1–2 mm beyond the fused tube of the staminal filaments. The vestiture of the ovary is usually constant within a species, be it glabrous or densely pubescent. Inside the ovary are 3 strips of ovules borne on a placenta. At the apex of the ovary are 3 styles (or 4, in the 4-carpellate species) that are nearly always free, to their bases and point of attachment.

The styles in *Passiflora* are solid inside and range from very slender and filiform, to stout and subcylindrical, to greatly expanded toward their tips and nail-shaped. They are white or green to dark purple or red, sometimes spotted with purple or red. Each style is tipped by a stigma that is either relatively large and hemispherical, depressed and obovoid, nearly spherical, or large and 2-lobed. The stigma is covered with microscopic multicelled papillae (Bernhard 1999). In many species, the styles move a long way after the flower opens in order to put the stigmas in potential contact with the pollinators, only to move back out of the way before the flower closes.

In a number of passifloras, some flowers on the plant may be female-sterile, especially at lower nodes along a shoot, and this may be especially true very early in the blooming season. These flowers function only as males. Sometimes the gynoecia of these flowers are not fully developed, appearing yellowish and smaller than normal. This is not a nutrient deficiency. Sometimes the ovary and styles look normal but do not follow the normal patterns of movement. If this is mixed with fully functional flowers, or if perfect flowers form later in the season, this reproductive condition is called andromonoecy. It has been clearly documented in only a few species, such as *Passiflora sicyoides*, *P. tarminiana*, *P. incarnata*, *P. vitifolia*, and *P. edulis* (Harms 1925, La Rosa 1984, MacDougal 1994, May and Spears 1988, Snow and Gross 1980). However, few of those working in the field have looked for andromonoecy, so it may well be a common condition that will be encountered more and more.

Fruit

The fruits of *Passiflora* are one-loculed and most are large or small berries with few to many seeds. Each seed is covered in a fleshy or gelatinous aril. A few species have dehiscent fruits, such as those of section *Xerogona* and some of supersection *Bryonioides*. The pericarp may be thick and rindlike, as with *P. quadrangularis*, to very thin and papery, as with *P. foetida*. The pericarps of subgenus *Astrophea* and supersection *Pterosperma* are thin and leathery (drying like "parchment," according to the older authors). The morphology and anatomy of the fruit is poorly known or unknown for most species, but there is considerable diversity in the genus.

The nature of the fruit is intimately tied to the mechanism of seed dispersal for each species. The seeds of nearly all species are dispersed by animals, mostly through the digestive tracts of birds or mammals.

There are a few main classes of fruit. One class includes the purple or blackish berries so common in subgenus *Decaloba*. These appear to be eaten by birds. Another class includes the yellowish fruit with an intense fruity odor and the taste passionfruit is famous for. These are common in subgenus *Passiflora* and are eaten by mammals, either in the canopy by fruit-eaters like monkeys, kinkajous, and bats, or on the ground by various omnivorous mammals like agoutis, tapirs, peccaries, coatis, skunks, and opossums. Some plants in subgenus *Passiflora* such as *P. variolata* and *P. maliformis* have superhard woody fruit. These are so hard that they cannot be opened without a hammer, and in nature they must be chewed open by a mammal. The astropheas usually have yellow fruit but are extremely poorly known, and it is not know whether the seeds are primarily dispersed by bird, mammal, or both. Another class includes the green, saclike, fragrant fruits that hang on the vine for a long time, such as those found in *P. menispermifolia*, *P. oerstedii*, and *P. microstipula*. These are probably eaten by frugivorous bats. Another kind of fruit that clings to the vine when ripe is the purple or red, sometimes dehiscent fruits with bright orange, odorless arils, like *P. lobata*, *P. morifolia*, and *P. exsudans*. The seeds of these fruits can be carried off one by one by birds, and the seeds may be extracted from the aril and discarded instead of swallowed. Some passionfruits native to dry areas, such as *P. arida*, *P. colimensis*, and *P. pilosa*, have small, greenish, fragrant, juicy fruits that fall to the ground and are eaten by small mammals. Then there are the bizarre 6-angled fruits of *P. goniosperma* and *P. pusilla*, which split open to release seeds with dry arils. Are these dispersed by ants or by ground-feeding birds?

Besides the main classes, the list goes on and on with each special exception, and truthfully, few people have ever really observed an animal eating a passionfruit or spreading its seeds. We urge anyone in the field to keep their eyes open and to report any findings.

Arils

Arils are the individual juice sacs that surround each seed. They are outgrowths of the tiny stalks that hold each seed to the placental wall of the fruit. Different species have arils of different lengths, from 1/2 the length of the seed to many times the length of the seed. Often, at full maturity the arils may disintegrate a little and form a slimy mass of pulp and seeds. The arils are open and sometimes incurved at their apices. The fact that they are open, however, can only be seen during their development or in the short-ariled species such as section *Xerogona* or supersection *Cieca*. The arils provide a food reward to the animal that carries the seed away. They are the edible part and the source of the juice from passionfruit, although some species have arils that are crisp instead of juicy, and some arils have no taste or are somewhat oily. In the species with colored arils, the arils become that color before the fruit wall ripens and changes color.

Seeds

The seeds of species of *Passiflora* are flattened, narrowly oblong to suborbicular in outline and usually have a beak at one end. They have an outer, thin membrane called the sarcotesta (still present after the aril is removed) and an inner, pitted or grooved, stony seed wall called the sclerotesta. The sarcotesta, being clear and only a few cells thick, rapidly degrades when the seed is eaten or planted but can often be seen as a whitish film over the pitting on the seeds. The size and especially the sculpturing of the sclerotesta present many good taxonomic characteristics in species of *Passiflora*. Most species have seed coats that are pitted, reticulate, or punctate, but supersection *Decaloba* has seeds with transverse grooves. The ridges between these grooves are often rugulose. In fact, the size and sculpturing are distinctive enough to be diagnostic for many of the species. Only John Vanderplank (1996a, 2000) has included photographs of seeds of various species in his books.

Inside the seed is abundant white endosperm, the starchy and oily food that will be consumed by the embryo before it is fully grown and ready to sprout. The embryo is relatively large and has a straight radicle. At maturity the seeds are usually dark brown, brownish black, or black. In some species, seeds are shed with the ripe fruit before the embryo is fully mature and ready to germinate, so there is at least a short period of post-dispersal maturation. Seed germination and dormancy have not been studied in most species.

Germination

Because the embryos are still growing inside the seed when first taken from the fruit, the seeds of many passionflower species need at least several weeks if not months of after-ripening before they will sprout. In nature this is done in the semimoist soil, so fully drying them may induce additional dormancy. It should be noted that the more than 525 known species of *Passiflora* come from every imaginable warm habitat. The seeds and germination have been little studied, so it is likely that there are many variations in the types of dormancy.

Seed germination in the genera *Passiflora, Tetrapathea, Deidamia,* and *Adenia* is epigeal—that is, the cotyledons emerge above the earth in the manner of the common bean plant. The cotyledons are elliptic or oblong-elliptic, green, and foliaceous. A very few species of *Passiflora* are hypogeal, meaning the cotyledons stay beneath the surface of the soil, trapped in the seed coat, and a naked shoot emerges toward the sun. Often the first true leaves are reduced in this kind of seedling. This type of germination is similar to that of the garden pea. To date, it is known only in *P. lutea, P. discophora, P. ovalis,* and an unnamed relative of *P. exsudans,* but most species have not been checked.

4

Cultivation

Different passionflowers have different demands corresponding to the conditions of the natural surroundings in each plant's native country. Many species of *Passiflora* are perfect as decorative container plants on a balcony or patio. Some are even suited for cultivation indoors on a windowsill. Several species can withstand short spells of slight frost and can be cultivated outdoors throughout the year in the appropriate climatic zones. A handful of passionflowers can even survive temperatures down to −15°C, and these can be regarded as hardy in much of Europe and the United States. On the other hand, a great many tropical species only flower when kept in a heated greenhouse. Because passifloras have such a broad range of cultivation requirements, they can be grown by virtually anyone.

A prerequisite for successful cultivation is the imitation of the conditions in the plants' natural surroundings. Try to obtain information regarding the origin of each species. Is it from hot and humid lowland jungles, or from cooler forests high on mountains? With the exception of many high-altitude species from the Andes, passionflowers generally prefer a sunny location sheltered from wind. Plants from high mountains, such as those of supersection *Tacsonia*, often do not tolerate temperatures of more than 25°C for a long period and must be kept partly shaded.

Another key to successful cultivation is correct watering. Passionflowers do poorly in soil that is too waterlogged or too dry, and it is important to keep the soil evenly moist. Sometimes it is advisable to water plants more than once a day according to their requirements rather than giving them too great a supply at one application.

Soil Requirements

Generally speaking, most passionflowers, particularly hybrids, can be successfully cultivated in standard commercial potting mixtures. To mellow and aerate such a mix, add perlite, vermiculite, expanded clay, or sand. Regular garden soil or compost can also be used, but these carry with them the danger of soil-borne diseases. Garden soils should be heated, or "pasteurized," in an 180°C oven for about 20 minutes to kill harmful bacteria, fungi, and especially pests like nematodes and weevil larvae. This method is rather troublesome and produces an unpleasant odor, but the pasteurized, nearly sterile mix will prevent any damage caused by the aforementioned organisms, and the trouble is certainly justified in view of the dramatic consequences of an infestation of nematodes. However, heat treatment has the disadvantage of removing all the beneficial soil organisms that would otherwise have a positive effect on the growth of the plant. In their natural surroundings, plants do not grow in sterile soil and flourish nevertheless. Perhaps the best advice is to try both garden compost and sterilized soil or commercial mix, and make a comparison between the two.

Mycorrhizal fungi live on the roots of more than 95% of all species of flowering plants. Plants in a symbiotic relationship with these helpful soil fungi are better able to absorb water and nutrients and are better protected from diseases. It is possible to take advantage of these positive effects by adding endo- and ectomycorrhizal fungal spores directly to potting mix. Various commercial brands are available in nurseries and by mail order, known by names such as mycorrhizal innoculant, beneficial fungus starter, and MF-tabs.

In their natural surroundings, passionflowers sometimes grow in relatively poor soils and may even occasionally be found growing along roadsides as weeds. For this reason, neglecting the composition of the potting mix may not have a considerable influence on the cultivation of most passionflowers. There are some exceptions, however. A number of the more sumptuous species from tropical lowlands, such as *Passiflora cirrhiflora*, *P. glandulosa*, and *P. cerasina*, will not grow well in a standard mix that contains much peat. In their natural surroundings these plants flourish in lateritic soil, a clayey residual product of rock decay that is red and has a high content in the oxides of iron and hydroxide of aluminum and a low proportion of silica. The nutrients of lateritic soil have been mostly washed out by years of tropical rain, and the remaining acidic soil is rich in iron and aluminum. Now the feat is to find the appropriate components for a comparable rooting medium. We have achieved good results with a formula of 1/3 standard commercial potting mix for cacti and succulents, 1/3 standard commercial soil-based potting mix, and 1/3 coarse sand. This, however, is not the final peal of wisdom; there may be a better formula.

The soil of the potted plant should be freshened from time to time. By watering the plant from the top, the soil is compressed, and in the course of time the potting mixture sags. The root ball must be loosened up and placed on a new layer of potting mixture in the same pot. The uppermost, often crusty and salty layer is also removed and replaced by fresh soil. This procedure, however, does not make up for repotting plants in a larger container, which should be done in regular intervals depending on the size of the plant. This particularly applies to young plants, which can be repotted twice in one year.

Passionflower roots rapidly rot in waterlogged soil, so encourage good drainage by adding gravel, pot shards, or sand to the base of the potting mixture. Do not allow the root ball to dry out, either. Most passionflower deaths among our collections resulted from fungal infections that developed after the great stress of drying out followed by overwatering.

A pH level in the acid range is favorable. On the commercial plantations in the tropics and subtropics these plants flourish optimally at a pH of 6–6.5. While the pH level of standard potting mixtures may be indicated on the label, homemade soils should be tested with indicator strips. A regular check is also advisable when limey tap water is used. When the pH reading is too high, add peat. Only in extreme cases should sulfur be used.

Fertilizing

Passionflowers need many nutrients if they are to grow well. During their growing period they must be fertilized regularly, and for the rest of the year feeding is adjusted to growth and reduced accordingly. Plants growing in pots have a greater demand for a supply of nutrients, since these are rapidly used up in a small container.

There are many fertilizers from which to choose. Mineral fertilizers are compound fertilizers containing the complete range of nutrients and trace elements. They are available on the market in either liquid or granulated form. Organic fertilizers such as horn meal and bone meal must be broken down by soil bacteria before the plants can use them. This occurs slowly over a long period of time, and thus an overdose of nutrients is unlikely. When plants are repotted, organic fertilizers are simply mixed with the soil or lightly worked into the top layer of potting mixture. Organic fertilizers, however, often do not contain all the necessary nutrients in a sufficient concentration, and so a mineral fertilizer should be added.

A prerequisite for healthy growth and development of sumptuous flowers is continuous fertilization at regular intervals so that the plant always has sufficient nutrients at its disposal. It is of no importance whether a low dose is administered together with daily watering, or the proprietary fertilizer is supplied each week or fortnight in accordance with the indications on the label. It need not be the most expensive product, but it does need to be a balanced fertilizer containing all important nutrients and trace elements in sufficient amounts. Nitrogen, phosphate, and potassium (N-P-K) should be contained in the following concentrations: nitrogen (N) at 10–12%, phosphate (P_2O_5) at 8–12%, and potassium (K_2O) at 10–15%. During the main growing season, 1–2 g of this type of fertilizer per liter of water can be applied with every watering, even if the label says that such a dose should be administered only once a week. In other words, the weekly ration can be multiplied. (This does not apply to young plants in small pots, however.)

Some gardeners use different fertilizers during the course of the year. At the beginning of the growing season a fertilizer for foliage plants with a high proportion of nitrogen (at least 20%) is administered to boost the formation of new shoots and leaves. After several weeks the content of phosphorus and potassium is increased to 18–25%, while the nitrogen content is reduced. This shift in nutrients favors flowering and fruiting. Several fertilizers suitable for flowering plants are offered on the market. Tomato fertilizer works well for fruiting vines.

If continuous fertilizing proves too tedious, or if fertilizing is forgotten now and again, the alternative is a slow-release fertilizer. A single application provides plants with the necessary nutrients for a complete growing season, and by releasing the nutrients slowly, the fertilizer will not burn the plants. This type of fertilizer is best mixed into the soil medium or at least scratched under the surface, as fertilizer placed right at the soil surface can lead to a harmful salt crust developing at the base of the stem. When selecting an appropriate slow-release fertilizer, ensure that its effect will last only for the desired timeframe.

Plants are also able to absorb nutrients through their leaves, though fertilizing through the leaves is only supplementary and can by no means replace the supply of nutrients from the soil. When fertilizing this way, one must see to it that the nutrients remain in liquid form on the leaves as long as possible in order to be optimally absorbed. Morning and evening hours are best suited for this. This kind of fertilizer is usually administered in a concentration of 0.05%. To avoid burning the leaves, make sure it does not exceed 0.1%.

Plants that exhibit an increasing number of yellow leaves are likely suffering from chlorosis, a condition caused by a lack of nutrients. Chlorosis may be caused by a pH level that is too high, often due to too much lime in the soil or water. Extreme pH levels cause nutrients and minerals to become bound in the soil and therefore unable to be absorbed by the plant. The red-flowering species in particular—*Passiflora vitifolia*, for example, and above all *P. racemosa*—tend toward chlorotic leaves, which can be traced back to a deficiency in iron. This deficiency can be controlled by administering chelated iron, which is available at nurseries.

When a large number of leaves turn yellow it is not necessarily caused by a deficiency in nutrients. Sometimes these symptoms are caused by overfertilizing, waterlogging, drying out followed by overwatering, lack of light, pest infestations, or drift from herbicides. Fertilizing with the belief that too much is better than too little is among the most harmful mistakes one can make. When the soil contains too much phosphorus or potassium, other important trace elements, such as iron, are bound in the soil and cannot be absorbed by the plant. Furthermore, overfertilizing can result in salt concentrations that reach such a degree that water is literally drawn away from the plant. This leads to wilting of the leaves, and the plant dries up in spite of good watering. A certain degree of sensitivity is needed to provide plants optimum nutrients.

Climbing Aids

In their natural surroundings, passionflowers climb shrubs and trees with the help of their tendrils. In a container they lack these necessary supports, which is why the gardener must give them a little help. A trellis is suitable and has a decorative effect when harmoniously matched with the container and plant. Plant stores offer a wide selection of trellises, from plastic-coated iron gratings and pyramids to wooden lattices and bamboo poles. Such a climbing aid can also be homemade at less cost. Bamboo poles are well suited for this. In a small pot, for example, 3 poles can be positioned in a triangle, with the upper ends bound by wire.

Pyramid structures such as this are best suited for the daintier passionflowers, such as *Passiflora sanguinolenta*, *P. citrina*, *P. conzattiana*, and *P. capsularis*. For more vigorous growers, like *P.* 'Amethyst', *P. ×belotii*, and *P. vitifolia*, a structure of 3 bamboo poles up to at least 1.5 m long is recommended, the upper ends kept apart by a strong wire. As the plants climb up the support, their blooming shoots hang down, creating a beautiful effect. A structure of 2 bamboo poles, horizontally connected with 2–5 shorter sticks, will also work. A word of caution, however: There is a maximum diameter that a tendril can grab, and for some species, there may be a minimum as well. A 7-cm pipe is outside the limit for most kinds of passionflowers, and fishing line is too small for some. Experience is the best teacher with each species, but trellis supports that are the diameter of a coat-hanger, thick twine, a pencil, or a finger will work just fine.

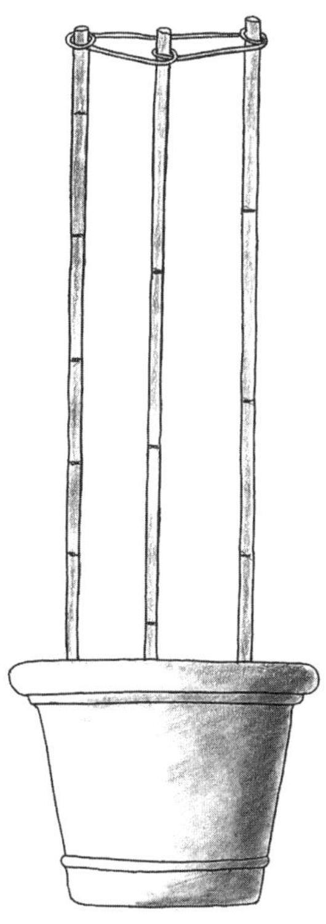

One suitable climbing aid.

A fence, an arbor, or the railing of a pergola can also be used as a climbing aid, and passionflowers can climb the walls or the downspouts of a house with the help of netting wire, plastic-coated mesh, or other small climbing aids. To induce abundant flowering, lead passionflowers toward the climbing aid at the beginning, lightly tying them to the support or leading them there with twine. Remember that passionflowers are tendril-climbers, not twiners whose stems twist around a support. These plants want their stems to be more or less straight and, in fact, resent being forced to adopt the twining lifestyle. Continuous twisting and turning of the shoots will inhibit the growth of the plants and their inclination to flower. Even if the plant looks rather unkempt, give it the opportunity to find its own way.

In flower shops passionflowers are mostly offered bound around a plastic or wire bow. This is not so much a climbing aid as it is a method to save space. Initially the passionflower grows and flowers willingly, but in the course of time it loses more and more of its strength. When enough room is available it is advisable to free the plant from this "circling." Though it can be difficult to unwind the shoots and fasten them to a suitable trellis, the effort will pay off. During such a procedure a kink in a shoot may well occur, but if straightened out gently it will not hinder the plant from growing on briskly. By cutting off the tip of a shoot, new lateral branches will develop with greater inclination to flower.

Overwintering

In winter, passionflowers suffer from lower temperatures, shorter days, and low light, and therefore this season is the most critical period for these plants.

Before night temperatures drop below 10°C, the more sensitive container plants, such as *Passiflora quadrangularis* and *P. vitifolia*, should be taken indoors. Depending on their resistance to cold, other species will need to be taken to their winter quarters later on; for now, though, these plants should just be cut back and thoroughly scrutinized for pests. Unlike many other decorative plants, passionflowers keep their foliage in winter, with the exception of certain herbaceous species such as *P. incarnata*, *P. lutea*, and *P. bryonioides*. Passionflowers must never be taken to a dark basement room; to do so would be to sentence them to death. The best possible winter location is a heated, lighted greenhouse or conservatory. Only a few enthusiasts can afford this, however, so compromises will most likely have to be made. Quite a number of robust species and many hybrids can be overwintered in an unheated or moderately heated room, in a light corridor, or in an attic with windows. As a rule, overwintering in a cozy living room is not so favorable, all the more when the heating is under a windowsill. Many garden centers and nurseries offer a winter service for container plants. Plants are provided professional care, but the service is relatively expensive.

With the help of supplementary light, the day can be prolonged artificially. Fluorescent tubes are a good choice for this and should be placed no more than 50 cm from the plants, since the intensity of light will otherwise rapidly diminish. The duration of lighting can be easily controlled by an inexpensive time switch; 12 or 14 hours of light per day is ideal. Even better is a dimming switch, which continually measures the light intensity from the window. As soon as the light falls below a fixed minimum, the switch turns the light on.

It is of great importance that passionflowers be kept at the right temperature, which can be very different for each plant. While 5°C or even slightly colder is sufficient for some species, many will not tolerate temperatures below 15°C. Hence it is of great importance to know the temperature needs of each plant. Few experiments have been conducted to determine exact minimum temperatures, and therefore the data given in this book is to be understood as a rule of thumb. In some cases temperatures may well be a little lower than stated, although the ideal temperature for overwintering is about 5–7 degrees higher than what we list. We have always endeavored to take into account the conditions for the plants in their natural surroundings. Furthermore, a plant's resistance against cold largely depends on the origin of the plant, not just the species. When a clone comes from a mild climate, it is more sensitive than a plant of the same species from extreme high altitudes.

How warm a greenhouse is to be kept will depend on the amount of insulation, the temperatures in winter, and how much money can be put into it. Overwintering in a heated living room is less expensive in this respect but does not offer these warmth-loving plants from the tropics the optimum conditions regarding light intensity and humidity. It makes good sense to limit oneself to the species that will get along quite well in whatever winter quarters are available.

Good air circulation is indispensable. In moist standing air, fungal infections are more likely to occur, and in dry standing air pests like spider mites may become a problem, which are very difficult to get rid of once they have infested a plant. To prevent the spread of disease, passionflowers should be regularly scrutinized, and at the beginning stage of any infestation, action should be taken immediately. Space plants far enough apart so that each gets enough light and air. A small oscillating fan on the lowest setting is helpful, although it does dry the pots more quickly. A fan coupled with daily misting and an occasional shower or misting to the point of run-off will help keep problems at bay. Yellow leaves and dead parts of the plant must be removed regularly. When these remain on the vine or are lying in the pot they are an excellent breeding ground for fungal infections.

During the whole of winter, passionflowers should be only moderately watered and scarcely fertilized, following their reduced winter growth. The root ball need only be a little moist. Waterlogging is strictly to be avoided.

As days get longer, passionflowers in greenhouses or conservatories put forth new growth. From this point onward the plants should be watered and fertilized regularly. When temperatures permit them to be taken outdoors, they will already be in top form and will therefore start flowering relatively early. As soon as day temperatures are regularly over 12°C—or, even better, over 15°C—the plants can be taken outdoors. We usually recommend waiting until the nights are consistently above 10°C, which corresponds with "melon-planting" weather. Extremely warmth-loving species should be kept indoors even a couple of weeks past this. Once in a while it happens that at exactly the time new growth has started, one plant or the other suddenly dies off. Although the plant has gotten through winter well and seems to be in good health, it suddenly wilts away from the top downward. The cause of this is often unclear, but in many cases the xylem is blocked by a fungal infection so that the transport of water and nutrients is interrupted. Unfortunately, there are no known countermeasures.

Pots and Containers

Selecting a pot is important, for a pot should both meet the demands of the plant and please the eye of the beholder. Pots are made of various materials and are available in many different forms. Beyond the usual terra cotta and plastic are pots made of stone, wood, clay, and other materials. Plastic pots have an advantage in that water evaporates very slowly from them, and thus watering is not needed as often. Plastic is disadvantageous, however, when solar radiation is intense, as the pot can become hot enough to damage the plant's roots. Hence it is advisable to avoid standing such containers in the blazing sun when they are used in a greenhouse.

Steadiness is an important characteristic to look for in pots. A container must withstand strong gusts and therefore should not be too narrow at the bottom. Heavy pots, too, are less likely to topple over. The size of the container is of some importance for certain passionflowers and can even be a decisive factor when it comes to whether or not the plant develops flowers. Many passionflowers blossom and flourish though the pot appears small in relation to the size of the plant, but *Passiflora actinia*, *P. holosericea*, *P. ligularis*, *P. maliformis*, *P. platyloba*, *P. serratodigitata*, *Tacsonia* species, and many others will not do so well in small pots. Only when planted in large tubs or allowed to root freely will they truly delight. Since these plants invest their complete strength in developing long shoots and an abundance of leaves, it often takes them longer to flower. The same holds true for passionflowers planted outdoors during summer. These plants often reach an undreamed-of size, and their leaves may grow to several times their usual dimensions. A sunny site is the standard precondition for successful cultivation. Apart from this, anyone wanting to experiment will have unlimited possibilities. Provided with a climbing aid, passionflowers climb everywhere, whether on garage roofs, pergolas, or the walls of a house. When planted in the open during summer, however, the question arises as to how the plants can be dug out before winter without being damaged. They will not tolerate the cutting off of large parts of the root ball. For this reason it is advisable to plant passionflowers in plastic pots provided with holes or in a wide-meshed wicker basket. When the plants are taken indoors in fall the roots that have grown out of the holes can then be cut off while the root ball in the pot is left undamaged.

Anyone who wants to test the hardiness of a plant, or who does not have enough room to overwinter a specimen that has reached a considerable size, should make cuttings so as to have young plants for the next year.

As a rule, passionflowers are moved to a larger container when the roots have completely filled the pot. An older plant, though its roots fill up its container, may well remain in it for several years, provided enough nutrients are supplied. A young plant, however, should be regularly moved into a container about 2–6 cm larger. The best time for potting is early spring, when new growth appears. At the same time, a slow-release fertilizer can be mixed into the fresh soil.

Pruning

When passionflowers are taken to their winter quarters, the shoots, which are often several meters long, are rather difficult to tame. Pruning in fall makes it easier to take plants indoors—and all the better, as indoor space is usually limited. However, plants should not be cut back so rigorously and indiscriminately that only the stem and a few nodes remain. This is particularly true for species that grow rather slowly; if cut back too much, a long time may pass before they regain their initial size. Furthermore, some shoots may dry out, which can result in the total loss of the plant. As a rule, plants are cut back to 1/2 or 1/3 of their original length. Always leave some green leaves. In principle, however, the available space is decisive. If there is enough space to overwinter larger plants, cut off only the tips of the shoots.

In winter, especially after a rigorous cutting-back in fall, new shoots suffering from insufficient light may appear weak or chlorotic and therefore unattractive. These cost the plant much strength, and most should be cut back regularly.

If one does not like pruning plants in fall, perhaps because the plants are still rather small or because they may continue flowering in winter quarters, pruning should take place in early spring. Both spring and fall have their advantages and disadvantages, so we advise a combination of both: a moderate cutting back in fall and a corrective pruning some months later in spring. Ailing, weakly growing plants always profit from a pruning, as do passionflowers infested with pests.

With the exception of *Passiflora amoena*, *P. discophora*, *P. racemosa*, and some other rarely cultivated species that can develop flowers on older branches or even on the stem, the overwhelming majority of passionflowers produce new buds only at the tip of the shoots. Thus, pruning supports the development of flowers by producing new lateral shoots.

Cultivation in Containers and Tubs

Many *Passiflora* species are excellently suited for cultivation on a balcony or terrace during summer. Whether the plan includes a solitary plant in a container or a group of plants situated as protection from the sun or as a visual screen, passionflowers always catch the eye. Enthusiasts have a large spectrum of flowers at their disposal, in all kinds of colors and forms. These plants generally

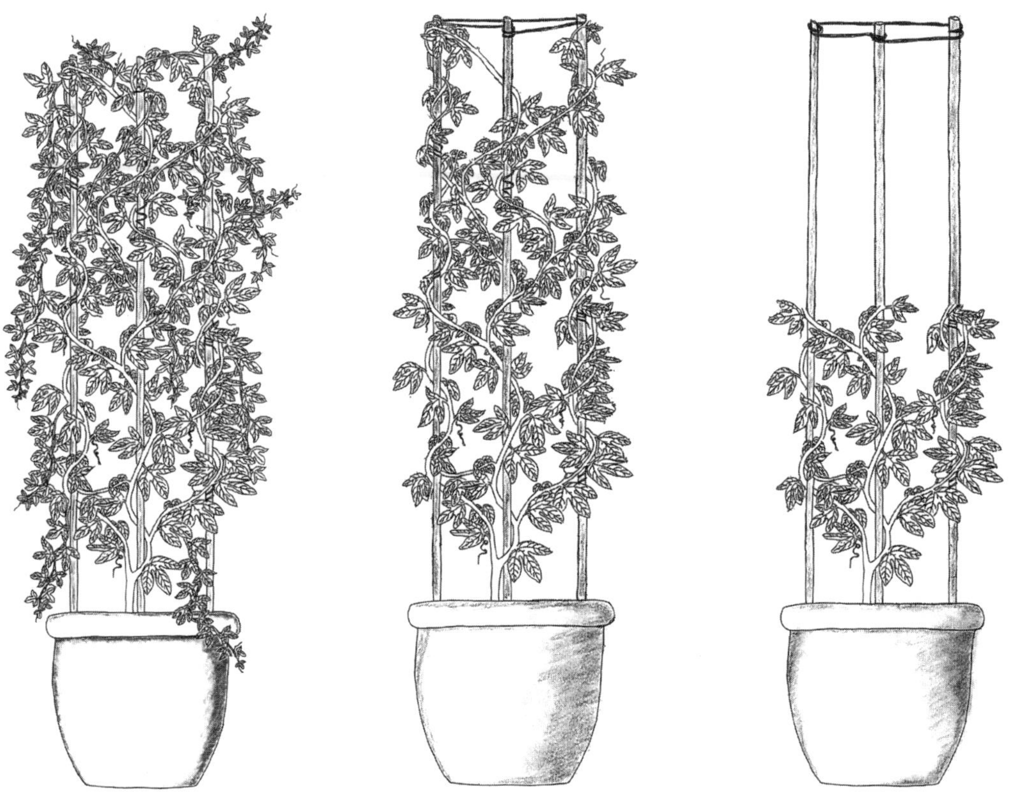

The stages of pruning: an unpruned plant, a light cutback, and a hard cutback.

prefer a sunny location sheltered from wind, with the exception of some species of supersection *Tacsonia*, which do not take well to extreme heat. In the open, passionflowers initially react to increased ultraviolet radiation with a sunburn, which appears as a whitish discoloration on the leaves. This is no reason for concern, however: the newly produced leaves, accustomed to the more intense sunlight, will not show these symptoms, and the damaged, unattractive foliage can simply be removed later.

If the plants have spent winter in a greenhouse or conservatory, a relatively early flowering can be expected. In most cases flowering time lasts until late fall. Plants are often still flowering when taken to their winter quarters. Try growing several varieties in order to enjoy flowers in every season.

Cultivation Under Glass

There is no doubt that in many regions in the United States and Europe, greenhouses and conservatories are the ideal locations for passionflowers—and not just for the fastidious and warmth-loving species that should only be cultivated in greenhouses but also for those recommended for growing in pots indoors or in the open.

In selecting species likely to prosper in a greenhouse or conservatory, the respective overwintering temperature is of great importance. Selection criteria such as flower color and size, fruit edibility, and plant shape should play a minor part initially.

If the facilities are available, plant certain passionflowers directly in the ground in a greenhouse or conservatory. More fastidious species, such as *Passiflora glandulosa*, *P. racemosa*, *P. nitida*, *P. laurifolia*, and *P. serratodigitata*, will only flower when they have reached a considerable size. Moreover, freely planted species have a very special charm because of their size and abundant flowers. Remember, in the wild these are canopy lianas. Regular pruning, however, is a must; otherwise a small greenhouse will become an impassable thicket.

When it comes to overwintering, the difference between the temperature in the room and the temperature on the floor is often underestimated. Even if the room temperature is 15°C, the floor temperature may sink to a level that is intolerable for quite a few species. In extreme cases ground frost may occur even in a heated greenhouse. At a permanent temperature below 12°C, the critical point for many warmth-loving species from tropical regions is already reached. Hence it is advisable to cultivate passionflowers on a table, or at least to protect them from contact with the cold floor by a polystyrene plate several centimeters thick. Installing a heating cable or a heating mat certified for use in moist places may procure the necessary floor temperature. Plants positioned close to the sidewalls or in the corner are also in danger of being too cold.

Due to the intense radiation of the sun in midsummer, the temperature in a greenhouse or conservatory may easily reach more than 40°C. In the long run this leads to inhibited flower formation and finally to an arrest in growth. The use of shading material and ventilators for air circulation can prevent an excessive rise in temperature.

Indoor Cultivation

Passiflora caerulea has been known as a houseplant for a long time, but several other passionflower species can also be cultivated on a windowsill. The insensitive species are well suited for this, satisfied with a relatively small pot and able to produce flower buds while still short. Except for some periods of short rest, many passionflowers blossom on a windowsill year-round. Some species, such as *P. capsularis*, *P. gracilis*, *P. herbertiana*, *P. morifolia*, *P. murucuja*, and *P. suberosa*, develop decorative but usually inedible fruit.

Regular feeding is as important for the development of flowers as is a bright location, preferably by a sunny window. In midsummer, when it might occasionally get too hot for plants at such a window, provide shade around midday.

If a store-bought passionflower is bowed around a wire hoop (a widespread bad habit in Europe), it should be removed from the hoop and adapted to a small window-sized trellis (see "Climbing Aids" at the beginning of this chapter). The plant will only need a little help in being directed toward the trellis and must only be cut into form occasionally. *Passiflora* 'Adularia', *P. allantophylla*, *P. karwinskii*, *P. sa-*

gasteguii, and some others are excellently suited as trailing plants and are very decorative with their long floriferous shoots against a window in the living room.

In many cases the heating source is just below the windowsill. This is very disadvantageous for passionflowers, especially in winter. The atmospheric humidity, which is already very low, is further reduced in such an area and may lead to an infestation with pests such as spider mites, which thrive under these conditions and reproduce very rapidly. Regular spraying with water is indispensable.

Outdoor Cultivation Throughout the Year

Increasingly popular, hardy passionflowers can remain in the same location for many years without needing to be moved inside each autumn. The stems often grow up to 8 cm thick. Such specimens can completely cover the wall of a house. Walls, pergolas, archways, visual screens, summerhouses, and fences offer an opportunity for planting passionflowers. Provided the vines get the necessary support, there are no limits to where they will grow.

Choosing a good location for a passionflower affords it the best chance for success. To develop properly, a plant needs a sunny sheltered position and the correct type of soil. Contrary to general belief, passionflowers planted outdoors generally do not perish due to low temperatures but rather due to soft rot. Therefore the soil should be double dug and mixed with sand or gravel before passionflowers are planted. Planting beneath a projecting section of roof is advantageous, as it will keep away some of the rain.

Sucker growths are an occasional nuisance, sometimes appearing within a radius of several meters around the plant. Such suckers should be removed, since they rarely produce flowers unless allowed to grow large.

In many regions of the United States and Europe only frost-resistant passionflower species are suitable for outdoor cultivation throughout the year. Many high-altitude species from the Andes, such as *Passiflora mixta*, *P. tripartita* var. *mollissima*, *P. jamesonii*, and *P. umbilicata*, will tolerate slight frost. However, these species are only adapted to nightly declines in temperature. They will not survive a long spell of cold weather.

On the other hand, some species are frost-resistant down to –15°C. Among these are *Passiflora caerulea* (including the white-blossoming 'Constance Eliott'), *P. incarnata*, *P. lutea*, and, with reservation, *P. tucumanensis*. Besides these, some hybrids—*P.* ×*colvillii*, for example, hardy to –12°C, or *P.* 'Incense', hardy to –8°C—have been developed from the aforementioned species and are regarded as frost-tolerant.

While *Passiflora incarnata* and *P. lutea* die down to the ground every year and build new growth the next spring, *P. caerulea* keeps its leaves at temperatures down to –5°C. Only at lower temperatures will it shed its foliage or lose a portion of its shoots to frostbite. A light protection in winter made of fir twigs, straw, or air-cushioned packaging is recommended. Passionflowers reach their absolute hardiness in the 2nd or 3rd year after planting. *Passiflora incarnata* only survives long freezes when fleshy roots have developed after about 2 years. The most suitable time for planting these is early in spring so that plants are able to develop strong roots before winter.

In order to increase the diversity of cold-tolerant plants, plant breeders have tried for years to develop frost-resistant hybrids through crossbreeding. This is still a very active front, so additional frost-resistant varieties may be found in the marketplace.

5
Propagation

PASSIONFLOWERS can be propagated either genetically or vegetatively with few problems. In most cases they are grown from seeds or cuttings. They can also be propagated vegetatively by layering, root cuttings, air layering, or grafting, but because these procedures are quite intricate, they are recommended only for those species whose cuttings root with difficulty.

Seeds

Young plants can be produced from store-bought seeds or from the seeds of fruit harvested from the garden. The seeds of most species need a soil temperature of at least 25°C in combination with a correspondingly high atmospheric or soil humidity. Sometimes even higher temperatures, such as 35°C, aid germination. While robust species such as *Passiflora caerulea*, *P. capsularis*, *P. morifolia*, *P. ligularis*, and *P. edulis* will germinate under less favorable conditions, growing more fastidious or tropical species from seeds will only be successful in a heated propagator or plastic-covered seed flat. To grow well and quickly, seedlings need at least 10 hours of daylight. When seeds are sown during winter, supplementary artificial light is necessary.

In most species, best results are achieved with dry fresh seeds. Older seeds may react unpredictably. Sometimes there is no difference to be noted in comparison with fresh seeds, but in most cases germination only occurs after some months, if at all. Seeds more than a few years old are rarely viable. However, agricultural experiments made in South America have revealed interesting results with *Passiflora tripartita* var. *mollissima*. While seeds several months old germinated only after a long time, if at all, the germination of seeds after a year's storage was rather good.

If the seeds are taken from a fruit, the arils must first be removed from the seeds—a rather tedious task. This can be achieved by rubbing the seeds in fine sand or washing them in a kitchen sieve. Under the moist, warm conditions in the propagator, mold may occur very rapidly, especially if small particles of pulp remain on the seeds. Some hobbyists dry the seeds on newspaper after washing; when completely dry, the seeds then separate easily from the remaining membranes. A preventive measure against mold is to treat seeds with a fungicide such as phthalimide or hydroxyquinoline sulfate, both tolerated well by plants (see chapter 6 for more information on using chemicals). Dry seeds can be soaked overnight in plain water or tea. In some texts it is advised to soak seeds in milk or passionfruit juice. This does not make sense and is even detrimental. The juice of passionfruit contains exactly those inhibiting substances that prevent the seeds from germinating in the fruit.

The soil mix should be rather poor in nutrients and germ-free or sterilized. Proprietary seed-growing mixtures are best suited for this, as are small peat pots filled with a mixture of sand, peat, and standardized potting soil in equal parts. Instead of sand, use pumice gravel or perlite to make the soil more porous. Avoid using soil from the garden. This can be the breeding ground for noxious microorganisms or fungi, which will prosper under the warm conditions of a propagator and jeopardize the barely resistant seedlings.

Slightly cover the seeds with soil and firm them down a little. Depending on how old the seeds are, germination will occur after 10–28 days. In rare cases the first seedlings will appear only after some months. It is important to remain patient, even if no seedlings can be seen after 8 weeks. To prevent the dreaded toppling-over disease or damping-off, spray the seedlings regularly with a fungicide and regularly air out the propagator. As the young plants develop further, carefully harden them off by occasionally taking the lid off the propagator for an hour or longer.

The plants should remain in the propagator until they have reached a height of about 5 cm. After this point, move them into small pots, being careful not to damage the roots. Spray them with water several times a day initially to make the transition easier. The first flowers may appear within 6–9 months on species such as *Passiflora capsularis*, *P. gracilis*, and *P. morifolia*. As a rule, most passionflowers produce flowers the 2nd year after sowing. Sometimes, however, several years may pass before the first flowers appear, as is the case with *P. serratodigitata* and other species that flower in the canopy of their native forests. There are rare cases in which plants raised from seedlings have not flowered even after many years of cultivation, despite the best care. This does not occur with vegetative propagation by cuttings taken from flowering plants of the same species.

Cuttings

Propagation from cuttings is the simplest, most widely used method. Vegetative propagation, a form of cloning, guarantees that the plants, especially hybrids, are multiplied true to seed. Passionflower vines grow rapidly and offer almost unlimited material for cuttings. Only select mother plants that are outstanding in every respect, especially when it comes to hybrids. Best results are achieved with cuttings from one-year-old growth. Terminal growth, provided it is not too soft, and growth from the middle of a shoot will grow roots well as a cutting. It is advisable to trim the cuttings just below a node, where root development seems to be more reliable than at any other spot. A cutting should be 5–12 cm long and have 2–3 nodes. First remove the tendrils, then any buds or old peduncles that may be present. The lowest leaf should also be removed or shortened so that it won't rot when it comes into contact with the soil. If desired, cut the remaining leaves in half, or remove every other leaf, to help keep the cutting from drying out.

Cuttings of *Passiflora* usually need a temperature of 25°C to root. Only the species of supersection *Tacsonia* and other representatives from cooler regions prefer 15°–20°C. Because a cutting can lose water to the air so much faster than it can absorb it, a very high atmospheric humidity must be guaranteed as well. A used aquarium with a plastic cover works well as a cutting propagator; set on a heating mat, it offers ideal conditions. Supplementary artificial light can be provided by fluorescent tubes inside the lid. Specialty stores offer propagators in various designs, from mini-greenhouses for a windowsill to fully controllable beds. Unfortunately, these devices are often of limited height so that only rather small cuttings can be accommodated. Insert the cuttings in a lightly fertilized, sterilized potting mixture. Most passionflower enthusiasts have their own special recipe. In our method, we begin by putting a layer of potting mixture about 5 cm high in a propagation box and setting small peat pots on it. The base of potting mix guarantees continuous moisture to the rooting medium, reduces the heat of the heating mat, and supplies the young plants with nutrients. A callus appears where the cutting was trimmed, and eventually roots will develop out of this. As soon as the roots have permeated the peat pot and extended into the potting ground for some centimeters, set the individual plants into pots that are not too large. This must be done very carefully so that the fresh, delicate roots are not damaged. The propagation box should be opened and aired out regularly and the cuttings sprayed with fungicide to prevent mold. It is important that the cuttings get sufficient light, but direct sun is to be avoided, particularly in summer.

Rooting hormones designed to make rooting easier or faster are widely available as talc-based powders or solutions. Judging from experience, however, these are no wonder drugs. We can only recommend preparations with indole-3-butyric acid as an active ingredient. These types of hormones can produce better results with difficult-to-root species. Treated cuttings can develop roots a few days earlier than normal. In Dutch nurseries a mixture of indole-3-butyric acid and naphtaleneacetic acid is applied to cuttings of reluctantly rooting species. When using hormones, make sure the temperature does not rise much beyond 25°C. The use of hormones is often unnecessary, however, as most species, especially the smaller climbers in subgenus *Decaloba*, root satisfactorily without such aids.

While rooting within 2–5 weeks is achieved at a success rate of 90% from spring to autumn, it may take up to 10 weeks in winter. At the same time, the success rate in winter is significantly lower than in summer. Enthusiasts often exchange or import cuttings. Wrapped in moist paper or cloth and put into a plastic bag, cuttings will last a number of days without damage. They must be stuck into the rooting medium with the correct end upward; when the tip of a shoot is missing, it is difficult to see which way is up. One hint is offered by the tendrils, which always grow from above the petioles and thus show which end is up and which end is down.

In order to safeguard one's own stock of plants, it is advisable to grow some young plants from rare or particularly beautiful specimens. Unfortunately, it happens once in a while that one or another passionflower does not survive winter or meets a sad fate. Rooting a few cuttings before they are needed will prevent any disappointment.

Layering

Experience shows that cuttings of some passionflower species root poorly or very slowly, as with *Passiflora cinnabarina*, *P. herbertiana*, *P. umbilicata*, *P. racemosa*, and the whole of subgenus *Astrophea*. For these species, propagation by layering is recommended.

On a flexible, not-yet-woody shoot, remove the leaves at a node. At this spot, make a cut about 1 cm long lengthwise with a sharp knife. If desired, treat the wound with a hormone rooting powder. Bend the shoot down and submerge it 5 cm deep into the soil in a bow-shaped form. This can be repeated several times when the shoot is very long.

This method is feasible in the open as well as in a greenhouse where pots are used. Root growth is expected only when the soil is moist enough and the temperature is 20°–25°C. To prevent the shoot from swinging back, peg it to the ground. After 4–6 weeks enough roots will have developed that the shoot can be separated from the mother plant.

Root Cuttings

Anyone who has planted *Passiflora caerulea*, *P. incarnata*, *P. lutea*, or *P. tucumanensis* in the open knows that these plants spread uncontrollably and extensively by producing suckers. This characteristic can be taken advantage of for propagation. Carefully expose the roots and underground rhizomes of the plant so that pieces 5–15 cm long can be cut off. Place the cuttings into a pot with permeable potting soil. Situate the pot in a warm place and keep it constantly moist. The new shoot will appear after about 2 months.

Air Layering

Another method that is successfully used with passionflowers is air layering, also called Japanese layering. To initiate root growth, remove all leaves, peduncles, or other attachments on a vigorous lateral shoot over a length of about 15 cm. Wound the stem by making a small slit at a node in this area. Make sure the cut has the right depth of about 2–3 mm. When the cut is made too deep (deeper than 1/2 the diameter of the stem), the shoot may not get enough nutrients and water from the roots and may therefore perish. On the other hand, if the cut is not deep enough, the wound may heal without any root formation. To prevent the wound from closing, wedge a small sterilized piece of wood or a pebble into the slit. Fix the shoot to a bamboo pole or similar support so that it cannot break off.

Place a plastic film around the wounded area, binding it together at the lower end, filling it with moist sphagnum moss, and then closing it at the upper end. If needed, cover the film with aluminum foil to protect the developing roots from light. When enough roots have developed, cut the shoot off below the roots and carefully plant it. Root development may take 8–12 weeks. Since the lateral shoot keeps on growing, new buds may have already appeared on it during this time. During root formation the moss must be kept moist.

Grafting

Passionflowers are grafted chiefly in commercial cultivation. Since different characteristics of the plants can be combined, this method can be a great economic advantage. Many passionflowers are susceptible to root nematodes and fungal infections such as fusarium wilt. In professional cultivation these problems can lead to enormous losses. Unlike *Passiflora edulis* f. *edulis, P. edulis* f. *flavicarpa* is resistant to nematodes and fusarium wilt. The pulp of the latter, however, does not compare to that of forma *edulis.* In such a case, the logical solution is to take advantage of the positive characteristics of both by grafting *P. edulis* f. *edulis* onto the rootstock of *P. edulis* f. *flavicarpa.* This method is also recommended for species that are difficult to propagate from cuttings, provided they can be propagated from cuttings at all. Good results have been obtained with *P. caerulea* as a stock for *P. umbilicata,* and with *P. tarminiana* as a stock for other species of *Tacsonia.*

There are several procedures to be distinguished in grafting. The most widely used form is called "whip and tongue grafting." Begin by shortening the rootstock to 10–20 cm with a sterilized sharp knife. On one side of the stock, make a slanted cut of about 2 cm toward the upper end. Make sure the cut surface does not become too wide. From another plant, take a piece about 15 cm long with internodes as short as possible. Shorten or remove the leaves. Now prepare this shoot, or scion, with a slant cut of 2 cm opposite a node. The cut should be congruent with the cut on the stock. Lay the edges of the cuts one on top of the other and bind the joint together with fiber or grafting rubbers. Close free parts of the wound with grafting wax. In a warm, moist climate the scion will rapidly set forth new growth, at which point the fibers can be removed.

6
Pests and Diseases

Klaus Balser

FOR MANY YEARS, numerous species and cultivars of passionflowers have thrived in parts of the world that are far away and completely different from their area of origin. Nevertheless, it is an enormous stress for a plant that is indigenous to tropical and subtropical regions to be forced to grow in surroundings so different from its habitat. To guarantee the well-balanced system in which each plant thrives in its ecological context, action must be taken to protect it from the damage caused by insufficient cultivation conditions and potential pests and diseases. The conditions of the plant's habitat must be imitated as accurately as possible, including temperature, light, atmospheric humidity, water supply, soil, and nutrition. As soon as one of these factors is out of balance, the plant becomes vulnerable to noxious organisms such as pests, fungi, bacteria, or viruses.

Although it is not possible to restore the biological equilibrium of the plant's natural habitat, natural predators can be introduced as a step in that direction. However, when plants are infested with several pests it is difficult to set up a parallel fauna, all the more as these organisms need suitable conditions, too. And once biological controls are introduced, the selection of effective pesticides is very limited, as these will often also eradicate the beneficial organisms. In the end it is left to the individual to decide. In many cases, pests may become so devastating that chemical warfare is needed to rescue a plant. When pesticides, fungicides, and other chemicals are applied judiciously and with the utmost caution, hazards can be kept to a minimum.

Before going into details regarding pest management tools, some words of warning must be given. Follow the instructions on the label accurately. Labels provide information about use rates, reentry period, application methods, precautions, and so on. Detect pests early on, and make sure the type of pest has been correctly identified; this is important when selecting the best pesticide. Rotate between chemical classes to reduce the risk of resistance, and rotate after 1 or 2 applications of the same product. At first use, try the product on only a small area of the plant, as some species might react susceptibly. Products containing oil or insecticidal soap are generally not the best choice; some species may react poorly by developing patches on the foliage, and some may even shed leaves.

Many effective insecticides, acaricides, and fungicides are available on the market. Information regarding proprietary names can be obtained through local suppliers or EPA extension services. Useful information can also be found on the World Wide Web.

Pests affected by selected insecticides and acaricides

Please note that the substances listed may lose their registration and become illegal to use at any time. Some substances or products are classified as restricted-use. Check with state or county authorities to confirm which products are approved for use.

	Aphids	Biting insects, caterpillars	Leaf miners	Mealy bugs, scale insects	Spider mites	Sucking insects	Thrips	Whiteflies
Abamectin					✘		✘	✘
Acephate	✘		✘	✘			✘	✘
Amitraz					✘			
Azadirachtine	✘				✘	✘	✘	✘
Bacillus thuringiensis		✘						
Chlorpyrifos			✘	✘			✘	✘
Clofentezine					✘			
(Beta-)Cyfluthrin		✘	✘	✘		✘	✘	✘
Cypermethrin		✘						
Deltamethrin		✘						✘
Diazinon	✘	✘	✘	✘	✘	✘	✘	✘
Dimethoate	✘	✘	✘	✘	✘	✘	✘	✘
Fenazaquin					✘			✘
Fenbutatin-oxide					✘			
Fenpropathrin	✘			✘	✘			✘
Fenpyroximate					✘			
Imidacloprid	✘		✘	✘		✘	✘	✘
Imidacloprid, methiocarb	✘	✘	✘	✘	✘	✘	✘	✘
Malathion	✘	✘	✘	✘	✘	✘	✘	✘
Methamidophos	✘	✘	✘	✘	✘	✘	✘	✘
Oxydemeton-methyl	✘				✘		✘	
Parathion	✘				✘		✘	
Permethrin		✘	✘					
Pyrethrine, piperonylbutoxid	✘	✘		✘	✘	✘		✘
Pyridaben					✘			
Tebufenpyrad					✘			

Aphids

Aphids, also called greenflies, blackflies, or plant lice, are winged or unwinged sucking insects. They are up to 0.5 cm long and can be different colors, such as green, pink, brown, or black. They feed on plant juices, causing deformed leaves and stunted growth. They also secrete a sticky substance called honeydew onto the leaves, stems, and buds of a plant, often leading to the development of blackish sooty molds. By tapping the plant's tissue through their threadlike mouthparts, aphids may transmit plant viruses. Overfertilization, resulting in high nitrogen levels in the plant, encourages aphid reproduction and can be avoided by using a low-nitrogen formula or slow-release fertilizer. Lady beetles, parasitic wasps, and green lacewings can keep aphids under control, and aphids can be washed off of plants with a strong stream of water.

Beetles and Bugs

These are not a big problem with passionflower vines, with the exception of the passionvine bug in Australia and the more than 200 species of stinkbugs active throughout the United States. Stinkbugs are 0.5–1.2 cm long and have an oval, broad, slightly convex shape. The head and prothorax form a triangle that points forward. They secrete a foul-smelling discharge that gives their resting place a nauseating odor that lingers for weeks. They range in color from green to brown and black, and adults as well as larvae feed on almost everything available. Some solitary stinkbugs may not be a problem and can be removed with an appropriate tool, such as long pincers (beware of the secretions!). However, severe infestations can only be controlled by using such chemicals as cypermethrin, deltamethrin, and propoxur.

Caterpillars

Leaves that are eaten away, rolled, or full of variously shaped holes indicate that the larvae of various butterflies and moths are at work. In most cases caterpillars are easily detected and removed;

only in cases of serious infestation is chemical control necessary.

Cutworms

The caterpillars of various moths usually feed on the roots of a plant during the day and come up to the lower leaves only in the dark. During this time, they can be searched for and destroyed. They are creamy white, dirty gray, or brown and up 4.5 cm long. Young larvae remain on the plant during the day and hide near the base of the plant at night. Action should be taken as soon as the first symptoms appear, either by collecting the young larvae from the foliage or by applying a suitable pesticide to the foliage, particularly at the base of the stem. Older larvae can eat up the rootstock of large plants within days. To find larvae in the soil, identify the entries to their hideouts, which will be holes about 0.5 cm diam. Larvae do not hide very deep in the ground, so they can also be found when the surface of the soil is turned over. As a last resort, baits can be prepared from a mix of bran, sugar, water, and parathion.

Leaf Miners

Within the leaf, white or brownish tunnels appear in the characteristic shape (linear, circular, or irregular, for example) of the particular leaf miner, caused by the larvae of various flies, moths, beetles, and sawflies. A slight infestation will not damage the plant too much, and affected leaves can be removed and burnt. Seriously infested plants should be treated with a suitable pesticide.

Mealybugs

Mealybugs are soft-bodied, grayish white, wingless insects with a white waxy covering, and are up to 0.7 cm long. They damage and weaken a plant by inserting their threadlike mouthparts into any part of it and sucking out the sap. An infestation usually starts at the base of the stem. A few solitary mealybugs are easily overlooked, as they hide in the leaf axils and near the base of the leaves. However, the infestation proceeds very rapidly, and soon mealybugs of all sizes can be found crawling around. A painstaking but effective method to fight an infestation in its beginning stages is to touch each individual with a small brush dipped in a 50% rubbing alcohol solution. This procedure must be repeated daily until no more bugs can be found. Biological controls include Australian ladybugs (*Cryptolaemus montrouzieri*), also called mealybug destroyers, and certain microscopic parasitic wasps. In extreme cases pesticides may be used.

Nematodes

Nematodes, also called eelworms or roundworms, are microscopic threadlike animals that utilize a stylet to puncture and feed from plant cells. Root nematodes cause short, stubby roots that may be brown and may develop swollen knots. Leaf and bud nematodes feed on leaves and cause brownish black patches. Such attacks cause discolored leaves and a stunted growth, shortening the life of the plant. Although there are several effective nematicides available, no single product is effective against all species of nematodes. Moreover, these substances are very toxic to humans and animals, and in many countries their use is restricted to certified applicators. If a cherished plant needs to be rescued, try to get professional help from an authorized person. Otherwise remove infested plants and clean the immediate surroundings so that no source of infection remains.

Scale Insects

Scale insects are white, yellow, gray, or brown insects with crusty shells, and are up to 0.7 cm long. They are closely related to mealybugs. By sucking sap they weaken the plant, and their honeydew serves as a medium for the growth of sooty molds. Scale insects are difficult to control. Like mealybugs, which are protected by their waxy covering, these pests are protected by their chitin shell. Pesticides with a systemic effect are recommended.

Spider Mites

Spider mites pierce the hostplant's epidermal cells with their mouthparts and extract plant juices, causing tiny chlorotic spots on the leaf. When many mites are feeding, the leaves turn yellow or brown and die off. If uncontrolled, spider mites may actually kill a plant.

Spider mites are not insects but are closely related to spiders, having 8 legs and a one-piece body. Their color is greenish to reddish, though overwintering females are orange. They are so small, less than 0.5 mm, that they can hardly be detected without a magnifying glass. When they appear in masses, they cover the leaves with fine webs, but at this stage it is almost too late to rescue the plant. Spider mites reproduce very rapidly, especially in dry, hot weather. Each adult female can lay up to 200 eggs (up to 20 per day), each hatching within a few days. The typical egg-to-adult cycle takes 7–14 days, depending on temperatures.

Immediate action must been taken at the first sign of infestation. An excellent biological control is another mite, the predatory *Phytoseiulus persimilis*, available from commercial biological control suppliers. If the right environmental circumstances can be provided, including temperatures above 18°C and atmospheric humidity above 60%, biological control is as effective as chemical control.

Serious infestations, however, can only be controlled by using appropriate miticides. Spider mites develop resistances very rapidly, so avoid applying the same product more than once a year. Several treatments at weekly intervals with products from different classes, and with different modes of action, are necessary to eradicate more resistant individuals and eggs. For example, at the first application, an organophosphorous pesticide such as dimethoate or methamidophos might be used, in the 2nd a pyrethroid (fenpropathrin or cyfluthrin), and in the 3rd a special miticide such as abamectin or pyridaben. The order can be reversed or changed. The issues of pesticide classes and different modes of action are a little complicated, however, so it is advisable to ask an expert for guidance.

Thrips

Thrips are small (about 1.25 mm long), hair-winged, yellowish brown to black insects. Their most notable characteristic is their featherlike wings. Thrips thrive in hot, dry conditions. They feed by piercing the leaf or petal surface and drawing sap from injured cells, which causes the leaves to discolor to silver-white. Serious attacks stop the flower buds from opening. Control of thrips is difficult, especially during summer months outdoors, because of constant migration from weeds, grass, flowers, and trees. In serious cases pesticides should be applied at short intervals.

Weevils

The adult black vine weevil is a grayish black, wingless beetle about 1 cm long with elbowed antennae and a short snout. It feeds at night, causing crescent-shaped notches along the leaf margins, and hides in the soil under the plants during the day. Search the plants for beetles with a flashlight as soon as damage is detected. They can also be trapped in pitfalls, which are made by placing a paper cup in the ground, with the rim at soil level, and bending a piece of paper to make a cone in the top so that the beetle slides down into the cup. The most common method of control is to place a poison on the foliage. Several sprays may be necessary, as the beetles are active for a long time.

Far more dangerous than the beetles themselves are their grubs. These legless, white, C-shaped, brown-headed larvae are up to 1.5 cm long. Found under the hostplant, they cause much more damage than the adults, eating away the roots and eventually killing the plant. Adults emerge from soil pupation in May or June. They must feed on plants for 3–5 weeks before they are ready to lay eggs. They may live 90–100 days and usually lay about 200 eggs during this time. The eggs hatch after 2–3 weeks and feed on plant rootlets. Control measures against grubs must be taken as soon as the eggs have hatched—that is, 5–7 weeks after the first beetles have appeared. While the adult population is kept under control by mechanical methods or foliar sprays, the larvae can be eradicated with the help of entomopathogenic nematodes (*Heterorhabditis megidis*), to be applied as soon as soil temperatures are at least 16°C. Although chemicals (methamidophos, parathion) are available to destroy the grubs by drenching the soil, these are only the 2nd best choice. If pesticides are applied to the soil, nematodes can only be inserted 6 weeks later, after the chemical has been degraded.

Whiteflies

Like aphids, these small, white-winged insects damage the plant by sucking sap. They can also transmit viruses, and the sooty mold that develops on their excretions further weakens the plant. They can be caught in yellow sticky traps hung just above the plants or sucked up with a vacuum cleaner. However, the immature, scalelike nymphs on the undersides of the leaves will remain there and build up new generations. Effective biological control can be achieved by the parasitic wasp *Encarsia formosa* when the infestation is not too severe. Otherwise, suitable pesticides must be used repeatedly.

Bacterial Diseases

Some of the most destructive diseases of plants, such as soft rots and wilts, are caused by bacteria. The symptoms caused by plant-pathogenic bacteria may include spots of various sizes and colors on leaves and stems. Often, however, such infections occur without obvious signs. Similar to the pathogens of fungal infections, bacteria invade the vascular tissue of the lower stem and the roots that conduct water and nutrients. They multiply and interfere with the transport of water and nutrients. The upper parts of the plant, cut off from their supply, wilt and die. Many unexplainable losses of plants are caused by such infections. Only when a cut is made through the stems and roots does it become obvious that the vascular tissue, the xylem that conducts water and nutrients, is damaged and has turned brownish. Histological analysis reveals that the cell walls are broken and toxins have built up so that the flow of water and nutrients is interrupted.

Preventive measures are the most important factors in fighting bacterial diseases. Dead plants, fallen leaves, and other debris must be removed immediately and disposed of in a garbage can, together with the whole pot and the rooting medium. Bacteria are ubiquitous; they are water- and windborne, transmitted by birds and insects, and dispersed by infected plant material. Healthy, well-kept plants that do not host sucking or biting pests will hardly be open to an attack by bacteria, which mostly invade a plant through wounds. Special attention should be paid to the rooting medium. To avoid root rot, keep the plant pest-free and make sure it never becomes waterlogged.

Fungal Diseases

Plants are often covered in a myriad of fungi, but let this come as no shock: well-kept passionflower vines are very resistant to fungal diseases. Nevertheless, the sudden inexplicable death of a cherished plant raises questions.

Just as there are both beneficial and noxious bacteria, there are both "good" and "bad" fungi. Bacteria and fungi are necessary to decompose the biosphere so that organic material is recycled in the ecosystem. A fungus does not make a distinction between a rotten tree on the forest ground and the leaf litter and debris covering the surface soil of a pot-grown plant. Preventive measures, such as favorable growing conditions and good sanitation, will keep fungal attacks at a minimum.

Fungal leaf spots, often brown or slate-gray, are caused by a wide range of fungi. All infected and fallen parts of the plant must be removed and burnt. Then the plant should be sprayed with triforine or a similar fungicide.

Phytophthora root rots and other root rots develop when the growing medium is too wet. The fungi, introduced in unsterilized soil mixes, build up in the soil and in plant remains over time. Symptoms of such a disease include enlarged roots that turn black from the stem downward although the fine roots appear healthy. The affected plants, pot, and growing medium must be removed and disposed of. Chemical control of *Phytophthora* is possible (fosetyl, propamocarb) but is mostly restricted to authorized personnel.

Black patches on the leaves and stems are symptoms of fusarium wilt, caused by the fungus *Fusarium*. The roots turn black and die off. There is no cure for this disease; the plant must be discarded and the vicinity thoroughly cleaned. *Verticillium* wilt is similar. Plants suffering from it exhibit a stained vascular system of stem and roots when a cut is made through them. Again, affected plants and soil must be removed and disposed of.

Powdery mildew caused by various fungi usually appears as a powdery white growth on the upper surface of the leaves, which may turn yellow and fall off. An infection is favored when plants are overcrowded and suffer from insufficient light and poor nutrition (as when they are given too much nitrogen). Greenhouse plants are especially endangered during winter, when light is insufficient and atmospheric humidity is high. Most unfavorable are drafts and abrupt changes in temperature. Affected parts should be removed and burnt. Among the numerous fungicides that are effective at battling powdery mildew are bitertanol, fenarimol, and kresoxim.

Downy mildew is caused by other species of fungi and must be treated with different fungicides, such as fosetyl, mancozeb, and metiram. It develops on the undersides of leaves, with the upper sides exhibiting yellow or brown blotches. Infections are worsened by humid conditions. Affected leaves must be removed.

When hygiene is not as good as it should be, gray mold caused by the fungus *Botrytis cinerea* may rapidly develop. This fungus enters through wounds and leaf scars. A gray fuzzy mold grows on areas of rotting tissue on leaves and stems and proceeds to healthy parts of the plant, causing these parts to die. All dead and rotting tissue must be removed, and the plants and neighboring areas should be sprayed with a suitable fungicide such as iprodione or dicarboximide.

At times, blackish, sooty molds develop on the upper surfaces of leaves. These are fungi from different genera that are growing on the excretions of pests such as aphids and whiteflies. These coverings can be washed off with a soft sponge and water.

Damping-off is a problem that occurs while plants are being propagated. The stems of seedlings turn black and shrivel at soil level, and then the seedlings topple over and die. Often a fluffy white fungal growth appears on the surface of the soil and on the died-off seedlings. Various fungi cause this disease. Again, prevention is the best weapon. Only sterilized soil mixes and clean propagating chambers, trays, or pots should be used. Another preventive measure is to spray the seedlings with an appropriate fungicide, such as captan.

Viral Diseases

A major problem with passionflowers is that many species, especially longtime cultivars, host a latent virus. In commercial fruit production, viruses are a serious economic drawback (the "woodiness" virus reduces yields considerably), but when it comes to the enthusiast, the problem is often aesthetic. In many cases the only symptoms produced by viral infections are reduced growth and dwarfing or stunting of the plant. The most obvious symptoms are the discoloration of the leaves, caused by the virus interfering with chlorophyll production and thus affecting photosynthesis. On the one hand, enthusiasts are fascinated by the enormous variety and beauty of species and cultivars they know to be infected with viruses. On the other hand, they face the risk that their hitherto healthy and immaculate plants might be infected with viruses by introducing dubious specimens.

Under favorable conditions in appropriate surroundings the symptoms of viral infections are hardly noticeable, but as soon as conditions get worse (less light, lower temperatures) the symptoms recur. Whether gardeners decide to keep their collections as virus-free as possible by not introducing potentially dangerous plants that already show symptoms or are known to have a latent virus, or whether they don't want to do without the gorgeous flowers of some virus-ridden cultivars, the approach is still basically the same. Even well-established, robust species and cultivars such as

Passiflora caerulea and *P*. 'Byron Beauty' may show symptoms of virosis if conditions are bad enough. Hence the only feasible approach is to take all preventive measures to keep viral infections at a minimum. These measures are similar to the conditions needed for successful cultural practice: enough light, adequate temperatures, sufficient atmospheric humidity, suitable rooting medium, balanced nutrition, and so on. Pests must be reduced to a minimum so that viruses cannot be transmitted by vectors such as sucking insects, and everything should be done to keep the plants as robust as possible. Any material that could pass the viruses on (cutting tools, for example) must be sterilized after each use. Smokers should clean their hands as well to avoid spreading the tobacco mosaic virus.

Even if some mosaic or ring spot patterns appear on a leaf every once in a while, in the final analysis it is the joy of growing these plants that characterizes the true enthusiast.

7
Medicinal Benefits

PASSIONFLOWERS have been used in folk medicine for ages and are finding an increasingly important place in modern medicine. Though many species are used in rural areas, *Passiflora incarnata* has emerged as the most economically important medicinal species on the international market. There are large areas of this species under cultivation in Belize, Brazil, Colombia, Guatemala, India, and particularly in the United States. Few species have been investigated chemically or clinically, and the preference of the pharmaceutical and herb industry for *P. incarnata* above others is mostly due to tradition and lack of bioprospecting.

In their native habitats, many passionflower species have been used in folk medicine for a very long time to combat various diseases. Throughout Central America an infusion of leaves from various *Passiflora* species with 2-lobed leaves is used as a diuretic. In India the unripe fruit of *P. foetida* is used as an emetic (don't try this at home!), and a decoction of dried herbage of *P. foetida* is said to have diuretic effects. The pulverized root of *P. quadrangularis* has been mixed with a grease and used as an analgesic ointment. In many parts of South America the leaves of *P. edulis* and other species are used as a diuretic; the leaves are also crushed and applied as a plaster to treat inflammations and asthma attacks. A concoction from fresh leaves of *P. laurifolia* is ingested as a treatment for tapeworms, and many other minor local uses are recorded.

Rather soon after its discovery, the passionflower was gradually introduced into European medicine. As early as the 17th century, a sedative and sleep-inducing effect was described. The first reports of the analgesic effects of passionflower preparations came from North America in 1867; reports from France were made about 30 years later.

Components and Effects

Many experiments on the effects of passionflower products have been made. A sedative effect has been found in a test with rats and mice. Further experiments have shown a paralyzing effect of the smooth muscle in the intestines of rabbits. The spasmolytic properties of passionflower preparations have been shown to relieve nervous stomach and intestinal discomfort. A bactericidal effect has been observed in substances contained in some species. Proof of a sedative and sleep-inducing effect in humans has also been established, as has a lowering of mild hypertension, even just from drinking the juice.

The most important components of the genus *Passiflora* can be divided into several chemical groups: alkaloids, cyanogenic glycosides, flavonoids, terpenes, sterols, and aromatic compounds of the fruit. It is not yet clear which chemicals, or interactions of chemicals, produce the various medicinal effects.

The greatest medicinal significance by far is attributed to "herba passiflora," a drug obtained from *Passiflora incarnata*. The following passages refer exclusively to this species. The drug is simply the pulverized aboveground parts of the plant (leaves, stalks, flowers), which are dried in a shaded location in the open air.

Until a few years ago the alkaloids and their derivatives were believed to be the main active agents in *Passiflora*. The alkaloid harman in *Passiflora incarnata* was isolated in 1954 (Neu). Lutomski (1959), Hultin (1965), and Ambühl (1966) confirmed the content of harman and discovered further alkaloids: harmin, harmol, harmaline, and harmanol. On the whole, their analyses of extracts from *P. incarnata* showed a concentration of alkaloids that was 0.01–0.09%. However, in more recent studies of "herba passiflora" only trace amounts (0.1 ppm) of harman-alkaloids could be detected (Rehwald and Sticher 1995). This concentration is too small to be responsible for the overall sedative effect of the drug.

Provided that the results in the studies from the 1960s are correct, the question arises as to why the concentration of alkaloids has been reduced so drastically since that time. Could this have been caused by commercial cultivation on giant plantations, or by sampling errors, or can other correlations be found? Is the herb less effective now? These questions deserve further investigation.

Thus the flavonoids gain more and more importance as the main active substances of passionflower extracts. Flavonoids are a group of plant color substances, widely spread among plants, that are found in the flowers and leaves and are responsible for the coloring of the heartwood of trees.

The exact effects of flavonoids like vitexine, saponarine, isovitexine, orientine, and isoorientine have not been determined completely, but there is no doubt about their importance. They have been used for years to strengthen capillaries and for their antioxidant properties. On the whole, flavonoids constitute about 0.4%–2% dry weight of *Passiflora incarnata*. There are several other components in passionflower extracts, including maltol (y-pyron-derivate), another pharmacologically active substance but one with poorly understood central depressant effects.

Application

The passionflower is most valued medicinally for its sedative and sleep-inducing effects. Although several species have been used for this effect, including *Passiflora quadrangularis*, there is actually only one species, *P. incarnata*, that is processed into medications and teas on a large and economically important scale. In comparison with other species, the sedative effect of *P. incarnata* starts more slowly but is stronger overall. This is one reason why it is preferred to other passionflower species as a mild sedative, although efficacy has been compared with few other species.

The use of passionflower extracts or teas is recommended particularly to treat the strain associated with stressful situations and the exacting demands of fast-moving contemporary life. There are no known side effects to these extracts, unlike most chemically produced sleeping and sedative drugs. Passionflower is mild enough for use by children, strong enough to help adults, and has no residual effects, such as lingering drowsiness. In homeopathic medicine an essence of the fresh herbage of passionflowers is used in either its pure form or as a combined preparation to treat insomnia and neuralgic pain.

Passionflower herbage, "herba passiflora," is usually offered in combination with other sedative medicinal herbs such as valerian, hop, Saint-John's-wort, and melissa, and can be purchased as dried tea leaves, either loose or in tea bags, or as a bottled tincture or extract.

8
Passionfruit

PASSIONFLOWER fruits come in a great variety of forms. Nearly all are berries, though a rare few are pulpy capsules, as with *Passiflora capsularis*. The fruits of different species differ considerably in size, shape, and color, as illustrated in Plate 12. While the smallest are about the size of a pea (*P. suberosa*), the fruits of *P. quadrangularis* are up to 15 cm diam. and weigh up to 4 kg (Plates 13 and 14). Shapes vary from globular to oval or spindle-shaped. They can be egg-shaped or ellipsoidal to nearly cylindrical. There are even a few passionfruits with 6 keel-like ridges, so that a cross section shows the form of a star. Fruits may be green to yellow, orange, red, purple-brown to dark blue, or almost black.

As mentioned in chapter 1, the Spanish conquerors called some of the larger fruits granadillas for their similarity to pomegranates, and the name is still used in Latin America.

Many species have edible fruit, but only 5 or 6 of these species are of major economic importance. Another 10–15 species are cultivated occasionally on a large scale. The rest are cultivated only locally on a small scale or are exploited as a natural resource. The edible part of the fruit is the juicy and aromatic aril, or juice sac, which surrounds each seed. Often the aril is whitish, grayish, yellow to orange, or strikingly red. This indicates that in their natural surroundings passionfruits are eaten not only by mammals but also by birds, which guarantees the dispersal of seeds over a larger area. People eat the juicy aril pulp fresh, spooning it out of the cut-up fruit together with the numerous seeds. With *Passiflora quadrangularis* and *P. alata*, the juicy thick wall of the fruit can also be eaten, though it is relatively tasteless.

Before trying fruit other than what is available at the market—fruit from the garden, for example—understand that not all edible fruit is a culinary delight. Some experiences in taste are rather insipid or disillusioning. There can be much variability in the taste of undomesticated species; growers should keep their eyes open for excellent variations. All species are likely to have fruit that is edible when ripe and mature—edible, but not necessarily tasty. That said, the immature fruit of all species should be considered poisonous due to the presence of cyanogenic glycosides, a source of cyanide.

In the temperate zone, some species also develop edible fruit when planted outdoors, including *Passiflora caerulea*, *P. edulis* f. *edulis*, *P. foetida*, and *P. incarnata*. Quite a number of other species, for example, *P. biflora*, *P. capsularis*, *P. cinnabarina*, *P. murucuja*, *P. rubra*, *P. sanguinolenta*, and *P. suberosa*, will produce fruit when grown outdoors during summer months, but the fruit will be barely edible. The fruit has quite a decorative effect, however, and the numerous seeds can be used for sowing. The majority of passionflowers will only produce fruit satisfactorily in a greenhouse or conservatory. Among these are *P. actinia, P. alata, P. laurifolia, P. maliformis, P. quadrangularis*, and *P. vitifolia*.

Unfortunately, many species that produce edible fruit are self-sterile, meaning they will only bear fruit with pollen from another clone. This kind of cross-pollination requires 2 nonrelated flowers of the same species that are open at the same time. Cross-pollination is also possible between different passionflower species, provided the number of chromosomes is identical (see chapter 17). The seeds of such fruit will bring about new hybrids with novel combinations of genes.

When passionflowers are cultivated under glass, the insects or hummingbirds needed to bring about pollination are absent, and so pollination must be brought about manually. To do this, the grower must know at what time the flowers are ready to receive the pollen. Immediately upon the opening of the blossom, the 3 styles with their swollen stigmas stand upright. As soon as the styles and their stigmas bend a little toward the base of the flower, the flower should be pollinated. At this time the stigmas are particularly receptive and can easily be spread with pollen. With forceps or thumb and index finger, the grower cautiously breaks off an anther from one flower and brushes the pollen onto the stigmas of the other plant. If it should rain within 90 minutes after the pollen has been applied, fruiting will most likely not take place. If the flower does not fall off and the ovary swells within 3 days, pollination has been achieved. Fruits mature in 40–90 days in most species.

The most important species with edible fruit are *Passiflora edulis* (including its best-known forms, *P. edulis* f. *edulis* and *P. edulis* f. *flavicarpa*), *P. laurifolia, P. ligularis, P. maliformis, P. pinnatistipula, P. quadrangularis, P. tarminiana, P. tripartita* var. *mollissima*, and *P. vitifolia*. Other species with edible fruits include *P. actinia, P. alata, P. ambigua, P. ampullacea, P. antioquiensis, P. arida, P. caerulea, P. coccinea, P. cumbalensis, P. edulis, P. foetida, P. herbertiana, P. incarnata, P. mixta, P. nitida, P. pergrandis, P. platyloba, P. popenovii, P. riparia, P. schlimiana, P. seemannii*, and *P. serratodigitata*.

Passiflora edulis

The best-known and most popular passionfruit comes from *Passiflora edulis*. There are about a hundred varieties, all differing in size and, in terms of fruit, color, quality, yield, resistance to nematodes, and ability to self-pollinate. Among the most important varieties are 'Common Purple' and 'Sevick Selection' (both grown in Hawaii), 'Frederick' (with large red fruit), 'Perfecta' (a commercial variety with large fruit), and 'Norfolk' (with white flowers and reddish purple fruit). These cultivars combine the properties of the wild purple *P. edulis* f. *edulis* and yellow *P. edulis* f. *flavicarpa* in varying degrees. In recent years systematic breeding programs have brought about some highly productive cultivars with either a higher pulp content (about 45% in one synthetic population, 48–59% in another), larger fruit (with a weight of up to 170 g), or a higher yield (55–65 tons per hectare).

Passiflora edulis f. *edulis*

In Europe the fruit of *Passiflora edulis* f. *edulis* is the best known of this species (see Plate 17). Apart from South America, the most important areas where crops are raised are in South Africa, India, Malaysia, Australia, New Zealand, and Hawaii.

Passiflora edulis f. *edulis* is the more cold-tolerant form and does very well at elevations of 600–2000 m. In Kenya it is successfully cultivated at an altitude of 2500 m. The seedlings are planted 3–4 m apart. The first crop can be harvested in the following year and will become continuously larger until the 5th and 6th years. At this point the yield will gradually diminish and the plantation must be renewed for economic reasons, though the plants can live much longer. On average, such a plantation has a life span of 7 years. In comparison with other cultivars, the annual yield per hectare is relatively low for typical forma *edulis*: 10–15 tons. As a rule the globular to ovoid fruit reaches 4–5 cm diam., the size of a hen's egg. The average weight of the purple to brownish fruit is 35 g; the very fragrant dark yellow pulp is about 35% of the fruit weight. It has a smooth leathery skin that begins to wrinkle after some days of storage. The wrinkles indicate that it has gained its full flavor and maturity, but nevertheless it can be stored for additional weeks and remain juicy. The taste of the pulp and the juice pressed out of it is slightly sour, with a touch of wine flavor and an exotic apricot-like aroma.

Many prepared foods are made with this fruit, including multivitamin juices, liqueurs, tropical punches, yogurts, ice creams, and confectioneries. The value of the juice and pulp is determined by their content of ascorbic acid, sugar, carotene, riboflavin, and minerals. The cold-pressed seeds yield a pale yellow oil with a mild, pleasant taste. This oil is used both in the kitchen and as a raw material in the paint and varnish industry. The compressed residues are used as fodder.

The fruit is also well suited for eating raw. To this end it is cut into halves and the refreshing pulp is spooned out of the skin. The numerous blackish seeds can be eaten too. The fresh pulp is also used to refine fruit salads, ice creams, and other sweets.

Passiflora edulis f. *flavicarpa*

The yellow-fruited *Passiflora edulis* f. *flavicarpa* is the other main cultivated stock (Plates 18–21). It is resistant to nematodes as well as to several fungal diseases, such as fusarium wilt. Sometimes called *maracuyá*, forma *flavicarpa* is particularly suited for warmer regions and thus grows best at elevations below 1000 m. The largest plantations are in Brazil, Colombia, Australia, Hawaii, and Peru. One hectare yields a crop of up to 50 tons of this yellow, ovoid fruit. Although the fruit does not measure up to the lovely taste of *P. edulis* f. *edulis*, it is significantly larger, at 7–12 cm long, and therefore the yield is considerably higher. The average weight of the fruit is around 90 g. In contrast to forma *edulis*, forma *flavicarpa* has brownish seeds. A major part of the fruit is used for juice production. *Passiflora edulis* f. *flavicarpa* is self-sterile.

As a rule, its flowers open around noon and close during the course of dusk. Natural pollination is accomplished by carpenter bees, and where there are no bees, pollination must be done by hand. Field-workers can pollinate about 600 flowers per hour with a success rate of about 70%. Fruit ripens in 70–80 days. Studies in Venezuela have shown that the highest yields are obtained when vines are planted 3 m apart. The average life span of a plantation of *Passiflora edulis* f. *flavicarpa* is 5–6 years.

Passiflora laurifolia

Passiflora laurifolia (Plate 22) is mainly cultivated in its tropical homeland, including the West Indies and South America. There are plantations in Venezuela, Guyana, Brazil, Peru, Trinidad, Barbados, Jamaica, Puerto Rico, and Cuba, where it is offered on the local markets. It is also cultivated on a smaller scale in Southeast Asia. In comparison with the other passionflower species presented here, its economic importance is rather insignificant and it is rarely exported. Its ellipsoid to ovoid orange-yellow fruit is 5–8 cm long; the pulp is whitish. The flowers do not open before afternoon. Cross-pollination is needed to produce good fruit. In its countries of origin, *P. laurifolia* is cross-pollinated by carpenter bees; in other regions, it must be done by hand.

Passiflora ligularis

Passiflora ligularis is cultivated worldwide for its fruit (Plate 23) and is nearly as economically important as *P. edulis*. Its original dispersal area is limited to the mountainous regions of 1000–2500 m in Central America and the mountain ranges of the northern Andes.

This delicious fruit is also available in the European market and is regarded as the tastiest of all passionfruits. When ripe it is yellowish orange with white speckles. It is 6–10 cm long with a diameter of 4–7 cm. The taste of the yellowish white aril pulp is reminiscent of gooseberries. The fruit resembles a small, smooth orange, elongated into the form of a drop. It is added to beverages, ice creams, and raw salads. Its average weight is around 50 g.

Contrary to popular belief, this fruit can only be stored for a few weeks. Despite its hard yet brittle shell, it should be eaten as soon as possible.

Passiflora maliformis

In Europe one can obtain the globular fruit of *Passiflora maliformis* only by growing it oneself. It is cultivated especially in northern Brazil, Ecuador, Colombia, and on the Antilles. The green fruit turns yellow-orange as it ripens. The diameter is 3–5 cm. The pulp must be extracted by machines, as the shell is very hard and difficult to break. It is possible, however, to extract small quantities of pulp with the help of a hammer or nutcracker. The taste of the sweetish, light orange pulp is similar to that of grapes. The juice is very refreshing and valuable due to its high content of ascorbic acid. It is usually mixed with other fruit juices. Though the vine is not self-sterile, it rarely fruits with its own pollen. Cross-pollination is necessary to secure fruit formation.

Passiflora pinnatistipula

Passiflora pinnatistipula belongs to supersection *Tacsonia*, a group of species that often produce edible fruit. Along with *P. antioquiensis* and *P. cumbalensis* (Plate 24), this species does not reach the economic importance of *P. tarminiana* or *P. tripartita* var. *mollissima*.

Passiflora pinnatistipula is indigenous to the high altitudes of the central Andes at elevations of 2000–4000 m. Its yellow-green fruit, called "cholupa" or "gulupa," is exported on a small scale to Europe. Large plantations can be found in Colombia, Chile, Peru, and central Bolivia. The globular fruit is up to 6 cm long. Its brittle, leathery shell contains yellowish white pulp that tastes sweetish to flavorless to slightly sour. It is eaten raw or used in the production of confectioneries, ice creams, and especially drinks.

Passiflora quadrangularis

Outside of Central and South America, where it can be found from Mexico to Brazil, *Passiflora quadrangularis* is cultivated in India, Thailand, Australia, the Philippines, and Hawaii (Plates 13 and 14). In Indonesia and Australia it fruits year-round. While varieties with large fruit produce 20–35 pieces per plant per year, varieties with medium and small fruit produce 70–90 pieces. Reports from Venezuela indicate that the main flowering time is May to October. The ellipsoid to narrowly ovoid, yellow fruit is the largest of the species *Passiflora*. It can reach a length of up to 30 cm and can weigh 2–3 kg. Some specimens are known to have weighed a little less than 4 kg. On average, the fruit ripens in 62–85 days.

Passiflora quadrangularis is cultivated in several varieties, with fruits that differ in quality and taste. One especially large-fruited form has sometimes been recognized under the synonym *P. macrocarpa*. There are also numerous hybrid and backcrossed forms with *P. alata*. On the whole, however, this species' melonlike taste does not measure up to the taste of the fruit of most other species. When ingested in large quantities the juice is said to have a slight narcotic effect. In addition to the white pulp, which contains large quantities of seeds, the rind of the fruit can also be eaten—a rather dubious pleasure.

Passiflora quadrangularis is able to produce fruit with its own pollen. In Europe and North America it is advisable to hand-pollinate the flowers 4–6 hours after they have opened.

Passiflora tarminiana

The fruit of this passionflower (Plates 15 and 25) is increasingly popular in the plant's homeland, from the Andes of Venezuela to Peru, due to its vigor and ease of cultivation. It is very similar to its more commonly cultivated close relative, *Passiflora tripartita* var. *mollissima*. These 2 forms are the only high-altitude passionflowers cultivated as fruit plants in all states of the Andes. *Passiflora tarminiana* is more vigorous and resistant to disease but does not taste quite as good. The narrowly ovoid to fusiform fruit is 8–14 cm long and 3.5–4.5 cm (rarely up to 5 cm) diam. When immature, the fruits are green with white dots. When ripe, the skin is yellow, sometimes orange-tinged, puberulent, and soft-leathery. The pulp is orange, mostly sweet but slightly sour, and of a firm consistency.

Passiflora tarminiana is able to fruit with its own pollen, differing in this respect from other species of supersection *Tacsonia* except *P. tripartita* var. *mollissima*. As these 2 species are pollinated by hummingbirds, enthusiasts in most temperate zones must do the job by hand.

Passiflora tripartita var. *mollissima*

Not to be confused with the very similar *Passiflora tarminiana*, this species is the basis for a large industry in the Andes mountains of South America. It is popular from Venezuela to Peru, as well as in New Zealand. *Passiflora tripartita* var. *mollissima* enjoys the reputation of a national fruit in Colombia. Although it is sometimes called *P. mollissima*, recent research has shown that it is better considered a variety of *P. tripartita*. On a vigorous plantation-grown plant there can be more than 100 fruit, collectively weighing up to 10 kg. In one year, 200–300 fruit can be harvested from one plant. At altitudes of 1800–2700 m there are yields of 20–30 tons per hectare on flourishing plantations. The yield is reduced at lower altitudes.

When immature, the fruit is uniformly green. The narrowly oblong fruit is similar to that of *Passiflora tarminiana* but is a paler yellow, 8–14 cm long, and 3–4.5 cm diam., rarely up to 5 cm. The skin is pale to light yellow, puberulent, and soft-leathery. The pulp is orange, aromatic, sweet and somewhat acid, and of a firmer consistency than that of the other fruits described here. The flavor of the pulp is superior to that of *P. tarminiana* and is more aromatic and tart. Because of its taste, rather than its shape, it is called banana passionfruit. The delicate skin discourages export shipping, so outside of the areas where it is grown the fruit is of less importance. In its homelands the slightly sour fruit, including the seeds, is eaten fresh. In addition, the pulp and juice are often used in ice creams, jellies, sorbets, cold soups, and even cakes. The fruit is regularly offered in Europe as "curuba" or "banana passionfruit." *Passiflora tripartita* var. *mollissima* is able to set fruit with its own pollen.

Passiflora vitifolia

The green, puberulent fruit has whitish to light green longitudinal stripes that sometimes appear only in traces or dots. It is 5–9 cm long, occasionally slightly longer, and 3–4(–6) cm diam. In the West Indies this species is cultivated for its fruit on a small scale. After 4–5 months the fruit is ripe and falls off. It is still rather sour at this point and needs at least another month before it is really sweet. The taste of the whitish, very fragrant pulp can be compared with sour strawberries. Since it is self-sterile in most cases, it must be cross-pollinated.

9
Passionflowers of Belize

Jan Meerman

GREEN HILLS is a small butterfly rearing and research station in the Cayo district of Belize. To any butterfly enthusiast, the mentioning of tropical America will immediately bring up associations with *Heliconius* butterflies. And anyone mentioning *Heliconius* butterflies will also mention passionflowers. So it is not surprising that at Green Hills, where we have bred 15 local species of the family Heliconidae, passionflowers play a critical role.

In order to feed a few thousand hungry caterpillars on a daily basis, Green Hills depends heavily on the natural vegetation of its 20-hectare property. Plant diversity in the tropics is high, and consequently many plants occur in too low a density to be harvested economically. Therefore we started small plantations of the butterflies' most desired foodplants, and among these were many local passionflowers. Thus the National Passionflower Collection of Belize was born.

Belize is one of the smallest nations in Central America (22,966 sq. km) but is home to 23 native and 2 cultivated species of *Passiflora*. Of these 23 native species, no less than 13 occur naturally at Green Hills. As a result, some plants in the collection are simply naturally occurring specimens that have been made accessible by leading a trail to them. The other species are grown along trellises in the vicinity of the butterfly flight area. This sounds easier than it is, however.

One of the benefits of living in a passionflower's natural habitat appears to be that one can grow it without really having to worry. After all, it is just a "weed." The reality is not so straightforward. Growing certain species proves to be a real challenge. I grow most plants outside in the natural soil, which has proven to be more effective than growing them in pots. Potted plants require much more attention, and where I live there is no access to commercial potting media or to piped water and electricity for fancy drip-irrigation systems. All plants need to be watered manually, and during our 4- to 5-month dry season, this really becomes a chore. Even then, some plants have a hard time in this seminatural habitat. One of the problem species is the beautiful *Passiflora mayarum* (Plate 26), which occurs as rare individuals on the property. Transplanting these to a more accessible spot has nearly always proven fatal. Pampered seedlings, once transplanted to an ideal spot, promptly die. My biggest success so far has been with a plant that was so badly damaged during collection that I discarded it. It took root where it was left and is now one of the most vigorous plants of my collection.

Other species that give me troubles are, less surprisingly, species that occur in habitats different than what Green Hills has to offer. Most Belizean species are associated with alkaline limestone-based soils. Areas with more acidic soils are often surprisingly poor in diversity of passifloras. Only

Passiflora costaricensis and *P. urbaniana* seem to prefer acidic soils, and needless to say these do not do well at Green Hills, which is on limestone.

Other problem species are the rare and elusive *Passiflora pittieri* (Plate 27) and *P. obovata*. Both required veritable expeditions to locate and collect them (Meerman 1993, 1998), and because of this they are among my most cherished species. These 2 species grow into enormous lianas and are quite long-lived and slow-growing. This slow growth is my biggest problem, since I also use these plants as caterpillar foodplants, and my plants need to produce a lot of growth to be used.

If long-lived, slow-growing species present a problem, fast-growing, short-lived species can be equally problematic. These include *Passiflora adenopoda* and most varieties of *P. foetida*. These species are typically associated with recently disturbed habitats. They grow up quickly, produce abundant fruit, and then simply disappear. This habit makes it difficult to maintain them in a more or less permanent display. This is a pity, since *P. adenopoda* produces acceptably edible fruits, and *P. foetida* is both attractively rich with flowers and a taxonomic puzzle.

More than 3500 people visit Green Hills annually, and for the benefit of these visitors we try to display species that either produce a lot of flowers or that have an interesting story to tell. Among the best species is *Passiflora seemannii*, with its stunning flowers and very pleasant scent. Other good examples are *P. serratifolia*, *P. biflora*, *P. guatemalensis*, and *P. lancetillensis*. The latter has the bonus of having an especially strong and pleasant scent. *Passiflora ambigua* is attractive as well but does not flower heavily, and worse, the flowers smell awful. This is why this plant has been banned to one of the more remote parts of our display. *Passiflora helleri* has lush, attractive foliage, which is just as well, because in the 5 years we have had it in the collection it has flowered only once. Unusual foliage is also the main attraction of *P. coriacea* and *P. xiikzodz*. Another delight is *P. rovirosae* (Plate 28). This slender vine is not widely known, yet the flowers are attractive and are borne in clusters on the stem. Perhaps the greatest attraction is the unusual fruit, which is star-shaped, bright orange-red, and dehisces from the apex up, showing the strongly contrasting white innards.

Research is an important function of the National Passionflower Collection at Green Hills. The first key to the passionflowers of Belize was produced using the plants grown in the collection (Meerman 1996). The collection also holds the living original type specimen of *Passiflora lancetillensis*, and growing this plant helped us discover a new butterfly species for Belize.

One group of passionflowers at Green Hills that requires more study is the *P. foetida* relatives. In Belize alone, there are no less than 8 named varieties (or are they species?) of section *Dysosmia*. One is still unnamed.

The vivid flowers of *Passiflora sublanceolata* are so striking that no one will confuse them with any of the *P. foetida* varieties (*P. foetida* var. *nicaraguensis*, for example, Plate 29), although the sterile plant can be nearly indistinguishable from some of them. With *P. urbaniana* we reach the edge of where we can be sure the taxon is a true and separate species. The remaining 6 taxa are open to debate. They were treated as varieties in Ellsworth P. Killip's monograph (1938), but in the field some of them seem to be good species.

One of them, *Passiflora ciliata*, is often considered a true species. Personally I have my doubts about this and prefer to treat it in Belize as *P. foetida* var. *nicaraguensis*. One variety that in my opinion is so different that it might be considered a distinct species is *P. foetida* var. *gossypiifolia*. In Belize all these taxa, whether they are species or varieties, seem to have distinct habitat preferences, and all come true from seeds. Green Hills may be able to play a role in solving some of these taxonomic questions.

10
Passionflowers of the Andes

Michael Schwerdtfeger

South America—most people imagine damp rain forests full of orchids, climbing plants, and ferns. Giant Brazil, with its centrally located Amazonian lowlands, is particularly responsible for this impression. Equatorial Amazonia, located just above sea level, is indeed the paragon of a region with "tropical conditions": the 4 seasons are unknown near the equator, and the differences between day and night temperatures (about 10 degrees) are greater than the differences between the warmest and coolest months. Temperatures are constantly high. The ground of the rain forest is about 25°C, but during the day, temperatures can rise considerably in the canopy of the forest or in deforested, sunny locations. There is also perpetual precipitation. Due to the constant low-pressure areas near the equator, almost every day brings thundery rain, and the average precipitation reaches several times the amount falling in North America—2000–3000 mm per year, compared with 385 mm in Los Angeles, 414 mm in Salt Lake City, 1050 mm in Vancouver, and 1053 mm in Washington D.C.

Generally speaking, one could say that "from the plant's point of view" such a permanently warm and humid climate provides the necessary conditions for the plant to reach its physiological optimum. The farther from the equator, the more the mean annual temperatures decrease, while the influence of the seasons increases. Outside the Tropics of Cancer and Capricorn, the permanently humid equatorial climate is superseded by a seasonal climate with a distinct succession of rainy periods and dry periods. The north of South America (northern Colombia, Venezuela), Mesoamerica, and the West Indies are located in this region, characterized by dry seasons, while the most southern states of the continent (Chile, Argentina), lying beyond the Tropic of Capricorn, reach into the temperate zone of the Southern Hemisphere. These regions experience cold winters with frost and ice in the Southern Hemisphere's winter months of July and August.

The climate of a region is determined not only by its distance from the equator but also by ocean currents and by its altitude above sea level. The Humboldt Current, or Peru Current, transports Antarctic cold to the western coast of Chile, Peru, and southern Ecuador, causing desertlike aridity in some regions. Conversely, humid Atlantic air masses cause annual precipitation of up to 4000 mm in the Brazilian coastal rain forest, so that relatively far to the south an evergreen rain forest can thrive. The coastal strip of southern Colombia and northern Ecuador, called the Chocó, experiences more than 10,000 mm of annual precipitation and is among the most humid and species-rich parts of the world.

Of decisive importance regarding the amount and distribution of temperatures and precipitation is the altitude above sea level. Alexander von Humboldt, during his expedition in South America, which led him from Venezuela through "Nueva Granada" (Colombia) to the Andes of Ecuador, was fascinated by the interrelationship between altitude and temperature. The famous scientist could take advantage of the fact that the volcanoes of the Andes outclass the highest European mountains by several thousand meters: Chimborazo, Ecuador, 6310 m; Aconcagua, Chile, 6960 m. Here, where grandiose formations of high mountains are directly under the equator, the decrease in temperature with regard to altitude can be studied without the interfering effect of the seasons. Besides this, the altitudes of the Andes—for example, in Ecuador—are areas where snow is assured year-round.

As far as the flora of the high Andes is concerned, there is also an "alpine zone" with small, hard "cushion plants" belonging to Asteraceae, Umbelliferae, and Ericaceae. This is similar to the vegetation that can be found in the high regions of the Alps or Rocky Mountains. However, in the Andes this zone begins at an altitude of about 4000 m, while at altitudes around 3500 m—quite surprising for a European or North American visitor—an andine forest thrives in which the genus *Polylepis* of the family Rosaceae plays an important role. However, wide parts of the andine forests were cut down, in many cases as early as during the times of the Inca culture, and became grasslands. By using these regions as pastures for sheep (formerly for llamas) and regular deliberate burning, this man-made landscape is maintained, and andine trees can only survive in steep, moist creek gorges. Nevertheless, these artificial grasslands, known in Venezuela, Colombia, and Ecuador as "páramo" and in Peru and Bolivia as the drier "puna," offer some of the most grandiose scenery in South America.

The páramo lies about 3500 m higher than the Amazonian lowlands. While in the Amazonian lowlands perpetual "warm greenhouse conditions" prevail, in the páramo there is a constant change between sunshine, fog, hail, and snow showers. Depending on the amount of sunshine, the day temperatures rise to 10°–15°C, but as soon as the sun sets, temperatures drop to near the freezing point.

Interestingly enough, genera and families of plants from tropical South America can be found reaching into the wet and cold heights of the andine páramo. In the brushland, orchids, epiphytic ferns, bromeliads, and species of Solanaceae and Melastomataceae can be found. At these altitudes the genus *Passiflora* is also represented with some of its most beautiful and very large-flowered species: the high Andes, at 2000–3500 m, are the home of supersection *Tacsonia* (see *P. harlingii*, Plate 30).

With their giant brilliant pink, violet, or scarlet-red flowers, tacsonias are among the most beautiful passionflowers (see Plate 31). A striking feature within this group is the cylindrical elongated floral tube, which can reach 2–4 cm long in some small-flowered species (*Passiflora mathewsii, P. schlimiana, P. gracilens, P. luzmarina*), as well as in *P. antioquiensis* and *P. insignis*, but is about 10 cm long in most species. *Passiflora mixta* var. *pilaloensis* can even reach a length of 13–15 cm. A corona can be clearly seen only in a few species (*P. pinnatistipula, P. insignis*); in most it is reduced to inconspicuous, tiny knots. This is due to the fact that tacsonias are adapted to a few flower-visiting hummingbird species with straight, long bills—above all the sword-billed hummingbird (*Ensifera ensifera*, Plate 32). The native range of this hummingbird covers exactly the same area that is home to supersection *Tacsonia*. The adaptation to this narrow circle of pollinators goes so far that other visitors to the flowers (small hummingbirds, insects) act as nectar robbers only but do not contribute to pollination.

Since all tacsonias but 2 species are completely self-incompatible, plants that grow in the wild rarely set fruit. The only tacsonias that often bear plentiful fruit are *Passiflora tripartita* var. *mollissima* and *P. tarminiana*. These are represented in almost all regions of the Andes, and one or both also occur escaped in the wild in Mexico, Hawaii, and even the Canary Islands. Both have been cultivated for centuries by the indigenous people of the Andes for their similar edible fruit, known as "curuba," "taxo," "banana passionfruit," or "puro puro." It seems that plants whose fruits were both plentiful and tasty were preferred over the ages, so that, unintentionally, self-fertile types were selected.

In *Passiflora mixta* and *P. mathewsii*, the flowers stand in a horizontal to upright position, while in most other species they are often pendent on long, thin peduncles, giving them the appearance of a luxurious, colorful lampshade (see, for example, Plate 33). These species are not only among the most beautiful passionflowers but with regard to color, size, and bearing are also among the most noble.

Why are such desirable plants so rarely found in gardens? Until the 1990s this question was easy to answer: no one was bringing them into cultivation. Now several species (particularly those from Ecuador) have come into cultivation via experienced enthusiasts, and the initial conjectures turned out to be right: under cultivation one can hardly provide the extreme conditions in which the plants grow in their natural surroundings. Hardly any tacsonias grow at altitudes below 2000 m. The habitats are on the outer slopes of the Cordilleras, which are covered with clouds for much of the day. Abundant rain falls every day, and temperatures are 0°–15°C year-round. Since in many parts of central Europe and North America similar temperatures prevail in summer, and since tacsonias develop quite well at temperatures around 25°C, it is not a problem to cultivate small plants in pots. However, in order to develop the often gigantic flowers, the plants must have shoots that are several meters long. In Europe, plants that are removed from their winter quarters in spring and spend the frost-free time of the year outdoors reach the necessary size to flower in fall when optimum conditions are provided. The inevitable overwintering indoors and the insufficient conditions in winter always mean a setback for the plants. When tacsonias are pot-grown in the usual way and the shoots are circled around a trellis or a similar climbing aid they will hardly ever flower.

The cool, humid altitudes of the Andes above 2000 m are also home to some passionflowers unrelated to *Tacsonia*—for example, *Passiflora exoperculata*, *P. indecora*, *P. alnifolia*, *P. viridescens*, and *P. ligularis*. Some of these species turn out to be quite free-flowering when cultivated in pots, while *P. ligularis*, like many tacsonias, must reach a considerable size before flowers can be expected. In regions where the plants must be taken indoors because of cold winters, pots must be wider than 20 cm. The location should be well aired, well lit, and spacious enough so that the plants must not be cut back too rigorously. When these conditions are provided, *P. tripartita* var. *mollissima*, *P. tarminiana*, *P. pinnatistipula*, *P. luzmarina*, and *P. umbilicata* can be induced to flower. *Passiflora* ×*exoniensis* and *P. manicata* (Plate 34) and its hybrids *P.* 'Coral Glow' and *P.* 'Coral Sea' will be more rewarding by setting more flowers. *Passiflora ampullacea* and *P. cumbalensis*, however, can hardly be induced to flower when cultivated in pots.

In regions with mild winters, where tacsonias can be planted freely outdoors (short spells of slight frost are tolerated), chances are far better that the plants will set many gorgeous flowers. The region in the United States from coastal San Diego to San Francisco is famous for having conditions that allow certain tacsonias to thrive. Other regions, such as coastal Chile, South Africa, New Zealand, and places with similar climates, might support more than the usual *Passiflora tripartita* var. *mollissima*. However, reports on experiences regarding such cultivation are still missing.

11

Passionflowers of the Atlantic Rain Forest of Brazil

Reto Gamma

THINKING of Brazil and the rain forest usually brings to mind the Amazon region. It is less well known that the Atlantic rain forest as a whole is richer in species than the forests that cover wide parts of the Amazon basin. The Atlantic rain forest reaches along the Brazilian coast from the state of Rio Grande do Norte to the state of Rio Grande do Sul, the most southern state of Brazil. It differs in several respects from the permanently humid forests of the Amazon basin. It is entirely located in the Southern Hemisphere, extending beyond the Tropic of Capricorn south of São Paulo into the subtropics that show marked seasonal changes. While the Amazon region is a lowland (Manaus, the capital of the state of Amazonas, is about 1400 km away from the mouth of the river near Belem, but only 40 m above sea level), the Atlantic rain forest reaches from sea level up to altitudes of more than 2000 m. Moisture-laden winds are constantly blowing from the Atlantic. When they reach the mountains they cool down, form clouds, and lose their moisture either by rain or fog. In many regions of the Atlantic rain forest, annual precipitation is 2000–4000 mm, significantly less than in the Amazon basin. Average temperatures are also lower, 14°–21°C. The highest elevation is Pico de Bandeira at 2890 m, once thought to be the highest mountain in Brazil before Pico de Neblina near the Venezuelan border turned out to be higher at 3014 m. Vegetation changes as altitude increases so that biodiversity is greater than in the Amazon region. At 60 million years old, the Atlantic rain forest is 20 million years older than the Amazon rain forest. The geographical border between the 2 regions is the Planalto Brazileiro, savanna-like elevated plains about 500 m above sea level. Less rain falls here than in the coastal regions and the Amazon basin, and many trees are deciduous. Due to this geographically separated development, the structure of vegetation in the Atlantic rain forest differs markedly from that in the Amazon basin. It is quite astonishing that there are some correspondences with the flora of the Andes. In the states of Rio de Janeiro and São Paulo there are some species of fuchsia whose closest relatives are at home in faraway Chile. Also, many other genera can be found that are at home in both the Andes and the southeast of Brazil, such as *Bomarea*, *Mutisia*, and *Hippeastrum*.

Very early, quite a number of European botanists—among them Carl Friedrich Phillip von Martius, Johann Christian Mikan, and Albert Loefgren—were interested in the Atlantic rain forest. They set up herbaria and introduced many species and genera into Europe, including *Schlumbergera truncata*, *Hippeastrum*, *Anthurium*, begonias, orchids, and—of course—passionflowers. Many of the species that were introduced then disappeared from cultivation, such as *Passiflora kermesina* (recently reintroduced), while others such as *P. racemosa* remained.

Passionflowers can be found in almost all vegetation zones, although they seem to be more frequent at altitudes above 800 m (see, for example, Plate 35). It might be that in lower regions they have to compete against other vines that spread from other areas of the tropics. Species such as *Passiflora capsularis, P. suberosa, P. foetida, P. organensis, P. alata, P. actinia, P. jilekii, P. amethystina*, and *P. haematostigma* have a wide native range, while extremely rare species such as *P. mendoncae, P. odontophylla, P. marginata*, and *P. ischnoclada* are endangered and face extinction because of deforestation of their habitat.

Restinga

The coastal forest, prevailing on sandy soil near the sea, is called *restinga*. It is characterized by trees that are rarely higher than 15 m and forms a band between the coastal dune vegetation and moister, taller upland forest of the hills. This vegetation zone is subject to considerable changes in weather: days are hot, rain is often sparse, and winds are frequent. In this vegetation zone many cacti and small palms, such as *Allogoptera* species, can be found—but also many passionflowers, including *Passiflora mucronata, P. galbana, P. farneyi, P. racemosa, P. pentagona*, and *P. kermesina*.

Coastal Rain Forest

In the dense rain forest of the lowlands, the trees reach a height of more than 30 m. The occurrence of many epiphytes makes this forest look similar to that of the Amazonian lowlands. Species like *Passiflora malacophylla, P. mucronata, P. suberosa, P. organensis, P. alata, P. actinia*, and *P. speciosa* (Plate 36) grow here.

Montane Rain Forest

The montane rain forest begins at altitudes above 800 m. It is very rich in species and the trees are densely covered with epiphytes like orchids, bromeliads, and leafy cacti. The clouds, coming from the sea, unload their rain here, contributing to its exuberance and its richness in species. The most interesting species of *Passiflora* can be found in these regions. *Passiflora speciosa*, closely related to *P. vitifolia* and at home in the states of Rio de Janeiro and Espirito Santo, is certainly among the most spectacular passionflowers. Its purplish pink blossoms reach a diameter of up to 22 cm. The extremely rare *P. ischnoclada* (Plate 37), recently rediscovered, grows in the state of São Paulo. Its pseudoracemose flowers often appear on old wood. Its bracts are purple, its flowers pink.

Some species from subgenus *Astrophea* are also found here, including *Passiflora haematostigma, P. rhamnifolia*, or *P. elliptica. Passiflora amethystina* is another species that does not grow below this altitude, and in various regions it proceeds to altitudes of up to 2000 m.

Montane Cloud Forest

At an altitude of about 1300 m, the montane rain forest changes into a forest that is mostly buried in clouds. The trees are smaller here but are nevertheless covered with many epiphytes. In the states of São Paulo and Paraná, the rare *Passiflora mendoncae* can be found in this vegetation zone. This beautiful species with purple bracts and flowers is pollinated by hummingbirds. *Passiflora deidamioides* (Plate 38) is also found at higher altitudes within this zone. The cloud forest often proceeds to altitudes of more than 2000 m and is then superseded by the mountain plains.

Mountain Plains (Campos de Altitude)

At altitudes beyond 2000 m, the climate is rougher. Gusty winds and lower night temperatures, falling to the freezing point in winter, determine the vegetation, which is dominated by sasa (small bamboos), grasses, small fuchsias, and hippeastrum. Some species of *Passiflora* are also found here, including *P. campanulata, P. vellozii, P. imbeana*, and *P. marginata*. Coming from such a rough climate, these species can be expected to withstand a slight frost.

At one time the Atlantic rain forest covered an area of about 1 million sq. km. Only 5% of that remains, and only 1/10 of what is left is a "primary" rain forest. Many regions, even Tijuca National Park in the city of Rio de Janeiro, have been deforested and reforested again. Of course, the resulting "secondary" forests are less rich in species. Many regions of the Atlantic rain forest are so-called hot spots. These are regions with a high density in species and a multitude of endemic plants. One such region is Macaé de Cima, Nova Friburgo, in the state of Rio de Janeiro, where *Passiflora odontophylla* is endemic. There are plans to protect the remaining parts of the Atlantic rain forest. This is all the more important in consideration of the fact that it is located, in contrast to the Amazon basin, in the most densely populated and most industrialized states of Brazil. It is hoped that these measures will be successful and that this rich biological universe can be preserved for future generations.

12
Subgenus *Astrophea*

ASTROPHEA is undoubtedly the most unusual subgenus within *Passiflora*. Some of these species do not look much like passionflowers at first glance. About 60 species of tree, shrub, and woody vine with unlobed leaves, mostly indigenous to the lowlands of northern South America, are included in this subgenus. Brazil holds a number of species, a few are Andean in distribution, and only 2 species grow in Central America: *Passiflora pittieri* (Belize to Colombia) and *P. tica* (Costa Rica and Panama).

The shrubs, and particularly the trees, in which the tendrils are entirely missing, differ considerably from the other climbing species by their growth habits (see Plate 39). The slender tree *Passiflora macrophylla*, for example, produces the largest leaves of the genus at a length of up to 95 cm.

Systematics—Subdivision in Sections

According to Killip (1938), Escobar (1994), and Feuillet (2002), the representatives of this subgenus were divided into 6 sections, with the 2 major, or most important, divisions being section *Astrophea* (trees) and section *Botryastrophea* (long-tubed lianas). A modern key to the 19 species of subgenus *Astrophea* of the Guianan area is provided by Feuillet (2002), and to all species of *Astrophea* by Escobar (1994). Feuillet and MacDougal (2004) organized the subgenus into 2 main divisions called supersections, each with 2 or 3 sections, and introduced a new section, *Capreolata*. We follow that classification of *Astrophea* here. A new and critical assessment of the sections is sorely needed.

The 10 species in section *Astrophea* are trees and shrubs without tendrils with a secondary xylem arranged in concentric rings. Their flowers are characterized by a short floral tube and a white coloring that forms a contrast to the often bright yellow corona of 2 or 3 series. Often the peduncles are dichotomous once or several times so that a multiflorous inflorescence is produced. Presumably these species—*Passiflora arborea*, *P. lindeniana*, and *P. macrophylla*, for example—are pollinated by large bees.

Section *Capreolata*, with 15 species, contains the close relatives of the tree passionflowers, with generally similar white and yellow flowers, but they are tendril-bearing vines or lianas, examples including *Passiflora citrifolia*, *P. cauliflora*, and *P. pittieri*. This includes Killip's invalid section *Dolichostemma*.

The 2 species of section *Leptopoda* possess a plumose corona, and *Passiflora leptopoda* itself has an unusual pendent flower on a long pedicel.

The 13 representatives of section *Botryastrophea* are vines with vigorous tendrils from the Amazon basin. Cross section of the woody stem shows anomalous secondary growth. Two series are recognized: series *Carnae* with petiolar nectaries on the adaxial surface of the petiole below the leaf blade, and series *Botryastrophea* with nectaries borne abaxially at the junction of the leaf blade, or on the base of the leaf midvein. All have orange or reddish to purple flowers with a conspicuous floral tube that is longer than the sepals, and a reduced corona. As a rule the flowers are produced on short or long leafless inflorescences that branch from old wood. The floral ecology (color, long tube, no odor) suggests that these vines (for example, *Passiflora amoena, P. pyrrhantha, P. spinosa,* and *P. balbis*) are pollinated by hummingbirds.

The section *Pseudoastrophea* is closely related to the reddish-flowered section *Botryastrophea* but has white flowers and is mostly found in eastern South America in drier habitats. It contains 17 scandent species of shrubs or woody vines with often well-developed tendrils. Their flowers appear singly or in pairs. The white to greenish white sepals and petals form the background for the greenish to yellow corona, which often shows purple speckles, as with *Passiflora candida, P. sclerophylla* (Plate 40), *P. haematostigma,* and *P. mansoi.* This section includes Killip's invalid section *Cirrhipes.*

The latest detailed revision dates from 1938, when Killip, in his monograph, based the description of some species on only a few or a single available specimen. Consequently, some descriptions are inaccurate or incomplete, which sometimes makes it difficult to classify the species. Many new species have also been discovered. Happily, the taxonomic revision of subgenus *Astrophea,* begun by Linda Escobar, is being completed by Anne Katherine Hansen within the project Flora Neotropica.

Cultivation

Many species of subgenus *Astrophea* produce very beautiful flowers that are admired and desired by many enthusiasts, but there are, unfortunately, only a few representatives under cultivation. They are particularly esteemed among lepidopterists because they are often the foodplants for the caterpillars of rare species of *Heliconius* butterflies.

Cultivation of the species of *Astrophea* is comparatively difficult. Flower buds are produced only when optimal conditions are provided. Very few species, such as *Passiflora lindeniana,* grow above 1500 m elevation in a temperate or subtropical climate. Most are from the tropical lowlands and need warmth, a high atmospheric humidity, and much light. Cultivation is most often successful only in a tropical house. Good drainage is strongly recommended. Growth of the treelike species is episodic, with spurts of new growth followed by months of apparent inactivity; this is normal for these plants. Sometimes the leaves of these trees are shed for a period. Ron Boender achieved excellent results in cultivating *Astrophea* species; in a lath house under Florida's sun, several species flowered and even set fruit. Cuttings, particularly those from treelike species, root with difficulty, so it is advisable to propagate plants from seeds when available.

Passiflora amoena Escobar

Syst. Bot. 19: 203 (1994)
amoenus: charming, lovely
Supersection *Pseudoastrophea,* Section *Botryastrophea*

This exceptional species (Plates 42 and 43), indigenous to the Guianas, was described by Linda Escobar only a few years ago, and it is a peculiarity within section *Botryastrophea.* The flowers are usually produced as multiflorous racemes on lignified shoots that are several years old, or even on the stem. In *Passiflora amoena* such a short raceme consists of up to 8 blossoms, each of which reaches a diameter of up to 4 cm. According to Feuillet (1998, 2002), the species thus differs from the very closely related *P. fuchsiiflora,* in which the inflorescences

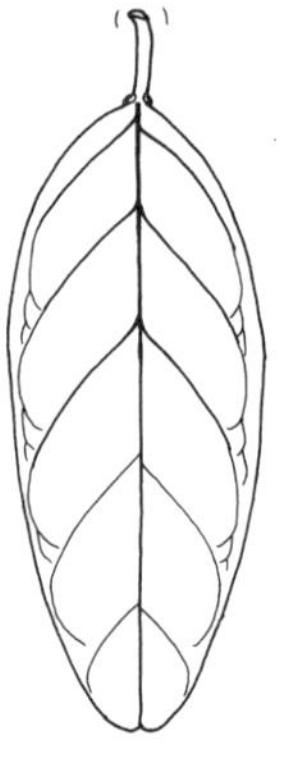

Passiflora amoena

consist of 10–20 blossoms. Moreover, the petioles of *P. fuchsiiflora* are 5–10 cm long—significantly longer than those of *P. amoena*, which are 1–5.5 cm long. The flowers are much more fleshy, almost rubbery, in *P. amoena*, and the corona is yellow rather than orange. *Passiflora amoena* is a real rarity and has been cultivated in Europe for only about 5 years, so experience regarding its cultivation is rather limited. In accordance with the conditions of its habitat, it is not easy to care for and must be cultivated in a heated greenhouse year-round.

Vine: glabrous with tendrils; stem: with anomalous secondary xylem; stipules: setaceous, ca. 0.3 cm long, deciduous; petioles: stout, 1–5.5 cm, with 2 glands at apex; leaves: unlobed, elliptic to ovate, 9.4–13 × 4–7.5 cm, obtuse to rounded at base, entire; pedicels: stout, 0.2–0.4 cm long; bracts: minute; floral tube: cylindrical, 3–5 cm long; flowers: pinkish red, salverform, 4.5–8.5 cm long, 3.5–4 cm diam., fleshy-coriaceous, inflorescences up to 4 cm long close to base of the plant; sepals: pinkish red, oblong, 1.4–1.6 × 0.8–1.6 cm; petals: pinkish red, 1.2–1.8 × 0.4–0.6 cm; corona: 3–5 series, fleshy, outer series laterally compressed, 0.7 cm long, pinkish red at base, at apex dilated and yellow; ovary: ellipsoid, glabrous, pink; fruit: ellipsoid, 6–9 × 3.5–5 cm, reddish green under a pink bloom; seeds: obovate, ca. 0.5 × 0.3 cm, reticulate.

Passiflora arborea Spreng.

Syst. Veg. 3: 42 (1826)
arboreus: treelike
Synonym: *P. glauca* Humb. & Bonpl.
Supersection *Astrophea*, Section *Astrophea*

Among the first tree passionflowers to be described was the type species of subgenus *Astrophea*, *Passiflora arborea* (Plate 44), growing at temperate altitudes in Colombia. At first sight, one would not think this plant to be a passionflower—it is a tree! *Passiflora arborea* produces inflorescences typical of section *Astrophea*, consisting of 2–8 white and yellow flowers up to 5 cm diam. Its leaves are remarkable, capable of reaching a length of 45 cm.

Many of the arborescent passionflowers have been called, and continue to be called, *Passiflora arborea* in collections. The true species is only found at 1400–2200 m in Colombia, and it is doubtful if it really is in cultivation. Specimens distributed in the 1990s under this name have been identified as *P. sphaerocarpa*. The few specimens under cultivation still called *P. arborea* must be evaluated when they bloom. Collections from Ecuador, declared to be *P. arborea* according to *Flora of Ecuador* (Holm-Nielsen et al. 1988), actually belong to 2 or 3 other species, including *P. putumayensis*.

Propagation has turned out to be very difficult in this group, and seeds are not available from Colombia for the time being. It seems to be comparatively difficult to get these species to flower, though few growers have any experience with them.

Passiflora arborea is also very similar to the closely related *P. lindeniana* from Venezuela, *P. macrophylla* from Ecuador, and *P. tica* from Costa Rica and Panama. The differences between *P. arborea* and these species are pointed out in the individual descriptions.

Tree: glabrous, 5–15 m high, without tendrils; stem: terete, younger parts slightly striate; stipules: setaceous, 0.1–0.2 cm long, soon deciduous; petioles: 2.5–4.5 cm long, glandless; leaves: unlobed, usually oblanceolate, or lanceolate to oblong, 10–45 × 5–15 cm, acute to acuminate at apex, obtuse at base, 2–4 glands on the midrib near base, membranous to subcoriaceous, entire; peduncles: 1- to 3-bifurcate; bracts: setaceous, 0.2–0.5 cm long; floral tube: campanulate to cylindrical-campanulate, 0.6–1.4 cm long; flowers: 3–5 cm diam., inflorescence 2- to 8-flowered; sepals: white inside, pale green outside, oblong, 1.8–2.5 × 0.5–0.9 cm; petals: white, subequal to the sepals; corona: 2 series,

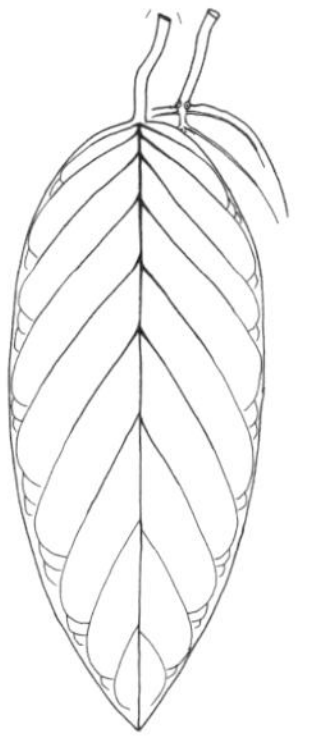

Passiflora arborea

base white, yellow apically, outer series 1–1.2 cm long, laterally compressed, upper 1/3 dilated and falcate, inner series 0.2–0.4 cm long, ligulate or filiform; ovary: narrowly ovoid, minutely tomentose; fruit: ellipsoid to ovoid, 3.5–4 × 2–2.5 cm, yellowish; seeds: ovate, 0.6 × 0.3 cm.

Passiflora candida (Poepp. and Endl.) Mast.

Trans. Linn. Soc. London 27: 629 (1871)
candidus: pure white
Synonym: *P. guedesii* Huber
Supersection *Pseudoastrophea*, Section *Pseudoastrophea*

This charming species (Plate 45), distributed in the tropical Amazon region in northern Brazil and in French Guiana, has been under cultivation, at least in Europe, since around 1999. Little information is known regarding the exact conditions needed for cultivation, such as the minimum temperature and the plant's willingness to flower. In view of the conditions in its natural surroundings, *Passiflora candida* is recommended for a heated greenhouse and should be offered optimum light conditions throughout the year.

The coriaceous leaves of this imposing woody vine with well-developed tendrils reach a length of up to 22 cm. Its fragrant flowers, about 6 cm diam., appear singly or in pairs and are typical of many species belonging to section *Pseudoastrophea*. The sepals and petals are white, and the outer rank of the corona is a rich yellow mottled with orange spots. A noteworthy characteristic is that the sturdy filaments of the outer corona series stand around the ovary in an erect-campanulate fashion. They are a little wavy in the upper 1/3 and remind one a little of *Passiflora tina*.

Vine: woody with tendrils; stem: terete or subangular, pubescent; stipules: setaceous, soon deciduous; petioles: stout, 2–2.5 cm long, with 2 glands at the apex; leaves: ovate to ovate-oblong, 8–22 × 7–14 cm, coriaceous, glabrous above, hirsute beneath; peduncles: stout, solitary or in pairs, 0.5–0.8 cm long, tomentulose; bracts: minute, soon deciduous; floral tube: 2.5–3.5 cm long, 1–1.5 cm diam. at throat, tomentulose; flowers: white, 5–7 cm diam.; sepals: white, oblong, 3–4.5 × 1–1.5 cm, fleshy; petals: white, subequal to the sepals, membranous; corona: 4 series, yellow, outer series liguliform, 1.5–2.9 cm long, bearing near middle of one side a short appendage, inner series 0.1–0.5 cm long; ovary: tomentose; fruit: ellipsoid, yellow-green, ca. 6 × 4 cm; seeds: ca. 0.7 × 0.4 cm, reticulate.

Passiflora citrifolia (Juss.) Mast.

Trans. Linn. Soc. London 27: 629 (1871)
citrifolius: having lemon-shaped leaves
Supersection *Astrophea*, Section *Capreolata*

The rarely grown *Passiflora citrifolia* (Plate 46), distributed from French Guiana to Peru, is among the few species from subgenus *Astrophea* that can be induced to flower under cultivation. It has a whitish floral tube, and the white sepals and petals form the background for the brownish purple corona, which is cream at the base and yellow at the tip. The orientation of the corona is an interesting, hitherto rarely used characteristic in subgenus *Astrophea*. In

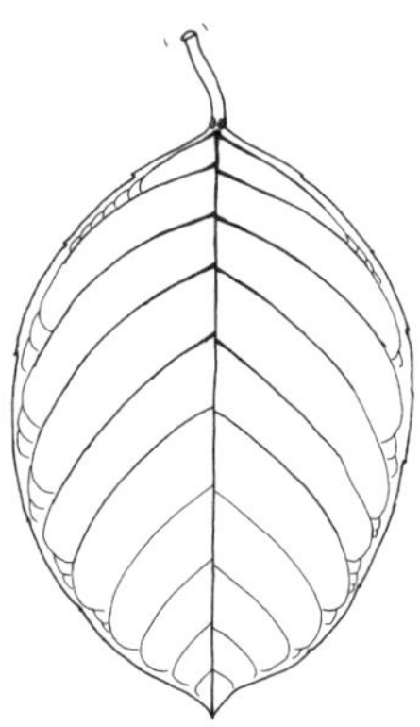

Passiflora candida

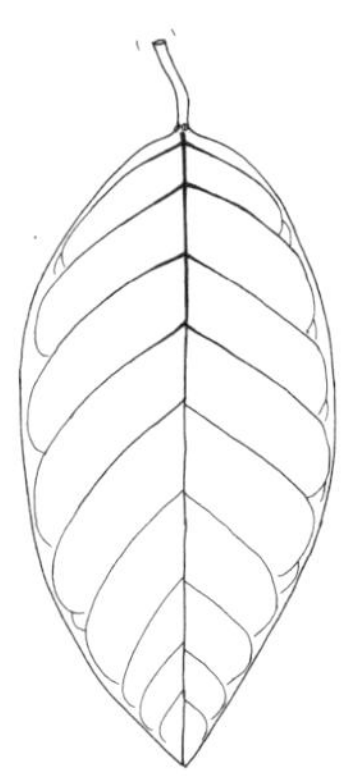

Passiflora citrifolia

this case the filaments of the outer coronal series, dilating abruptly in the upper 1/3 and getting thinner toward the tip, stand nearly upright in a wide campanulate fashion. The diameter of the flowers is about 4.5 cm.

However, the plants under cultivation differ in several respects (for example, in the length of the peduncles) from the lacunary description in Killip's monograph, and the species is suspiciously similar to *Passiflora cauliflora*. Perhaps analysis of the many more recent collections will shed light on these problems.

Vine: woody, glabrous throughout except the ovary; stem: terete, strongly striate; stipules: minute, soon deciduous; petioles: stout, 1.5–5 cm long, with 2 sessile glands at apex; leaves: unlobed, normally oblong, 10–20 × 5–10 cm, abruptly short-acuminate at apex, rounded or shallowly cordate at base, coriaceous, entire; peduncles: 0.4–3 cm long; bracts: minute, soon deciduous; floral tube: cylindrical to funnel-shaped, 1–2 cm long, ca. 0.5 cm at throat, white; flowers: white, 3.5–4 cm long, 4–5 cm diam.; sepals: white inside, greenish white outside, oblong, 2–2.8 × 0.5–0.8 cm; petals: white, subequal to the sepals; corona: 4–5 series, outer series 1.2–1.5 cm long, erect, orange to purplish brown at base, yellow at tips; ovary: tomentose; fruit: ellipsoid, 4–5 cm × 3 cm; seeds: ca. 0.7 × 0.35 cm, reticulate.

Passiflora haematostigma Mart. ex Mast.

Fl. Bras. 13 (1): 574 (1872)
haematostigmus: having blood-red spots (in reference to the mottled corona)
Synonym: *P. platystyla* Mast.
Supersection *Pseudoastrophea*, Section *Pseudoastrophea*

Depending on the color and structure of the corona, the representatives of section *Pseudoastrophea* can be subdivided into at least 2 groups. The larger group includes species with a yellow, partially orange, or reddish purple mottled corona, including *Passiflora candida* and *P. sclerophylla*. The other group includes a few species with a greenish corona that mainly grow in eastern and southern Brazil. The most common and hence best-known species from this group might well be *P. haematostigma*. Its almost recumbent outer corona filaments are green with purple mottles. This species occasionally does not produce tendrils. Its flowers are 4–6 cm diam. and do not appear in inflorescences but singly or in pairs, which is a characteristic of this section.

The coriaceous, unlobed leaves are 4–13 cm long. They are mostly glabrous on the upper surface and more or less pubescent with brown hair on the underside. Our investigations have shown that *Passiflora haematostigma* does not seem to be under cultivation. Cultivation of some species of subgenus *Passiflora* from southeastern Brazil has shown that these plants are far more unpretentious and free-flowering than their relatives from the Amazonian lowlands. *Passiflora haematostigma* should also be easier to cultivate than many other species of *Astrophea*.

Vine: woody, or shrub, puberulent, younger branches villous, tendrils slender or well developed, occasionally absent; stem: bark brown; stipules: setaceous, soon deciduous; petioles: 2–4 cm long, with 2 scarlike, brown glands at the apex; leaves: unlobed, variable, elliptic or oblong-ovate to oblong-lanceolate, 4–13 × 2–11 cm, coriaceous, glabrous or rarely puberulent above, yellow-villous underneath; peduncles: solitary or in pairs, 1.2–4 cm long; bracts: setaceous, soon deciduous; floral tube: cylindrical-campanulate, 1–1.7 cm long, 0.6–1 cm wide at throat; flowers: white, 3–4 cm long, 4–6 cm diam.; sepals: white, linear-oblong, 2–3 × 0.5–0.7 cm; petals: white, linear-spatulate, 1.5–2 × 0.3–0.4 cm; corona: 2 series, outer series 0.8–1.5 cm long, dilated just above middle, inner series linear-clavate, ca. 0.2 cm long; ovary: oblong, densely

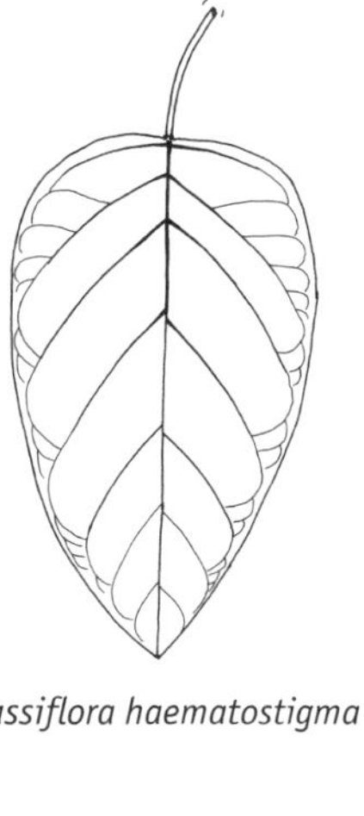

Passiflora haematostigma

puberulous; fruit: ellipsoidal, 4–6 × 3–3.5 cm, yellow; seeds: ovoid, 0.5–0.7 × 0.4 cm, reticulate.

Passiflora kawensis Feuillet

Novon 4: 236–238 (1994)
named after Montagne de Kaw, French Guiana, the type location
Supersection *Pseudoastrophea*, Section *Pseudoastrophea*

Passiflora kawensis (Plate 47) was discovered by Feuillet in Montagne de Kaw, a relatively small area in northeastern French Guiana, when he was working on the revision of the Passifloraceae for the "Flora of the Guyanas" project. Many new species of *Passiflora* from various subgenera have been found in this area during the last several years.

A peculiarity of the species, pointed out by Feuillet in his first description, is a reddish trochlea that appears as an umbrella-shaped ring on the androgynophore. This feature can otherwise be found only in *Passiflora pentagona* (Brazil), *P. rhamnifolia* (Brazil), *P. tessmannii* (Peru), *P. trochlearis* (Ecuador), and *P. venosa* (Bolivia). It is an important diagnostic feature. This ring, located on the androgynophore above the operculum, seems to be an additional barrier to keep small insects from getting to the nectar. The white flowers can reach a diameter of up to 7 cm, and the yellow outer corona filaments are erect-campanulate and conspicuously dilated at the tip. A few collectors have cultivated *P. kawensis* in the last few years. So far it has turned out to be a difficult plant that will flower only in a heated greenhouse.

Vine: pubescent with long, thick tendrils; stem: terete; stipules: minute; petioles: 1–2.2 cm long, without glands; leaves: unlobed, oblong, 10–15 × 4–5.5 cm, acute to acuminate at apex, asymmetrically rounded at base, with 2 glands on the midrib close to the base, entire; peduncles: solitary or in pairs, ca. 1.8 cm long; bracts: minute; floral tube: ca. 1.3 cm long, 0.7–1.1 cm diam., green outside, yellowish inside; flowers: white, ca. 3.5 cm long, 6–7 cm diam.; sepals: white inside, green outside, oblong, 3.5–4 × 0.7 cm; petals: white, 3.5 × 0.5 cm; corona: several series, outer 2 series yellow, the outermost subdolabriform, 1 cm long, verrucose at apex, inner series capitate, 0.5 cm long; ovary: angled, densely rufo-tomentose; fruit: mature fruit unknown.

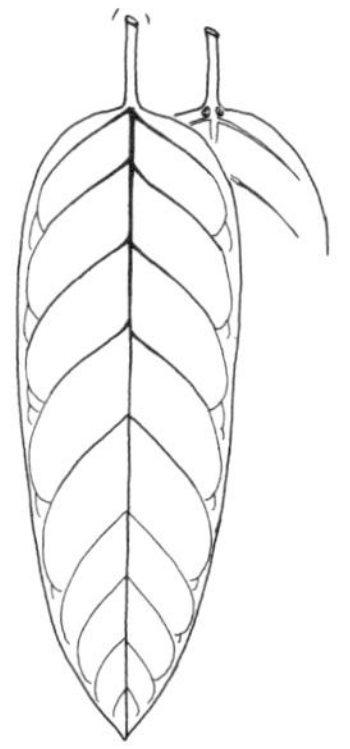

Passiflora kawensis

Passiflora lindeniana Triana & Planch.

Ann. Sci. Nat. Bot. 5 (17): 182 (1873)
named after Jean Jules Linden, collector of the type plant
Supersection *Astrophea*, Section *Astrophea*

While most representatives of subgenus *Astrophea* grow in the tropical lowlands, *Passiflora lindeniana* (Plate 48) advances to altitudes of up to 2720 m in Venezuela. Following the conditions of its natural surroundings, it flourishes when night temperatures are around 12°C and day temperatures are 20°–25°C. Although it tolerates a wider temperature spectrum in comparison with other species of *Astrophea*, it cannot be called easy to cultivate. The old foliage is regularly shed before new growth appears. Moreover, the plant may easily become stunted in winter, which for the most part is due to insufficient light conditions.

Because the rooting of cuttings is not very promising, this plant should be propagated from seeds. Fortunately, in recent years collectors have been able to get hold of seeds from Venezuela.

Passiflora lindeniana is distinguished from the closely related *P. arborea* by its campanulate floral tube, which is as long as it is wide. It can be distinguished from *P. emarginata* and *P. sphaerocarpa* of Colombia by its larger, ovate fruit. Another species that can be confused with *P. lindeniana* as a herbarium specimen is the Venezuelan *P. araguensis*. Although the latter has comparable flowers, there is no question that it is a woody vine with weak tendrils.

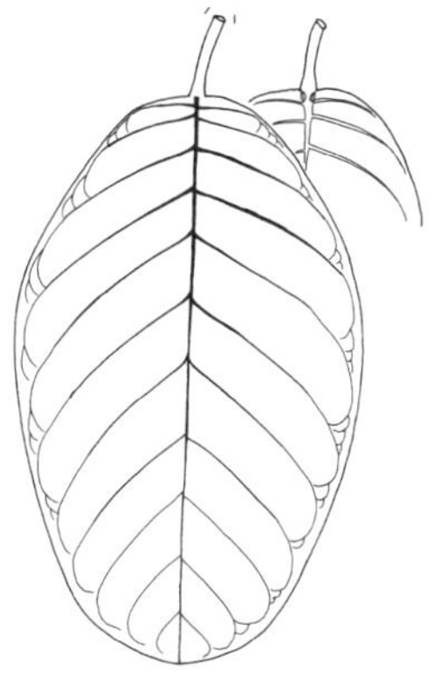

Passiflora lindeniana

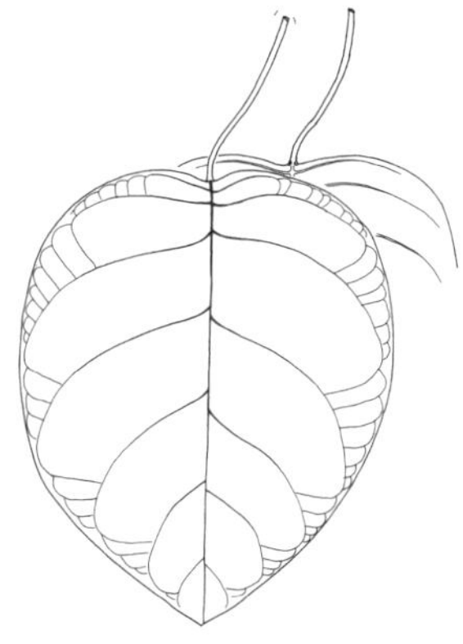

Passiflora longiracemosa

Tree: glabrous except ovary, 3–18 m high, without tendrils; stem: terete, woody, up to 25 cm diam. at very base; stipules: minute, soon deciduous; petioles: stout, 2–3 cm long without glands; leaves: unlobed, oblong-ovate, 10–25 × 5–15 cm, subacute or slightly emarginate at apex, midnerve biglandular at base; peduncles: up to 4 cm long, 1- to 2-, rarely 3-bifurcate; bracts: minute; floral tube: broadly campanulate, 1–1.2 × 0.8–1 cm; flowers: white, 5–6 cm diam., inflorescence 2- to 6-flowered; sepals: greenish white, lanceolate, 2.5–3 × 0.8–1 cm; petals: white, 2–2.5 × 0.8 cm; corona: ca. 3 series, yellow, outer series ca. 1.3 cm long, upper 1/3 dilated, filaments of inner series broadly linear, 0.2 cm long; ovary: tomentose; fruit: ovoid, 3–6 × 2.5–3 cm, yellow, sometimes red-splotched; seeds: ca. 0.5 × 0.4 cm, reticulate.

Passiflora longiracemosa Ducke

Archiv. Jard. Bot. Rio de Janeiro 3: 221, pl. 23 (1922)
longiracemosus: having long racemes
Supersection *Pseudoastrophea*, Section *Botryastrophea*

Passiflora longiracemosa (Plates 49 and 50), distributed in the tropical regions of Guyana, Amazonian Brazil, and Venezuela, is known for its gigantic inflorescences, each up to 60 cm long. These are among the longest within the whole genus, exceeded only by a couple of close relatives, such as *P. balbis*. However, some specimens have significantly shorter racemes and are similar to the closely related *P. fuchsiiflora*. Both species can be distinguished by their corona. Three series are found in *P. fuchsiiflora*, with the outermost series about 1 cm long and subdolabriform, while in *P. longiracemosa* there are 2 series with the outermost 0.2–0.5 cm long and filamentous.

Its very beautiful flowers with long floral tubes reach a diameter of up to 3.5 cm. They abound at the base of the cauliflorous raceme. Large, leathery leaves on long petioles are often a distinguishing feature in comparison with other species of section *Botryastrophea*. Unfortunately, this rare woody vine is not yet in cultivation.

Vine: glabrous, woody with well-developed tendrils; stem: terete, woody; stipules: minute; petioles: stout, 2–9 cm long, with 2 glands at apex; leaves: unlobed, variable, broadly ovate to lanceolate-oblong, 7–15 × 3.5–14 cm, obtuse and emarginate at apex, shallowly cordate at base, thick-coriaceous, entire; pedicels: 1.5–3 cm long; bracts: minute; floral tube: cylindrical, 3–5 × 0.8–1 cm; flowers: reddish to coral pink, ca. 3.5 cm diam., inflorescence 10–60 cm long, floriferous mostly near base; sepals: reddish to coral pink, linear-oblong, 1–1.5 × 0.4–0.5 cm; petals: subequal to the sepals; corona: 2 series, outer series filiform, 0.2–0.5 cm long, yellowish, violet at tips, inner series minute, dark violet; ovary: glabrous; fruit: obovoid, ca. 5 × 3 cm, 6-angled, greenish rose; seeds: 0.7 × 0.4 cm.

Passiflora macrophylla Spruce ex Mast.

J. Linn. Soc. Bot. 20: 31 (1883)
macrophyllus: large-leafed
Synonyms: *P. gigantifolia* Harms, *P. lorifera* Mast.
Supersection *Astrophea*, Section *Astrophea*

Like *Passiflora arborea* and *P. lindeniana*, *P. macrophylla* (Plate 51) is a treelike species, and it is culti-

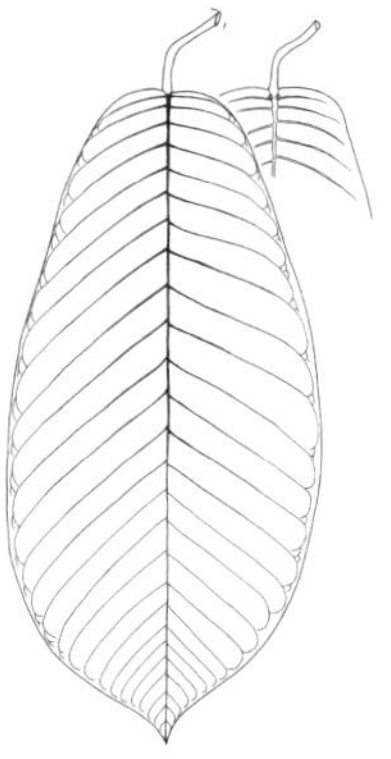

Passiflora macrophylla

vated only occasionally. Only the tip of the stem bears leaves. This warmth-loving tree of the western lowlands of Ecuador has almost no lateral shoots and reaches a height of 3–6 m.

On a mature plant, the leaves can reach a length of up to 95 cm. Thus this species has the largest leaves of the genus and can be easily identified. Other books make reference to *Passiflora gigantifolia*, but this is just a synonym (Holm-Nielsen et al. 1988).

An interesting phenomenon of movement can be observed in this species. As soon as the flowers open, the otherwise pendent leaves rise to give a view of the flowers lying beneath them. The plant produces pretty inflorescences with 2–12 single flowers that are white and yellow and about 6 cm diam. Provided tropical conditions and time, this species may flower in cultivation. Remember that this is one of those species that can appear lifeless between growth spurts. As is true for all trees among passionflowers, cuttings root only with great difficulty. Better chances of success are offered by sowing seeds or by air layering.

Passiflora macrophylla is exclusively suited for cultivation in a heated greenhouse year-round.

Tree: 1–6 m high, glabrous except the ovary, without tendrils; stem: terete, woody; stipules: setaceous, normally soon deciduous; petioles: stout, 3.5–7 cm long, glandless; leaves: unlobed, oblong or oblong-ovate, 23–95 × 10–38 cm, acute to acuminate at apex, obtuse to slightly cordate at base, midnerve biglandular at base, entire; peduncles: 1- to 3-bifurcate, inflorescence 8–25 cm long; bracts: minute; floral tube: cylindrical, 1.6–2.2 cm long, ca. 0.5 cm diam.; flowers: white, 6–7 cm diam., inflorescence 2- to 12-flowered; sepals: white inside, greenish white outside, oblong-lanceolate, 3.5–4 × 0.6–1 cm; petals: white, 3.3–3.9 × 0.6–1 cm, white; corona: 2–3 series, outer series 1.6–2.2 cm long, filiform and white at base, upper 1/3 dilated, triangular or falcate, yellow, inner series 0.1–0.3 cm long; ovary: narrowly ovoid, glabrescent; fruit: globose, 3.1–4 × 2–4 cm, glabrous; seeds: obovate, 0.4–0.5 × 0.2–0.3 cm, reticulate.

Passiflora maguirei Killip

Bull. Torrey Bot. Club 75: 415 (1948)
named after Bassett Maguire, collector of the type specimen
Supersection *Astrophea*, Section *Capreolata*

Unfortunately, *Passiflora maguirei* (Plate 52) is not under cultivation, as is true of many other species presented in this chapter. It is included because it is of special interest with regard to systematics. The assignment of this rare species within the subgenus turns out to be difficult, since it incorporates attributes from different sections. Its flowers, borne in dense fascicles, and the floral tube, which is up to 2 cm long, suggest a relationship with section *Botryastrophea*. The whiteness of the flowers and the coronal structure resembles certain species from section *Pseudoastrophea*, such as *Passiflora citrifolia* and *P. costata*.

The native range of *Passiflora maguirei* reaches from Guyana to Amazonian Brazil and southern Venezuela. Günter Gerlach discovered the plant near Puerto Ayacucho, an area in which the species was previously unknown.

Besides the characteristic inflorescences, the oblong, acuminate leaves, with relatively few principal lateral nerves, make a definite identification easier.

Vine: woody with tendrils, glabrous except the ovary; stem: terete; stipules: microscopic; petioles: 1.5–5 cm long, stout, with 2 glands at apex; leaves: unlobed, oblong, 12–35 × 5–12.5 cm, acuminate or abruptly short-acute at apex, rounded or cuneate at base, coriaceous, entire; pedicels: up to 1 cm long; bracts: subulate, 0.2–0.3 cm long; floral tube: funnel-shaped, 1.5–2 cm long, ca. 0.8 cm at throat, lon-

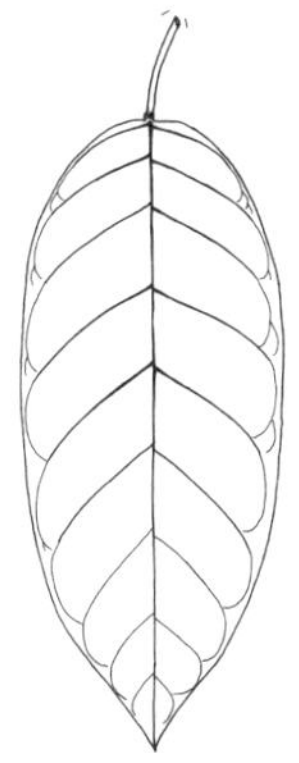

Passiflora maguirei

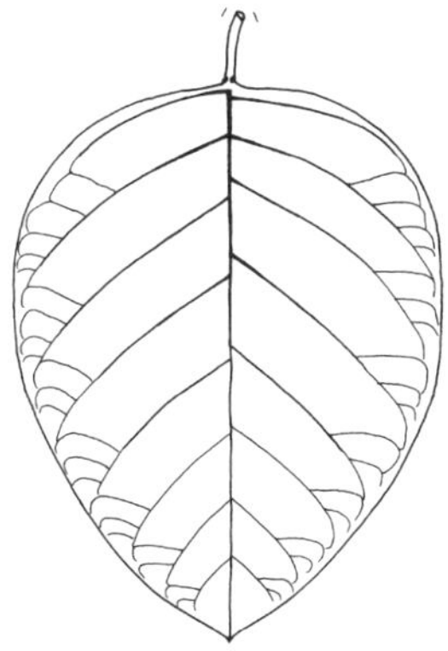

Passiflora mansoi

gitudinally striate; flowers: greenish white, 5–6 cm diam., ca. 4.5–6 cm long, in dense fascicles; sepals: white inside, greenish outside, linear-oblong, 2.5–3.5 × 0.6–0.8 cm; petals: white, as long as sepals, ca. 1.2 cm wide; corona: 4 series, yellow, outer series 2–2.2 cm long, linear-dolabriform, subulate at the apex, inner series decreasing in length from 0.1 to 0.6 cm; ovary: ovoid, puberulent; fruit: unknown.

Passiflora mansoi (Mart.) Mast.

Trans. Linn. Soc. London 27: 629 (1871)
named after Manso, collector of the type specimen
Supersection *Pseudoastrophea*, Section *Pseudoastrophea*

While the majority of species presented in this chapter have glabrous leaves, species with pubescent leaves are also found, particularly in central and southern Brazil. Among these is *Passiflora mansoi* (Plate 53), clearly distinguished from other species distributed in the region, such as *P. haematostigma*, *P. pentagona*, and *P. rhamnifolia*, by its yellow corona. The filaments of the outer series surround the androgynophore in a wide bell-shaped form. Besides the pubescent leaves and the color and structure of the corona, the broadly ovate leaves are another reliable distinguishing feature.

The reasons why this and many other species of *Astrophea* are not under cultivation are evident. Fruit with ripe seeds are found too rarely in the known habitats. Cuttings suffer during transport and hardly develop roots. And even if one succeeds in bringing such a species into cultivation, it turns out to be a difficult child and is soon ailing.

Shrub: subscandent with few tendrils on upper branches, partly pilosulous; stem: terete; stipules: setaceous, soon deciduous; petioles: 1–2 cm long, with 2 glands at apex; leaves: unlobed, broadly ovate, 8–15 × 5–12 cm, acute or acuminate at apex, rounded at base, subcoriaceous, pubescent beneath, entire; peduncles: 0.5–2.5 cm long, solitary; bracts: minute; floral tube: cylindrical-campanulate, 1–1.2 cm long, 0.6–0.8 cm at throat; flowers: white, pilosulous outside, ca. 4.5 cm diam., ca. 3.5 cm long; sepals: white inside, greenish outside, ovate-oblong, 1.9–2.2 × 0.8–0.9 cm; petals: white, oblong, ca. 1.5 × 0.5 cm; corona: 2 series, yellow, outer series 1–1.3 cm long, dolabriform in upper 1/4, subulate at apex, inner series up to 0.4 cm long; ovary: ovoid, densely pilosulous; fruit: nearly globose, 3–4.5 × 2.5–3.5 cm, yellow; seeds: 0.6–0.7 × 0.4 cm, reticulate.

Passiflora pittieri Mast.

Bot. Gaz. 23: 246 (1897)
named after Henri François Pittier, collector of the type plant
Supersection *Astrophea*, Section *Capreolata*

Passiflora pittieri (Plate 54) has one of the larger ranges of species in *Astrophea*, reaching from Belize to western Colombia. Its coevolved butterfly herbivore, *Heliconius sapho*, has been collected in southeastern Mexico as well, so it may well be discovered there someday. The species is a low canopy liana with tendrils, but it has a shrubby juvenile stage. In the north of its range, the flowers differ somewhat; it is not clear if these deserve taxonomic recognition. In 1993, Jan Meerman began a study

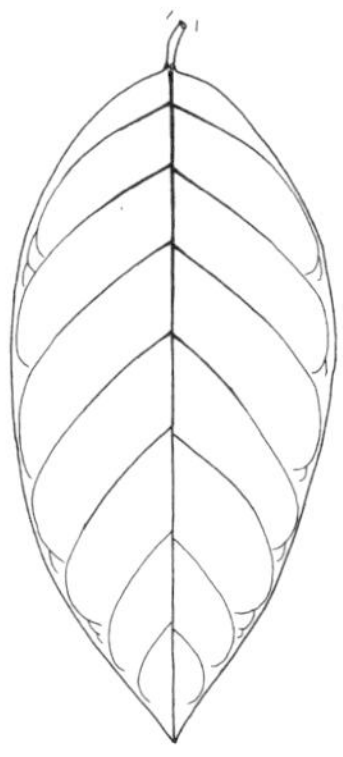

Passiflora pittieri

of *P. pittieri* in Belize. The flowers there are 4–5 cm diam. and whitish; the corona in 3 series is yellowish to orange and bowl-shaped. Further south the corona is 4–5 seriate and often spreads open like a saucer. A characteristic of this species, and of closely related species like *P. haughtii*, is the long tubular operculum that is exerted beyond the floral tube, possibly an adaptation for hummingbird pollination. Killip used this as the basis for his invalid section *Dolichostemma*. *Passiflora tica* and *P. pittieri* are the only representatives of subgenus *Astrophea* in Central America.

Passiflora pittieri resembles *P. tina* from Ecuador but differs by having a nonwavy, more or less recumbent outer corona series, fewer filaments, and usually fewer flowers per node. *Passiflora pittieri* is represented in a few collections now, perhaps because it is relatively easier to obtain, and it is possible to induce flowering under cultivation.

Vine: glabrescent, tendrils well-developed or occasionally absent; stem: terete; stipules: linear-subulate, 0.1–0.4 cm long, soon deciduous; petioles: 1–2.5 cm long, grooved, 2 glands at apex; leaves: unlobed, oblong to oblong-ovate, 8–32 × 4–14 cm, acuminate at apex, obtuse to rounded at base, subcoriaceous, entire; peduncles: solitary or in pairs, normally 1- to 2-bifurcate; bracts: 0.1–0.2 mm long, deciduous; floral tube: cylindrical-campanulate, 0.8–1 cm long, 0.6–0.7 cm diam. at throat; flowers: white, 5–8 cm diam.; sepals: white, oblong, 2.5–4 × 0.7–0.9 cm; petals: white, subequal to or shorter than the sepals; corona: (3–)4–5 series, filaments of outer series 1.3–2.3 cm long, yellow to yellow-orange, sometimes with red mottling, dilated near apex, tapering gradually to base, filaments of inner series 0.1–0.5 cm long, yellow to orange or red; ovary: densely tomentose; fruit: narrowly ovoid to ellipsoid, 6–8(–9) × 3–4 cm, greenish or yellow; seeds: large, 0.7–0.9 × 0.6–0.8 cm, reticulate.

Passiflora putumayensis Killip

Publ. Field Mus. Nat. Hist., Bot. Ser. 19: 532 (1938)
named after Putumayo, Colombia, the type location
Supersection *Astrophea*, Section *Astrophea*

Passiflora putumayensis (Plate 55) is among the smallest species of the treelike passionflowers. In its natural surroundings, the eastern foothills of the Andes from southern Colombia through Ecuador to Peru, it can be found at elevations of 1100–1800 m. It reaches a height of only 1.5 m. This species is by no means rare and occurs in masses in its habitats. It seems to be a foodplant for *Heliconius wallacei* butterflies.

In 1999, Jacob Olander succeeded in bringing this species into cultivation. He found it in the foothills around the Cascada de San Rafael (San Rafael waterfalls), Ecuador, in a tropical forest. He passed it to Ron Boender, and at his place it started to bloom at a height of 30 cm with large amounts of flowers. Its growth is somewhat episodic, with spurts of new growth followed by a short period of apparent inactivity. When *Passiflora putumayensis* has reached a height of about 1 m under cultivation it starts to deteriorate.

Passiflora putumayensis is among the most free-flowering species from subgenus *Astrophea* currently under cultivation. Its inflorescence consists of 4–9 flowers, each reaching 4 cm diam. The blossom with its short cylindrical-campanulate floral tube, the white sepals and petals, and the yellow corona are absolutely typical of the treelike representatives.

In *Flora of Ecuador* (Holm-Nielsen et al. 1988), collections of *Passiflora putumayensis* and at least one other species, presumably a new one, were erroneously assigned to *P. arborea*, whose native range is restricted to Colombia, according to the most recent findings.

Treelet: 1–1.5 m high, without tendrils, glabrous; stem: terete, younger parts slightly striate;

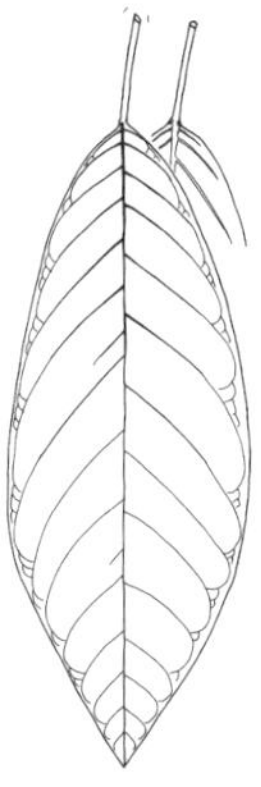

Passiflora putumayensis

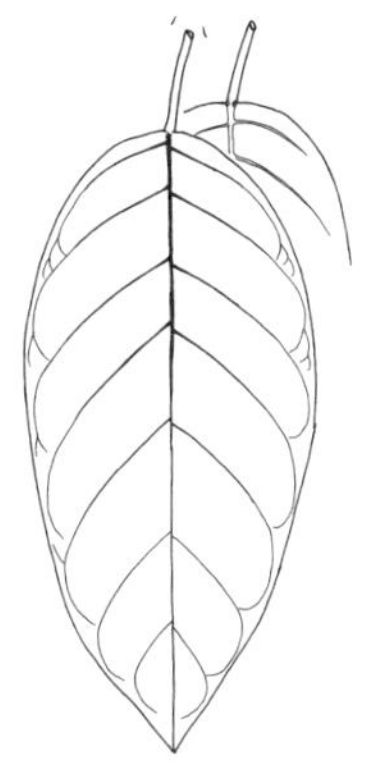

Passiflora pyrrhantha

stipules: setaceous, soon deciduous; petioles: 2–3.8 cm, without glands; leaves: unlobed, elliptic to oblong-ovate, 16–27 × 6–11 cm, 2 scarlike glands on the midrib near base, membranous, cuneate to rounded at base, entire; peduncles: 4–6.5 cm, 2- to 4-bifurcate; bracts: minute; floral tube: cylindrical-campanulate, 0.7–1 cm long; flowers: white, 3.6–4.3 cm diam., inflorescence 4- to 9-flowered; sepals: green to greenish white outside, whitish inside, oblong, 1.6–2 × 0.5–0.7 cm; petals: white, oblong, 1.4–1.7 × 0.4–0.5 cm; corona: 2 series, outer series 1.1–1.4 cm long, yellow with white base, narrowly linear in basal 1/2, dilated above middle, attenuate at apex, inner series ca. 0.3 cm long, yellow, filiform; ovary: ovoid to oblong-ovoid, tomentose; fruit: unknown.

Passiflora pyrrhantha Harms

Notizbl. Bot. Gart. Berlin 9: 977 (1922)
pyrrhanthus: having fire-colored flowers
Supersection *Pseudoastrophea*, Section *Botryastrophea*

The brilliant orange flowers of *Passiflora pyrrhantha* (Plate 56) are unique within subgenus *Astrophea*, and there is hardly another species in the entire genus with a comparable flower color. All parts of the flower are vivid orange, even the pollen.

The flowers appear in groups, characteristic of section *Botryastrophea*, on inflorescences 9–20 cm long. At a maximum 8 cm diam., the flowers might well be the largest within the whole subgenus.

Passiflora pyrrhantha is distributed in the Amazonian lowlands of Venezuela, Colombia, Ecuador, and Peru, where it occasionally advances to the moderate altitudes of about 1000 m. It has a remarkable characteristic in common with the closely related *P. spinosa* and *P. spicata*: the lower, slightly widened part of the floral tube is divided into 5 segments. This peculiarity is due to membranes that connect the 5-angled androgynophore with the floral tube. A higher rate of pollination is thus achieved, because the pollinating hummingbirds must slowly move around the flower to empty the individual nectar chambers.

Unfortunately, though *Passiflora pyrrhantha* is exceptional in many respects, it has rarely been cultivated. Because of its exacting requirements, this very beautiful species is reserved for enthusiasts who own a heated greenhouse.

Vine: woody with few tendrils, partly puberulent; stem: young stems puberulent, striate, bark of older stems grooved; stipules: minute, absent or soon deciduous; petioles: 1–4 cm long, striate with 2 glands at junction to blade; leaves: unlobed, ovate to lanceolate-oblong, 10–25 × 6–15 cm, acuminate at apex, rounded at base, subcoriaceous, glabrous; pedicels: stout, 0.2–0.4 cm long; bracts: minute; floral tube: cylindrical to slightly funnel-shaped, 3.2–4.5 cm long, 1.5–1.8 cm wide at throat; flowers: fiery orange, 5–9 cm diam., inflorescence 7–20 cm long; sepals: orange, linear-oblong, 3–3.6 × 0.6–1 cm; petals: orange, subequal to the sepals; corona: 2 series, orange, outer series 0.5–0.6 cm long, slightly dilated, filaments of inner series 0.2–0.4 cm long, filiform; ovary: ovoid, tomentose; fruit: 6–9 × 3.5–5 cm, greenish with tough, coriaceous pericarp, 0.3–0.7 cm thick; seeds: 0.7 × 0.45 cm, reticulate.

Passiflora quelchii (Juss.) Mast.

Trans. Linn. Soc. London 27: 629 (1871)
named after John Joseph Quelch, collector of the type plant
Supersection *Astrophea*, Section *Capreolata*

Only 3 specimens of the extremely rare *Passiflora quelchii* (Plate 57) have ever been found in its habitat of savannas in Guyana. It was last observed there by James Mallet.

The typical tendrils are missing in this small tree or shrub. It is conspicuous for its narrow leaves, short cauliflorous inflorescences, comparatively long floral tube, and outer yellow corona filaments in a bell-shaped form. Killip placed this species in section *Botryastrophea*, but it is clearly related to *Passiflora maguirei* and *P. cauliflora* in section *Capreolata*. Unfortunately, this unusual species is not under cultivation and is of conservation concern.

Shrub or small tree: without tendrils; stipules: minute, soon deciduous; petioles: 1.5–2 cm long, with 2 raised glands at apex; leaves: unlobed, oblanceolate to narrowly oblong, 8–12 × 2–4 cm, obtuse or short acute at apex, tapering from the upper part to a narrowly cuneate base, with 12–15 lateral nerves on each side, margins strongly undulate, entire; pedicels 0.3–0.5 cm long; bracts: minute, soon deciduous; floral tube: cylindrical, 2.5–3 cm long, 0.4–0.7 cm at throat, greenish outside; flowers: 4–5 cm diam., white, inflorescence very short; sepals: white inside, greenish outside, oblong-lanceolate, 2–2.5 × 0.5–0.7 cm; petals: white, subequal to the sepals; corona: 2 series, outer series yellow, ca. 1.5 cm long, dilated above middle, subdolabriform, inner series ca. 0.1 cm long; ovary: oblong, minutely pubescent; fruit: unknown.

Passiflora rhamnifolia Mast.

Fl. Bras. 13 (1): 575 (1872)
rhamnifolius: having buckthorn-shaped leaves
Supersection *Pseudoastrophea*, Section *Pseudoastrophea*

Like *Passiflora kawensis*, *P. rhamnifolia* (Plate 58) has a reddish trochlea that appears as an umbrella-shaped ring. For more details on the function of this unusual ring, see the description of *P. kawensis*.

Unfortunately, *Passiflora rhamnifolia* is not in cultivation. It belongs to a small group of species around *P. haematostigma* that stand out because of their green coronal filaments, which are partially mottled with purple spots. This species differs only slightly from the closely related *P. haematostigma*, which has no umbrella-shaped ring, and from *P. pentagona*, which has pendent flowers and leaves that are glabrous above. The distinguishing criteria in the literature are not always tenable, and it is difficult to separate the species, even by subtle foliar venation characteristics. They are all indigenous to southeastern Brazil and may be sorted ecologically by habitat. For example, *P. pentagona* appears to be restricted to coastal restinga thickets. Nevertheless, the species are very close, and it is possible that in a new, detailed revision of subgenus *Astrophea*, 2 of these species might be subsumed under the same name.

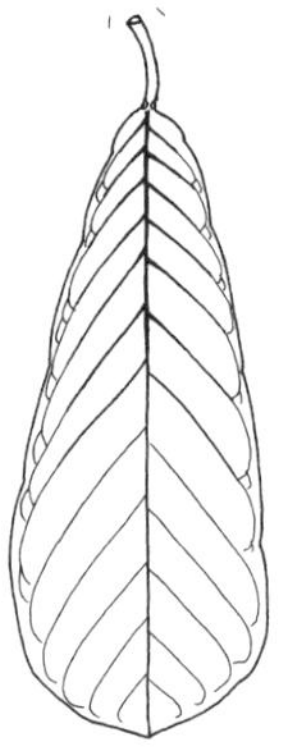

Passiflora quelchii

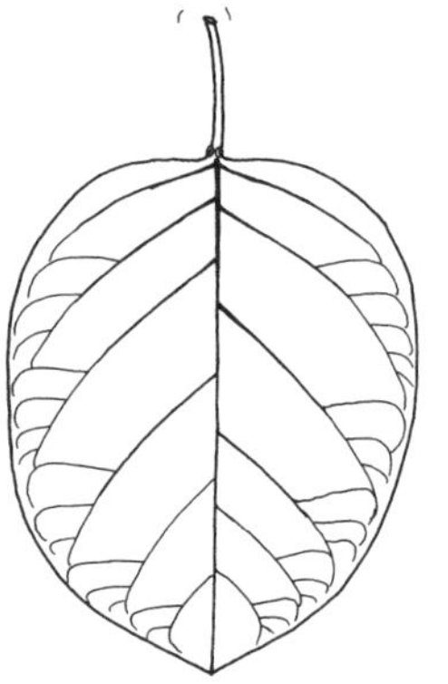

Passiflora rhamnifolia

Shrub: scandent with slender or stout tendrils, puberulent in some parts; stem: subterete; stipules: setaceous, soon deciduous; petioles: 1–2.5 cm long, with 2 glands at apex; leaves: unlobed, ovate-oblong to ovate-elliptic, 5–11 × 4–7 cm, subacute at apex, rounded at base, entire; peduncles: 1–2 cm long, solitary, ascendant; bracts: setaceous, soon deciduous; floral tube: cylindrical-campanulate, ca. 1 × 0.6 cm, greenish outside; flowers: white, 3.5–4 cm diam.; sepals: white inside, greenish outside, oblong-lanceolate, 1.5–2 × 0.5–0.6 cm; petals: similar to the sepals; corona: 2 (rarely 3) series, greenish, mottled with purple spots, outer series 1–1.5 cm long, subdolabriform, dilated near apex, inner series 0.3–0.4 cm long, filiform below, broadly dilated at apex; ovary: ovoid, tomentulose; fruit: obovoid to subglobose, 3–4 × 2–3 cm, yellowish; seeds: 0.5–0.6 × 0.3–0.4 cm, reticulate.

Passiflora sclerophylla Harms

Notizbl. Bot. Gart. Berlin 6: 347 (1917)
sclerophyllus: having hard leaves
Supersection *Pseudoastrophea*, Section *Pseudoastrophea*

This lovely species (Plate 59) differs from other representatives of subgenus *Astrophea* by its hard foliage. The venation stands out distinctly on the underside of the leaf.

Passiflora sclerophylla is at home in southern Venezuela and in Guyana at altitudes of up to 2500 m, but unfortunately, it is not under cultivation. Its handsome flowers reach a diameter of 5 cm. The verrucose, yellow corona and whitish sepals and petals are usually covered with purple mottles, although specimens without mottles can occur. The intensity of this mottling is quite variable. The petals can be mostly purple or mostly white. The coronal filaments, standing upright in a campanulate form, do not show a typical swelling in the upper 1/3; in this regard, they are quite similar to those of *P. candida*.

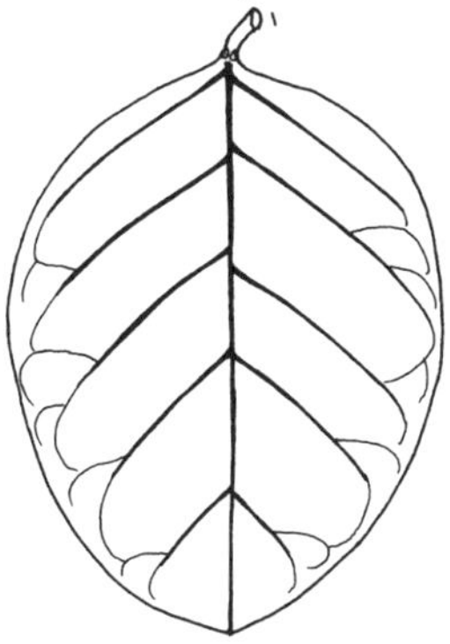

Passiflora sclerophylla

Shrub: scandent with tendrils; stem: younger parts angulate and puberulent; stipules: minute; petioles: 0.5–1 cm long, with 2 glands at apex; leaves: unlobed, more or less broadly ovate, 4–8.5 × 3–7 cm, emarginate or subacute at apex, rounded at base, thick-coriaceous, glabrous and shining above, short-villous beneath, veins elevated, entire; peduncles: solitary or in pairs, 1–1.5 cm long; bracts: minute; floral tube: broadly funnel-shaped, ca. 1 cm long, 1 cm at throat; flowers: white, usually mottled with purple, 4–5.5 cm diam.; sepals: white with purple spots inside, rarely plain white, greenish outside, oblong, 2.3–2.7 × 0.6–0.8 cm, fleshy; petals: similar to the sepals; corona: 2–3 series, outer series ca. 2 cm long, yellow with purple spots, verrucose; ovary: tomentose; fruit: 5–6 × 4 cm; seeds: unknown.

Passiflora securiclata Mast.

Kew Bull. 1893: 12 (1893)
securis: hatchet, ax
Synonym: *P. retrorsa* Killip
Supersection *Pseudoastrophea*, Section *Botryastrophea*

Passiflora securiclata (Plate 60) is another species from section *Botryastrophea* that is closely related to *P. longiracemosa* and *P. holtii*. These 3 species have in common a pseudoracemose inflorescence that is at least 10 cm long and that, in *P. securiclata*, can reach a length of 50 cm. This species is at home along the Río Orinoco in Venezuela and in northeastern Brazil. In contrast to *P. longiracemosa*, the flowers of which are mostly at the base of the inflorescence, flowers abound at the tip in *P. securiclata*.

Like other representatives of section *Botryastrophea*, *Passiflora securiclata* produces scentless flowers that are pollinated by hummingbirds and that are characterized by a long floral tube. The diameter of the handsome red flowers is about 4 cm. The conspicuous corona is widened and verrucose.

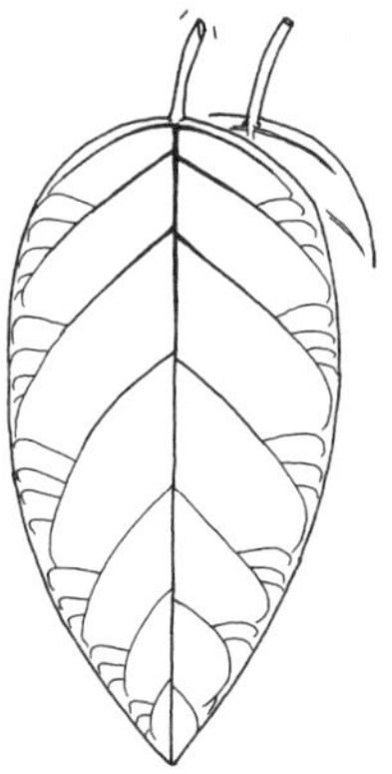

Passiflora securiclata

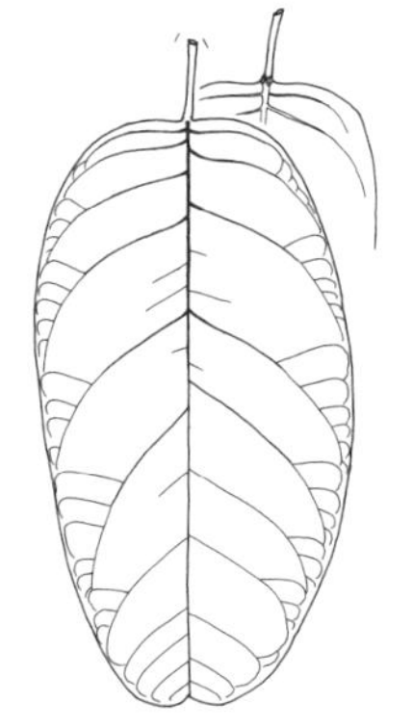

Passiflora sphaerocarpa

Unfortunately, *P. securiclata* has not yet been cultivated.

Shrub: subscandent or woody vine, glabrous, base of deciduous tendrils reduced to recurved spines; stem: angulate to terete; stipules: setaceous, soon deciduous; petioles: 1–1.5 cm long, with 2 glands at junction to blade; leaves: unlobed, variable, from ovate-oblong to oblong-lanceolate, 6–15 × 2–6 cm, obtuse, occasionally short acuminate at apex, rounded at base, entire; pedicels: 0.5–1 cm long; bracts: minute; floral tube: cylindrical, 2–4 cm long, 0.3–0.5 cm diam. at throat, reddish; flowers: reddish, ca. 4 cm diam., 4–6 cm long, flowers in pseudoracemose inflorescence, 15–50 cm long; sepals: reddish, oblong, 1.5–2 × 0.3–0.4 cm; petals: similar to the sepals; corona: 2 series, reddish orange, outer series 0.3–0.5 cm long, dolabriform, verrucose, inner series ca. 0.1 cm long; ovary: ellipsoid, minutely puberulent; fruit: ovoid, 3–4 × 2–3 cm; seeds: 0.6 × 0.4 cm, reticulate.

Passiflora sphaerocarpa Triana & Planch.

Ann. Sci. Nat. Bot. 5 (17): 184 (1873)
sphaerocarpus: round-fruited, having a round fruit
Supersection *Astrophea*, Section *Astrophea*

This is another medium-sized tree in which the tendrils are missing. *Passiflora sphaerocarpa* (Plates 39 and 61) is the most frequently found species on the Eastern Cordillera of Colombia. It mainly grows at altitudes of 500–1700 m on the foothills directed toward the Amazonian lowlands.

At 4–5 cm diam., its white flowers are unspectacular, but its corona has some particularities. It is not light yellow to yellow, as with most treelike representatives of the subgenus, but orange-yellow. In addition, it has comparatively few filaments, which lie more or less flat and are not positioned upright as is normally the case.

Passiflora sphaerocarpa has been cultivated in the United States for some years and has flowered regularly there. However, for many years the plants under cultivation were regarded and distributed as *P. arborea*. While there is a similarity with regard to the narrow leaves, the 2 species can be distinguished unequivocally by their height and by their coronal structure. Since the genuine *P. arborea* has not yet been brought into cultivation, all plants cultivated under this name should be checked to find out whether they are *P. sphaerocarpa* or another species.

Tree: 2–4 m high, without tendrils, glabrous; stem: terete, younger parts slightly striate; stipules: setaceous, soon deciduous; petioles: 1–2.7 cm, without glands; leaves: unlobed, elliptic to oblong-ovate, 7–20(–28) × 4–9(–16) cm, 2 scarlike glands on the midrib near base, membranous to subcoriaceous, obtuse to rounded at base, entire; peduncles: 2.5–3.5 cm, bifurcate; bracts: minute; floral tube: campanulate to cylindrical-campanulate, 0.6–0.9 cm long; flowers: white, 4–5 cm diam., inflorescence 2-flowered; sepals: white, oblong, 1.8–2.5 × 0.7–1 cm; petals: white, oblong, 1.6–1.8 × 0.4–0.6 cm; corona: 2 or 3 series, outer series 1– 1.4 cm long, orange-yellow with white base, laterally flattened, upper 1/3 dilated and falcate, just 28–36 filaments, inner series 0.1–0.2 cm long, orange-yellow, slightly dilated at apex; ovary: ellipsoid to oblong-ovoid, tomentose; fruit: spherical, ca. 2 cm diam.; seeds: ca. 0.6 × 0.5 cm, reticulate.

Passiflora spinosa (Poepp. and Endl.) Mast.

Trans. Linn. Soc. London 27: 630 (1871)
spinosus: full of spines
Supersection *Pseudoastrophea*, Section *Botryastrophea*

The range of the hummingbird-pollinated *Passiflora spinosa* (Plate 62) reaches from Colombia and Ecuador to Peru, where it grows in the eastern lowlands. To re-create the Amazonian habitat for this handsome but rarely cultivated species, a gardener must provide much warmth, a high atmospheric humidity, and optimum light conditions year-round. Hence, *P. spinosa* is suited for a heated greenhouse where 20°C is guaranteed in winter. A remarkable feature of this woody vine, which was described as the first species of section *Botryastrophea*, is the deciduous tendrils that, when fallen away, leave a spine at their base. Undoubtedly this serves to ratchet the vines up through thickets and tree branches during winds. Furthermore, there are 5 nectar chambers at the base of the floral tube (for more information, see the description of the closely related *P. pyrrhantha*). The reddish flowers are produced on pseudoracemose inflorescences that can be up to 25 cm long. Its sepals are significantly shorter than the floral tube, a typical characteristic of the species from this section. The diameter of the flower is 4–5 cm.

Vine: woody, glabrous except the ovary, tendrils reduced to spines; stem: terete; stipules: minute, soon deciduous; petioles: 1.3–3 cm long, with 2 glands at junction to blade; leaves: unlobed, oblong-ovate to oblong-lanceolate, 9–18 × 2.8–9 cm, acute to acuminate at apex, truncate at base, margin entire, coriaceous; pedicels: 0.3–0.5 cm long; bracts: minute; floral tube: cylindrical, 4.5–6 × 0.6–0.8 cm long, bright red; flowers: bright red, 4–5 cm diam., 6–7 cm long, inflorescence 7.5–25 cm long; sepals: bright red, oblong-lanceolate, 1–2.2 × 0.3–0.7 cm; petals: similar to the sepals; corona: 2 series, yellow, outer series 0.2–0.5 cm long, filiform or falcate-dolabriform, inner series ca. 0.1 cm long; ovary: linear-ovoid, tomentulose; fruit: yellow with longitudinal green stripes; seeds: 0.6 × 0.4 cm, reticulate.

Passiflora tica Gómez-L. & L. D. Gómez

Phytologia 49: 56 (1981)
tica: slang for "Costa Rican woman"
Supersection *Astrophea*, Section *Astrophea*

Passiflora tica (Plate 63) is a tree or large shrub without tendrils. It is one of the few species of subgenus *Astrophea* from Central America. It differs above all from the closely related *P. arborea* in having leaves that smell like burnet saxifrage (*Pimpinella saxifraga*). Morphologically, the leaves of *P. tica* are proportionately wider in shape, and the coronal filaments are long-tapering to the apex, not just suddenly pointed. The flowers of *P. tica*, pollinated by bees, are absolutely characteristic of section *Astrophea*. The white sepals and petals form a background contrast to the rich yellow dolabriform outer corona filaments.

Passiflora tica is the foodplant for *Heliconius eleuchia* butterflies and hence of particular impor-

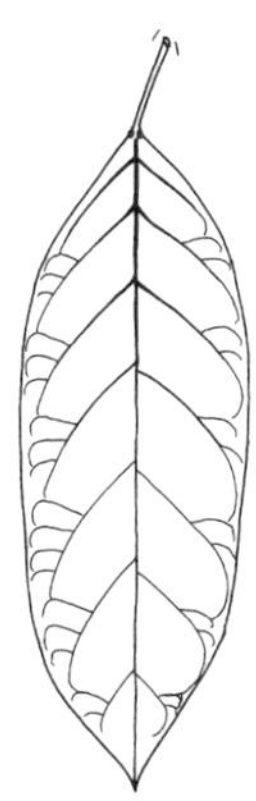

Passiflora spinosa

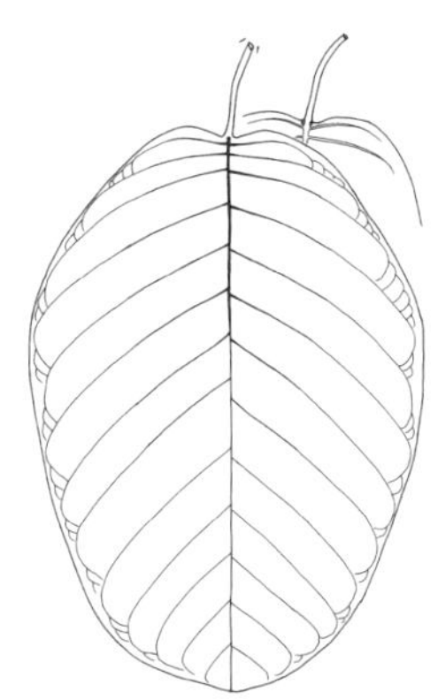

Passiflora tica

tance for lepidopterists. Many species of *Passiflora*, among them several species from subgenus *Astrophea*, which many enthusiasts can't imagine being without, were first cultivated as foodplants for butterflies.

Tree or shrub: 3–15 m high, without tendrils; stem: terete; stipules: 0.15–0.4 cm, soon deciduous; petioles: 2.5–4.5 cm long, with 2 sessile glands on base of midrib; leaves: unlobed, elliptic-oblong to elliptic-obovate, 12.5–66 × 6.5–30 cm, slightly acuminate and shortly mucronate at apex, rounded to obtuse at base, glabrous above and beneath, entire; peduncles: 1- or rarely 2-bifurcate, 1.5–12 cm long, pedicels 1.5–6.5 cm long; bracts: minute, soon deciduous; floral tube: cylindrical, 0.6–0.9 cm long, 0.4–0.5 cm diam. at throat; flowers: white, 4.5–5.5 cm diam., 3.7–4 cm long; sepals: white, oblong, 2.7–3.5 × 0.8–1.2 cm, white; petals: similar to the sepals; corona: 2 (rarely 3) series, yellow, filaments of outer series 1.4–2 cm long, dolabriform, inner series 0.1–0.2 cm long; ovary: densely tomentose; fruit: widely ellipsoid to subglobose, 3–4.5 × 2.5–4.3 cm, yellowish or cream; seeds: 0.4–0.5 × 0.3–0.35 cm, reticulate.

Passiflora tina R. Boender & T. Ulmer

Sendtnera 7: 5 (2001)
named after Tina Garzon, builder of the Tinalandia Hotel, near the type collection
Supersection *Pseudoastrophea*, Section *Capreolata*

Passiflora tina (Plate 64) is a new species from northwestern Ecuador. The number of known representatives belonging to subgenus *Astrophea* that are indigenous to Ecuador has more than doubled since 1988, and more are expected to be found.

The essential characteristics of *Passiflora tina* are its flowers, which are grouped in dense fascicles, and the structure of the outer corona filaments. The diameter of a single flower reaches the considerable width of 6–7.5 cm. The purplish orange to purplish brown corona contrasts with the white sepals and petals.

This new species is a very close relative of *Passiflora pittieri* and at times has even been thought to be identical with it. However, the 2 species can be distinguished easily. While *P. tina* has a campanulate corona, *P. pittieri*, in the strict sense, is characterized by a recumbent corona. The number of filaments in the outer corona series and the number of flowers per node are also good distinguishing features.

In 1996, Ron Boender, during one of his visits to the province of Pichincha, discovered 3 juveniles of this new species while observing *Heliconius sapho candidus* search for a hostplant. He succeeded in cultivating *Passiflora tina*, and the treelets turned into climbing vines with well-developed tendrils. Under cultivation it has flowered from old wood on shady parts of the plant 2–3 times per year. Though this species is under cultivation, and Boender succeeded in getting it to flower in the Florida sun, *P. tina* is a difficult species. Only a few enthusiasts will have the necessary facilities to cultivate it.

Vine: woody, glabrous, with well-developed tendrils on younger stems; stem: terete; stipules: setaceous, 0.1–0.2 cm long, soon deciduous; petioles: 0.9–1.5(–1.9) cm long, with 2 sessile glands at junction of blade; leaves: unlobed, oblong to oblong-ovate, (7.5–)12.5–25(–35) × (3.5–)5–9(–13.5) cm, acute to abruptly acuminate at apex, cuneate at base, subcoriaceous, entire; pedicels: 0.5–0.7 cm long; bracts: subulate, 0.2–0.3 cm long; floral tube: cylindrical to campanulate-cylindrical, 0.8–1.2 × 1–1.5 cm, green; flowers: white, 6–7.5 cm diam., 3.5–5.5 cm long, grouped in dense fascicles; sepals: whitish inside, green outside, narrowly oblong, 2.9–4.1 × 0.8–1.3 cm; petals: white, nearly as long as sepals, 1–1.5 cm wide; corona: 5 series, outer series 2–2.8 cm long, cream at base, purplish orange to purplish brown above middle, yellow at apex, 2nd series 0.9–1.3 cm long, 3rd series 0.6–0.8 cm long; ovary: ovoid, tomentose; fruit: unknown.

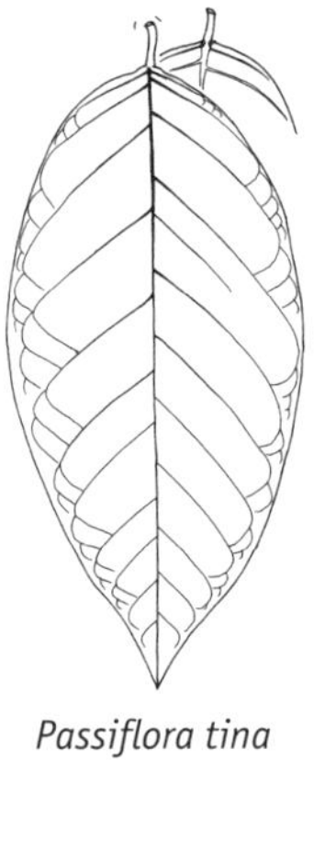

Passiflora tina

13
Subgenus *Deidamioides*

THE SMALLEST subgenus, *Deidamioides,* is something of a hodgepodge, containing 13 relatively primitive species of *Passiflora* not assignable to the other 3 subgenera. All species of the subgenus, no matter how different they may be, have tiny stipules, petiolar glands usually near the junction to the leaf blade, unlobed or 3- to 9-foliate leaves with entire margins, small bracts, a flower stalk jointed well below the base of the flower, a more or less plicate operculum, and a glabrous or nearly glabrous body. As juveniles, many have tendrils with distal adhesive disks that allow them to climb smooth-trunked trees deep in the primary rain forest. Except for the 2 representatives from section *Mayapathanthus,* all species have flowers borne on the tendrils, a phenomenon that is seen very rarely in the genus *Passiflora* but that is the rule among other genera of the passionflower family. This characteristic is evidence of a more primitive stage in the evolution of passionflowers. The primitive inflorescence in the genus can be conceived as a branched one, with the middle flower transformed into a tendril, all borne on a peduncle or stalk. Reduction of the common stalk has resulted in a flower or 2 plus a tendril at each leaf axil. One would hardly imagine that eons ago these were stalked groups of flowers; yet this is what can be seen today in subgenus *Deidamioides.* This group also includes the only passionflowers with branched tendrils. The former genus *Tetrastylis,* now a section within *Deidamioides,* was regarded by Killip (1938) and Benson (1975) as a genus more primitive than *Passiflora* within Passifloraceae but is now placed here in *Passiflora* in this relatively primitive subgenus.

Subgenus *Deidamioides* is primarily a South American group: besides the Central American section *Mayapathanthus* with 2 species, only one other species comes north into Panama and Costa Rica from the Andes of Colombia. Although the origin of passionflowers has been obscured by time, northwestern South America can rightly be called the cradle of diversity for *Passiflora,* both for the sheer number of species and for being home to a number of these primitive varieties.

Systematics—Subdivision in Sections

Despite the relatively small number of species within this subgenus, a division is made into 5 sections, 2 of which are monotypic. In the systematics set up by Killip (1938) these sections often had the rank of subgenera.

The section *Tryphostemmatoides,* containing about 7 species, is the largest within subgenus *Deidamioides.* In their juvenile stage, and even later, these species are characterized by a dainty growth and small leaves. The rather small flowers appear in pairs on the tendrils, though this is not true for *Passiflora discophora*, a species found in Ecuador. Moreover, the tendrils are furcate or bifurcate and often develop tiny disks on the tendril tips that help it climb the trunks of trees.

The type species of this subgenus is found in the monotypic section *Deidamioides*. It is characterized by compound 3-foliate leaves, flowers that are borne on the tendrils, and a whitish corona in several series.

Passiflora cirrhiflora (Plate 65) is the only representative of section *Polyanthea*. Characteristics of this species from the Guianas are the compound 5- to 9-foliate leaves, the flowers that are borne on the tendrils, and the yellow-orange corona in several series.

In the remaining 2 sections, 3 or 4 more species are found that, like *Passiflora discophora*, do not produce their flowers on the tendrils but on peduncles of their own. The only representatives of section *Tetrastylis* are *P. ovalis* and a very similar species that is being named by Fabio Augusto Vitta. These Brazilian species are characterized by their racemose inflorescences and zygomorphic bat-pollinated flowers with 4 stigmas.

The species from section *Mayapathanthus* are woody canopy vines that reach a gigantic size. Their seedlings resemble species of section *Tryphostemmatoides* and have tendrils with adhesive disks. Glands can often be found on the base of the sepals.

Cultivation

Subgenus *Deidamioides* contains species that grow in the tropical lowlands and species from very moist mountain forests, some of which grow at moderate altitudes and some of which flourish at higher altitudes up to 3000 m. The conditions in the species' natural surroundings must be taken into account when the plants are cultivated. The species from section *Tryphostemmatoides* may well be compared with the many free-flowering and less demanding representatives of subgenus *Decaloba*. *Passiflora cirrhiflora*, however, must be compared with the exacting species from subgenus *Astrophea*.

The 2 representatives of section *Mayapathanthus* can become quite large. Few enthusiasts will have the space for them in their greenhouse or conservatory.

Deidamioides species can be easily propagated from seeds.

Passiflora arbelaezii Uribe

Caldasia 7 (35): 335 (1957)
named after Enrique Pérez Arbeláez, collector of the type
Section *Tryphostemmatoides*

Passiflora arbelaezii (Plate 66), at home from the moist lowlands of eastern Nicaragua to western Colombia, belongs to a group of species with small leaves and short furcate or bifurcate tendrils. These species constitute section *Tryphostemmatoides*, named thus because the inflorescence looks much like those found in the African genus *Tryphostemma*, another member of the passionflower family. Because of its dainty growth, *P. arbelaezii* is easily confused with the closely related *P. discophora*, *P. gracillima*, and *P. tryphostemmatoides*. We can only confirm that *P. arbelaezii* and *P. discophora* are in cultivation. It is possible that all species currently cultivated in Europe under the name *P. gracillima* are actually *P. arbelaezii*.

While *Passiflora discophora* produces flowers on peduncles of its own, the significantly smaller flowers of the other species presented in this chapter appear in pairs on the tendrils. *Passiflora arbelaezii* can be easily distinguished from *P. gracillima* and *P. tryphostemmatoides* by its larger flowers and longer outermost coronal series.

Passiflora arbelaezii is relatively easy to grow and likes a protected, sunny location, preferably under

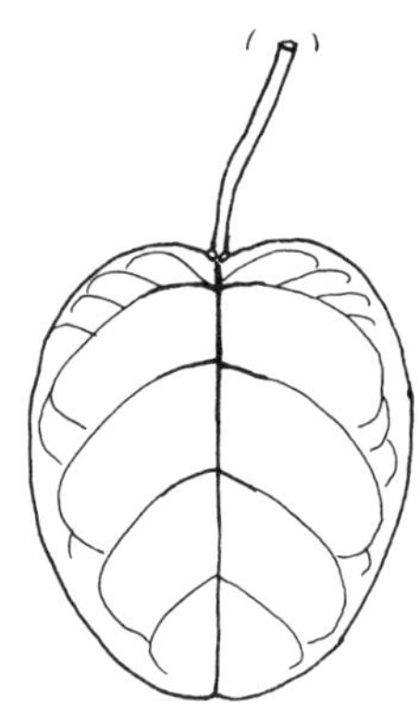

Passiflora arbelaezii

glass. As this dainty species is satisfied even with a 15-cm pot, it may also be grown on a windowsill. Its whitish and greenish yellow flowers are mainly produced on shoots that have failed to find a support. A tendril usually bears 2 flowers that reach 3.5 cm diam. Before flowering, this species often produces small bright yellow aborted buds that persist for weeks on old flower stalks. These are considered to be mimics of heliconid butterfly eggs and are a normal feature of most species in this section. While *P. gracillima* and *P. tryphostemmatoides* may proceed to altitudes of up to 3000 m, *P. discophora* and *P. arbelaezii* are lowland species, only rarely found as high as 1000 m. Accordingly, during overwintering, temperatures should not drop below 12°C.

Vine: glabrous; stem: terete, striate; tendrils: 3- to 5-parted; stipules: setaceous, ca. 0.1 cm long, deciduous; petioles: 1.8–2.3 cm, with 2–4(–6) glands at apex; leaves: unlobed, ovate, 1.8–6(–7) × 1.8–5(–6) cm, usually with one laminar nectary borne at apex of midvein, membranous, entire; peduncles: solitary, 1.1–8 cm long, pedicles 1.3–3.5 cm long, 2-flowered cyme terminating in a tendril, tendril shriveling or sometimes aborted; bracts: minute; flowers: greenish white to whitish and greenish yellow, 3–3.5 cm diam.; sepals: greenish white to whitish, linear-lanceolate, 1.5–2.5 × 0.4–0.8 cm; petals: greenish white to greenish yellow, linear-lanceolate, 1.4–2.1 × 0.3–0.5 cm; corona: 2–3 series, outer series 1–1.3 cm long, basally light green to cream or light yellow, distally pale yellow to white, inner series 0.2–0.5 cm long, greenish yellow, white at apex; ovary: ovoid, glabrous; fruit: 3–4.5 × 3–3.5 cm, subglobose to widely ellipsoid, yellowish to red-orange; seeds: ca. 0.4–0.7 × 0.3–0.6, reticulate or rugose to irregularly transversely sulcate.

Passiflora cirrhiflora Juss.

Ann. Mus. Natl. Hist. Nat. 6: 115 (1805)
cirrhiflorus: *cirrhus* (tendril), *florus* (flowered)
Synonyms: *P. jenmanii* Mast., *P. septenata* DC.
Section *Polyanthea*

Passiflora cirrhiflora (Plates 65 and 67), outstanding in several respects, is an object of desire among many enthusiasts because of its yellow-orange flowers. When the first richly illustrated books on passionflowers were published in the early 1990s, John Vanderplank's photo of *P. cirrhiflora* was a real sensation. No such flower had been seen before. Greenish yellow sepals, alternating with rich yellow-orange petals, are paired with an outer corona that is orange-red at its base and rich yellow in the upper part, and these surround the androgynophore in an upright campanulate fashion. The extremely wavy coronal filaments are also fringed, a feature otherwise known of only in a few species belonging to subgenus *Astrophea*. The flowers, appearing in pairs midway along the tendrils, reach a diameter of 6–8 cm. The stigmas are truly sensational and may well be the largest within the whole genus.

Even without flowers, *Passiflora cirrhiflora* can be distinguished from other species at first sight. Its attractive leaves are 5- to 9-foliate. Besides that, the base of the individual leaflets shows 2 or rarely 3 teeth that end in a filiform awn. The function of these conspicuous projections is a complete mystery. Because of its various morphological features, *P. cirrhiflora* is the only species within section *Polyanthea*.

Not surprisingly, many enthusiasts visited the home countries of this species after having seen its picture. It is endemic not only in the frequently visited French Guiana but also in the tropical lowlands of Suriname and Guyana. It likes to climb on bushes and small trees, and when these are missing, as is often the case along roads, the shoots of this gigantic species often cover the ground.

The enthusiasts that went off in search of *Passi-*

Passiflora cirrhiflora

flora cirrhiflora sometimes brought seeds back to Europe, and later to the United States. Quite a number of specimens are under cultivation by now, and these can be easily propagated from cuttings. This warmth-loving species may be called quite exacting. Strictly speaking, it is only recommended for a heated greenhouse with winter temperatures above 15°C and additional lighting. The quality of the rooting medium is an important issue. A mix that contains too much peat will not be well suited. Administering some lateritic soil with a high content in iron and aluminum will be of great help making the mix correspond to the conditions in the species' natural surroundings. Loam, perlite, or potting mixes that contain very coarse sand and are used for cacti may also be used. Further essentials are excellent drainage and a very deep pot.

Vine: large, nearly glabrous; stem: terete; stipules: setaceous; petioles: stout, 5–10 cm long, with 2 paired, in general sessile, scarlike glands near base; leaves: 5- to 7-foliate or occasionally 9-foliate, 7.5–14 × 10–20 cm, leaflets aristate-bidentate near base, entire; peduncles: stout, 1–4 cm long, 2-flowered at apex, terminating in a stout tendril, pedicles 2–5 cm long, base of tendril retained as climbing hook; bracts: linear-subulate, ca. 1 cm long, with 2 glands at base; floral tube: broadly campanulate, 0.5–1 cm long; flowers: greenish yellow to orange-yellow with reddish corona base, 6–8 cm diam.; sepals: green outside, greenish yellow to yellow inside, 2.5–3 × 0.7–1 cm; petals: yellow, mottled with red speckles, oblong, about 2.5 × 0.6–0.8 cm; corona: 3 series, outer series ca. 3 cm long, liguliform, deep red to reddish orange in lower 1/3, deep yellow above, wavy, filaments fimbrillate along the margin, inner series linear, about 1 cm long, whitish with red apex, capitate; ovary: ovoid, whitish, tomentulose; fruit: globose, grayish brown with white mottles, pericarp coriaceous; seeds: broadly obovate, ca. 0.7 × 0.4–0.5 cm, transversely 5- or 6-sulcate.

Passiflora deidamioides Harms

Repert. Sp. Nov. 19: 57 (1923)
deidamioides: resembling the genus *Deidamia*
Section *Deidamioides*

Passiflora deidamioides (Plate 68) is another species that occupies a special taxonomic position. Harms (1925) was well aware of the peculiarity of this species when he assigned it to its own subsection, *Deidamioides*, instead of to supersection *Decaloba*. Killip (1938) took over Harm's classification but raised the subsection to the monotypic subgenus *Deidamioides*. Following the revision of the systematics, this subgenus was acknowledged and enlarged by Feuillet and MacDougal.

Passiflora deidamioides, as a typical representative of the subgenus named after it, produces its flowers on the tendrils. Less than 5% of the more than 525 species of *Passiflora* produce their flowers on the tendrils. The greenish white flowers of *P. deidamioides* reach a diameter of 5–6 cm. Immediately after each flower opens, the sepals and petals reflex completely.

The native range of *Passiflora deidamioides* is restricted to southeastern Brazil. It is very rarely cultivated, if at all. Plate 68 was taken by Lawrence E. Gilbert, who succeeded in bringing a single specimen to flower in his collection at the University of Texas at Austin.

Vine: medium to large, glabrous; stem: subterete; stipules: minute, soon deciduous; petioles: stout, 1.5–5 cm long, with 2–4 paired, sessile glands, below and above middle; leaves: 3-foliate, 5–8 × 7–12 cm, entire; peduncles: 4–6 cm long, 2-flowered, terminating in a long tendril, pedicles slender and 2–3 cm long; bracts: setaceous, 0.2–0.3 cm long; flowers: greenish white and pale yellow to cream, 5–6 cm diam.; sepals: green outside, greenish white inside, oblong, ca. 2.5 × 1 cm; petals: greenish white to white, oblong, subequal to sepals; corona: 7 series, outer series 1.5–2 cm long, white,

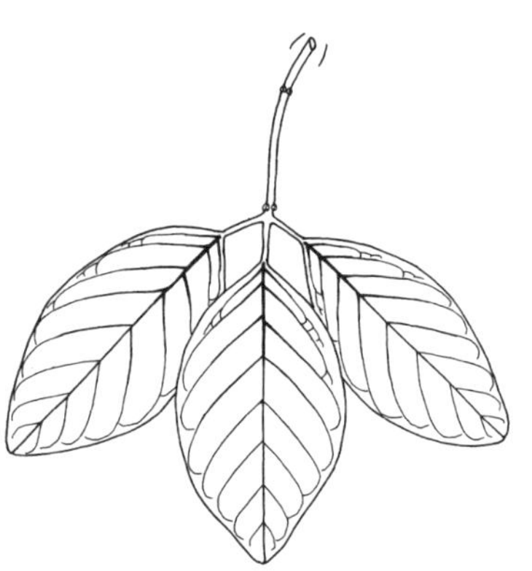

Passiflora deidamioides

2nd series ca. 1–1.5 cm long, white, inner series 0.3–0.1 cm long, pale yellow to greenish yellow; ovary: ovoid, glabrous; fruit: unknown.

Passiflora discophora Jørgensen & Lawesson

Nord. J. Bot. 7: 127 (1987)
discophorus: bearing a disc (in reference to the tendril tips)
Section *Tryphostemmatoides*

Passiflora discophora (Plate 69), another exceptional species, is at home in north-central to central Ecuador, where it grows at altitudes of 200–1000 m. Because of its unlobed leaves, only 2–6 cm long, it looks very dainty at first sight. Nevertheless, under ideal conditions it can develop into an unexpectedly large plant. The diameter of the lower lignified stem may well reach 1 cm or more. To guarantee a vigorous growth, *P. discophora* should be planted directly in the ground in a greenhouse. It prefers summer temperatures around 25°C and winter temperatures around 12°C. Under these conditions it will flower regularly.

The white flowers with a yellow and white corona effuse a pleasant scent and reach a diameter of up to 4 cm. They are produced either on the ends of the shoots or on old wood. The flowers, borne on peduncles of their own, have an unambiguous characteristic that distinguishes *P. discophora* from other representatives of section *Tryphostemmatoides*, which also have small leaves and which are similar in their vegetative appearance: the androgynophore is curved, with its stigmas and anthers grouped into one position. In the other species, such as *P. arbelaezii*, *P. gracillima*, and *P. tryphostemmatoides*, the flowers are produced on the tendrils and are more radially symmetric. Further distinctive features include the diameter of the flower, the form and color of the fruit, and the seeds. In *P. discophora*, as in the other species in this section, the furcate or bifurcate tendrils bear terminal discs that reach a diameter of up to 3 mm. With the help of these small holdfasts, each resembling a discus, the vine is able to climb even smooth surfaces. The seeds have hypogeal germination, which is extremely rare among passionflowers.

Vine: glabrous; stem: terete, striate; stipules: setaceous, 0.1 cm long, soon deciduous; petioles: 0.6–1.5 cm without glands; leaves: unlobed, oblong to ovate, (1–)2–6 × 0.9–3 cm, with 2 sessile glands near base beneath, entire; peduncles: 0.7–1.5 cm long, solitary or in pairs; bracts: minute, soon deciduous; flowers: white and yellow, 3.5–4 cm diam.; sepals: whitish green outside, white inside, oblong-ovate, 2.4 × 0.9 cm; petals: white, oblong, 1.7 × 0.5 cm; corona: 3 series, outer series up to 1.8 cm long, yellow to yellow-orange at base, white in upper 1/2, inner series 0.1–0.6 cm long; ovary: ovoid, glabrous; fruit: subglobose to ovoid, 3.8 × 3.2 cm, bright red to bright orange; seeds: ovate, ca. 0.4 × 0.3 cm, slightly transverse-sulcate, black.

Passiflora macdougaliana Knapp & Mallet

Ann. Missouri Bot. Gard. 71: 1068 (1984)
named after John M. MacDougal
Section *Mayapathanthus*

Passiflora macdougaliana (Plate 70) is a large, woody canopy liana that flowers only in treetops in its natural surroundings. It is found only in the province

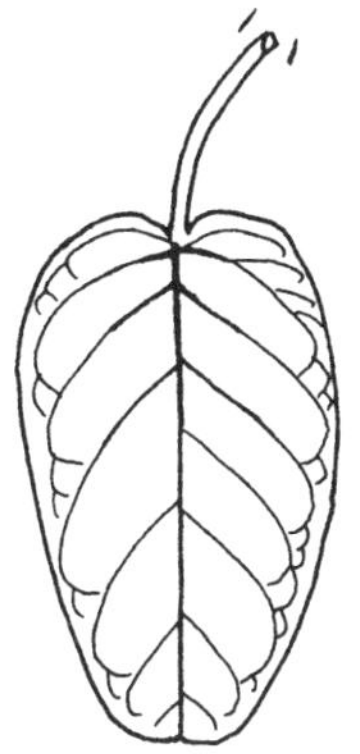

Passiflora discophora

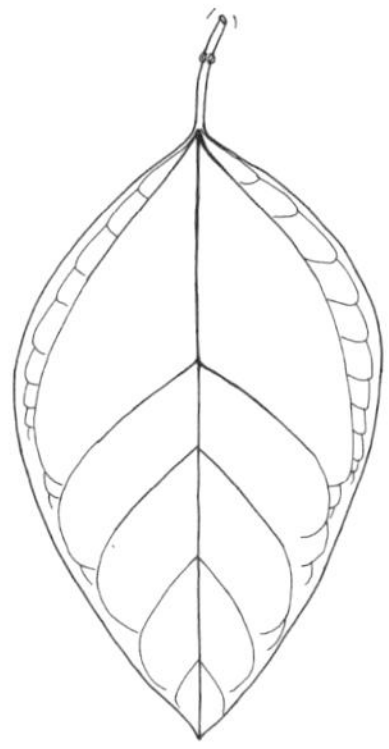

Passiflora macdougaliana

of Colón in central Panama, where it grows in mature rain forests from near sea level up to 500 m. It is fairly common on the Santa Rita Ridge.

This species, along with its sister species, *Passiflora obovata*, forms the small section *Mayapathanthus* in subgenus *Deidamioides*. The species are very poorly known large to giant vines with white flowers and unlobed leaves. There are 2 petiolar nectaries, the stipules and bracts are tiny, the buds often have external nectaries, and the flower has a plicate operculum. The juveniles of *P. obovata*, at least, are very similar to the adult vines of section *Tryphostemmatoides* and thus suggest a close relation between the sections.

The 2 species may be distinguished by their leaves. Those of *Passiflora obovata* have 3 major veins at the base, versus 5 in *P. macdougaliana*. Furthermore, the white, unmarked flowers of *P. macdougaliana* are larger and are characterized by an undulate corona in 8–10 series. Its flowers effuse a sweetish scent and reach a diameter of 5.5 cm. The glands outside the flower at the base of the sepals are very unusual, similar to those seen in *P. edulis* f. *flavicarpa* or *P. variolata*.

Passiflora macdougaliana is not under cultivation, and it might be difficult to grow if it were, as it can reach more than 35 m long.

Vine: woody, canopy, nearly glabrous; stem: terete; stipules: linear, 0.1–0.2 cm long, soon deciduous; petioles: ca. 2.5 cm, with 2 paired glands just beyond the middle; leaves: unlobed, ovate, ca. 13.5 × 8 cm, shiny above, glaucous below, entire; peduncles: ca. 4.5 cm long, solitary; bracts: deltoid, ca. 0.1 × 0.1 cm, pale brown, scattered along peduncle; flowers: white to cream, 5–6 cm diam.; sepals: white inside, green outside, narrowly triangular, ca. 3 × 1–1.5 cm, with 0, 1, or 2 glands outside, without awn; petals: white, narrowly triangular, ca. 3.5 × 1.5 cm; corona: 10 series, white to cream, densely packed, outer 5 series up to 3.5 cm long, next 2 series ca. 0.7 cm long, inner series 0.2–0.3 cm long; ovary: ellipsoid, glabrous; fruit: unknown.

Passiflora obovata Killip

Publ. Carnegie Inst. Wash. 461: 308, pl. 1 (1936)
obovata: inversely ovate (in reference to the leaf shape)
Section *Mayapathanthus*

Passiflora obovata (Plate 71) is one of the most vigorous passionflowers. Its shoots can climb to a height of 18 m and the base of the stem can reach a diameter of 20 cm. An abundance of flowers appear in the leaf axils as soon as the plant has reached the tree tops, the greater light intensity inducing flower bud formation. *Passiflora obovata* can hardly be perceived from the forest ground, because its flowers are almost out of sight. Only the fallen whitish flowers, having a purplish corona base, evidence the existence of a passionflower in the treetops. It is hard to believe that such an imposing plant produces such small flowers, each only about 4 cm diam.

Together with its sister species, *Passiflora macdougaliana*, *P. obovata* constitutes the small section *Mayapathanthus*. When Killip described this species, he realized it had an exceptional status within subgenus *Decaloba* because of its petiolar glands, the venation of its leaves, and its smaller bracts, and assigned it to a new section, *Mayapathanthus*, where it remains. However, this section is now included in subgenus *Deidamioides* instead of *Decaloba*.

Passiflora obovata has a wide native range but is rare within that range. It can be found, spottily, in wet primary forest from Oaxaca, Mexico, to Costa Rica at altitudes of 850–1400 m. Only in parts of Belize can it be considered common, where it is found in the wet limestone hills of the southern

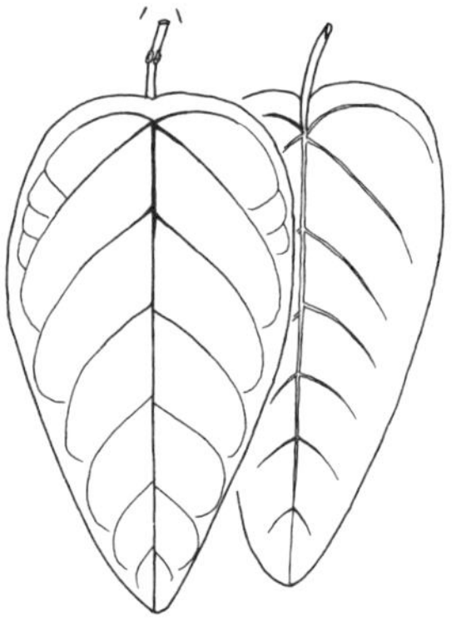

Passiflora obovata

Maya Mountains. Except in Belize, it is not under cultivation.

Vine: woody, canopy, nearly glabrous; stem: subangular, older stems dark; stipules: linear, soon deciduous; petioles: ca. 2 cm, with 2 paired, scarlike glands above middle; leaves: unlobed, obovate to oblong-ovate, 6–12 × 3–7 cm, subcoriaceous, entire; peduncles: slender, 1–3.5 cm long, in pairs; bracts: minute, up to 0.15 cm long, closely appressed to the peduncle; flowers: white to cream, and pale purple, ca. 4 cm diam.; sepals: whitish green outside, whitish inside, oblong, ca. 1.5 × 0.8 cm; petals: white, linear-oblong, ca. 1.6 × 0.4 cm; corona: (2–)3 series, outer series 1.5–2.7 cm long, pale purple in lower 1/2, white above, wavy in upper 1/3, inner series 0.1–0.5 cm long; ovary: globose, glabrous; fruit: 5–7 × 4–6 cm, yellow-green; seeds: ca. 0.6 × 0.6–0.7 cm, irregularly deeply rugose with 5–7 radial and transverse ridges.

Passiflora ovalis Vell. ex Roemer

Fl. Flumin. 9: pl. 75 (1827); *Fam. Nat. Syn.* 2: 168 (1846)
ovalis: egg-shaped (in reference to the leaves)
Section *Tetrastylis*

This species (Plates 72 and 73) has 4 stigmas and styles instead of the usual 3, as indicated in the sectional name. After *Passiflora lobata*, another species with 4 stigmas, was recently assigned to genus *Passiflora* by John MacDougal, it was only a matter of time before the controversial genus *Tetrastylis* within the family Passifloraceae had to be given up. This was done in the wake of the new systematics, and the only remaining species, *P. ovalis*, was assigned to subgenus *Passiflora*. A 2nd Brazilian species in this group, very similar to *P. ovalis* and also with 4 styles, is now being described by Fabio Augusto Vitta. The characteristic of 4 stigmas proved to be inappropriate in delimiting this genus because even within various *Passiflora* species, flowers with 4 stigmas instead of the usual 3 may occur, and the generic characteristic of a well-defined androgynophore is present.

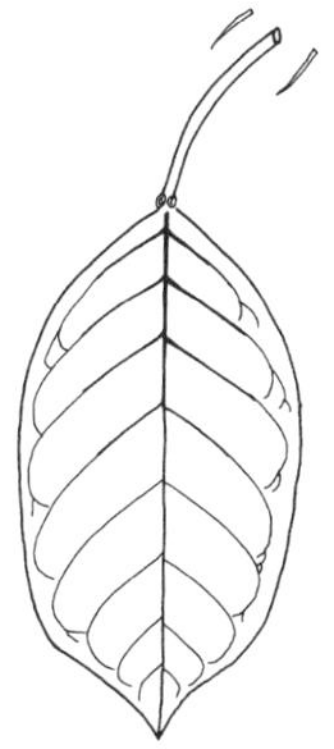

Passiflora ovalis

Passiflora ovalis, at home in southern Brazil, was the 2nd passionflower that was proven to be pollinated by bats. Investigations revealed that the flowers of these woody vines, growing in the edge of the forest, were visited by 2 phyllostomid bat species: *Glossophaga soricina* and *Rhinophylla pumilio* (Buzato and Franco 1992). However, it might be possible that the latter visited the plants because of their fruit.

Most *Passiflora* species are pollinated by bees and other insects or by hummingbirds. Not more than about 2% of the known species are pollinated by nocturnal bats. Taking into account the features of other species pollinated by bats, such as *P. galbana*, *P. macropoda*, and *P. trisecta*, the characteristics of "bat-flowered" species can be summarized: the flowers are mainly white, often with sepals and petals that reflex soon after the flower opens; the flowers, which open at dusk, are pendent on long peduncles or stand out of the foliage; the stigmas and anthers are inclined toward the side the bats are landing on; the nectar has a characteristic sugar concentration; the anthers bear a large amount of pollen; the corona is reduced; the androgynophore has shifted from the center to one side; and the floral tube is campanulate-cylindrical and wide open at its tip.

Passiflora ovalis has all of these characteristics. Further peculiarities are the position of the 4th stigma on the opposite side and the racemose inflorescences that distinguish it from other bat-flowered species. The flowers, several of which can be open at the same time, are about 5.5 cm diam. They are borne on pendent racemes that can reach a length of 75 cm and may consist of up to 20 singular flowers. While all other bat-flowered species are scentless or almost scentless, *P. ovalis* gives off a strong garlic odor. Alas, at present this rarity is not under cultivation. Lawrence E. Gilbert (personal communication) discovered that this species also has seeds with hypogeal germination (see *P. discophora*).

Vine: glabrous; stem: terete, longitudinally sulcate, corky near base; stipules: setaceous, 0.8–1 cm long, soon deciduous; petioles: 2.5–4 cm, with 2 paired, sessile glands at base; leaves: unlobed, elliptic to elliptic-oblong, 6–10 × 3–5.5 cm, coriaceous, entire; peduncles: up to 1 cm long, 2-flowered, pedicels 1.5–4 cm long, stout; bracts: setaceous, 0.1–0.2 cm long, soon deciduous; floral tube: 0.3–0.5 cm long, green outside; flowers: greenish white to white, 5–6 cm diam.; sepals: greenish outside, whitish inside, lanceolate-oblong, 2.5–3 × 0.5–0.7 cm; petals: white, lanceolate-oblong, 1.3–2 × 0.3–0.5 cm; corona: 2 series, outer series ca. 1 cm long, white, slightly wavy, inner series ca. 0.5 cm long; ovary: oblong, glabrous; fruit: ovoid, 6–10 × 3.5–8 cm; seeds: obovate, 0.5–0.7 × 0.4–0.5 cm, reticulate.

14

Subgenus *Decaloba*

Species from subgenus *Decaloba* have the reputation of being small but easy to grow. More than 200 species are classified here, including most of the small-flowered and small-fruited species, as well as those with 2-lobed or variegated leaves. Most are small vines or forest-edge climbers rather than forest lianas. Petiolar nectaries are usually absent, and bracts are usually tiny and inconspicuous. What the decalobas may lack in the size and impact of their flowers is more than made up for by their interesting leaves and by the ease with which they can be cultivated and propagated. Many species of *Decaloba* are very well known and have been cultivated for many decades, among them *Passiflora biflora*, *P. coriacea*, and *P. trifasciata*.

The most beautiful and colorful displays of foliage may be found among the decalobas, as well as some of the most bizarre leaf shapes—leaves said to resemble swallowtails, shields, sausages, duck feet, grasses, boomerangs, kitten heads, coins, chicken feet, and half moons. Supersection *Decaloba*, in particular, has many variations based on a 2-lobed ground plan, a leaf shape most unusual in flowering plants. The juveniles of many species in subgenus *Decaloba* have whitish variegation on the upper surface of the leaves, especially along the 3 main veins. Sometimes the variegation is retained all the way to reproductive age, and when this is combined with a purple leaf, the appearance is especially striking. We recommend growing many species of this subgenus just for the curious and attractive foliage (see, for example, Plate 74).

The vast majority of decalobas are pollinated by bees and wasps, but there are a number of hummingbird-adapted species, in shades of red and with long androgynophores, such as the *Passiflora murucuja* group, and at least one species is known to be pollinated by nocturnal bats: *P. penduliflora* (Kay 2001). The flowers of most species are 3–4 cm diam. and white or pale greenish yellow, but the palette of floral colors is nearly complete in the subgenus, with every color of the rainbow represented except pure deep blue. The filamentous corona is usually in 1 or 2 series, the outer series often yellow or banded with reddish purple. The operculum is plicate. Fruits of most species are small purple-black berries, eaten by birds, but a few species have green or red fruits. Not all fruits are small berries, however: some are up to 20 cm long, and a few even have capsulelike fruits that split open to release the seeds. Some species have seeds with a reticulate or pitted seed coat; the more advanced species have seeds that are transversely grooved or sulcate. The type of seed sculpturing is closely correlated with the presence of glands appearing as dots on the leaf blade and with the shape or posture of the growing shoot tip (MacDougal 1994): species with grooved seeds usually have glands on the leaf blade and have cernuous new growth.

The main native range of *Decaloba* species is distinctly farther to the north than the range of other subgenera. No other subgenus occurs so often in Mesoamerica, where so many species are endemic. The northern border is marked by *Passiflora lutea*, which proceeds as far as Illinois, Indiana, Ohio, and Pennsylvania. The farther one gets to the south, the more the number of species endemic in the United States rises, reaching 5 species as far south as the Mexican border. In Mexico the number of species increases considerably and remains consistently high throughout Mesoamerica and the Greater Antilles, where decalobas drastically outnumber passionflowers from other subgenera. There are also very many species in Colombia and Ecuador. From there the percentage of decalobas in the flora decreases in absolute number and relative importance progressively toward the east and south. *Decaloba* species grow from near sea level to altitudes of about 3000 m.

This is the only subgenus containing species that are not at home in the New World. Approximately 20 species are indigenous to Southeast Asia, Oceania, and Australia. The best known among them are the Australian *Passiflora aurantia*, *P. cinnabarina*, and *P. herbertiana*.

Systematics—Subdivision in Sections

The name *Decaloba* was coined by de Candolle in 1822 when he divided passionflowers into 7 sections. The name refers to the 10 lobes of the perianth, the 5 sepals and 5 petals. Of course, nearly all passionflowers have this, so he further specified that these species are the ones with small, tiny, or no bracts. In 1828, Heinrich Gottlieb Ludwig Reichenbach raised de Candolle's section *Decaloba* to the rank of subgenus. This is the first mention of a subgenus *Decaloba*, which, however, was subsequently overlooked or became forgotten until cited by John MacDougal in 1994. In 1846, Max J. Roemer, one of the great students of passionflowers, decided that the variety of flowers and fruits was so great in the genus that *Passiflora* should be split up into a number of genera. We do not follow that today, but *Decaloba* became a full genus for a while. Working at Kew in 1871 and 1872, the great Maxwell T. Masters published a complete monograph of all passionflower species and their close relatives. His system of classification employed the rank of subgenus, which is higher than and contains sections, and his subgenus *Plectostemma* contained the species we put in *Decaloba*. He had overlooked Reichenbach's earlier use of *Decaloba* as a subgenus. Therefore, subgenus *Plectostemma* and subgenus *Decaloba* are basically the same thing, but we are obligated to use the older name, *Decaloba*. In the 20th century, Harms (1925) did not use the rank of subgenus but split *Passiflora* into many sections and subsections, inventing many new sections to recognize special floral characteristics. Killip's famous system (1938) closely approximated that of Harms but raised most of the distinctive groups in the genus to the subgenus level. Killip also overlooked the old subgenus *Decaloba*, which is why *Plectostemma* is the name used in his book. Modern field-workers, especially Feuillet and MacDougal (2004), lumped together many sections and subgenera based more on their leaves and seeds than on their flowers, and we now have a large subgenus *Decaloba* split into 8 supersections.

The core of the subgenus is based around the type species, *Passiflora biflora*. This is therefore in supersection *Decaloba*, the largest group in the subgenus, with about 120 species. These are the 2-lobed passionflowers, based on reduction of the central lobe of a 3-lobed leaf. The laminar nectaries appear as ocellae between the main veins, well away from the margin of the leaf, and there are no petiolar nectary glands. These are advanced characteristics in the genus and, as noted above, are correlated with sulcate seeds and new stem growth shaped like a shepherd's crook. In a few species, such as *P. lutea* and *P. rubra*, the laminar nectaries have been lost, and in species such as *P. bicornis* the bracts are not setaceous but are the size of a fingernail. There are many groups of related species, and here can be found the old subgenera *Murucuia*, *Pseudomurucuja*, *Psilanthus*, and the so-called *Astephia*. The chromosome number for this whole group is basically $n = 6$ (Snow and MacDougal 1993).

Two sections are recognized in supersection *Decaloba*: section *Decaloba* proper, with 106 species, and the small section *Xerogona*, with about 13 species with capsular fruit, including *Passiflora capsularis, P. rubra*, and *P. citrina* (Plate 75). Besides the unusual dehiscent fruit, the primarily Mesoamerican *Xerogona* is characterized by the absence of both bracts and laminar nectaries, and the grooves of the seeds are often smooth and shiny. Species in both sections are usually very easy to cultivate and propagate.

The 20 species native to the Old World, in southern and Southeast Asia, Australia, and Oceania, comprise supersection *Disemma*. Except for the 3 Australian species—*Passiflora aurantia, P. cinnabarina*, and *P. herbertiana*—these are poorly known in the West. These species are discussed in 3 monographs (Cusset 1967, de Wilde 1972a, Green 1972), the one by Willem J. J. O. de Wilde being particularly helpful for its range maps and illustrations of leaves. The Chinese and Southeast Asian species are mostly white-flowered, and a few are 2-lobed or have variegation on the leaves. Modern genetic studies are underway by Shawn E. Krosnick that may soon illuminate the genealogy and evolution of this group. Few of the Old World species are presently available for cultivation in the West.

Supersection *Multiflora* is a bit of a hodgepodge, including about 20 decalobas with 2 petiolar nectaries and either branched inflorescences or primitive conditions of the laminar nectaries and seeds (for example, *Passiflora holosericea* and *P. sodiroi*). The type species is *P. multiflora*, a most unusual plant with extremely reduced flowers—the smallest in the genus, perhaps the whole family. This species of coastal scrub has mass flowering with hundreds or sometimes thousands of tiny flowers all open at one time. The other main species group placed here is the Andean *P. lobbii* group, with 9 species. New growth is straight, and chromosome numbers are unknown in the supersection except for one count of $n = 7$ from *P. holosericea*. Variegation appears to be absent except in the *P. lobbii* group.

A group of 8 species centered in the Amazon basin is recognized as supersection *Auriculata*. These have petiolar nectaries and straight new growth but also have laminar nectaries well onto the leaf blade, though not in 2 neat rows as in the higher decalobas. The petals are very narrow and reduced and are easily overlooked, as with *Passiflora auriculata*. The seeds are partway between being reticulate and sulcate. Leaf variegation is absent, and new growth is straight.

The most observable characteristic of supersection *Cieca* is the absence of petals. This is primarily a northern Central American and Caribbean group and contains 17 species, among them such well-known passionflowers as *Passiflora suberosa, P. coriacea, P. xiikzodz, P. juliana*, and *P. viridiflora*. Studies underway by Kristen Porter-Utley (2003) are clarifying the messy situation around *P. suberosa* and *P. pallida*, and we look forward to the publication of those results. Variegation is usually present, at least in seedlings, and new growth is straight. The chromosome number of this group is $n = 6$ or a tetraploid based on that.

Supersection *Bryonioides* is centered in southern Mexico and northern Central America, and many of the species have leaves that resemble those of species in the cucumber family, such as the genus *Bryonia*. Since many of the 20 species of this supersection are adapted to the seasonally dry temperate forests of Mexico, their stems often die back to the roots each year. This supersection has also been known as section *Pseudodysosmia*, "the hooked trichome group." All but 2 species are covered in tiny hairs that are curved over at the end to form a sharp hook. There is no leaf variegation, even in seedlings, and the shoot tip is straight, although tendrils are often coiled during development. Stipules are usually conspicuous, which is unusual in subgenus *Decaloba*. Fruits vary here from fragrant pale green berries (as with *P. karwinskii* and *P. pilosa*) to purple berries with orange arils (as with *P. morifolia* and *P. adenopoda*) to very long pods that split open when mature (as with *P. dioscoreifolia* and *P. lobata*). The chromosome number is $n = 6$ except for one count of $n = 7$ in *P. lobata*.

Two supersections are peripheral and basal to the rest of the supersections of subgenus *Decaloba* and could conceivably be recognized as supersections of their own. These groups of large, quite woody vines are centered in northern Central America. Both groups have leaf variegation, crook-tip growth, and a plicate floral operculum. The first and perhaps most primitive section is supersection *Pterosperma*, including the species *Passiflora eueidipabulum* and *P. lancetillensis*. Named for its distinctive, broadly winged seeds, this group has flowers borne off the tendril, a primitive condition (MacDougal and Hansen 2003). The leaves are unlobed, with one main vein, and there may be many petiolar nectaries. The chromosome number is n = 9 in *P. microstipula*. The other basal group in subgenus *Decaloba* is supersection *Hahniopathanthus*. The name means "Hahn's passionflower," and besides *P. hahnii*, the group is comprised of at least 4 other species, including *P. guatemalensis* and *P. membranacea*. These vines have a leaf with 3 main veins, marginal laminar nectaries, and greatly exaggerated stipules and bracts. Several species have a peltate leaf, where the petiole is inserted not at the edge of the blade but at some distance from the margin. Two important trends characterizing the supersection are the loss of petiolar nectaries in most species and the large leafy bracts reduced in number to only 2. The fruits and seeds are more like those of subgenus *Passiflora*.

Cultivation

The great majority of *Decaloba* species are recommendable because of their ease of culture and their willingness to flower. Many species produce flowers even on small plants, so that they can be grown in smaller pots and placed on the windowsill. Nevertheless, some species must become 2–3 m high to flower. Because of the modest size and predominantly white or greenish flowers of many *Decaloba* species, some enthusiasts do not consider them particularly attractive. Large, sumptuous flowers seem to be preferred, such as can be found in many species of subgenus *Passiflora*. Nevertheless, we would like to speak up here for growing *Decaloba* species. What is the use of having in cultivation the most sumptuous species if they cannot be brought to flower, or if they flower only rarely? A plant in full bloom is more enjoyable. Although the flowers of species from *Decaloba* are rather small, they often appear in abundance. A few red, pink, yellow, and violet species are available, and more hybrids are being created each year. Moreover, these plants are sometimes self-fertile, so that decorative fruits are set, as, for example, with *P. herbertiana*, *P. morifolia*, *P. gracilis*, *P. suberosa*, and forms of *P. rubra* and *P. capsularis*.

This subgenus contains various species whose flowers offer some pretty spots of color, and the curiously shaped or variegated foliage of many species is very decorative even without flowers. When selecting plants, try to avoid being dazzled by the size and color of the flowers; instead select reliable species of different forms that can provide a sense of achievement, especially when first beginning to grow passionflowers. This is quite feasible with *Decaloba* species.

Because of their varying origins, the species of this subgenus have different cultural demands, and it is difficult to set up generally valid rules. The whole range of possibilities becomes apparent when one regards the minimum temperatures that can be tolerated by a certain species. While *Passiflora lutea* from the United States stands temperatures down to –15°C, representatives from the tropical lowlands need temperatures above 12°C or even 15°C. Between these 2 extremes, all other species must somehow be assorted in accordance with their demands. In the majority of cases the plants can be overwintered at temperatures of 7°–10°C. Generally speaking, the overwintering of *Decaloba* species is unproblematic.

Cuttings from decalobas, with the exception of a few species, easily develop roots within 10–14 days, and high losses are not to be feared. Hormones are not always needed on these, and many will root in a glass of water. Sometimes field collectors find that coils of stems of these species have started to root right in their plastic bag before they have even arrived home from their trip. Even growing plants from seeds is quite successful. Plants often flower 1/2 a year after sowing the seeds.

Passiflora adenopoda DC.

Prodr. 3: 330 (1828)
adenopodus: having glands on stalks
Synonyms: *P. acerifolia* Schlecht. & Cham., *P. aspera* Sessé & Mociño, *P. scabra* Sessé & Mociño
Supersection *Bryonioides*

Passiflora adenopoda (Plate 76) is native from Mexico to Peru, where it usually grows at altitudes of 1000–2000 m above sea level, although it can be found as low as 300 m, especially in Mexico.

This species features 2 stalked nectar glands on the upper 1/3 of the petioles, and it is due to these glands that the plant was given its Latin name. The glands can have a diameter of 0.2–0.4 cm. The usually 5-lobed leaves are also remarkable: bristly, hooked hairs on their surface cause them to stick to clothing like strips of Velcro. In the plant's natural surroundings, this dense hairiness of miniature fishhooks is an effective protection against damage done by small caterpillars. Forms from Venezuela have an unusually short corona.

Unfortunately, cultivated plants—at least the clones cultivated in Europe—have so far turned out to be reluctant to flower. This is regrettable, since *Passiflora adenopoda* is otherwise an easy-to-handle, rapidly growing species. Min. temp.: 5°C.

Plant: short-pubescent throughout with sharply hooked hairs; stem: terete, woody at base, up to 2.5 cm wide; stipules: suborbicular to reniform, 0.8–2.5 × 0.6–1.8 cm, cuspidate; petioles: 5–13 cm long, with 2 stipitate glands in upper part of petiole; glands 0.4–1 × 0.2–0.4 cm; leaves: 5-lobed, rarely 7-lobed, juvenile leaves 3-lobed, 7.5–19 × 9–24 cm, bearing a few submarginal laminar nectaries, cordate at base, denticulate or nearly entire; peduncles: solitary or in pairs, 2–4.8 cm long; bracts: ovate-oblong, 1–1.8 × 0.8–1.5 cm, borne near middle of peduncle, lacerate; flowers: white with violet-purple bands, 3.5–5.5 cm diam.; sepals: white to whitish green, oblong, 1.8–3.5 × 0.8–1.5 cm, with awn 0.2–1.1 cm long; petals: white, 0.8–1.5 × 0.4–0.8 cm; corona: 1 series, white with 4–7 violet to purple bands, filaments 0.8–2 cm long; ovary: densely tomentose, ellipsoid; fruit: obovate or widely ellipsoid, 4–6 × 3–5 cm, purple; seeds: obovate, ca. 0.5–0.6 × 0.4 cm, reticulate.

Passiflora affinis Engelm.

Boston J. Nat. Hist. 6: 233 (1850)
affinis: related
Supersection *Decaloba*, Section *Decaloba*

Passiflora affinis (Plate 77), indigenous to southern Texas and northeastern Mexico, is mostly found in river valleys. The specific epithet suggests a close relationship to *P. lutea*, with which it is often confused. The 2 species are easily distinguished by examining the undersides of the leaves: those of *P. affinis* feature dots of nectary glands, while those of *P. lutea* have no glands at all. The 2 species can be also be distinguished by the presence or absence of bracts on the flower stalk: *P. affinis* has 3 setaceous bracts, but bracts are absent in *P. lutea*. Remarkably, in *P. affinis* the apex of the outer corona filaments sports a more or less distinct knob. These knobs are missing in *P. lutea*. Euminid wasps are important pollinators of both species.

Passiflora lutea continues to advance further northward in the United States and has proven to

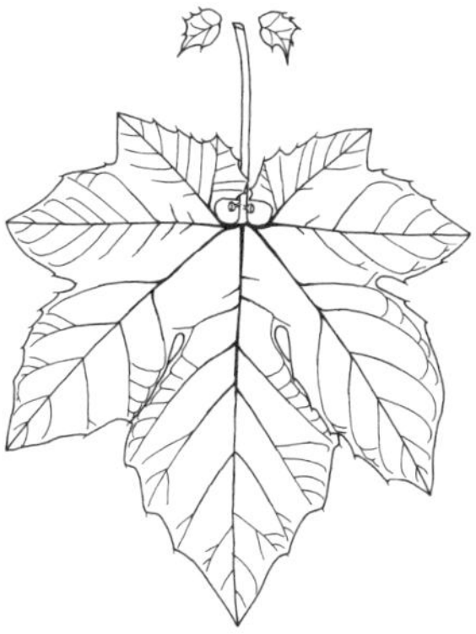

Passiflora adenopoda

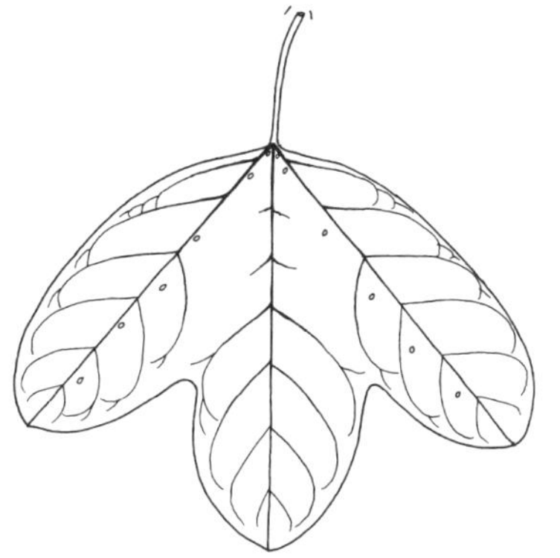

Passiflora affinis

be hardy to Zone 5. *Passiflora affinis*, however, needs higher temperatures, and this should be taken into account when cultivating it. Despite its very attractively lobed leaves, *P. affinis* species has been cultivated only occasionally. This may be due to its rather inconspicuous pale yellow flowers, which are relatively small at a diameter of 2–2.5 cm. In this it shares its lot with other species of subgenus *Decaloba*, all of which deserve to be cultivated more often. Min. temp.: 7°C.

Plant: glabrous throughout; stipules: setaceous, up to 0.2 cm long, falcate, purple; petioles: slender, 1–3.5 cm long without glands; leaves: 3-lobed, middle lobe slightly longer than lateral lobes, 2–10 × 3–14 cm, bearing laminar nectaries, slightly cordate to subtruncate, entire; peduncles: solitary or in pairs, 1–3 cm long; bracts: setaceous, 0.1–0.3 cm long; flowers: greenish yellow, 2–2.5 cm diam.; sepals: oblong-lanceolate, greenish yellow, 1–1.2 × 0.2–0.3 cm; petals: linear-lanceolate, greenish yellow, 0.6–0.8 × 0.1–0.2 cm; corona: 2 series, filaments of outer series filiform, 0.7–1 cm long, knobbed at apex, yellowish green with purplish brown base, filaments of inner series ca. 0.2 cm long; ovary: globose or ovoid, glabrous; fruit: subglobose, 0.8–1 × 0.8 cm, purplish black; seeds: 0.3 × 0.2 cm, transversely sulcate.

Passiflora allantophylla Mast.

Bot. Gaz. 16: 7 (1891)
allantophyllus: having sausage-shaped leaves
Supersection *Decaloba*, Section *Decaloba*

The weakly growing *Passiflora allantophylla* (Plate 78) is particularly suited as a trailing plant. This species produces numerous greenish flowers on pendent shoots. They can appear singly or in pairs; in rare instances the peduncle is forked, in which case 3 flowers may be open at a leaf axil. With a diameter of about 1 cm, the flowers are among the smallest of the genus.

The leaves, which are also very small, are shallowly 2-lobed and give the plant a very decorative effect. The undersides are purple-brown, and the upper sides feature yellowish nectar glands near the base of the leaves.

This species comes from the cool, moist oak forests of the mountains of Guatemala, Honduras, and El Salvador, at 1400–2000 m above sea level.

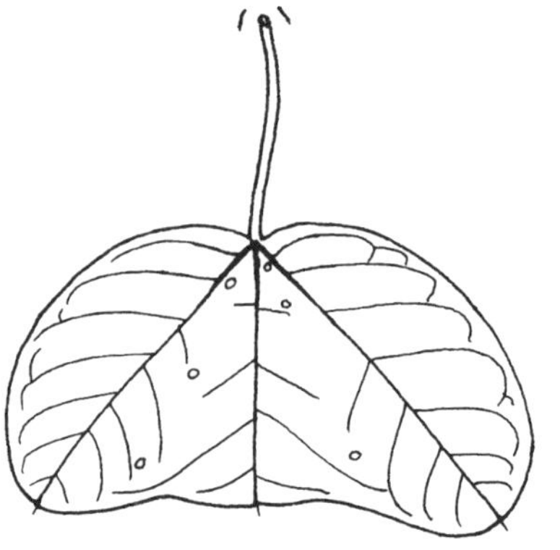

Passiflora allantophylla

When it is overwintered at temperatures around 15°C, it keeps flowering without problems. Otherwise it is content with lower temperatures. However, 7°C should be the lowest temperature.

Occasionally *Passiflora allantophylla* is confused with the closely related *P. ornithoura*. The latter differs considerably, however, by having significantly longer leaves that are 2-lobed for more than 1/2 of their length. *Passiflora ornithoura* also has slightly larger flowers, with petals 4–6 mm long.

Plant: glabrous throughout; stem: angulate to flattened, sulcate-striate; stipules: setaceous, ca. 0.2 cm long; petioles: slender, 0.5–1 cm without glands; leaves: 2-lobed or 3-lobed with short middle lobe, 1–4.2 × 1.5–7 cm, bearing 1–3 pairs of laminar nectaries, rounded at base, entire; peduncles: slender, solitary or in pairs, 0.5–1.5 cm long, occasionally 3-flowered, pedicels nearly as long as the peduncles; bracts: minute; flowers: greenish with purple and yellow corona, 1–1.2 cm diam.; sepals: greenish oblong, 0.5–0.8 × 0.2–0.4 cm; petals: greenish, oblong, 0.3–0.4 × 0.2 cm; corona: 1 (rarely 2) series, 0.2–0.4 cm long, clavate, purple-brown at base, yellow at apex, inner series (if present) less than 0.1 cm long; ovary: glabrous, subglobose; fruit: subglobose, 0.8–1.3 × 0.7–1 cm, blackish purple; seeds: ca. 0.3–0.4 × 0.2–0.3 cm, transversely sulcate.

Passiflora alnifolia Kunth

Nov. Gen. Sp. 2: 136 (1817)
alnifolius: alder-leafed
Synonym: *P. bauhinifolia* Kunth
Supersection *Decaloba*, Section *Decaloba*

This species (Plate 79) is not so common among enthusiasts. It originates from Ecuador, Colombia, and possibly Peru, where it is often found at elevations of 1500–3000 m. Since it grows side by side with tacsonias, it is recommended for cultivation outdoors in summer or for a cold greenhouse. Like many high-elevation species, *Passiflora alnifolia* ails during short, hot spells of more than 30°C. An infestation of spider mites and a premature drop of flower buds often follow.

Given day temperatures of 20°–25°C and nightly drops to 10°–15°C, this species will produce countless flowers of about 5 cm diam.

According to Holm-Nielsen and others (1988), *Passiflora alnifolia* and *P. bauhinifolia* may be regarded as the same species. There are few collections of the variant originally described as *P. bauhinifolia*, a form with unusual deeply, obtusely lobed leaves, and none are in cultivation. Min. temp.: 2°C.

Plant: puberulous to tomentose; stem: subtriangular, striate or rarely grooved; stipules: linear, 0.3– 0.7 cm long; petioles: 1–3.5 cm long without glands; leaves: variable, 2- or 3-lobed, 2.5–11 × 1.5–8.5 cm, bearing laminar nectaries, truncate or slightly cordate at base, puberulous to tomentose beneath, glabrous above, entire; peduncles: solitary or in pairs, 2–4.5 cm long, tomentose to villous; bracts: minute; flowers: white, 4–5 cm diam.; sepals: oblong-lanceolate, 1–2.4 × 0.3–0.9 cm, inside white or occasionally white with bluish violet base, outside brown, purple, or green; petals: oblong, 0.7–1.4 × 0.1–0.5 cm, white or occasionally white with bluish violet base; corona: 2 series, filaments of outer series 0.4–0.8 cm long, purple-brown with some lighter bands; ovary: subglobose, woolly; fruit: globose to ovoid, 1.3–2 × 1–2 cm, green.

Passiflora anfracta Mast. ex André

J. Linn. Soc. Bot. 20: 38 (1883)
anfractus: anfractuous, full of windings
Supersection *Decaloba*, Section *Decaloba*

Passiflora anfracta (Plate 80), endemic in the districts of Guayas and Los Ríos in the western lowlands of Ecuador, can hardly be distinguished from the closely related *P. vespertilio*. A comparison of the flowers and leaves of both species does not show any significant difference. Holm-Nielsen and others (1988) regard the cleft bracts as a distinguishing feature. However, it is questionable whether this criterion justifies the separation into 2 species. It seems necessary to check in further studies whether it makes sense to put *P. anfracta* and *P. vespertilio* together into one species. The geographical separation of the 2 native ranges by the Andes must be taken into account.

The warmth-loving *Passiflora anfracta* is recommended for cultivation under glass. Given a sheltered location, it can also spend the summer outdoors, although significantly fewer flowers will be produced. This species often ails when light intensity and temperatures are low during the overwintering period. Min. temp.: 10°C.

Plant: nearly glabrous or puberulous; stem: angulate to terete; stipules: linear, 0.2–0.5 cm long; petioles: 0.3–4.5 cm long without glands; leaves: 2- or 3-lobed, 1.4–10.3 × 5–18.3 cm, bearing laminar nectaries, coriaceous, entire; peduncles: solitary or in pairs, 0.5–1.6 cm long; bracts: linear-lanceolate, 0.2–0.3 cm long, cleft or occasionally entire; flowers: white, 3.5–4.5 cm diam.; sepals: white, ovate-oblong, 1.2–2 × 0.6–1 cm; petals: white, ovate-ob-

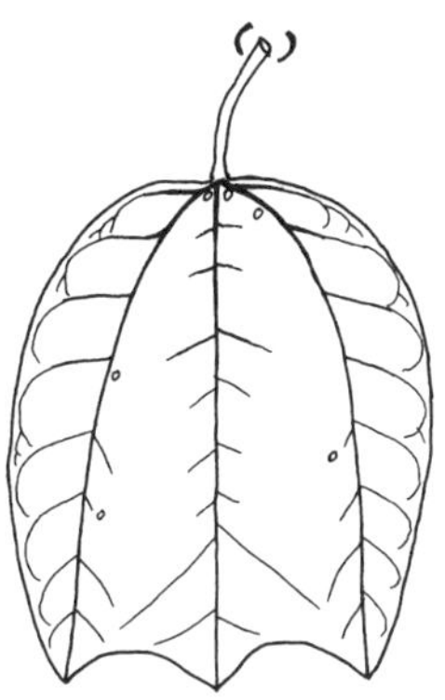

Passiflora alnifolia

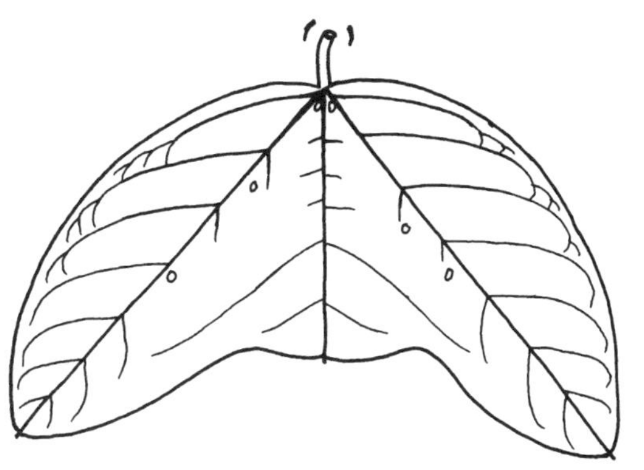

Passiflora anfracta

long, 1–1.7 × 0.3–0.7 cm; corona: 2 series or occasionally 1 series, white, filaments of outer series 0.9–1.3 cm long, filiform, curled at apex, filaments of inner series 0.4–0.6 cm long, filiform; ovary: ovoid, glabrous; fruit: globose, 2–2.5 × 2–2.4 cm, blackish blue; seeds: ovate, 0.3–0.4 × 0.2 cm, transversely rugulose.

Passiflora apetala Killip

J. Wash. Acad. Sci. 12: 255 (1922)
apetalus: lacking petals
Supersection *Decaloba*, Section *Decaloba*

Passiflora apetala (Plate 81) stems from Costa Rica and Panama, where it grows at elevations of 1300–2200 m. Despite its colorful leaves, it is not so often cultivated. The deeply cut, 2-lobed, decorative leaves are variegated light green along the main nerves on the upper side. Pale yellow to whitish green flowers appear in pairs and are rather small, 1.2–2 cm diam.

As the specific epithet suggests, the flowers of this plant do not generally have petals, although sometimes there are a few vestigial ones less than 3 mm long. Within subgenus *Decaloba* there are about 18 species in which the petals are also missing and hence have only 5 sepals. Among these are such well-known species as *Passiflora coriacea*, *P. gracilis*, and *P. suberosa*, which in addition have 2 glands on the petioles. *Passiflora apetala* and the somewhat similar *P. poeppigii* can easily be distinguished from these species, as they do not have glands on the petioles.

Passiflora apetala is a rather robust, vigorous species. It may be grown on a windowsill, outdoors, or in a greenhouse or conservatory. Min. temp.: 5°C.

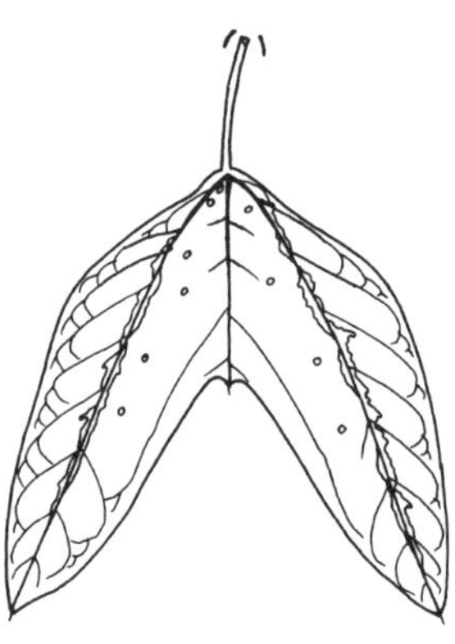

Passiflora apetala

Plant: glabrous throughout; stem: obtusely angled, striate; stipules: setaceous, 0.2–0.6 cm long; petioles: 1.5–3 cm without glands; leaves: 2-lobed, 3–9 × 2–11.5 cm, bearing laminar nectaries, cuneate to subrotund at base, adaxially with light green variegations along the main nerves, entire; peduncles: in pairs, rarely solitary, 1–3 cm long, slender; bracts: setaceous, 0.1–0.3 cm long; flowers: greenish white or pale yellow, 1.2–2 cm diam.; sepals: greenish white, oblong-lanceolate, 0.7–1.2 × 0.2–0.3 cm; petals: absent or rarely vestigial to 2.5 mm long; corona: 1 (rarely 2) series, white or light yellow, often dull purplish at base, filaments 0.2–0.5 cm long, filiform, inner series (if present) less than 2 mm long; ovary: subglobose, glabrous; fruit: subglobose, 0.7–1.5 cm diam., purplish black; seeds: broadly ovate, ca. 0.3 × 0.2 cm, transversely sulcate.

Passiflora aurantia G. Forster

Fl. Ins. Austr. 621 (1786)
aurantiacus: orange-red
Synonyms: *P. adiantifolia* Lawrence ex Ker-Gawl., *P. adiantum* Willd., *P. banksii* Benth., *P. baueriana* (Endl.) Mast., *P. brachystehanea* (F. v. Muell.) Benth., *P. coccinea* Banks & Sol. ex DC., *P. coerulescens* (Seem.) Harms, *P. glabra* Wendl., *P. samoensis* Exell, *P. storckii* (Seem.) Drake, *P. vitiensis* (Seem.) Mast.
Supersection *Disemma*, Section *Disemma*

Passiflora aurantia (Plate 82) can be distinguished from other representatives of subgenus *Decaloba* even at first sight, due to its characteristic flowers, which do not open completely. As the flower begins to open, the sepals and petals are cream-yellow, but they gradually darken to orange-red before the flower finally falls off. As flowers usually remain open for 3 or 4 days, blooms of different colors may appear on the same plant.

This species' native range reaches over southern and eastern Australia, Fiji, Malaysia, New Guinea, and other small islands in that area. Thus, *Passiflora aurantia* belongs to the small group of species from the Old World. The free-flowering and robust *P. aurantia* is a favorite both in Europe and in the

Passiflora aurantia

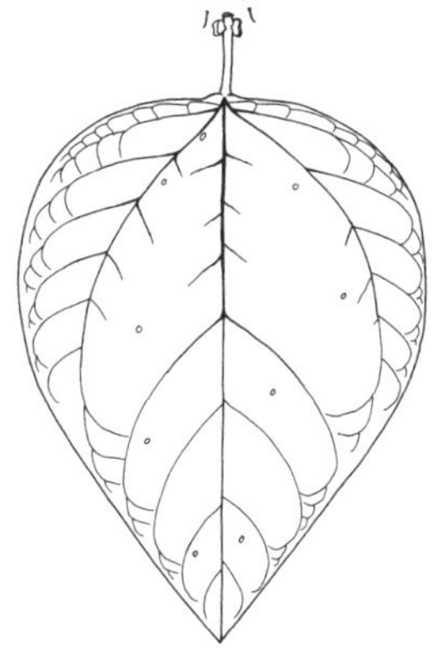

Passiflora auriculata

United States. It will flourish as well on a windowsill as it will outdoors or under glass. Min. temp.: 7°C.

Plant: glabrous throughout (except in forma *pubescens*); stem: subterete; stipules: minute, soon deciduous; petioles: 1–4 cm long, with 2 glands in the upper part, glands occasionally missing; leaves: 3-lobed to about 1/2 of their length, 2–9.5 × 2–10 cm, rounded or truncate at base, entire; peduncles: solitary or in pairs, 1–5 cm long; bracts: filiform, about 0.4 cm long, deciduous; flowers: cream-yellow at first, darkening to orange-red, 5–8 cm diam.; sepals: cream-yellow to orange-red, oblong-lanceolate, 2.5–5 × 1–1.5 cm, abaxially with fleshy keel; petals: cream to yellow to orange-red, oblong-lanceolate, 1.5–2.5 × 0.7–1 cm, membranous; corona: 1 series, red, filaments 1.2–2 cm long, filiform; ovary: subglobose to ellipsoid, glabrous; fruit: subglobose, slightly purple when ripe, 3–5.5 × 2–4.5 cm; seeds: reticulate.

Passiflora auriculata Kunth

Nov. Gen. Sp. 2: 131 (1817)
auriculatus: ear-shaped
Synonyms: *P. appendiculata* G. F. W. Mey., *P. cayaponioides* Rusby, *P. cinerea* Poepp. & Endl., *P. cryptopetala* Hoehne, *P. cyathophora* Ham., *P. kegeliana* Garcke, *P. rohrii* DC., *P. torta* Mast.
Supersection *Auriculata*

Passiflora auriculata (Plate 83) owes its name to the 2 characteristic ear-shaped nectar glands on its petioles. Its native range reaches from Nicaragua in the north over wide parts of Central and South America to Brazil and Bolivia. It advances from the coastal lowlands to the moderate elevation of 1400 m.

This variable species has not yet made its way into commercial cultivation, but it is represented in the collections of dyed-in-the-wool *Passiflora* and *Heliconius* enthusiasts. This is because although *P. auriculata* is a good foodplant for some species of butterfly caterpillars, its greenish yellow flowers are rather inconspicuous at only about 2 cm diam. Moreover, this plant needs a lot of space. Small plants in a pot rarely flower, so plant *P. auriculata* out in a spacious greenhouse or conservatory. Under these ideal conditions it can rapidly develop into an imposing plant with mass flowering—numerous flowers opening en masse in batches over a few days, with several weeks passing before the next batch appears. Min. temp.: 10°C.

Plant: glabrous to puberulent; stem: terete or angulate, striate; stipules: setaceous, ca. 0.1–0.2 cm long, deciduous; petioles: 1.4–4 cm long, with 2 auricular glands in lower part, glands 0.1–0.3 cm wide, often yellow; leaves: variable, unlobed to more or less 3-lobed, 5–15 × 2–12 cm, bearing laminar nectaries, truncate to cordate at base, entire; peduncles: solitary or in pairs, 0.6–2 cm long; bracts: minute; flowers: light greenish yellow to greenish white with a purple-brown or reddish center, 2–2.5 cm diam.; sepals: green to greenish white, oblong-lanceolate, 0.6–1.5 × 0.2–0.3 cm; petals: greenish white, linear, 0.4–1.1 × 0.1 cm; corona: 2 series, filaments of outer series greenish yellow with purple-brown base, filiform, undulate, 0.6–1.7 cm long, inner series green, 0.1–0.3 cm long; ovary: ovoid, puberulent; fruit: globose, black, 1–1.7 cm diam.; seeds: obovate, ca. 0.3–0.4 × 0.2–0.3 cm, transversely rugulose.

Passiflora bicornis Houston ex Mill.

Gard. Dict., 8th ed., Passiflora 13 (1768)
bicornis: 2-horned (in reference to the 2-lobed leaves)
Synonyms: *P. divaricata* Griseb., *P. pulchella* Kunth
Supersection *Decaloba*, Section *Decaloba*

Passiflora bicornis (Plate 84) is a 2-lobed *Decaloba* with large bracts, an unusual feature for the section. The bright yellow corona stands out strongly from the background of blue or violet in the flower, much like *Passiflora standleyi*, though it is stronger and larger in *P. bicornis*. Formerly widely known as *P. pulchella*, the correct name was discovered only recently. With a native range from the Isthmus of Tehuantepec, Mexico, to northern Venezuela, this is a lowland species mostly found at sea level to 900 m elevation. It inhabits sunny, often seasonally dry areas such as tropical dry forest, dry coastal forest, and beaches. Min. temp.: 10°C.

Plant: nearly glabrous throughout; stem: angulate, striate; stipules: linear-triangular, 0.3–1 cm long; petioles: slender, ca. 1–3 cm, without glands; leaves: 2-lobed or 3-lobed with short middle lobe, 2–8 × 3–10 cm, bearing laminar nectaries, cuneate to obtuse at base, lobes obtuse, rounded, or often truncate, entire; peduncles: solitary, rarely in pairs, 2.5–9 cm long, 1-flowered; bracts: foliaceous, 0.7–2.2 × 0.4–2 cm, olive to brownish, sometimes flushed with purple, lustrous, oily; flowers: pale blue to strong purplish blue or violet, with yellow corona, ca. 3–3.5 cm diam.; sepals: greenish outside, violet, bluish, or bluish purple inside, oblong-lanceolate, 1.7–2.8 × 0.7–1 cm; petals: violet to purplish blue, oblong-lanceolate, 0.8–1.5 × 0.5 cm; corona: ca. (3–)4–5 series, outer series 0.5–1.1 cm long, clavate and dilated to lobed at apex, yellow, often reddish purple at base, inner ca. 2–4 series 0.3–0.7 cm long, yellow, dilated at apex; ovary: glabrous, subglobose; fruit: subglobose 1.2–2.2 cm diam., purplish black; seeds: ca. 0.3–0.4 × 0.2 cm, transversely sulcate.

Passiflora biflora Lam.

Encycl. 3: 36 (1789)
biflorus: 2-flowered
Synonyms: *P. brighamii* S. Wats., *P. glabrata* Kunth, *P. lunata* J. E. Sm., *P. spathulata* Mast., *P. transversa* Mast.
Supersection *Decaloba*, Section *Decaloba*

Passiflora biflora (Plate 85) is undoubtedly among the most free-flowering, rewarding species of subgenus *Decaloba*. It will flower as a small, compact plant on a windowsill, or in a larger pot on a terrace, on a balcony, or under glass. The flower buds appear in pairs in the leaf axils. Sometimes, however, the flowers open one after the other. The whitish sepals and petals form a beautiful contrast with the corona, which is a rich yellow and markedly thickened in the upper 1/3.

In central Europe, potted *Passiflora biflora* starts to flower immediately after it is taken outdoors toward the end of May. When continuously provided with temperatures around 15°C, high atmospheric humidity, and good lighting conditions, it will flower year-round with only a few short breaks.

This plant's 2-lobed (rarely 3-lobed) leaves may have varying forms, which sometimes leads it to be confused with its near relatives *Passiflora cuneata*, *P. micropetala*, and *P. yucatanensis*. By comparing

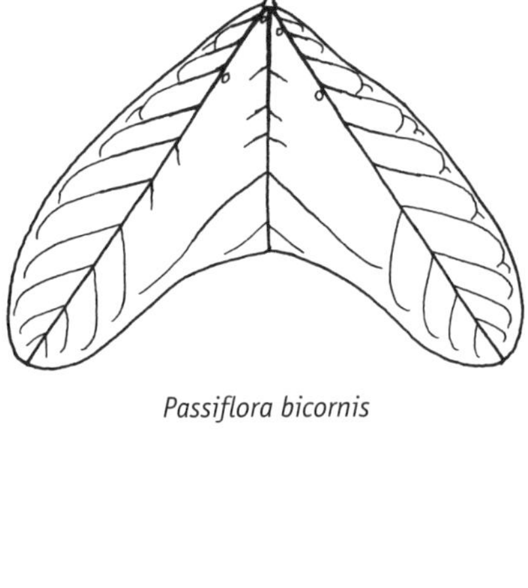

Passiflora bicornis

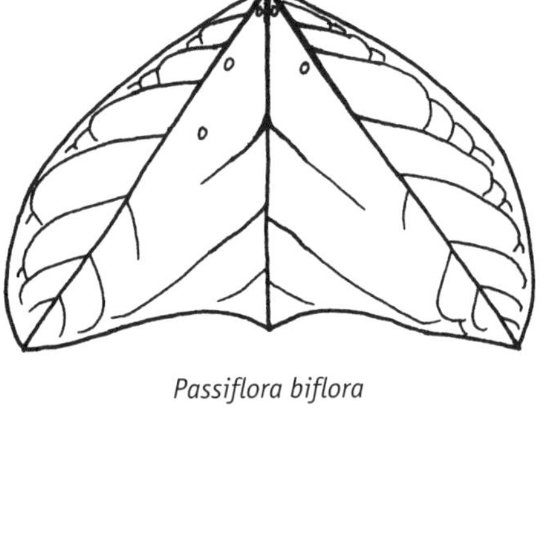

Passiflora biflora

the flowers, the differences become obvious. In *P. biflora*, the peduncle is usually jointed 0.6–1.2 cm below the flower, sometimes resulting in a joint below the middle. Admittedly, there are distinctive variations in this widespread species. A coastal race, for example, has more colorful flowers and leathery leaves with conspicuously elevated leaf glands.

The native range of *Passiflora biflora* reaches from Mexico over wide parts of Central America to the northeast of South America, where it proceeds to elevations of up to 1500 m. Min. temp.: 5°C.

Plant: glabrous or partly puberulent; stem: 5-angulate, strongly grooved; stipules: linear to setaceous, 0.1–0.4 cm long; petioles: 0.5–3 cm long without glands; leaves: quite variable, 2-lobed, occasionally with a short 3rd lobe at middle, 0.8–10 × 2.5–10 cm, bearing laminar nectaries, entire; peduncles: in pairs, 0.7–3 cm long; bracts: minute, necrescent, 0.1–0.2 cm long; flowers: white and yellow, 2.5–3.5 cm diam.; sepals: oblong-ovate, white, 1–1.7 × 0.5–0.7 cm; petals: oblong-ovate, white, 0.7–1.4 × 0.5 cm; corona: 2 series, filaments of outer series 3- to 4-angled, yellow, often purplish brown or dull reddish near base, dilated above middle, 0.7–0.9 cm long, inner filaments ca. 0.3 cm long, greenish; ovary: subglobose or ovoid, glabrous to tomentose; fruit: globose to ovoid, 2–3 × 1.5–2 cm, blackish blue; seeds: obovoid, ca. 0.3 × 0.2 cm, transversely sulcate.

Passiflora boenderi MacDougal

Novon 13 (4): 454–458 (2003)
named after Ronald Boender, founder of Passiflora Society International
Supersection *Decaloba*, Section *Decaloba*

On one of his many travels to Costa Rica, Ron Boender learned of a peculiar passionflower growing by the house of university student Andrés Vega. Together, Boender and Vega collected a 2nd specimen in its habitat. *Passiflora boenderi* (Plate 86) is very rare, found only at 740–800 m elevation at 2 sites in the Caribbean watershed of Costa Rica. Convinced that what they had found must be a new species, Boender passed the material on to John MacDougal, who had worked extensively on *Decaloba* species. Only after MacDougal saw living flowers did he finally agree that it was new. In fact, it corresponded with century-old herbarium material he had on hand, which had been insufficient for an unambiguous definition. In addition to the Passiflora Society International, Boender founded Butterfly World in Florida, and has great merit for preserving and providing knowledge about many *Heliconius* and other butterflies.

Passiflora boenderi produces greenish flowers that are about 2 cm wide. Its amazing 2-lobed leaves are far more attractive, however, and are the reason to cultivate the plant. On the upper side they are strongly variegated in a yellowish green along the 2 outer nerves, and on the underside they are purplish. Golden yellow leaf nectaries dot the upper surface. This beautiful species needs abundant moisture and is quite touchy, as Boender himself has explained: "Any duress, the plant will drop its leaves and pretty much be invisible." Patience and moisture will be repaid with new growth after some weeks. Min. temp.: 7°C.

Plant: nearly glabrous; stem: angulate; stipules: linear-triangular, 0.2–0.4 cm long; petioles: 1–4 cm long, without glands; leaves: 2-lobed about 1/3–1/2 of their length, occasionally with a short middle lobe, 4–17 × 4–9.5 cm, bearing orange-yellow laminar nectaries, cuneate to obtuse at base, variegated above, purplish beneath, entire; peduncles: slender, solitary or usually in pairs, 1–4 cm long; bracts: minute; flowers: light yellowish green with yellowish corona, 2 cm diam.; sepals: oblong-lanceolate, light yellowish green, 0.7–1.4 × 0.3–0.5 cm; petals: ovate-oblong, greenish white, 0.3–0.6 × 0.2–0.3 cm; corona: 2 series, filaments of outer series

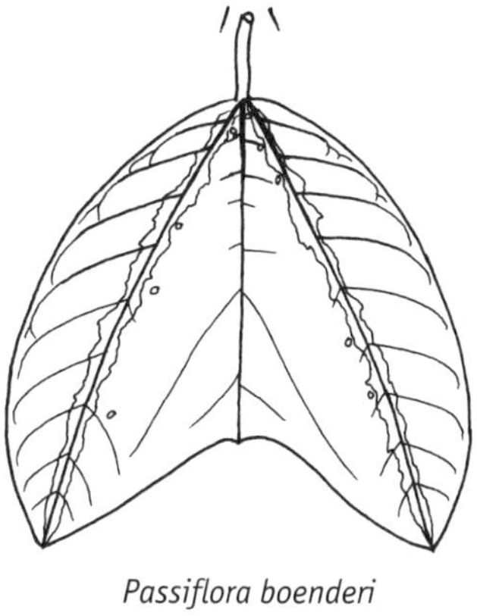

Passiflora boenderi

0.3–0.5 cm long, yellow, purple near base or lower 1/2 and sometimes also at apex, dilated above middle, filaments of inner series few, 0.1–0.2 cm long, purple with yellow at apex; ovary: subglobose, puberulent or rarely glabrous; fruit: subglobose, 1–2 cm diam., purplish black; seeds: ca. 0.4 × 0.3 cm, transversely sulcate.

Passiflora bryonioides Kunth

Nov. Gen. Sp. 2: 140 (1817)
bryonioides: similar to the genus *Bryonia*
Synonym: *P. inamoena* A. Gray
Supersection *Bryonioides*

Passiflora bryonioides (Plate 87), at home in southern Arizona and wide parts of dry Mexico, is among the few passionflowers that die back to the ground each year. It belongs to a small group of passionflowers that flourish in seasonally dry habitats at elevations of 1100–2500 m.

Unfortunately, this species is often confused with its close relatives, especially *Passiflora karwinskii* and *P. morifolia*. One distinctive feature is the leaves. While several laminar nectaries occur in *P. bryonioides* and *P. karwinskii*, these are almost always missing on the leaves of *P. morifolia*. *Passiflora bryonioides* and *P. karwinskii* can be distinguished by the former being evenly and finely pubescent under the leaf, while the latter is pubescent mostly or only on the veins under the leaf. The corona is significantly longer in *P. karwinskii* and curved at the tip. *Passiflora morifolia* differs from both by the pubescent ovary and purple fruit with red-orange arils.

Passiflora bryonioides is occasionally cultivated in the United States but is presently cultivated only rarely in Europe. It appears to require cross-pollination early in summer but tends to produce numerous edible fruits by self-pollination toward the end of the flowering period (MacDougal 1994). This species has few needs and can be successfully cultivated in pots or under glass. Corresponding to the conditions in its natural surroundings, the plant will tolerate a dry spell during overwintering and may be cut back hard just before the beginning of the next season's growth. Min. temp.: 7°C.

Plant: pubescent to hispidulous throughout; stem: subterete, striate; stipules: semiovate to sub-

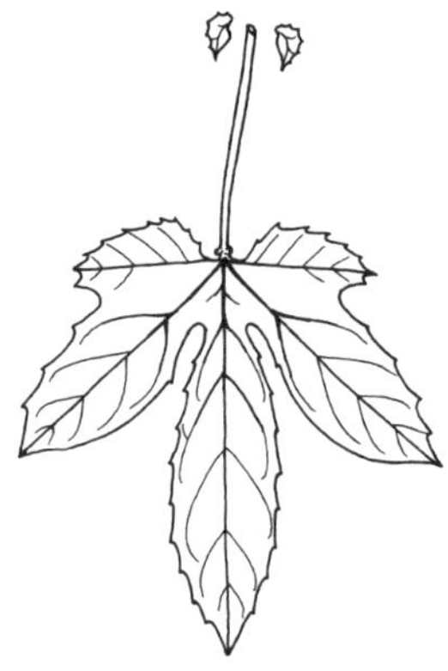

Passiflora bryonioides

reniform, 0.4–1.3 × 0.1–0.9 cm, more or less ciliate; petioles: 2–6.5 cm long, with 2 glands near apex, glands 0.1–0.2 × 0.1 cm; leaves: 5-lobed or occasionally 3-lobed, 2.5–10 × 4–11.5 cm, bearing 1–3 laminar nectaries per sinus, serrate to serrate-incised; peduncles: solitary, 1.5–3.5 cm long; bracts: linear-lanceolate, 0.2–0.7 cm long; flowers: whitish 2.5–3 cm diam.; sepals: ovate-lanceolate, 1.3–2 × 0.4–0.8 cm, greenish yellow to whitish; petals: oblong-lanceolate, 0.6–1.3 × 0.2–0.4 cm, white; corona: 1 (rarely 2) series, filaments of outer series filiform, 0.7–1.2 cm long, purple at base, white above, filaments of inner series (if present) less than 0.4 mm long; ovary: ellipsoid, glabrous; fruit: ovoid or ellipsoid, 3.5–4.5 × 2–2.5 cm, green with white speckles; seeds: ca. 0.4 × 0.2 cm, finely reticulate.

Passiflora capsularis L.

Sp. Pl. 957 (1753)
capsularis: having capsular fruit
Synonyms: *P. hassleriana* Chod., *P. paraguayensis* Chod., *P. piligera* Gardn., *P. pubescens* Kunth
Supersection *Decaloba*, Section *Xerogona*

Passiflora capsularis (Plate 88), widespread in many parts of Central and South America, is a robust, free-flowering species that will flourish in a pot on a windowsill, outdoors, or under glass. Its white flowers reach a diameter of up to 4 cm. The flowers of the widespread 'Vanilla Cream' effuse a pleasant vanilla-like scent. The odor is most intense in the first hours after the flower has opened and lasts until midday.

Some clones of *Passiflora capsularis* are self-compatible, so the reddish brown fruit might be ex-

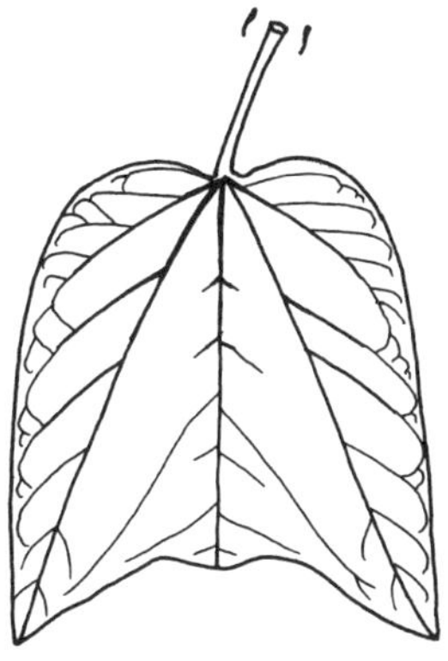

Passiflora capsularis

pected even indoors. The fruit is ribbed and has the form of a star in cross section. All species of section *Xerogona*, including such well-known species as *P. citrina*, *P. sanguinolenta*, and *P. rubra*, have these capsular fruits in common, although the fruits are not recommended for consumption. Sometimes it is difficult to distinguish between *P. capsularis* and *P. rubra*. A reliable distinguishing feature is the color of the corona. While it is generally pink at the base in *P. rubra*, the corona of *P. capsularis* is pure white in clones from Central America and in nearly all forms from South America. According to Holm-Nielsen and others (1988), the pubescence of the ovary and the fruit are further important distinguishing features. Min. temp.: 5°C.

Plant: pubescent; stem: 3- or 5-angulate, striate; stipules: setaceous, 0.2–0.7 cm long, more or less falcate; petioles: 1–6 cm long without glands; leaves: 2-lobed or occasionally with short middle lobe, 3–10 × 2.5–16 cm, laminar nectaries absent, cordate at base, entire; peduncles: solitary or occasionally in pairs, slender, 1–3 cm long; bracts: absent; flowers: white, 2.5–4 cm diam.; sepals: oblong-lanceolate, white, 1.3–2.4 × 0.3–0.5 cm; petals: oblong-lanceolate, white, 0.8–1.6 × 0.1–0.4 cm; corona: 1 (rarely 2) series, filaments of outer series filiform, 1–1.6 cm long, white, very rarely with pink base, filaments of inner series (if present) few; ovary: ellipsoid, glabrous to minutely puberulent; fruit: ellipsoid to fusiform, 3–7 × 1–2.5 cm, reddish brown, dark purplish red, or red, 6-keeled, glabrous, dehiscent; seeds: ovate, 0.3–0.4 × 0.1–0.2 cm, transversely grooved.

Passiflora chrysosepala Schwerdtfeger

Haussknechtia 6: 41–46 (1997)
chrysosepalus: having golden yellow sepals
Supersection *Decaloba*, Section *Decaloba*

Thanks to Giselher Behnken and Michael Schwerdtfeger, quite a number of the most beautiful passionflower species indigenous to Ecuador are now cultivated in Europe. Unfortunately, Behnken and Schwerdtfeger were unable to succeed with *Passiflora chrysosepala* (Plate 89). It is hoped that one day this desirable species will be brought into cultivation.

With a diameter of 6–7 cm, its fragrant flowers reach a considerable size for a species of subgenus *Decaloba*. The white sepals are yellowish brown on the reverse side and give the buds a golden tinge. Its light green leaves are 3-lobed at the tip and similar to those of the closely related *Passiflora alnifolia*, but are overall significantly larger. In accordance with the conditions of its natural surroundings, *P. chrysosepala* might prove to flourish under cultivation without problems.

Plant: subglabrous to minutely pubescent; stem: angulate, striate; stipules: setaceous, 0.5–0.8 cm long, falcate; petioles: up to 5 cm long without glands; leaves: 2- or 3-lobed with short middle lobe, 8–17 × 6–12 cm, truncate to cordate at base, bearing 4–8 big laminar nectaries near base and some less conspicuous nectaries in the distal part, entire; peduncles: solitary or in pairs, 5–9 cm long; bracts: setaceous, ca. 0.4 cm long, scattered above the middle of peduncle; flowers: white, 6–7 cm diam.; sepals: ovate-oblong, white inside, golden brown-

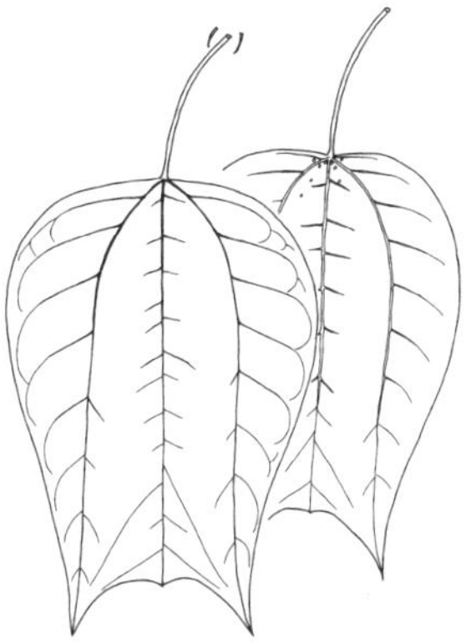

Passiflora chrysosepala

ish outside, 2.8–3.2 × 1 cm, cupped; petals: white, subequal to the sepals; corona: 2 series, filaments of outer series ca. 1 cm long, dark bluish purple with a narrow green base and a dilated green or white tip, filaments of inner series ca. 0.4 cm long, greenish white; ovary: nearly glabrous; fruit: subglobose, ca. 4.5 cm diam., bright yellow; seeds: unknown.

Passiflora cinnabarina Lindley

Gard. Chron. 1855: 724 (1855)
cinnabarinus: vermilion
Synonyms: *P. banksii* Benth. sensu Mast., *P. muelleriana* (Reg.) Mast.
Supersection *Disemma*, Section *Disemma*

Originating in the subtropical areas of southeastern Australia, where it grows near the coastal regions of New South Wales and Victoria, *Passiflora cinnabarina* (Plate 90) can be cultivated with success outdoors during summer in wide parts of Europe and the United States. In a sunny location it will produce its first flowers about 4 weeks after being taken outdoors. The flowers are about 6 cm diam. and feature the particularly attractive and rare combination of red sepals and petals with a yellow, upright corona. When cultivated in a greenhouse or conservatory, this species flowers nearly year-round.

Compared with other passionflower species, propagation by cuttings turns out to be very difficult with *Passiflora cinnabarina*. Though the cuttings callus in 1–2 weeks, root growth often stagnates for several weeks.

Passiflora cinnabarina can be easily propagated from the fresh seeds of the regularly produced but not very edible fruit. Seeds are adapted to the extensive fires that often break out in Australia, and so germination is positively influenced by the short-time application of heat. Min. temp.: 7°C.

Plant: glabrous; stem: terete, slender; stipules: linear-subulate, 0.3–0.8 cm long, curved; petioles: 1.5–6 cm long without glands; leaves: 3-lobed for about 2/3 of their length, 4–12 × 6–13 cm, glabrous, truncate to cordate at base, entire; peduncles: solitary, 1.5–3 cm long; bracts: linear, near middle of peduncle; flowers: red, 4.5–7 cm diam.; sepals: oblong-lanceolate, red inside, greenish red outside, 2.5–3.5 × 0.4–0.7 cm, slightly keeled outside; petals: lanceolate, red, 1–1.5 × 0.3–0.4 cm, membranous; corona: 1 series, yellow, filiform, 0.8–1 cm long; ovary: ellipsoidal, glabrous; fruit: subglobose to ellipsoid, 2.5–4 × 2–3.7 cm, grayish green; seeds: reticulate.

Passiflora citrina MacDougal

Ann. Missouri Bot. Gard. 76: 354 (1989)
citrinus: lemon-yellow
Supersection *Decaloba*, Section *Xerogona*

Passiflora citrina (Plates 75 and 91) may well be among the most rewarding, most free-flowering passionflowers, and it is especially attractive for its yellow blossoms, a rarity among passionflowers. For these reasons it is represented in many collections and is regularly offered by flower shops and garden centers. It will flower in a small pot and so is suited for indoor cultivation. It will also make a good impression in a container outdoors. The

Passiflora cinnabarina

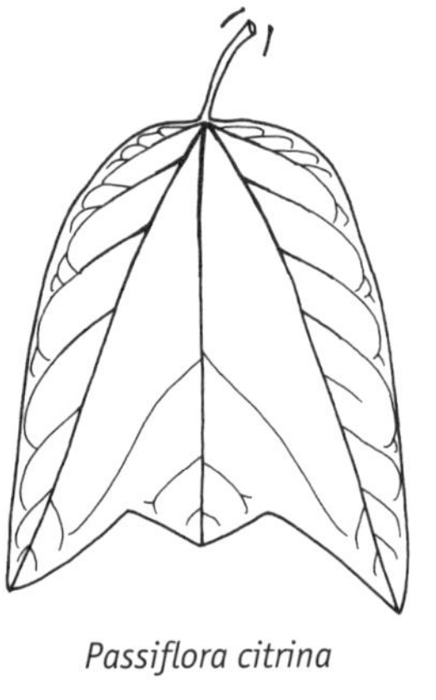

Passiflora citrina

lemon-yellow flowers, 3.5–5.5 cm diam., appear early in summer. Until it is taken to its winter quarters in fall, *P. citrina* remains a flowering eye-catcher. When cultivated under glass it produces flowers year-round with only short interruptions.

The shape and size of the flowers are similar to those of *Passiflora sanguinolenta*, its close relative. A salmon-colored hybrid of the 2 species was brought onto the market under the name *P.* 'Adularia' by John Vanderplank in 1992.

The native range of *Passiflora citrina* stretches from eastern Guatemala to central Honduras, from where it was introduced to the trade by John MacDougal in 1989. In habitat it grows in moist pinewoods at elevations of 600–1500 m. Min. temp.: 5°C.

Plant: pubescent; stem: ca. 5-angular, striate; stipules: setaceous, 0.4–1 × 0.1 cm, falcate; petioles: 0.3–1.4 cm occasionally to 2 cm long, without glands; leaves: variable, 2-lobed or occasionally 3-lobed, 2–12 × 1.5–10 cm, laminar nectaries absent, rounded to truncate at base, occasionally with light green variegation along main veins, entire; peduncles: solitary or occasionally in pairs, 0.5–4.5 cm long; bracts: absent; flowers: yellow, 3.5–5.5 cm diam.; floral tube: 0.5–1.3 cm long, 0.5–0.9 cm diam., yellow; sepals: narrowly oblong-lanceolate, yellow, 2–3 × 0.3–0.5 cm; petals: narrowly oblong-lanceolate, yellow, 1.9–2.8 × 0.2–0.3 cm; corona: 1 (rarely 2) series, outer series pale yellow with yellow apex, erect, 0.9–1.4 cm long, inner series (if present) hidden in tube, less than 2 mm long; ovary: obovoid-ellipsoid, densely pubescent; fruit: ellipsoid, 6-keeled, reddish, 2.5–4.5 × 1.2–2.5 cm, dehiscent; seeds: 0.3–0.4 × 0.2 cm, transversely grooved.

Passiflora cobanensis Killip

J. Wash. Acad. Sci. 14: 111 (1924)
named after Cobán, Guatemala, near the type location
Supersection *Decaloba*, Section *Xerogona*

Passiflora cobanensis (Plate 92) is another previously little-known representative of section *Xerogona*. It has one peculiarity in common with its sister, the doubtfully distinct *P. brevipes*: contrary to the other species of this section, the leaves of both are unlobed. This facilitates a quick distinction.

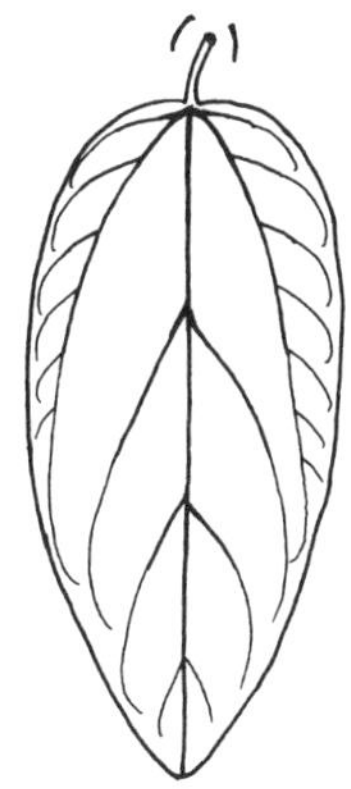

Passiflora cobanensis

Killip (1938) only knew this species from its type location between Chamá and Cobán in northern Guatemala, at about 950 m. It is now known to be rather common farther to the north, on calcareous soils in the foothills of the Atlantic drainage of Chiapas, Mexico, up to 1700 m. Its greenish flowers reach a diameter of about 2 cm. The short corona with 2 series is purplish, and yellowish at the apex. Comparable flowers can be found in *Passiflora conzattiana*, which also belongs to section *Xerogona*. Specimens of *P. cobanensis* can be found in the collections of only a few specialists. Its demands can be compared with those of *P. citrina*, *P. capsularis*, and *P. conzattiana*, which are cultivated more often. Min. temp.: 7°C.

Plant: pubescent; stem: 3- to 5-angulate, slender; stipules: lanceolate, 0.4–1 × 0.1–0.5 cm; petioles: 0.8–1 cm long without glands; leaves: unlobed, ovate-lanceolate, 6–14 × 3–7 cm, laminar nectaries absent, rounded or slightly cordate at base, glabrous above, pubescent beneath, entire; peduncles: usually in pairs, 0.5–1.5 cm long; bracts: absent; flowers: greenish, 2–2.5 cm diam.; sepals: lanceolate, greenish, 1–1.3 × 0.2–0.3 cm; petals: lanceolate, greenish, 0.5–0.7 × 0.2–0.3 cm; corona: 2 series, outer series purplish brown and yellowish at the apex, filaments linear-clavate, 0.3–0.5, inner series 0.1–0.2 cm long; ovary: obovoid to slightly angular, densely tomentose; fruit: obovoid or ellipsoid, 6-keeled, 2–5 × 1.5–2 cm, reddish; seeds: ca. 0.4 × 0.2 cm, transversely grooved.

Passiflora colimensis Mast. & Rose

Contr. U.S. Natl. Herb. 5: 181, pl. 20 (1899)
named after Colima, Mexico, the type location
Supersection *Bryonioides*

Passiflora colimensis (Plate 93) is another representative from supersection *Bryonioides*, revised by John MacDougal in 1994. (Similar revisions are needed for many more supersections of genus *Passiflora*). This species is native in southwestern Mexico, where it grows mainly in the interior regions of Colima, Michoacán, and Guerrero in more or less dry forests at elevations of 400–1100 m. It is distinguished from the closely related *P. bryonioides* and *P. morifolia* by its shallowly dentate leaves and by its stems, which are nearly hairless between the nodes. The center lobe of the leaf is not narrowed at the base in *P. colimensis.*

Like most species from supersection *Bryonioides*, *Passiflora colimensis* deserves to be cultivated. Like many species in this group, it is adapted to seasonally dry areas and has more or less annual stems from perennial roots. Its sweetly scented flowers can reach a diameter of up to 4 cm. Min. temp.: 7°C.

Plant: nearly glabrous except for hooked trichomes at nodes; stem: angulate; stipules: ovate-auriculate to widely falcate, 0.3–1 × 0.1–0.6 cm, entire or with 1–3 teeth; petioles: 2.5–11 cm long, with 2 glands at the apex, glands 0.1–0.3 × 0.1–0.2 cm; leaves: 3-lobed or occasionally 5-lobed, 4–12 × 4.5–17 cm, laminar nectaries usually absent, rarely 1 per sinus, cordate at base, denticulate at margin; peduncles: solitary, 3–5.5 cm long; bracts: linear-lanceolate, 0.2–0.6 × 0.1 cm; flowers: white, 3.5–4 cm diam.; sepals: more or less triangular, white, 1.3–2.2 × 0.6–1.2 cm; petals: oblong-ovate, white, 0.8–1.2 × 0.4–0.7 cm; corona: 1 series, very rarely some vestigial filaments of a 2nd series seen, filaments 1–1.3 cm long, white with 2 purplish bands in lower 1/2 or these merged into one wide band; ovary: ellipsoid, glabrous; fruit: ellipsoid to ovoid, 6–8 × 2.5–3.5 cm, yellowish green with white speckles; seeds: widely obovate, 0.4–0.5 × 0.3 cm, reticulate.

Passiflora conzattiana Killip

J. Wash. Acad. Sci. 17: 425 (1927)
named after Cassiano Conzatti, a Mexican botanist
Supersection *Decaloba*, Section *Xerogona*

The dainty *Passiflora conzattiana* (Plate 94) is at home in the wet montane or cloud forests of east-central Mexico. Introduced by John MacDougal in 1980, it is cultivated only rarely, since its flowers, about 2 cm diam., are not among the most conspicuous of the genus. It is exactly this unpretentiousness that makes it so charming in comparison with other more sumptuous representatives of subgenus *Passiflora*. *Passiflora conzattiana* is adapted to shade and produces abundant, attractive foliage.

This passionflower thrives as well outdoors as in a greenhouse or on a windowsill. Since it sometimes tends to run along the ground, it makes a good impression as a ground cover or trailing plant. In an ideal location it will flower year-round. Otherwise, numerous flowers with greenish white sepals and petals appear throughout summer. The short corona is deep purple, almost brown at the base, and yellowish in the upper 1/3. Its flowers

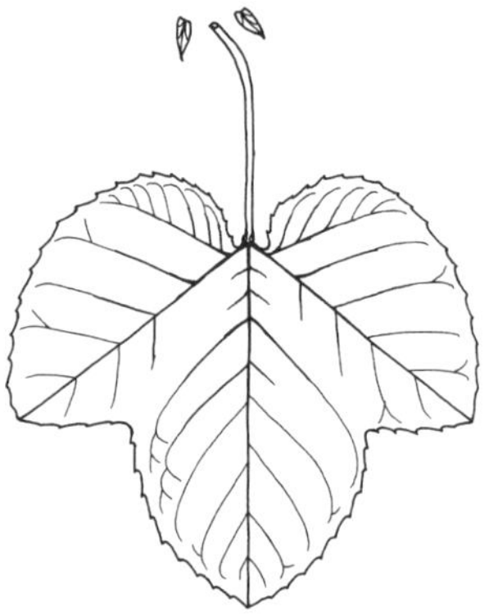

Passiflora colimensis

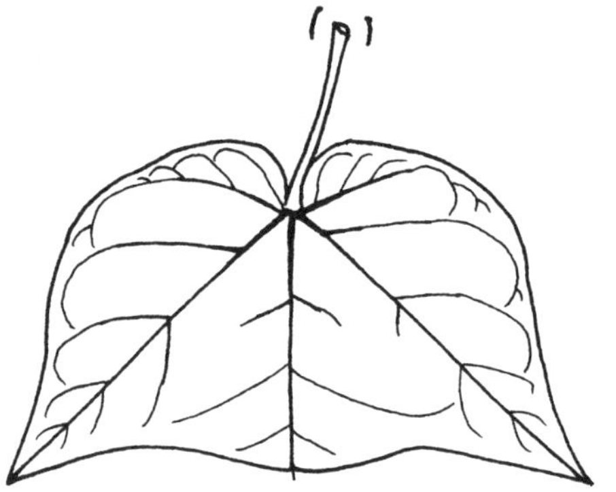

Passiflora conzattiana

look very similar to the closely related *P. pusilla*, *P. cobanensis*, and *P. goniosperma*. These can be distinguished from one another by the form of their leaves and seeds and by the number of coronal series.

The stiff 2- or 3-lobed leaves of *Passiflora conzattiana* are pubescent, as is the stem, which becomes brown with age. Min. temp.: 5°C.

Plant: pubescent; stem: angled, slender, turning reddish brown; stipules: linear-lanceolate, ca. 0.4 cm long; petioles: 0.8–2 cm long without glands; leaves: 2-lobed or with a small 3rd lobe in the middle, 2–5 × 2.5–8 cm, laminar nectaries absent, cordate at base, entire; peduncles: solitary or in pairs, slender, ca. 2 cm long; bracts: absent; flowers: yellowish green, 1–2 cm diam.; sepals: lanceolate, yellowish green, 0.8–1.2 × 0.2–0.4 cm; petals: lanceolate, yellowish green, 0.4–0.6 × 0.2 cm; corona: 1 series, filaments relatively few, 0.3–0.4 cm long, purplish brown in the lower 2/3, yellowish in the upper 1/3; ovary: ellipsoid to ovoid, densely pubescent to tomentulose; fruit: fusiform to ellipsoid, nearly 5 × 1 cm, 6-keeled, dark purplish red, dehiscent; seeds: 0.2 × 0.2 cm, transversely grooved.

Passiflora coriacea Juss.

Ann. Mus. Natl. Hist. Nat. 6: 109, t. 39, fig. 2 (1805)
coriaceus: leathery (in reference to the leaves)
Synonyms: *P. clypeata* J. E. Sm., *P. difformis* Kunth
Supersection *Cieca*

Passiflora coriacea (Plate 95) is cultivated mainly for its decorative leaves, which are unlobed, leathery, and much wider than long. On the upper side they are often irregularly splotched in pale green or yellowish green. This plant, found at elevations up to 2000 m in many parts of Central and South America, is suited for a greenhouse and can also be cultivated on a windowsill when young. When cultivated outdoors it should be given a sunny location protected from too much rain.

The flowers of this robust species appear singly or in pairs in the leaf axils. *Passiflora coriacea* often produces leafless pseudoracemes with many flowers at the ends of the shoots. A single flower reaches 2.5–3.5 cm diam. The absence of petals is characteristic of supersection *Cieca*. There are more than one species among the plants now called *P. coriacea*, according to recent, as-yet-unpublished studies by Kristen Porter-Utley (2003). Plate 95 shows the commonly cultivated form native to Mexico and northern Central America, probably better called *P. sexocellata*. The original *P. coriacea* is from South America; the plants from Costa Rica and Panama will receive a new name. Min. temp.: 5°C.

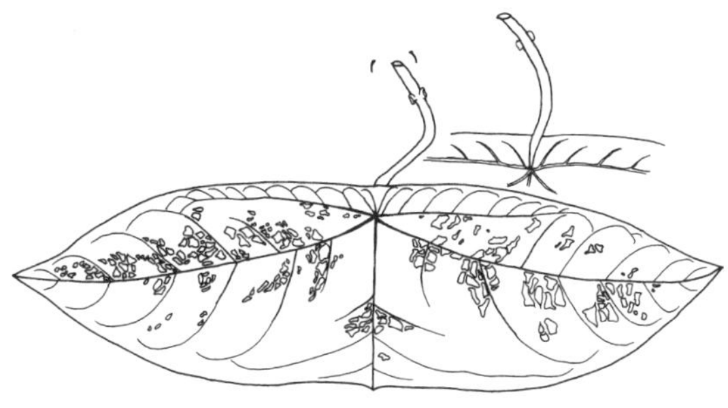

Passiflora coriacea

Plant: glabrous with exception of stem and petioles; stem: angulate; stipules: narrowly linear, 0.3–0.6 cm long; petioles: 2–4.5 cm long, with 2 (or extremely rarely 4) sessile or subsessile glands near or below the middle; leaves: 2- or obscurely 3-lobed, peltate, 2.5–7 × 7–30 cm, bearing laminar nectaries, coriaceous, entire; peduncles: solitary or in pairs, 0.3–2 cm long, or in a terminal reduced raceme 4–25 cm long; bracts: absent; flowers: yellowish green or (in South America) cream, 2.5–3.5 cm diam.; sepals: oblong-lanceolate, yellowish green or cream, 0.7–1.5 × 0.2–0.5 cm; petals: absent; corona: 2 series, filaments of outer series filiform, 0.6–0.8 cm long, purplish brown at base and yellowish above or (in South America) whitish with violet basally, inner series ca. 0.2 cm long; ovary: ovoid, glabrous; fruit: globose, 1–2 cm diam., bluish black; seeds: obovate, curved, ca. 0.4 × 0.2 cm, brown, reticulate.

Passiflora costaricensis Killip

J. Wash. Acad. Sci. 12: 257 (1922)
costaricensis: originating from Costa Rica
Supersection *Decaloba*, Section *Xerogona*

The native range of *Passiflora costaricensis* (Plate 96) reaches from near sea level in Mexico through wide parts of Central America to Colombia and Ecuador in the northeast of South America. Its white flow-

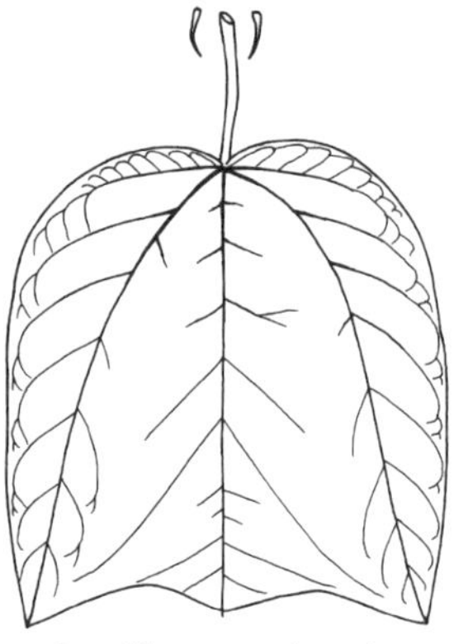

Passiflora costaricensis

ers, which reach a diameter of 4–5 cm, resemble those of the very closely related *P. rubra*. However, the clone distributed in Europe has turned out to be rather unwilling to flower, which could explain why the species is cultivated only rarely. Regardless, we recommend this species on account of its leaves and thick triangular stem. The attractive leaves, acutely 2-lobed at the tip, are relatively large and conspicuously hirsute. This makes the plant easily distinguishable from other representatives of section *Xerogona*, such as *P. rubra* and *P. escobariana*.

Passiflora costaricensis develops into an imposing plant that is better suited for a conservatory or greenhouse than for containers left outdoors during summer. It becomes too large for cultivation on a windowsill. This plant prefers a warm, protected location and will suffer when temperatures dip too low in winter. Min. temp.: 10°C.

Plant: densely hirsute; stem: sharply 3-angulate; stipules: linear, 0.5–1.2 cm long; petioles: 1.5–4 cm long, without glands; leaves: 2-lobed near apex of lamina, 9–16 × 7–15 cm, laminar nectaries absent, rounded to cordate at base, hirsute especially beneath, entire; peduncles: solitary or in pairs, 0.9–2 cm long; bracts: absent; flowers: white, 4–5 cm diam.; sepals: lanceolate, white inside and sometimes flushed purple, green and hirsute outside, 1.5–2.5 × 0.4–0.6 cm; petals: lanceolate, white, 1–1.8 × 0.2–0.4 cm; corona: 1 or 2 series, outer filaments narrowly ligulate, 1–1.5 cm long, usually pink or purple at base and cream-white above, inner series (if present) 0.1–0.5 cm; ovary: ellipsoid, minutely puberulent; fruit: fusiform-ellipsoid, 5.5–11 × 1.5–2.5 cm, 6-keeled, wine-red, or red on ridges, cream between, dehiscent; seeds: narrowly ovate, 0.3–0.4 × 0.1–0.2 cm, transversely grooved.

Passiflora cuneata Willd.

Enum. Hort. 696 (1809)
cuneatus: wedge-shaped (in reference to the base of the leaf)
Synonyms: *P. bifurca* Triana & Planch., *P. flexicaulis* Killip, *P. furcata* Mast.
Supersection *Decaloba*, Section *Decaloba*

Native to the Venezuelan and Colombian mountains, this species (Plate 97) is suited for any well-lit location. When cultivated outdoors in a container it produces numerous flowers, 3–4 cm diam., throughout summer. Under glass the flowering season is even longer, by several months. Under favorable conditions *Passiflora cuneata* may well flower in a greenhouse year-round. The sepals and petals of the species are usually white, but those of *P. cuneata* 'Miguel Molinari', described by John Vanderplank, are increasingly purple toward the middle. This variety was initially introduced under the tentative name *P. lourdesae*, which was never validly published. The 2-lobed, light green leaves vary somewhat in form. The 2 lobes can be more or less distinct. Along the outer main nerves there are conspicuous yellowish laminar nectaries that act as egg mimics.

Overall, *Passiflora cuneata* is a recommendable, free-flowering representative of subgenus *Decaloba*. It is represented in more and more collections. Min. temp.: 5°C.

Plant: nearly glabrous; stem: angulate; stipules:

Passiflora cuneata

linear or setaceous, up to 0.6 cm long; petioles: slender, 1.5–3 cm long, without glands; leaves: variable, 2-lobed for 1/5–2/3 of their length, occasionally with an additional middle lobe equaling or shorter than lateral lobes, 4–9 × 3.5–6 cm, bearing yellow laminar nectaries, cuneate or rounded at base, hirsute especially beneath, entire; peduncles: solitary or in pairs, slender, up to 5 cm long; bracts: minute; flowers: white, occasionally purplish white, 3.5–4 cm diam.; sepals: lanceolate, white inside, green outside, 1–1.6 × 0.3–0.5 cm; petals: oblong-lanceolate, white, 0.5–0.7 × 0.3–0.4 cm; corona: 2 series, filaments of outer series 0.3–0.6 cm long, dilated above middle, cultrate or spatulate, yellow or yellow-green with or without purple bands; ovary: subglobose, glabrous or pilose; fruit: globose, ca. 1.5 cm diam.; seeds: ovoid, up to 0.3 cm long, transversely sulcate.

Passiflora cupraea L.

Sp. Pl. 955 (1753)
cupreatus: copper-colored
Synonym: *P. cavanillesii* DC.
Supersection *Decaloba*, Section *Decaloba*

Passiflora cupraea (Plate 98), indigenous to the coastal regions of eastern Cuba, the Bahamas, and Haiti, has entered the collections of many enthusiasts in recent years. Its reddish or reddish brown to copper-colored flowers reach a diameter of about 5 cm. The long, narrow sepals and petals bend outward and sometimes twist, giving a rather asymmetric impression. *Passiflora cupraea* was once incorporated in subgenus *Pseudomurucuja*, along with the closely related *P. perfoliata* and other species. In accordance with the systematics by Feuillet and MacDougal, it is now assigned to the species-rich subgenus *Decaloba*. Elma Kay recently completed a study of this group of West Indian red-flowered decalobas.

Passiflora cupraea

Because of its flower structure and atypical unlobed leaves, *Passiflora cupraea* can hardly be confused with other species under cultivation. On the whole, it is a comparatively dainty species. Because it is warmth-loving, this species is particularly suited for a greenhouse or conservatory, where it will flower reliably. Min. temp.: 10°C.

Plant: glabrous; stem: angulate, striate; stipules: setaceous, up to 0.3 cm long, soon deciduous; petioles: slender, 0.5–1.3 cm long, without glands; leaves: unlobed, ovate to oblong, 2.5–7 × 1.5–5 cm, occasionally with some laminar nectaries, more or less rounded at base, entire; peduncles: solitary or in pairs, slender, up to 2.5 cm long; bracts: minute, soon deciduous; flowers: reddish or red-brown, 4.5–5.5 cm diam.; sepals: linear-oblong, red-brown, greenish outside, 1.5–2.5 × 0.3–0.4 cm; petals: linear-oblong, red-brown, 1–1.8 × 0.2–0.3 cm; corona: 1 series, filaments 0.3–0.5 cm long and up to 0.1 cm wide, purplish brown and orange at the apex; ovary: ovoid, glabrous; fruit: globose, blackish, ca. 1 cm diam.; seeds: nearly cuneate, 0.4 × 0.2 cm, ovoid, transversely sulcate.

Passiflora dioscoreifolia Killip

J. Wash. Acad. Sci. 14: 108 (1924)
dioscoreifolius: having leaves like *Dioscorea*
Supersection *Bryonioides*

Passiflora dioscoreifolia (Plates 99 and 100) is remarkable in several respects. Of particular interest are its flowers, which reach a diameter of 5–6 cm. At the tip of its laterally inclined androgynophore, it bears 4 stigmas instead of the usual 3. These are not distributed evenly around the ovary. Three hang down together with the zygomorphic androecium, and one reaches the opposite direction. A comparable flower structure can only be found in the sister species *P. lobata*, which, however, can immediately be distinguished from *P. dioscoreifolia* by its 3-lobed leaves with laminar nectaries. *Passiflora dioscoreifolia*, endemic in the lower montane

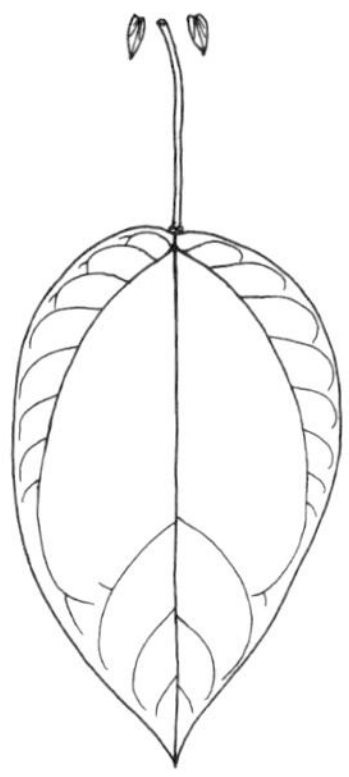

Passiflora dioscoreifolia

wet forests and cloud forests of Costa Rica, grows at elevations of 1100–1600 m. In habitat it flowers and bears fruit year-round.

Cultivation of this unusual species is restricted to a small number of enthusiasts, mainly in the United States. It is hoped that in the future more collectors will be able to enjoy it. Min. temp.: 7°C.

Plant: glabrous except for hooked trichomes on petioles, top of leaves, peduncles, and basal part of sepals; stem: subterete, sulcate; stipules: semiovate, 0.7–1.7 × 0.3–0.8 cm; petioles: 2–5 cm long, with 2 glands at apex of petiole; leaves: unlobed, ovate to narrowly ovate, 7–18 × 3.5–8 cm, laminar nectaries absent, cordate at base, sparsely pubescent above with hooked trichomes, glabrous beneath, entire; peduncles: solitary or in pairs, 2–5 cm long; bracts: linear-lanceolate, 0.2–0.4 cm long; flowers: white, 5–6 cm diam.; sepals: oblong-lanceolate, white, outside greenish, sometimes spotted with purple, 2–3.3 × 0.7–1.1 cm; petals: oblong-lanceolate to lanceolate, white, 1.5–2 × 0.5–0.7 cm; corona: 1 series, filaments 1.6–2.8 cm long, filiform, completely white or white with a broad purple band near base; ovary: ellipsoid, glabrous; fruit: ovoid-oblong, long-stipitate, reddish, 13–21 × 3.5–5 cm, dehiscent; seeds: almost ovate, ca. 0.9 × 0.6 cm, reticulate.

Passiflora escobariana MacDougal

Novon 2: 365, fig. 6 (1992)
named after Linda K. Escobar, passionflower expert and author
Supersection *Decaloba*, Section *Xerogona*

John MacDougal named this species after Linda Escobar, renowned as a top expert on passionflowers, especially tacsonias and astropheas. Her early death in 1993 meant a great loss for all of us who could profit from her enormous knowledge.

Passiflora escobariana (Plates 101 and 102), known from only 2 places in the lowlands and foothills of northern Colombia and Panama, belongs to section *Xerogona*. It is closely related to *P. costaricensis* and *P. rovirosae*. All 3 species tend to produce leafless, flowering axillary short-shoots (MacDougal 1992).

By comparing the pubescence of the upper leaf surfaces, the pubescent *Passiflora escobariana* can be distinguished from the hirsute *P. costaricensis*. It differs from *P. rovirosae* by its poorly developed or absent inner coronal series and by its leaves. In contrast to *P. escobariana*, *P. rovirosae* has a well-developed inner coronal series and nonrugose leaves.

There have only been a few specimens of *Passiflora escobariana* under cultivation, all cuttings of the type collection. On the whole this species needs more attention than most others from section *Xerogona*. Its flowers, about 4 cm diam., close near midday and effuse a discreet, sweetish scent. The flowers range from white to pale pink. Min. temp.: 10°C.

Plant: minutely puberulent; stem: 4- to 5-angular, reddish; stipules: linear-lanceolate or slightly falcate, 0.5–0.7 cm long; petioles: 2.5–6 cm long without glands; leaves: 2-lobed, occasionally with additional middle lobe, which is shorter than the lateral lobes, 7–25 × 6–12 cm, laminar nectaries absent, cordate at base, rugose, entire; peduncles: solitary or usually in pairs, 0.5–2.5 cm long, red;

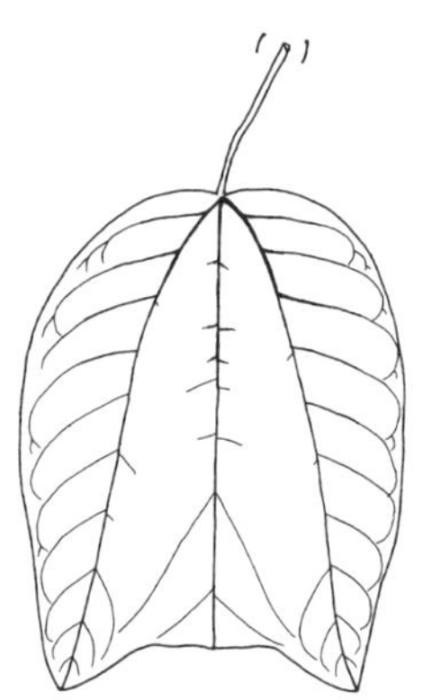

Passiflora escobariana

bracts: absent; flowers: cream-white, usually tinged pink near center, 3.5–4.5 cm diam.; sepals: oblong-lanceolate, white, usually tinged pink on the edges and basal part, 1.8–2.3 × 0.5–0.8 cm; petals: narrowly ovate to lanceolate, more or less pinkish, 1.5–1.7 × 0.4–0.5 cm; corona: 1 or 2 series, filaments of outer series 1–1.3 cm long, filiform, white and pale yellow above middle, filaments of inner series (if present) 0.2–0.3 cm long; ovary: ellipsoid, puberulent; fruit: fusiform, 6-keeled, ca. 7 × 2.5 cm, red on ridges, cream between, dehiscent; seeds: unknown.

Passiflora eueidipabulum Knapp & Mallet

Ann. Missouri Bot. Gard. 71: 1070 (1984)
named after the butterfly *Eueides lineata*
Supersection *Pterosperma*

Many new species of *Passiflora* were discovered when lepidopterists investigated the biology of certain butterflies whose evolution and existence are closely connected to passionflowers. *Passiflora eueidipabulum* (Plate 103), for example, was found during the investigation of the native range of the butterfly species *Eueides lineata*. This butterfly's hostplant in Mexico is *P. microstipula*. However, populations of *E. lineata* were also found in more southern regions of Central America, so the question arose as to which passionflower species could be a hostplant in those areas. As early as 1982, James Mallet and Jack Longino predicted the occurrence of such a species. Their conjecture proved to be correct when a short time later all stages of *E. lineata* were found in abundance on old and new growth of a *Passiflora* species (later named *P. eueidipabulum*) in the province of Coclé, Panama. The native range of this woody vine reaches from Costa Rica to central Panama. Its occurrences are exclusively found in very wet habitats and in low cloud forests at altitudes below 1000 m.

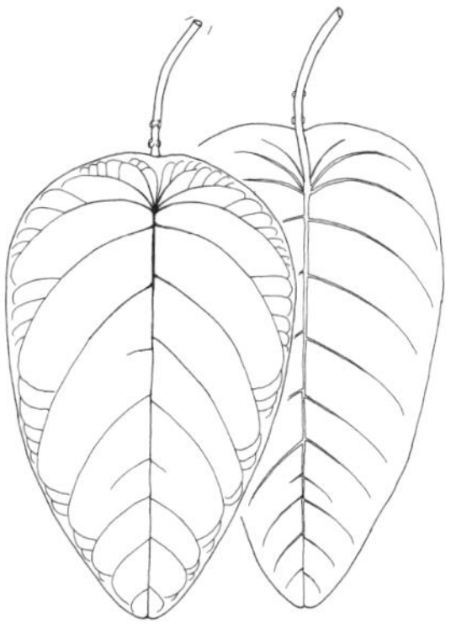

Passiflora eueidipabulum

Because of its characteristic winged seeds and vegetative morphology, *Passiflora eueidipabulum* was assigned to supersection *Pterosperma*. This supersection includes *Passiflora lancetillensis* and *P. microstipula* as well, which are also characterized by large, unlobed leaves and flowers that are borne on the tendrils. In *P. eueidipabulum*, usually 2 flowers of about 5–6 cm diam. appear on the tendrils. They are greenish white to maroon and effuse a sweetish scent. *Passiflora eueidipabulum* has an unmistakable characteristic that distinguishes it from its close relatives: peltate leaves. The petiole ends 1–2 cm below the base of the leaf in the midrib. This species has not yet come into cultivation.

Vine: glabrescent; stem: terete, waxy; stipules: setaceous, 0.2–0.3 cm long; petioles: 4.5–8 cm, with 4–6 mostly paired glands above middle; leaves: unlobed, ovate, 9–18 × 5–15 cm, peltate, with some laminar nectaries, entire; peduncles: 4–5.5 cm long, 2-flowered cyme terminating in a tendril; bracts: lanceolate, 0.1–0.3 cm long; flowers: white, green, and maroon, 5–6 cm diam.; sepals: whitish, oblong-ovate to triangular, 2.2–2.5 × 1.6 cm; petals: white, oblong-ovate, 2.2–2.5 × 1.5 cm; corona: ca. 4 series, outer series 1.3–1.5 cm long, base mottled green and maroon, bright lemon-yellow above, inner series 0.2–0.8 cm long, mottled green and maroon; ovary: ovoid, pubescent; fruit: ovoid, 8.5–10 × 5 cm, light yellow-green; seeds: elliptic, ca. 1 × 1 cm, winged, reticulate.

Passiflora exoperculata Mast.

Fl. Bras. 13 (1): 556 (1872)
exoperculatus: lacking a lid
Supersection *Multiflora*

The disjunctive native range of *Passiflora exoperculata* (Plate 104) is remarkable. This species is common in one region of southern Ecuador and then skips to another spot in northern Bolivia at elevations of 2200–3200 m. In Peru, lying in between, the species has been found only twice: once in the north, once in the south. The distance between the

2 native ranges is more than 1000 km. Jürgen Skrabal and others (2001) point out that *P. exoperculata* is a very heterogeneous species that varies considerably with regard to pubescence, leaf shape, position of petiolar glands, and length of the androgynophore. These characteristics, in combination with the disjunctive distribution, might suggest that the specimens placed in *P. exoperculata* represent different taxa.

The plants growing in Europe were raised from seeds Michael Schwerdtfeger collected in Ecuador. Consequently the data given here refer to the Ecuadorian representatives. The very short androgynophore of the flowers, which are about 2.5 cm wide, is remarkable.

The native habitat of *Passiflora exoperculata* is cool to cold mountains, so this species should be cultivated like a tacsonia and protected from temperatures that are too high. Cultivation in a cold greenhouse or outdoors in summer will suit it well. Min. temp.: 0°C.

Plant: nearly glabrous; stem: younger parts angulate, older stems terete; stipules: linear, 0.3–0.8 cm long; petioles: 0.7–2.4 cm long, with 2 glands in upper 1/3, glands 0.1–0.2 cm diam.; leaves: 2-lobed, occasionally with additional middle lobe, which is shorter than the lateral lobes, 1–4 × 3.3–9 cm, laminar nectaries absent, slightly cordate to truncate at base, entire; peduncles: solitary or in pairs, 1.5–2.5 cm long; bracts: linear; flowers: greenish yellow, 2.3–2.6 cm diam.; sepals: ovate, greenish yellow inside, reddish violet outside, 0.6–1.1 × 0.3–0.5 cm; petals: oblong, greenish yellow, 0.5–0.9 × 0.3 cm; corona: 2 series, filaments of outer series 0.5–0.9 cm long, filiform, falcate, purple at base, yellowish above, filaments of inner series 0.1–0.3 cm long, purple or green; ovary: globose, glabrous; fruit: subglobose, nearly 1 cm diam., purple; seeds: ovate, 0.4–0.5 × 0.3 cm, reticulate.

Passiflora fanchonae Feuillet

Candollea 41: 175, fig. 2 (1986)
named after Marie-Françoise Prévost, nicknamed Fanchon, who first realized the species was distinct
Supersection *Auriculata*

French Guiana seems to have become a favorite destination for passionflower enthusiasts. Despite its relatively small size, the country accommodates an unusually large number of *Passiflora* species. A growing number of species from French Guiana are making their way into collections after being taken home as souvenirs.

Among these is *Passiflora fanchonae* (Plate 105), often found in disturbed habitats as a ground cover. Its flowers, which are up to 3 cm diam., usually appear in pairs in the leaf axils. At first sight they remind one of *P. auriculata* because of their size and color. Upon closer inspection one notices the petals, the inner coronal series with a thickened tip, and the strange operculum, suggesting a relationship with the group around *P. misera*. It is particularly conspicuous for the elongated base of its leaf blade, which extends in a winglike lamina along the upper part of the petiole. Surprisingly, the berries are red.

This warmth-loving species is recommended for a greenhouse, although in warmer climates it may also flourish outdoors in summer. It is a rather

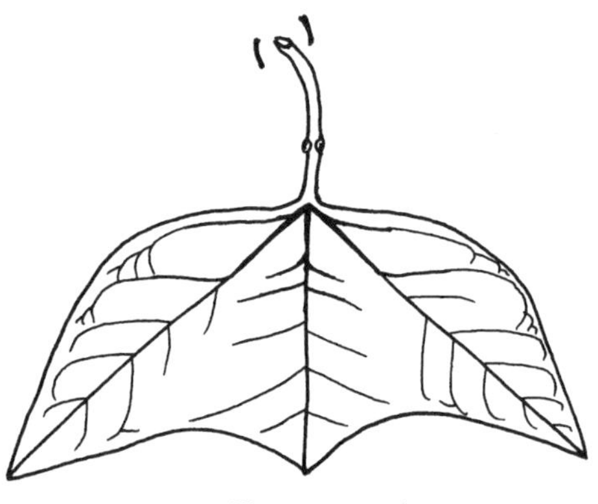

Passiflora exoperculata

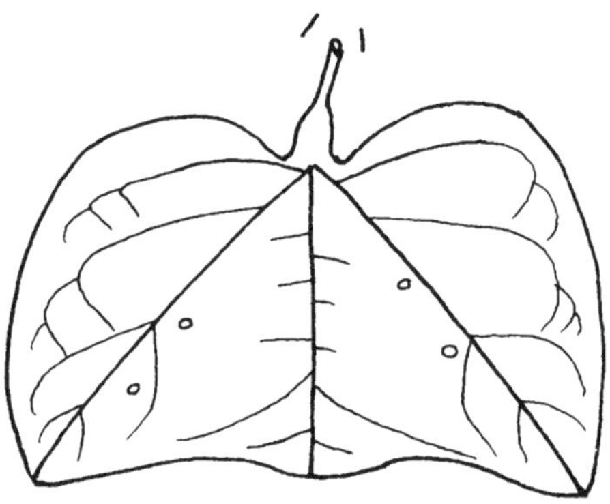

Passiflora fanchonae

finicky representative from subgenus *Decaloba*. Min. temp.: 10°C.

Plant: nearly glabrous; stem: terete; stipules: setaceous, soon deciduous; petioles: 1–1.5 cm long without glands; leaves: slightly 3-lobed at apex, 2.5–3.5 × 5–6 cm, bearing 2 pairs of laminar nectaries, base subcordate, lamina lengthened along upper part of petiole, entire; peduncles: solitary or usually in pairs, 0.6–0.8 cm long; bracts: setaceous; flowers: greenish yellow, 2.5–3 cm diam.; sepals: oblong-lanceolate, greenish, 1–1.2 × 0.6–0.7 cm; petals: linear-lanceolate, greenish, 0.6–0.7 × 0.1–0.2 cm; corona: 2 series, filaments of outer series 1–1.5 cm long, purplish in lower 1/2, yellowish in upper 1/2, curled near apex, filaments of inner series 0.3–0.4 cm long, purple at base, white at apex, broadly capitate; ovary: subglobose, glabrous; fruit: subglobose, up to 1.5 cm long and wide, red; seeds: obovate, 0.4–0.5 × 0.3 cm.

Passiflora gilbertiana MacDougal

Ann. Missouri Bot. Gard. 76: 610, figs. 2, 3b (1989)
named after Lawrence E. Gilbert, expert on passionflower butterflies and passifloras
Supersection *Decaloba*, Section *Decaloba*

Through his excellent publications, Lawrence Gilbert has contributed considerably to the understanding of the ecology of Passifloraceae and *Heliconius* butterflies. During his travels to Central and South America, a butterfly has often shown him the way to hitherto unknown passionflower species, many of which have been brought into cultivation. In central Costa Rica in 1977, Gilbert discovered the pretty *Passiflora gilbertiana* (Plate 106) at an elevation of about 1500 m.

Particularly remarkable in this species are its yellowish flowers, which are about 2 cm diam. and exude an unpleasant odor. The leaves are extremely attractive, usually 2-lobed at the apex, or 2-lobed to the middle in some presently uncultivated clones. On the upper surface a narrow band along the 3 main nerves is set off in light green, while the leaves are purple on the underside. These 2 characteristics of *Passiflora gilbertiana* were transmitted to *P.* 'Sunburst', a well-known hybrid with *P. jorullensis* that is distributed far more widely than its parent plants. *Passiflora gilbertiana* is a vigorous,

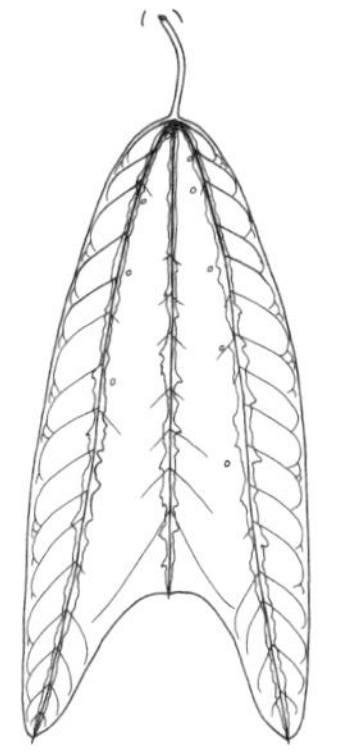

Passiflora gilbertiana

free-flowering species that will flourish as a container plant, similar to 'Sunburst'. Min. temp.: 7°C.

Plant: minutely puberulent except for the leaves; stem: subterete, striate; stipules: linear-lanceolate, 0.2–0.6 cm long, falcate; petioles: 1–5 cm long without glands; leaves: 2-lobed, 4–17.5 × 2–8 cm, bearing 5–17 yellow laminar nectaries, cuneate to obtuse at base, variegated above, purplish red beneath, entire; peduncles: solitary or in pairs, 2.5–4.5 cm long; bracts: linear-triangular, 0.1–0.5 cm long; flowers: greenish yellow, ca. 2 cm diam.; sepals: oblong-lanceolate, greenish yellow, 1.1–1.5 × 0.3–0.6 cm; petals: ovate to ovate-oblong, greenish yellow, 0.2–0.4 × 0.1–0.2 cm; corona: 1 or 2 series, filaments of outer series 0.5–0.7 cm long, yellow, filaments of inner series (if present) ca. 0.1 cm long; ovary: ellipsoid to subglobose, glabrous to densely puberulent; fruit: subglobose, 1.2–1.4 cm diam., purplish black; seeds: obovate, 0.3 × 0.2 cm, transversely sulcate.

Passiflora gracilis J. Jacq. ex Link.

Enum. Hort. Berol. Alt. 2: 182 (1822)
gracilis: slender, tender
Supersection *Bryonioides*

Passiflora gracilis (Plate 107) grows wild spottily from Chiapas, Mexico, to Costa Rica, Venezuela, Guyana, Brazil, and Ecuador. It is nearly unique among passionflowers, for it is an annual. Every year the plants die off in late fall. Numerous brilliant red fruits, mostly set after self-fertilization, provide enough seeds to grow new plants by germinating early the next spring.

The dainty *Passiflora gracilis* will do well in a rel-

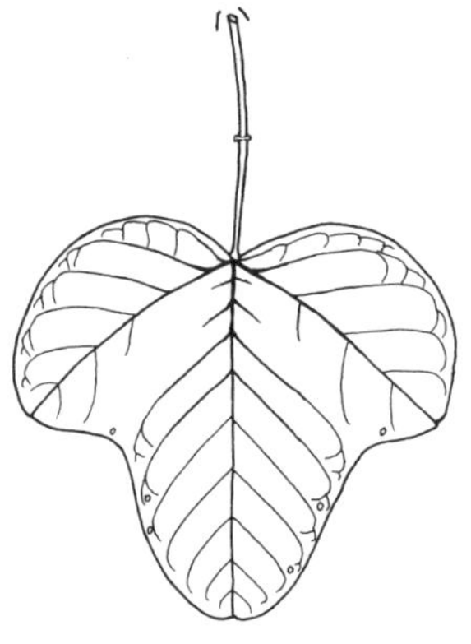

Passiflora gracilis

atively small pot and is suited for a windowsill, although it will also flourish under glass or outdoors during summer. It prefers a sunny location where it will produce masses of whitish flowers, each about 2.5 cm diam., in which the petals are missing.

On the whole, *Passiflora gracilis* is a very recommendable passionflower from subgenus *Decaloba*. It grows quickly, blooms reliably, and always sets pretty fruit. The disadvantage is that new plants must be grown from seeds every year. It seems that many enthusiasts shun this effort, as this species is cultivated only rarely. Min. temp.: 10°C.

Plant: annual, glabrous; stem: angulate, striate; stipules: linear, 0.1–0.4 cm long, falcate; petioles: slender, 2.5–9.5 cm long, with 2 glands in lower quarter, rarely an additional gland above middle, very rarely glands missing; leaves: 3-lobed for usually 1/2 of their length, 2.5–10 × 2.5–13 cm, bearing laminar nectaries near margin, cordate at base, entire; peduncles: solitary, 1.5–5 cm long; bracts: 1–3 in number, setaceous; flowers: white or occasionally whitish green, 2–3 cm diam.; sepals: oblong, white or occasionally whitish green, 1–1.4 × 0.2–0.4 cm; petals: absent; corona: 2 series, filaments of outer series 0.6–0.9 cm long, filiform, white with violet bands, filaments of inner series 0.1–0.3 cm long, filiform, greenish, slightly capitate; ovary: ovoid, slightly zygomorphic, glabrous; fruit: ellipsoid, 2.5–4 × 1.2–2.5 cm, red to orange-red, at least apically, often softening apically or irregularly dehiscent; seeds: oblong-ovate, 0.4 × 0.2– 0.3 cm, coarsely reticulate.

Passiflora guatemalensis S. Wats.

Proc. Amer. Acad. Arts 22: 473 (1887)
guatemalensis: coming from Guatemala
Synonym: *P. hahnii* in the sense of Killip and other authors, not of Mast.
Supersection *Hahniopathanthus*

Gardeners delight in the decorative leaves of *Passiflora guatemalensis* (Plate 108), which are usually purple on their undersides. This may be the only consolation for growing this species, however, as it will only flower satisfactorily under glass and as a large plant. Under optimum conditions it produces dozens of flowers at a time, each up to 5 cm diam. and among the most beautiful of the subgenus. When the plant is cultivated outdoors during summer, flowers appear only now and then.

Passiflora guatemalensis, common from Mexico to Venezuela, looks very similar to *P. hahnii*. For a long time they were thought to be one and the same species, and Killip (1938) erroneously mentioned *P. guatemalensis* as a synonym for *P. hahnii*. It may be distinguished from *P. hahnii* by its stipules and shorter peduncles, and by the usual lack of extrafloral nectaries at the base of the leaf blade. While in *P. guatemalensis* the toothed stipules end in filiform appendages, in *P. hahnii* these are reduced to tiny teeth at most, or a single central tooth. John Vanderplank (1996a) points out that the flowers of *P. hahnii* appear singly, while in *P. guatemalensis* they mainly appear in pairs. This is generally true but not as dependable as the structure of the stipule. Min. temp.: 7°C.

Plant: glabrous throughout; stem: terete, wiry; stipules: reniform, 0.6–2.5 × 1.2–4.5 cm, long-ciliate; petioles: 2–5 cm long without glands, slender; leaves: inconspicuously 3-lobed at apex, middle lobe the largest, 6–11.5 × 5–10 cm, laminar nectaries absent, or very rarely 2 on margin of laminar base, peltate, entire; peduncles: solitary or usually in pairs, 1–3 cm long, slender; bracts: 3 or only 2, cordate, 1.7–3 × 1.3–2.6 cm, green, rarely tinged pink marginally, the 3rd bract (if present) much smaller, inconspicuous, early deciduous; flowers: white with yellow corona, 4–5 cm diam.; sepals: oblong to oblong-ovate, white, 1.3–2 × 0.8–1 cm; petals: oblong to oblong-ovate, white, 1.2–2 × 0.7–0.9

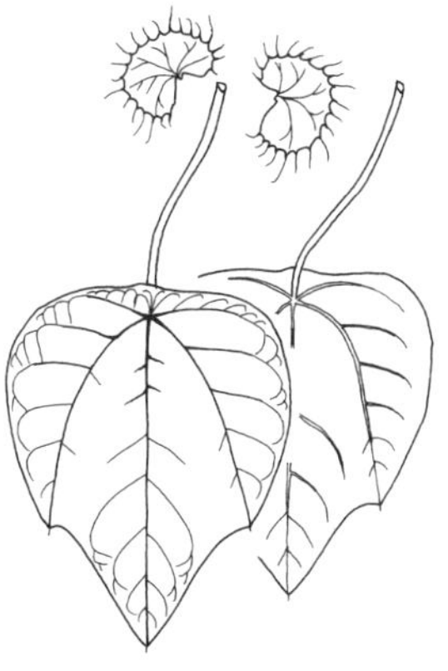

Passiflora guatemalensis

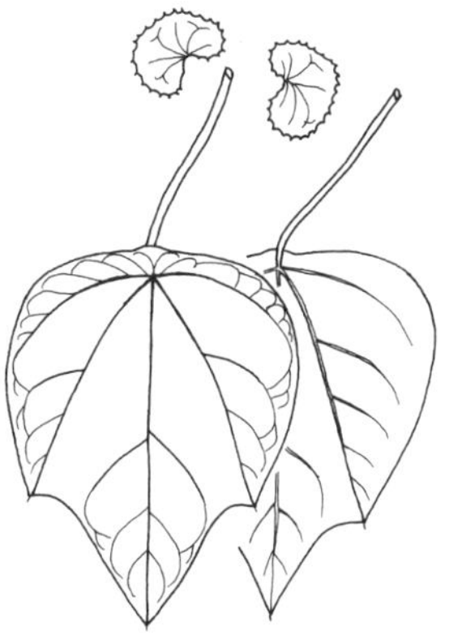

Passiflora hahnii

cm; corona: 2–3 series, filaments of outer series 0.3–1 cm long, dilated at apex, yellow or orange-yellow, filaments of inner series 0.2–0.5 cm long, yellow, clavate; ovary: ovoid, glabrous; fruit: subglobose, 3–4.5 cm diam.; seeds: ovate-oblong, curved, ca. 0.5–0.6 × 0.3–0.4 cm, reticulate.

Passiflora hahnii (Fourn.) Mast.

Fl. Bras. 13 (1): 569 (1872)
named after botanist Gotthold Hahn, who introduced the species in France
Synonym: *P. cookii* Killip
Supersection *Hahniopathanthus*

Passiflora hahnii (Plate 109) and *P. guatemalensis* are morphologically and hence also systematically interesting species. They differ in several respects from the typical phenotype of *Decaloba* species. Normally the bracts and stipules in species of subgenus *Decaloba* are rather small and inconspicuous. This is not the case with *P. hahnii*, *P. guatemalensis*, and *P. membranacea*, in which the large, bud-enveloping bracts and large, reniform stipules are conspicuous from a distance. Particularly remarkable is the trend of reduction of the number of bracts to merely 2; nearly all other species from genus *Passiflora* have 3 bracts or none at all. *Passiflora hahnii* and *P. guatemalensis* are similar in appearance but can be distinguished in several ways (see *P. guatemalensis*).

The native range of *Passiflora hahnii* reaches from southeastern Mexico through Central America to Colombia in northwestern South America. This is mainly a species of wet mountains, found at 1000–2100 m elevation; it always grows higher than *P. guatemalensis*. There is much poorly documented variation in bract size and number of coronal rows in the north of its range, and there could be more than one species represented. Bracts larger than 6 × 6 cm have been seen in Chiapas, Mexico. Under cultivation this species have proven quite reluctant to flower; it probably needs to achieve a large size. This is all the more regrettable since its fragrant flowers, up to 6 cm diam., are very handsome. *Passiflora hahnii* is very suited for cultivation under glass or outdoors during summer. Min. temp.: 7°C.

Plant: glabrous throughout; stem: terete, wiry; stipules: reniform, 0.6–2.7 × 1–4.2 cm, entire to crenate or minutely toothed; petioles: 1.5–3 cm long without glands, slender; leaves: inconspicuously 3-lobed at apex, middle lobe the longest, 5–14.5 × 3.5–13 cm, bearing laminar nectaries on margin of laminar base, peltate, entire; peduncles: solitary or occasionally in pairs, 2.5–7 cm long, slender; bracts: only 2, cordate, 2–6.2 × 1.3–6.4 cm, light green to light pink; flowers: white with yellow corona, 4–6 cm diam.; sepals: oblong to oblong-ovate, white or cream, 1.7–3 × 0.7–0.9 cm; petals: oblong to oblong-ovate, white, nearly as long as the sepals; corona: 1–5 series, usually ca. 3–4, filaments of outer series 0.6–1.5 cm long, dilated at apex, yellow, filaments of inner series 0.2–0.5 cm long, yellow at apex, clavate; ovary: ovoid, glabrous; fruit: subglobose, ca. 4–5 cm diam.; seeds: ovate-oblong, curved, ca. 0.6 × 0.4 cm, reticulate.

Passiflora helleri Peyr.

Linnaea 30: 54 (1859)
named after Carl Heller, discoverer of the type plant
Synonym: *P. fuscinata* Mast.
Supersection *Decaloba*, Section *Decaloba*

Passiflora helleri (Plate 110) is attractive because of its beautiful glossy foliage. There are numerous nectar-producing gland-spots along the 2 outer main nerves of the leaves, which are 4–12 cm long. The leaves are nearly always longer than wide and are never variegated. *Passiflora talamancensis* can have nearly identical leaves, but its peduncles are usually longer and its flowers have 2 coronal rows.

Unfortunately, this vigorous species is a little lazy in flowering. In spring it develops very fragrant white flowers, each 3–4 cm diam., only on shoots that are several, even many, meters long.

Its habitats are the moist and moderate altitudes from Puebla, Mexico, to Costa Rica. Although in Europe and North America *Passiflora helleri* will grow on a windowsill or even outdoors in summer, the ideal location is in a greenhouse. When the plant is overwintered, temperatures should not drop below 10°C.

Plant: glabrescent throughout; stem: terete, grooved-striate; stipules: falcate, 0.3–0.7 cm long; petioles: 2–3 cm, without glands; leaves: very shallowly 3-lobed with short middle lobe, 4–12.5 × 2.5–8 cm, bearing laminar nectaries, entire; peduncles: in pairs, 2–5 cm long, 1-flowered; bracts: setaceous; flowers: whitish with yellow and purplish brown corona, ca. 3 cm diam.; sepals: greenish outside, white inside, oblong-lanceolate, 1.5–2.2 × 0.6 cm; petals: white, oblong-lanceolate, 1.2–1.6 × ca. 0.6 cm; corona: 1 series, 0.5–0.8 cm long, laterally compressed, dilated distal to middle then tapering, light green to yellowish with adaxial and basal flush of red-purple, purple, or brown; ovary: puberulent, often in rows, subglobose; fruit: subglobose, 1.5–2.3 cm diam., blackish purple; seeds: ca. 0.3–0.4 × 0.2–0.3 cm, transversely sulcate.

Passiflora herbertiana Ker-Gawl.

Bot. Reg. 4: 737 (1823)
named after collector Herbert Lord Caernarvon, who introduced the plant in Britain
Synonyms: *P. biglandulosa* Caley, *P. verruculosa* Weinm.
Supersection *Disemma*, Section *Disemma*

Passiflora herbertiana (Plate 111) is one of 3 species indigenous to Australia. This recommendable species will flourish in a greenhouse, on a windowsill, or outdoors. It can be induced to flower reliably even in a 15-cm pot. It flowers untiringly and in a greenhouse will produce flowers almost year-round if appropriate temperatures are provided. Outdoors and on a windowsill, flowering time usually lasts from June to late September. Flowers are 5–8 cm diam. and stay open 3–4 days. Over the course of time the intensity of their color increases, which is also the case with *P. aurantia*, another Australian species. Immediately upon opening, the flowers are greenish yellow to cream, becoming darker and increasingly yellow before they wither.

Passiflora herbertiana produces numerous fruits with its own pollen. Australian Aborigines like very much to eat these fruits, which are green with white mottles. The whitish pulp contains many

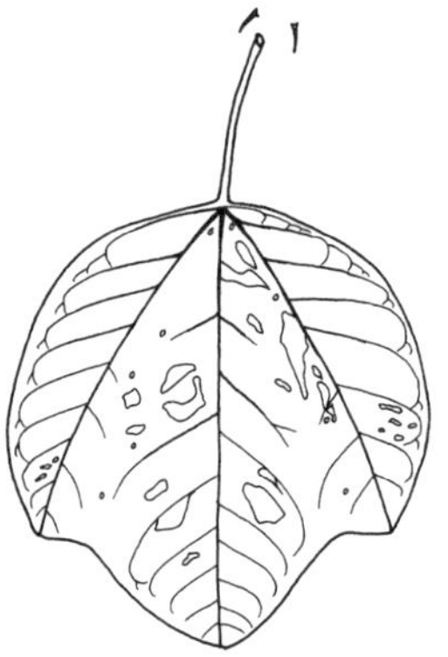

Passiflora helleri

Passiflora herbertiana

tiny black seeds, and these may be used to propagate the species quite easily. Cuttings, however, will often not develop roots, although they form callus very rapidly.

Passiflora herbertiana is remarkably resistant to cold. Planted-out specimens are said to have survived slight frost. Nevertheless, a minimum temperature of 7°C should be provided to get this plant through winter without difficulties.

Vine: pubescent; stem: terete; stipules: setaceous, 0.1–0.4 cm long; petioles: (1.5–)2.5–5(–7.5) cm long, with 2 large glands near apex; leaves: 3-lobed, rarely 5-lobed, 4–8(–11) × 4–8(–12) cm, entire; peduncles: 2–3.5 cm, solitary; bracts: minute, in lower 1/2 of peduncle, free; flowers: greenish yellow, yellow, or orange-yellow, (5–)6–8 cm diam.; sepals: outside greenish, inside greenish yellow, yellow, or orange-yellow, oblong-lanceolate, 2.5–3.7 × 0.6–0.8 cm; petals: greenish yellow, yellow, or orange-yellow, oblong-lanceolate, 1.5–2.2 × 0.4–0.5 cm; corona: 1 series, (0.5–)1–1.5 cm long, erect, yellow; ovary: ovoid, glabrous; fruit: ovoid, 4–7 cm long, obscurely 3-sided, green with white speckles.

Passiflora hirtiflora Jørgensen & Holm-Nielsen

Nord. J. Bot. 7: 132 (1987)
hirtiflorus: having bristly flowers
Supersection *Decaloba*, Section *Decaloba*

This species with bristly flowers, endemic at moderate altitudes of the Andes in Ecuador, was introduced to the United States by John MacDougal in 1992. The rather dainty *Passiflora hirtiflora* (Plate 112) owes its attractiveness to its beautifully colored leaves. They are deep purple-brown with a paler variegation, and the intensity of these colors exceeds that of most other passionflowers with colorful leaves. At a length of 4.5–8 cm, the leaves, 3-lobed at their tips, are comparatively small.

The unusually large bracts, quite unusual in *Decaloba* species, are another conspicuous feature. *Passiflora hirtiflora* owes its specific epithet to the flowers, which are hirsute on the back, as is the whole plant. The flowers are about 4.5 cm diam. Although this species originates near the equator, those we have planted always flower during winter, when days are shorter and temperatures lower—or, in other words, when conditions worsen. The question arises: do they flower because of these conditions or despite them?

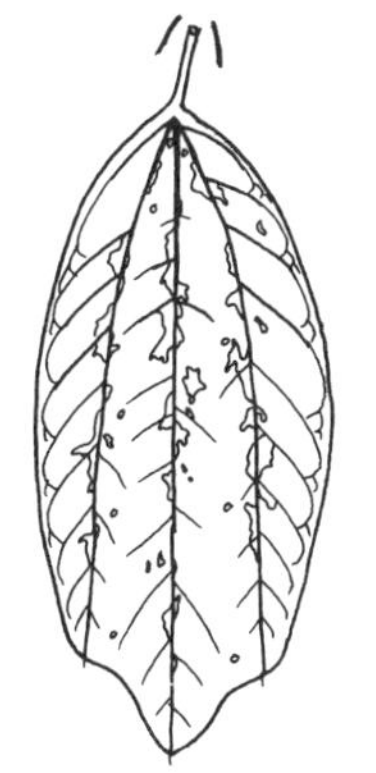

Passiflora hirtiflora

Like tacsonias, this species dislikes temperatures above 30°C and should therefore spend summers outdoors. In addition, it likes a stony rooting medium better than a potting mix that contains peat. Peat retains moisture too long for a plant that does not develop a vigorous rootstock.

Passiflora hirtiflora will get through winter without difficulties at temperatures above 7°C.

Vine: white-hirsute; stem: angulate; stipules: linear, falcate, 0.8–0.9 × 0.1–0.2 cm; petioles: 1.5–1.8 cm without glands; leaves: 3-lobed at apex, 4.5–7.2 × 3.1–5.1 cm, bearing laminar nectaries, greenish purple above, purplish beneath, entire; peduncles: 2–2.6 cm, solitary or in pairs; bracts: ovate, 1.6–2 × 0.9–1.3 cm, entire or somewhat undulate, free; flowers: whitish green to white and purple, 4.5 cm diam.; sepals: pale green outside, greenish white to white inside, oblong-ovate, ca. 2 × 0.7 cm; petals: white, oblong, 1 × 0.4–0.5 cm; corona: 2 series, outer series 0.4–0.7 cm long, white banded with purple, inner series 0.3–0.4 cm long, green with white apex, slightly capitate; ovary: ellipsoid, hirsute; fruit: globose to subglobose, hirsute.

Passiflora holosericea L.

Sp. Pl. 958 (1753)
holosericeus: covered with dense, silky hair
Synonym: *P. reticulata* Sauv.
Supersection *Multiflora*

Passiflora holosericea (Plate 113) is among our favorite passionflowers. Hardly any other species flowers so abundantly. Its clusters of flowers at each node ensure weeks of uninterrupted bloom. The flowers, 3.5–4.5 cm diam., are very beautiful. The sepals and petals are white, the corona is dark yellow and often reddish purple at the base, and the flowers exude a pleasant sweet scent. However, the plant can only display all of its beauty when planted in a large greenhouse, as it needs bright light. In most cases small plants produce only 1 or 2 flowers on one peduncle, but there are usually 3 or even 4 flowers per peduncle on long pendent shoots. In July, hundreds of flowers hang down from the greenhouse roof on shoots several meters long.

The entire plant is covered with silky hair, which is especially noticeable on the large 3-lobed leaves. As a rule the leaves are not indented very deeply; only the rounded tips of the 3 lobes are visible. This species always sports 2 curious small cusps or teeth at the base of its leaf blade.

The native range reaches from Cuba and coastal Mexico through coastal and drier parts of Central America to Colombia and Venezuela in northwestern South America. In these tropical regions it usually grows below altitudes of 1000 m. In accordance with the conditions of its natural habitats, the warmth-loving *Passiflora holosericea* feels at home when temperatures above 15°C are guaranteed during winter months. Specimens that can root freely may well tolerate temperatures below 10°C for a short while.

Plant: densely soft-pubescent throughout; stem: almost terete; stipules: setaceous, ca. 0.2–0.5 cm long; petioles: 1–3 cm, with 2 subsessile glands near or below the middle; leaves: 3-lobed, or obscurely so, with central vein the longest, 4.5–16.5 × 3–14 cm, often bearing laminar nectaries near the margins, entire except basal part of leaf with 2 cusps or teeth; peduncles: in pairs, 0.5–3(–4.7) cm long, (1–)3(–4)-flowered; bracts: linear-lanceolate, 0.1–0.8 cm long; flowers: white with yellow and dark reddish corona, 3.5–4.5 cm diam.; sepals: greenish outside, white inside, ovate-oblong, 1.3–1.8 × 0.6–0.8 cm; petals: white, elliptic-oblong, 1.3–1.5 × 0.6–0.7 cm; corona: 2 series, outer series 0.8–1.4 cm long, laterally compressed, dark yellow and often brick-red or purple-brown basally, inner series 0.3–0.6 cm long; ovary: densely pilose, subglobose; fruit: subglobose, 1.8–3 × 1.7–2.5 cm, purplish; seeds: ca. 0.4–0.5 × 0.3 cm, reticulate.

Passiflora indecora Kunth

Nov. Gen. Sp. 2: 134 (1817)
indecorus: unadorned
Synonym: *P. involucellata* Harms
Supersection *Decaloba*, Section *Decaloba*

Passiflora indecora (Plate 114) comes from the province of Loja, southern Ecuador, where it grows at altitudes up to 2600 m in locations where tacsonias also flourish. Though in other ways undemanding, this species dislikes heat spells and so should be cultivated outdoors as long as weather conditions allow. However, as soon as temperatures drop to near the freezing point, *P. indecora* must be taken to its winter quarters, with temperatures around 5°C.

As a rule this species produces numerous flowers of about 4 cm diam. before and after midsum-

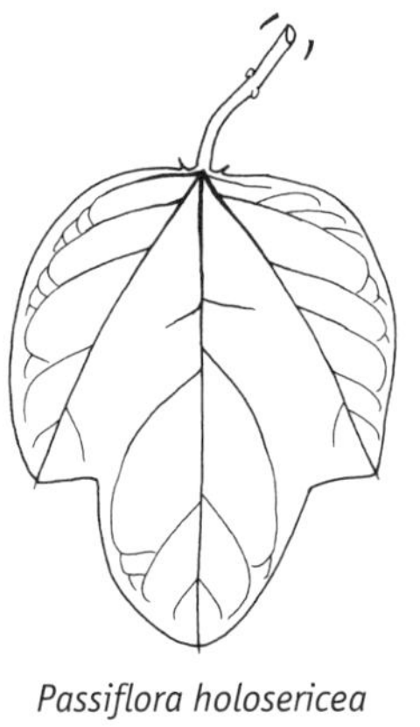

Passiflora holosericea

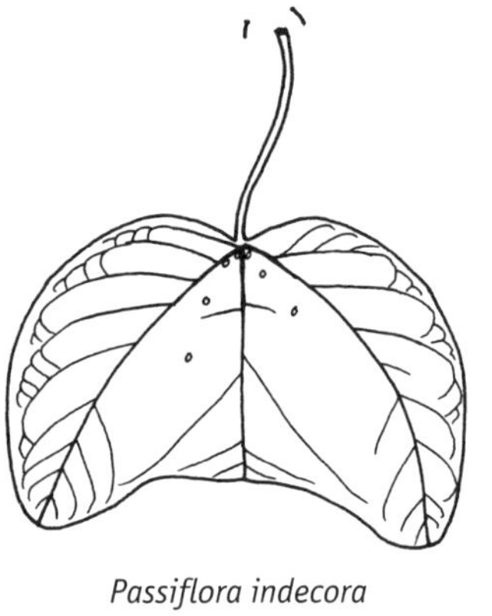

Passiflora indecora

mer. The sepals and petals are whitish, often with a bluish violet tinge toward the center of the flower. While the corona is purple-brown in its center, the filaments of the outer series, 0.5–0.6 cm long, are banded white and purple-brown. The unusually large bracts and short petiole are remarkable for a *Decaloba* species. Since *Passiflora indecora* is not so free-flowering as a small plant, it should be grown in a pot of at least 20 cm diam. Its 2-lobed leaves are covered with several yellowish glands that are most frequent near the base of the leaf. In cooler conditions the undersides of the leaves have a slight purple tinge.

Passiflora indecora is sometimes confused with *P. colinvauxii*, endemic in the Galápagos Islands, which has come into cultivation quite recently. *Passiflora colinvauxii*, however, has considerably wider leaves, much smaller bracts, and a flower that is very similar to *P. punctata*.

Vine: mostly glabrous; stem: angulate, striate; stipules: broadly subulate, 0.4–0.8 × 0.1 cm, purplish; petioles: 0.8–2.6 cm without glands; leaves: 2-lobed, 2.5–7.5 × 2.5–7.2 cm, bearing laminar nectaries in basal 1/3, entire; peduncles: 1.8–3 cm, solitary or in pairs; bracts: ovate, petiolate, 0.5–0.8 × 0.2–0.5 cm, free; flowers: white and purplish brown, 3–4 cm diam.; sepals: green outside, greenish white to white inside often with bluish violet area at base, ovate-oblong, 1.3–1.4 × 0.6–0.8 cm; petals: white, narrowly oblong, 0.8–1 × 0.2–0.4 cm; corona: 2 series, outer series 0.5–0.6 cm long, white with at least one purple band, inner series 0.3–0.4 cm long, greenish or purplish brown; ovary: ovoid, densely villous; fruit: ovoid, ca. 1.5 × 1.1 cm, pubescent; seeds: ovate, 0.3 × 0.2 cm, transversely rugulose.

Passiflora jatunsachensis Schwerdtfeger

Haussknechtia 6: 43 (1997)
named after Fundación Jatun Sacha, Ecuador, location of the type plant; in Quichua, *jatun sacha* means "large forest"
Supersection *Auriculata*

Passiflora jatunsachensis (Plate 115), endemic in the tropical lowlands of the province of Napo, northeastern Ecuador, was recently described by Michael Schwerdtfeger. This species combines some char-

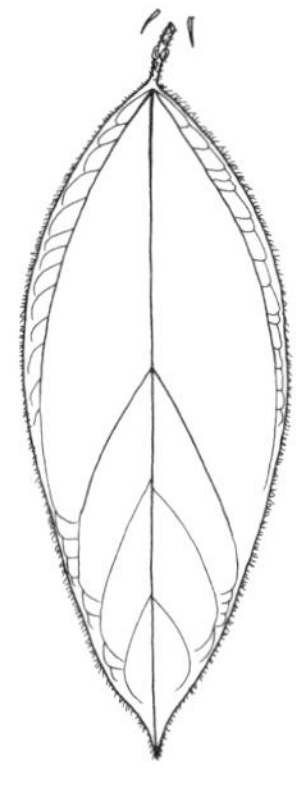

Passiflora jatunsachensis

acteristics of *P. suberosa*, *P. auriculata*, and *P. ferruginea*. It is distinguished from *P. ferruginea* by its large, unlobed leaves with an entire margin. In contrast to *P. suberosa*, the outermost corona series is undulate and considerably larger. The longer, denser pubescence of *P. jatunsachensis* is another distinguishing feature.

Specimens that are planted out in a heated greenhouse develop into large plants with shoots that can reach a length of several meters. The internodes may well be 20–25 cm long. This plant's mass flowering is also remarkable, and similar to others in supersection *Auriculata*. Within 1–2 days up to 50 flower buds that have been developing over previous weeks may open. In cultivation such batches can occur up to 3 times a year. The sepals of the flowers, about 2.5–3 cm diam., are greenish yellow, while the petals are very reduced or sometimes absent.

Vine: large, densely pubescent; stem: terete; stipules: setaceous, about 0.8 cm long; petioles: 1–3 cm, with 2 large, auricular glands near middle; leaves: unlobed, elliptic-oblong, 14–25 × 6–9 cm, bearing 6–12 laminar nectaries, entire; peduncles: 1.5–2 cm, in pairs; bracts: setaceous, 0.2–0.3 cm long; flowers: greenish yellow, 2.5–3 cm diam.; sepals: green outside, greenish yellow inside, oblong, ca. 1.5 × 0.5 cm, pubescent outside; petals: 0.3 × 0.1 mm or absent; corona: 2 series, outer series ca. 1.8 cm long, greenish yellow, wavy in upper 1/3, inner series ca. 0.4 cm long, greenish yellow, white at apex, capitulate; ovary: ovoid, densely puberulous; fruit: unknown.

Passiflora jorullensis Kunth

Nov. Gen. Sp. 2: 133 (1817)
named after Mount Jorullo, Mexico, location of the type plant
Synonym: *P. medusae* Lemaire
Supersection *Decaloba*, Section *Decaloba*

The typical *Passiflora jorullensis* (Plate 116) is from the temperate mountainous regions of central to southwestern Mexico and grows at altitudes of 1300–1800 m. It is rarely represented in European collections. *Passiflora hahnii* is occasionally cultivated as *P. jorullensis*, though the two can easily be distinguished at first glance by their leaves.

The flowers of *Passiflora jorullensis*, 3.5–4.5 cm diam., give off an unpleasant odor. However, the flamelike color of the orange-red corona is pleasing, and unlike most passionflowers, the flowers stay open 2 days, changing to a purplish pink the 2nd day. This plant flowers late in summer when cultivated outdoors but will flower as early as June when grown in a greenhouse. Since in its natural surroundings it advances to altitudes up to 1800 m, it is quite possible that it might do well at temperatures below 12°C in winter.

What was previously known as *Passiflora salvadorensis* has recently been reduced to a variety of this species, *P. jorullensis* var. *salvadorensis*. The flower is the same, but the leaves resemble (or mimic) *Bauhinia* and are more deeply 2-lobed, with obtuse lobes. The habitat is mostly tropical dry forest from Oaxaca, Mexico, to El Salvador, and although this variety is not in cultivation, it might be expected to require warmer temperatures.

Patrick J. Worley crossed *Passiflora jorullensis* var. *jorullensis* with *P. gilbertiana*, which resulted in the well-known and much-cultivated *P.* 'Sunburst'.

Passiflora jorullensis, together with *P. apetala*, *P. boenderi*, *P. gilbertiana*, and some others, belongs to an independent group of Mesoamerican species that has yet to be named. Characteristic features of this group from section *Decaloba* are 2-lobed, colorful, patterned leaves, yellowish to greenish sepals and petals, a filiform outer series of the often unbanded corona, and significantly smaller petals than sepals. Killip (1938) did not recognize the connection and thus classified the species known at his time into different series.

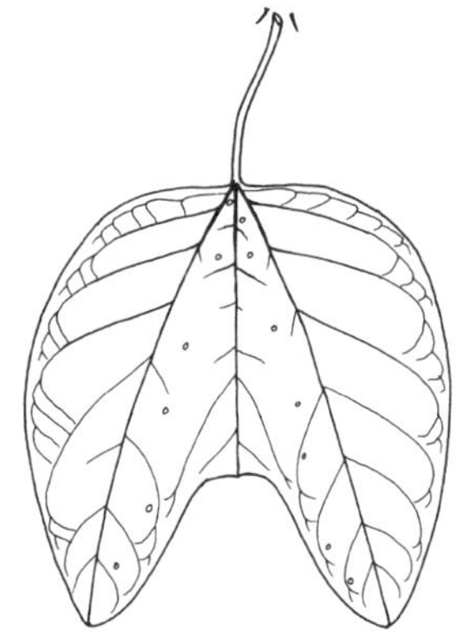

Passiflora jorullensis

Plant: lightly puberulent to glabrescent; stem: subangulate, sulcate-striate; stipules: setaceous, subfalcate, ca. 0.2–0.4 cm long; petioles: 2–7 cm, without glands; leaves: very shallowly 2-lobed or 3-lobed with short middle lobe, 3.5–10 × 3–11.5 cm, bearing laminar nectaries, almost glabrous, entire; peduncles: in pairs, 2–5 cm long, 1-flowered; bracts: setaceous, 0.2–0.6 cm long; flowers: light yellowish green with orange-red or reddish corona, turning purplish, 3.5–4.5 cm diam.; sepals: light yellowish green, narrowly oblong-lanceolate, 1.5–2.5 × 0.4–0.5 cm; petals: pale green to cream, inconspicuous, 0.3–0.8 × 0.1–0.3 cm; corona: 1 series, 0.8–1.5 cm long, filiform, orange-red, red-orange, or reddish, turning purplish or purplish pink the 2nd day; ovary: slightly puberulent, rarely glabrous, subglobose to ellipsoid; fruit: subglobose, 1–1.6 cm diam., blackish purple; seeds: ca. 0.3 × 0.2 cm, transversely sulcate.

Passiflora juliana MacDougal

Novon 2: 358 (1992)
named in honor of Julianna Miley
Supersection *Cieca*

Passiflora juliana (Plate 117) is yet another species that was grown and appreciated before being given a scientific name. Seedlings from the 1979 type collection were already cultivated by some enthusiasts before this species was described by John MacDougal in 1992. Unsurprisingly, in the meantime it made its entry into many collections in Europe.

Passiflora juliana inhabits southwestern Mexico, where it grows in coastal regions at altitudes up to

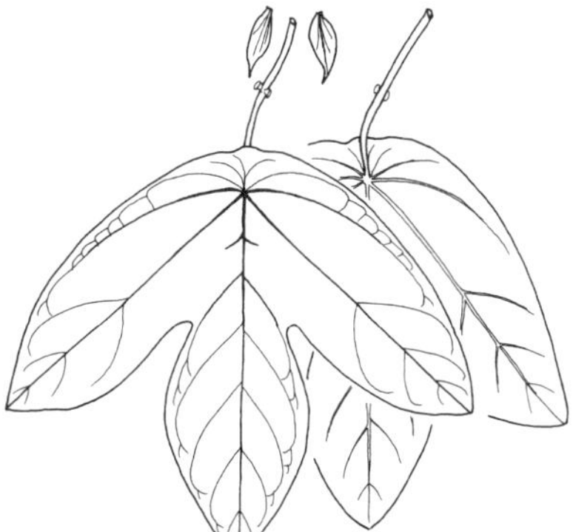

Passiflora juliana

800 m. The petals of the greenish flower are missing. The flowers are 2 cm diam. and appear in pairs, or, in older plants, in leafless terminal spikes up to 20 cm long at the ends of the shoots. The leaves are considered by many to be the most beautiful feature, being peltate, 3-lobed with the central lobe narrowed at the base, and an unusual matte soft green or sea-green.

Although this warmth-loving species will flourish on a windowsill or at a sunny location outdoors, it is better suited for cultivation in a greenhouse. Overwintering temperatures should be around 15°C, and special care must be given during this time to avoid overwatering.

Plant: glabrescent; stem: almost terete; stipules: foliaceous, elliptic-obovate to ovate, 0.6–2.3 × 0.3–1.5 cm; petioles: 1–5.5 cm, with 2 glands near or below the middle; leaves: 3-lobed with the middle lobe narrowed at base, 3–10 × 6–16 cm, bearing laminar nectaries near margin, peltate, entire; peduncles: in pairs, 0.7–3.3 cm long, 1-flowered, often in condensed terminal inflorescences to 20 cm long with aborted leaf blades; bracts: absent; flowers: light yellow-green with dark purple center, ca. 3 cm diam.; sepals: light yellow-green, lanceolate, 0.9–1.2 × 0.4–0.6 cm; petals: absent; corona: 2 series, outer series 0.6–0.8 cm long, tapering to tips, light greenish yellow to yellow apically, sometimes with a touch of purplish basally, inner series 0.3–0.4 cm long, capillary; ovary: glabrous, ellipsoid; fruit: subglobose to prolate, 1.3–2.5 × 1.3 × 2 cm, blackish purple; seeds: ca. 0.4 × 0.2–0.3 cm, reticulate.

Passiflora karwinskii Mast.

Fl. Bras. 13 (1): 555 (1872)
named after Baron Wilhelm Friedrich Karwinski von Karwin, discoverer of the type plant
Synonyms: *P. platyneura* Eastw., *P. pringlei* Robinson & Greenm.
Supersection *Bryonioides*

In older plants the formation of tendrils is occasionally suppressed, which characterizes this species as rather bushy, trailing, and only slightly climbing. Hence *Passiflora karwinskii* (Plate 118) is particularly suited for a hanging basket, as a trailing plant, or as a ground cover when planted out.

Passiflora karwinskii is indigenous to the mountains of Mexico in seasonally dry forests with pines. In a greenhouse it flowers untiringly from April to September. Outdoors or on a windowsill, flowering starts 2–3 months later and is not so abundant. The white petals of the fragrant flowers, which are 4–5 cm diam., contrast with the cream-and-violet-zoned corona. *Passiflora platyneura*, a very short and nearly glabrous variant that is here treated as a synonym, is a link between the typical species and the Sonoran *P. quercetorum*, and perhaps deserves species recognition. *Passiflora karwinskii* is overwintered at temperatures above 10°C. The perennial woody taproot is among the largest of the whole genus.

Plant: glabrescent except pubescent with hooked trichomes on stems, petioles, and veins under leaf; stem: almost terete, sulcate-striate, almost annual; stipules: lanceolate-falcate, 0.2–0.5 cm long; petioles: 0.6–4 cm, with 2 glands near apex; leaves: 3-lobed or nearly 5-lobed, middle lobe narrowed at

Passiflora karwinskii

base, 2.5–8 × 2.5–9 cm, sometimes bearing laminar nectaries near margin, almost entire or deeply serrate-lobed; peduncles: solitary or in pairs, 1.9–5.3 cm long, 1-flowered; bracts: setaceous, 0.2–0.6 cm long; flowers: whitish and violet, ca. 4–5 cm diam.; sepals: greenish or tinged red-purple outside, white inside, lanceolate-triangular, 1.7–2.1 × 0.7–0.9 cm; petals: white, narrowly oblong-lanceolate, 0.9–1.4 × 0.2–0.5 cm; corona: 1 series, 1–1.8 cm long, filiform, whitish with violet band or bands, often pale yellow toward tips; ovary: glabrous, ellipsoid; fruit: ovoid, conical at both ends, 3–4.5 × 1.8–2.6 cm, light green with pale apex when mature; seeds: ca. 0.4 × 0.25 cm, finely reticulate.

Passiflora lancearia Mast.

J. Bot. Brit. and For. 23: 114 (1885)
lancearius: lanceolate (in reference to the leaves)
Supersection *Decaloba*, Section *Decaloba*

Passiflora lancearia (Plate 119) is a real beauty within this subgenus. Its flower is dominated by green, yellow, and purple tints arranged alternately around the center. These tints are often very bright, so that the coloration is rich in contrast. A further dash of color is provided by the androgynophore, which is dark purple, as are the ovary and the anthers in the center of the flower. Sometimes the ovary is pure bright red or white. The diameter of the flower is 4–5 cm.

This species can be found from sea level up to 2900 m but is most commonly found at 1000–1900 m. It is common in Costa Rica and Panama but is also known farther north, with one site known in Guatemala and another in Chiapas, Mexico. The oblong-lanceolate leaves are sometimes purplish on the underside and have 3 parallel nerves, quite an unusual feature among *Decaloba* species. The species has diagnostically large seeds and unusually small stipules, and the fruits are particularly thick-walled.

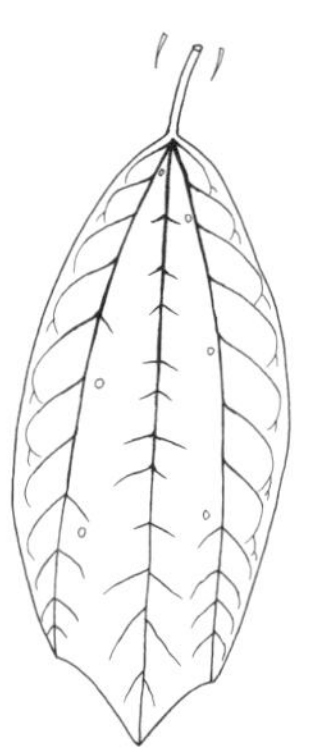

Passiflora lancearia

This species is rarely cultivated by enthusiasts. Its demands are similar to those of other species from moderate altitudes. It prefers constant day temperatures of 20°–25°C year-round, and night temperatures should be 10°–15°C. During overwintering, however, lower night temperatures may well be tolerated.

This species was previously thought to occur also in Ecuador and is listed in the *Flora of Ecuador* (Holm-Nielsen et al. 1988). However, comparative studies have shown that the species in Ecuador is *Passiflora smilacifolia*, recently described by John MacDougal. While the 2 species are similar in their vegetative appearance, especially as herbarium specimens, the flowers show marked differences. *Passiflora smilacifolia* produces white flowers with a pink touch. A photo of such a flower, listed erroneously as *P. lancearia*, can be seen on page 172 of *Passionsblumen* (Ulmer and Ulmer 1997).

Vine: glabrous; stem: stout, terete, striate; stipules: minute, less than 0.2 cm long, soon deciduous; petioles: 1–1.5 cm, without glands; leaves: unlobed or obscurely 3-lobed at apex, oblong-lanceolate, 4–11 × 2–6 cm, bearing 4–8 laminar nectaries, coriaceous, entire; peduncles: 0.5–2.4 cm, in pairs; bracts: setaceous, 0.15–0.4 cm long; flowers: green, yellow, purple, 4–5 cm diam.; sepals: green, ovate-lanceolate, 1.5–2.4 × 0.7–1.2 cm; petals: greenish to white, ovate, 1–1.5 cm; corona: 2 series, outer series 0.4–1.1 cm long, greenish mottled with purple speckles in lower 1/2, purple band above middle and yellow in upper 1/3, inner series up to 0.4 cm long, purplish green with green apex, capillary; ovary: ovoid, glabrous; fruit: subglobose, 3–5 cm diam.; seeds: broadly ovate, 0.5–0.6 × 0.4–0.6 cm, transversely sulcate, with 7–8 ridges.

Passiflora lancetillensis MacDougal & Meerman

Novon 13 (4): 460–464 (2003)
named after Lancetilla Valley, Honduras
Supersection *Pterosperma*

Many enthusiasts will welcome information about this unusual species (Plate 120), which was in cultivation for several years and established in many collections before being officially named. It was distributed in the trade under the name *Passiflora lancetillensis* and thankfully has retained that name.

Passiflora lancetillensis was first discovered in the 1930s in Honduras. Alas, flowers could not be found. Sixty years later it was also discovered in Belize, again without flowers. In 1994 Jan Meerman succeeded in bringing seedlings of *P. lancetillensis* into cultivation. The plants grew excellently and flowered for the first time in 1996. In addition to *P. eueidipabulum* and *P. microstipula*, *P. lancetillensis* belongs to supersection *Pterosperma*. The representatives of this section are characterized by large unlobed leaves, numerous petiole glands, minute stipules, flowers that are borne on the tendrils, and large flattened seeds with short "wings" on their margins.

Up to 6 of its very fragrant flowers, reaching a diameter of up to 5 cm, appear near the base of the tendrils. The pale apricot of the outermost coronal series contrasts with the otherwise cream flower, giving the plant a very distinctive appearance.

The vigorous and robust *Passiflora lancetillensis* usually sets its flower buds in late summer and fall. As soon as days grow shorter, numerous flowers begin to open. A 25-cm pot is large enough for a plant to reliably flower, although a location under glass is preferable. This species should be overwintered at temperatures above 10°C.

Vine: large, lightly pubescent, except stems and upper surface of leaf glabrous at maturity; stem: terete; stipules: minute, soon deciduous; petioles: stout, 3–5.5 cm, with 7–12 glands scattered along petiole, glands 0.1–0.2 cm long, depressed spheroid, narrower in juveniles; leaves: unlobed, ovate to ovate-cordate, 8–22 × 7–19 cm, glabrous above, pubescent beneath, never variegated; peduncles: solitary, minute to 3 cm long, 1- to 6-flowered cyme terminating in a tendril, the pedicels 2.8–5 cm long; bracts: minute to 0.2 cm long; flowers: white to cream with very pale apricot, 4.5–5 cm diam.; sepals: green and pubescent outside, white to cream inside, narrowly ovate-oblong, 1.8–2.5 × 0.6–0.9 cm; petals: white, oblong, 1.9–2.7 × 0.6–0.8 cm; corona: 3–4 series, outer series white to cream or very pale apricot becoming more intense near base, 1.5–2 cm long, filiform, slightly wavy, inner 2–3 series white, 0.3–0.6 cm; ovary: ellipsoid, densely pubescent; fruit: widely ellipsoid to subglobose, 6.5–8 cm long including 2.5–3 cm stipe, 4–4.8 cm diam., light yellowish green; seeds: 0.9–1.2 × 0.7–0.8 cm, reticulate in center or the pits in a few wavy rows, with 4 conspicuous striate erose marginal wings.

Passiflora lobata (Killip) Hutchinson ex MacDougal

Phytologia 60: 446 (1986)
lobatus: lobed (in reference to the leaves)
Synonym: *Tetrastylis lobata* Killip
Supersection *Bryonioides*

Passiflora lobata (Plates 121 and 122), indigenous to Nicaragua, Costa Rica, Panama, and Colombia, originally belonged to the genus *Tetrastylis*. In 1967, John Hutchinson stated that *Tetrastylis lobata* should be included in the genus *Passiflora*, and about 20 years later John MacDougal incorporated it formally into subgenus *Decaloba*.

Passiflora lobata has 4 styles, as indicated by its

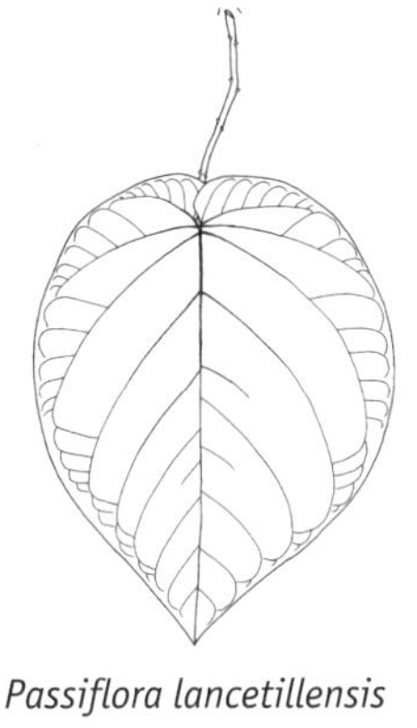

Passiflora lancetillensis

Passiflora lobata

old generic name, and consequently also has 4 stigmas. The 5 stamens are grown together at their base, and all 5 anthers face one direction: downward. The androgynophore is slightly curved and the flowers are presented laterally. Its white flowers reach a diameter of up to 6 cm. In its natural surroundings of wet tropical forest, this extraordinary species is not pollinated by bats, as one might suspect, but by large bees of the genus *Epicharis*. The species advances to altitudes of up to 1600 m in its countries of origin, and it needs much moisture.

Passiflora lobata is increasingly cultivated, but since it is self-incompatible, few collectors will see its amazing fruit. Each fruit hangs like a long red sack up to 18.5 cm long before splitting open to reveal up to a hundred or more large seeds covered by bright reddish orange pulp. In view of the conditions prevailing in its natural surroundings, the plant should be kept neither too warm nor too cold. Day temperatures around 25°C and nightly drops to 14°C will have a positive effect. During the overwintering period, temperatures of more than 10°C will suit the plant very well, although even 5°C will be tolerated for a short time. When kept in a greenhouse, plants flower July to September.

Plant: lightly pubescent throughout with hooked trichomes, except one side of leaf often glabrescent; stem: almost terete to slightly 3-keeled; stipules: foliaceous, ovate-auriculate, 0.4–1.8 × 0.2–1.1 cm; petioles: 2–6.5 cm, with 2 (rarely 4) glands; leaves: 3-lobed, 4–17 × 4–17 cm, bearing laminar nectaries near margins, cordate, entire; peduncles: in pairs, 1.3–5.3 cm long, 1-flowered; bracts: setaceous, 0.1–0.4 cm long; flowers: white with purplish red, ca. 4.5–6 cm diam.; sepals: green outside, white inside, lanceolate-triangular, 2.2–3.5 × 0.7–1.1 cm, with slight apical horn; petals: white, lanceolate, 1.2–2.3 × 0.5–0.9 cm; corona: 1 series, 1–2.6 cm long, filiform, white or white with purplish red bands, very rarely with a 2nd vestigial inner series, the operculum usually dark wine-red; ovary: glabrous, ellipsoid; fruit: narrowly obovoid, long-stipitate, 11–18.5 × 3.5–6 cm, reddish or purplish red, irregularly dehiscent; seeds: ca. 0.7–0.9 × 0.4–0.6 cm, coarsely reticulate.

Passiflora lobbii Mast.

Fl. Bras. 13 (1): 53 (1872)
named after William Lobb, plant collector
Synonyms: *P. niorbo* Planch., *P. obtusiloba* Mast., *P. obtusiloba* var. *glandulifera* Harms
Supersection *Multiflora*

After 30 weeks of field studies, Jürgen Skrabal and others (2001) presented an excellent revision of the *Passiflora lobbii* group. Among other things, the authors noted that the presence or absence of petiolar glands should not always have a major taxonomic importance. Ellsworth P. Killip (1938), for example, assigned *P. obtusiloba*, in which petiolar glands are absent, to section *Decaloba*, while he put *P. lobbii* (Plate 123), which has 2 petiolar glands, into section *Cieca*. As a matter of fact, both are so closely related that *P. obtusiloba* has been declared a subspecies of *P. lobbii*.

The *Passiflora lobbii* group consists of 8 species whose native range extends along the Andes from northern Ecuador through Peru to northern Bolivia. As a consequence of the field studies, several of these species have come into cultivation. Among them is *P. lobbii*, which is divided into 3 subspecies, all endemic in Peru. Because of its small 3-lobed leaves, *P. lobbii* is basically a rather dainty passionflower, although it can produce shoots of several meters in length in its natural surroundings. Under cultivation it will flower without problems even in a 20-cm pot. The flowers, 2–3 cm diam., appear in pairs. Because of their size and their yellowish green coloration they are not very spectacular.

Passiflora lobbii is found in the Andes at high el-

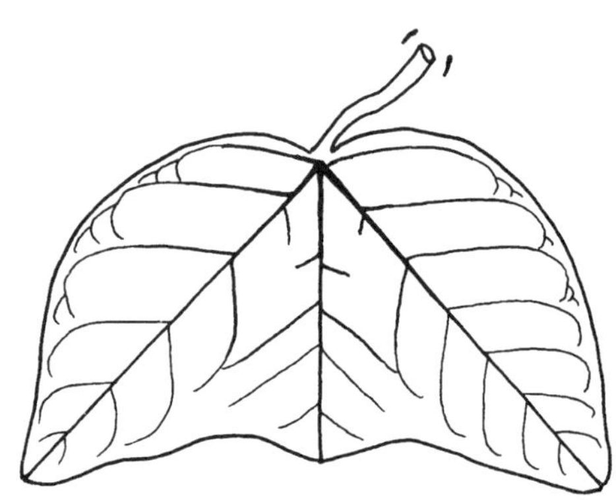

Passiflora lobbii

evations of 2500–3300 m and is therefore quite resistant to cold. During overwintering, it will easily tolerate night temperatures just above freezing.

Vine: nearly glabrous; stem: angulate, striate; stipules: setaceous, 0.3–0.8 cm long, soon deciduous; petioles: 0.6–2 cm, with 0–2 (rarely 4) glands; leaves: 3-lobed, 1.5–4.4 × 2–8(–9.5) cm, coriaceous, entire; peduncles: 1.5–4 cm, generally in pairs; bracts: minute, 0.2–0.4 cm long; flowers: green, greenish yellow, 1.7–3 cm diam.; sepals: red, greenish red or green outside, green inside, triangular, 0.6–1.2 × 0.3–0.7 cm; petals: green to greenish white, narrowly ovate, 0.5–0.9 × 0.2–0.3 cm; corona: 2 series, outer series 0.5–1 cm long, purplish at base, white, whitish green or pale yellow above, inner series 0.1–0.4 cm long; ovary: ovoid, glabrous; fruit: globose, ca. 1.5 cm diam.

Passiflora lutea L.

Sp. Pl. 958 (1753)
luteus: yellow
Supersection *Decaloba*, Section *Decaloba*

This species (Plate 124) is at home in forested eastern North America, advancing in the north to Missouri, Illinois, Indiana, Ohio, and Pennsylvania—farther than any other species of *Passiflora*. In the west the border runs through Texas, Oklahoma, and Kansas. Judging from the conditions in its natural surroundings, *P. lutea* might well be the hardiest of all passionflowers, although it is closely approached by *P. incarnata*. It regularly survives winters with temperatures of –15°C and short cold snaps to –30°C. Thus, *P. lutea* can also be overwintered outdoors in Europe and North America when provided the additional protection of fir twigs, straw, or foliage. As with *P. incarnata*, the shoots of this herbaceous species die down to the ground in winter. The annual new growth can reach a height of 3 m. With a diameter of about 1.5–2 cm, its greenish yellow flowers are rather small. It runs underground extensively by thick white rhizomes and can emerge on the other side of the garden path from where it was seen the year before.

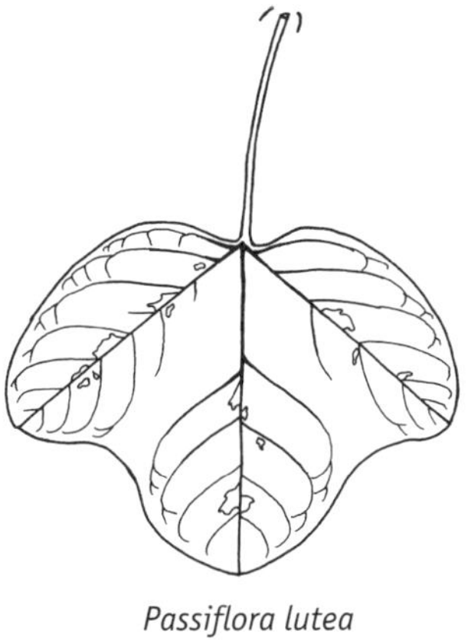

Passiflora lutea

This species has hypogeal germination. Only 4 species in the entire genus are known to have this condition, though most have not been checked. *Passiflora lutea* can be successfully propagated from root cuttings, as well as from cuttings and seeds.

Plant: glabrous to sparsely pubescent throughout; stem: almost terete, sulcate-striate, annual; stipules: setaceous, falcate, ca. 0.2–0.5 cm long; petioles: 1–5 cm, without glands; leaves: 3-lobed with shorter middle lobe, 2–10 × 2.5–15 cm, laminar nectaries absent, entire; peduncles: slender, in pairs, 1.5–6 cm long, 1-flowered; bracts: absent or rarely 1 or 2, microscopic; flowers: light greenish yellow, ca. 1.5–2 cm diam.; sepals: greenish yellow, oblong-lanceolate, 0.6–1.2 × 0.2–0.3 cm; petals: pale greenish yellow, almost whitish, narrowly lanceolate, 0.5–0.8 × 0.15 cm; corona: 2 series, outer series 0.5–1 cm long, filiform, pale yellow or pale greenish yellow, usually with dull violet at base, inner series ca. 0.2 cm long, capillary; ovary: glabrous, ellipsoid; fruit: subglobose, 0.7–1.2 × 0.7–1 cm, blackish purple; seeds: very widely obovate, ca. 0.4 × 0.3 cm, transversely sulcate.

Passiflora membranacea Benth.

Pl. Hartweg. 83 (1841)
membranaceus: thin-skinned
Supersection *Hahniopathanthus*

This extraordinary passionflower has thus far remained a rarity. Although *Passiflora membranacea* (Plate 125) was cultivated in Europe for some years, reports about flowering specimens were very uncommon, and it was regarded as reluctant to flower. Fortunately, a different clone from Chiapas was introduced by Dennis Breedlove and grown at the Strybing Arboretum in San Francisco. Both this

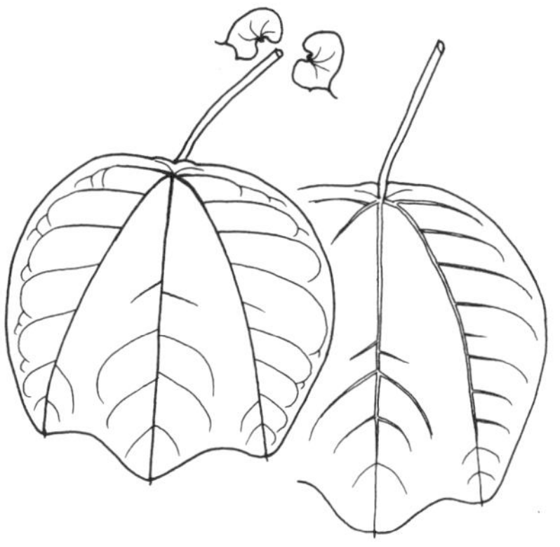

Passiflora membranacea

and another clone from Guatemala introduced by John MacDougal came into general circulation, and both turned out to be much more free-flowering.

Passiflora membranacea is native to cloud forests from the mountains of Mexico to Panama at altitudes of 1200–3350 m. It needs abundant moisture and moderate to slightly cool conditions. The bell-shaped, pale greenish to yellow flowers hang down on long, thin peduncles. Conspicuous and characteristic of the species are its 2 (very rarely 3) reddish bracts, which give the flower a distinctive effect. Moreover, being hummingbird-pollinated, *P. membranacea* has a long androgynophore, an unusual feature in subgenus *Decaloba*. The flowers open only about halfway and reach a diameter of up to 5 cm. The ovoid or ellipsoid fruit is edible but so highly perfumed as to be cloying.

In coastal California this species is cultivated outdoors year-round because it tolerates cooler periods with temperatures near the freezing point. It is advisable to overwinter *Passiflora membranacea* at rather low temperatures, 7°–10°C, to secure abundant flowering the following summer. Although it is very advantageous for this species to be planted directly in the ground or cultivated in a spacious container, even a medium-sized plant in a 20-cm pot may set some flowers.

Plant: glabrous throughout, except Breedlove clone with puberulous underside of leaf; stem: terete; stipules: foliaceous, clasping, 0.5–2 × 1.5–3.8 cm, entire to crenate, glandular; petioles: 2–4 cm, without glands, or rarely with 1–2 small glands near or above middle; leaves: very shallowly 3-lobed, almost orbicular, 5–9 × 6–9 cm, bearing laminar nectaries at margin, subpeltate 1–3 mm from margin, often reddish purple beneath, almost entire; peduncles: solitary or in pairs, 7–17.5 cm long, 1-flowered; bracts: 2, 2.7–4.6 × 2.5–6.3, foliaceous, pink to vivid purplish red, rarely a 3rd bract present up to 1 × 1.2 cm, free; flowers: pale green to light greenish yellow or cream, campanulate, ca. 3.5 cm diam.; sepals: pale green to light greenish yellow or cream, oblong-lanceolate, 3.5–6 × ca. 1–1.3 cm; petals: pale green to pale greenish yellow or whitish, oblong, 2–5.5 × ca. 1 cm; corona: 1 or 2 series, hidden, outer series 0.7–2.5 cm long, cream, inner series (if present) 0.1–0.6 cm long; ovary: glabrous, ellipsoid; fruit: ellipsoid, 7–10.5 × ca. 4 cm, greenish yellow to yellow; seeds: ca. 0.6 × 0.4 cm, reticulate.

Passiflora microstipula Gilbert & MacDougal

Lundellia 3: 1 (2000)
microstipulus: having small stipules
Supersection *Pterosperma*

The peculiar *Passiflora microstipula* (Plate 126) has delighted enthusiasts for more than 20 years. In the 1970s, Lawrence E. Gilbert collected seeds from a specimen in Mexico and raised plants from them at his institute in the University of Texas at Austin. Even at first glance it was evident that this species was unique in several respects and would not fit into any existing subgenera. Accordingly, Gilbert and John MacDougal assigned it to section *Pterosperma* of subgenus *Decaloba* (MacDougal and Hansen 2003), and it was later raised to the rank of supersection (Feuillet and MacDougal 2004).

This species is outstanding first of all for its large seeds, which are surrounded by large striate wings with erose margins. Gilbert also noticed different stages in the development of the plants. Juveniles have pilose, heart-shaped leaves that show a white variegation along the midrib. As the plants age, the leaves become more elongate, glabrous, and philodendron-like. Furthermore, while in young plants the petiolar nectaries are small, green, and filiform, in older specimens they are conspicuous, ivory-colored, and globose. When the typical leaf form has become established, the first flowers are produced. As is typical of this section, they appear in 1 or 2 pairs on the tendril. Flowers are white with a reddish to purplish coronal base, reach a diameter of 4.5–5.5 cm, and effuse a sweet fragrance faintly like wilted rose petals. The greenish yellow fruits contain sweetish, transparent arils with a faint vanilla odor. It is not known how the seeds are spread, but the fragrant fruits hang on the vine at maturity and are probably eaten by frugivorous bats.

Passiflora microstipula is at home in very humid tropical forests at elevations of 20–330 m from Veracruz to Oaxaca, Mexico. In its natural surroundings it develops from a medium-sized woody vine into a canopy liana with shoots up to 15 m long. Accordingly it needs much space when cultivated. Planted directly in the ground, it can reach the dimensions of *P. quadrangularis* or *P. alata*. When grown as a container plant, it requires a pot at least 30 cm diam. to flower reliably. Because of its tropical origin, it prefers a protected location under glass. It will ail when temperatures drop below 15°C for a long time, especially in winter. Alas, cuttings of *P. microstipula* root very badly. Months may pass before new growth appears.

Vine: glabrous or puberulent; stem: terete with posture of shoot tip cernuous, waxy; stipules: narrowly triangular, 0.2–0.4 cm long, soon deciduous; petioles: 4–8(–10) cm, with 6–8 subsessile, large glands; leaves: unlobed, ovate to ovate-elliptic, 7–22 × 5.5–16 cm, bearing laminar nectaries; pedicles: 1.5–2.5 cm long, 2- to 4-flowered cyme terminating in a tendril; bracts: linear-lanceolate, 0.1–0.3 cm long, soon deciduous; flowers: white, yellow, green, and reddish, 4.5–5.5 cm diam.; sepals: greenish outside, white inside, oblong-lanceolate, 1.7–2.4 × 0.7–1 cm; petals: white, oblong-ovate, 1.8–2.5 × 1–1.4 cm; corona: 2–3(–4) series, outer series 0.9–1.3 cm long, green at base, banded reddish purple to dark red near the middle, pale to bright yellow at apex, inner series 0.2–0.6 cm long; ovary: ellipsoid, minutely puberulent; fruit: ovoid, 6–8 × 4–5.5 cm, yellowish green; seeds: strongly compressed, 0.8–1.1 × 0.7 × 1 cm, obscurely reticulate, large striate wings at margin.

Passiflora morifolia Mast.

Fl. Bras. 13 (1): 555 (1872)
morifolius: mulberry-leafed
Synonyms: *P. erosa* Rusby, *P. heydei* Killip, *P. warmingii* Mast., *P. weberiana* Mast.
Supersection *Bryonioides*

Passiflora morifolia (Plate 127) is an easy species cultivated by many enthusiasts. It will grow abundantly even on a windowsill and does not require winter temperatures greater than 7°C. When 15°C is guaranteed throughout the year and sufficient light is provided, this plant will flower continually with only a few short breaks. The main flowering time is June to October.

The fragrant white and purple flowers, about 2.5–3 cm diam., are self-pollinating. The dark purple to black-violet fruits are similar in size and color to small plums and hang on the plant alongside flowers. Contrary to assertions made in some books, this plant's orange to brick-red arils are not tasty. They might be digestible, but they can hardly be called mouth-watering.

The main regions of the large native range of *Passiflora morifolia* are Bolivia, Argentina, Paraguay,

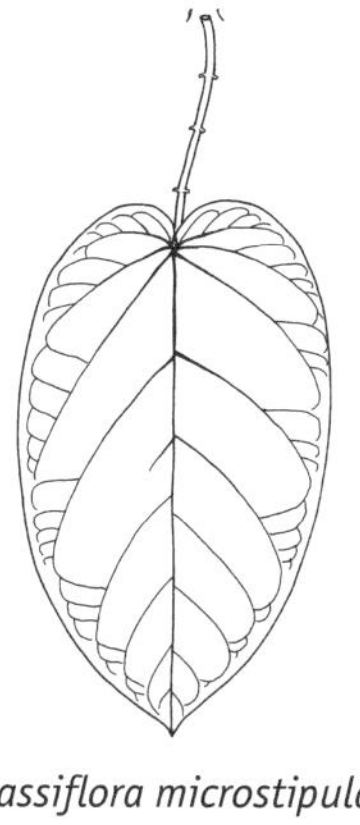

Passiflora microstipula

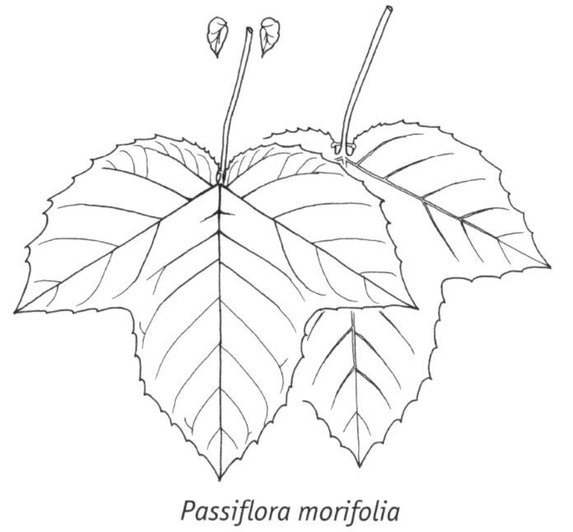

Passiflora morifolia

and Brazil. This plant was temporarily introduced into the southern United States years ago but apparently did not survive. It also grows in isolated pockets in Guatemala, Colombia, Venezuela, Ecuador, and Peru in seasonally dry forests, mainly at altitudes of 400–2800 m. In wild habitat the stems of this species die back each dry season to near the base, but in a house or greenhouse this may not occur. *Passiflora morifolia* is very well suited as a foodplant for the caterpillars of heliconid butterflies, and so can often be found in butterfly gardens.

Plants cultivated under the name *Passiflora bryonioides* are almost always *P. morifolia*. These can easily be distinguished by the central lobe of the leaf, always narrowed at the base in *P. bryonioides*, and by the dark violet fruit with orange pulp that is characteristic of *P. morifolia*. Sometimes *P. morifolia* is cultivated under the name *P. sicyoides*, but that species differs by the petiolar nectaries in the lower 1/2 of the petiole. Min. temp.: 10°C.

Plant: pubescent throughout with hooked trichomes; stem: almost terete, sometimes annual; stipules: ovate-auriculate, 0.4–1 × 0.2–0.7 cm; petioles: 2–7 cm, with 2 glands toward the apex; leaves: 3-lobed, 3.5–13 × 4.5–15.5 cm, rarely bearing 1–2 laminar nectaries near margin, denticulate to serrate; peduncles: solitary or usually in pairs, 1–3 cm long, 1-flowered; bracts: setaceous, 0.2–0.6 cm long; flowers: white with dark purple center and white and purple corona, 2.5–3 cm diam.; sepals: greenish outside, white inside, lanceolate, 1.3–2 × 0.5–0.8 cm; petals: white, lanceolate, 0.7–1.1 × 0.3–0.5 cm; corona: 1 series, 0.5–1.1 cm long, filiform, white with few purple or violet zones; ovary: hispidulous, subglobose; fruit: subglobose to ellipsoid, 2.5–4 × 2.2–3.5 cm, blackish purple; seeds: ca. 0.4–0.5 × 0.3–0.4 cm, coarsely reticulate.

Passiflora murucuja L.

Sp. Pl. 957 (1753)
Murucuia: former name of the genus; *murucuja*: vernacular name for "passionflower"
Supersection *Decaloba*, Section *Decaloba*

Passiflora murucuja (Plate 128) has rapidly become a favorite passionflower. Relatively little known in the early 1990s, it is now offered even at garden centers. Its flowers, 4–6 cm diam., are particularly charming. The light red to blood-red sepals and petals reflex more and more when the flower has opened. The red or orange-red corona is connate into a tubular membrane of about 1 cm long, enclosing a large part of the androgynophore. Because of the unusual coronal structure, the "hummingbird-flowered" *P. murucuja* was first regarded as a separate genus from passionflowers; later it was made the type species of subgenus *Murucuia* in *Passiflora*. It is now recognized as a member of subgenus *Decaloba*, a conclusion made obvious by the leaves though initially obscured by the exquisitely modified flower. Many enthusiasts hold that this is the most beautiful species within subgenus *Decaloba*.

Alas, this persistently flowering species occasionally aborts some of its numerous flower buds. Nevertheless, more than 6 flowers may be open in one day on a relatively small plant in a 15-cm pot. This species fruits with its own pollen, quite often producing globose, black fruit.

Passiflora murucuja produces tiny greenish "pseudoflowers," or aborted flowers, especially during winter—a hitherto unexplained phenomenon. Due to its unpretentiousness and its easily restrained growth, this species is excellently suited for cultivation on a windowsill, but it will also flourish outdoors or in a greenhouse. Indigenous to Puerto Rico, Haiti, and the Dominican Republic, it is satisfied with an overwintering temperature above 10°C.

As with almost all red-flowering species, *Passiflora murucuja* runs the risk of developing chlorotic leaves and paler flowers. Regular application of an additional fertilizer containing sequestered iron is recommended.

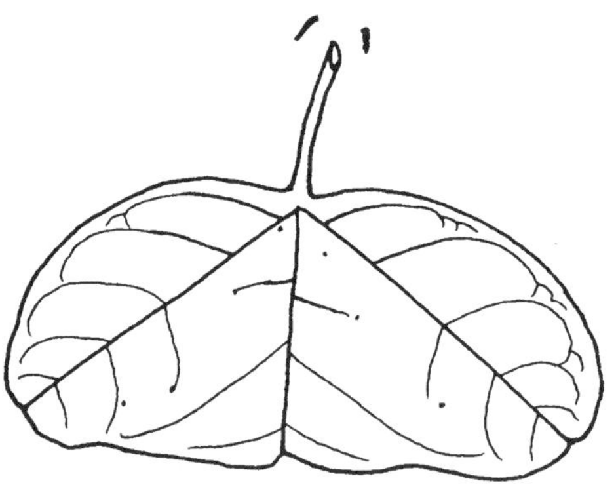

Passiflora murucuja

Vine: glabrous; stem: angulate, deeply grooved, wiry; stipules: linear-setaceous, 0.2–0.4 cm long; petioles: up to 1.5 cm long without glands; leaves: 2-lobed, variable, 1–2(–3) × 2.5–4(–7) cm, bearing laminar nectaries, entire; peduncles: 1–2.5 cm, solitary or often in pairs; bracts: setaceous, 0.2–0.4 cm long, below middle of peduncle, free; floral tube: 0.5–0.7 cm long, red outside; flowers: red, 4–6 cm diam.; sepals: red, lanceolate-oblong, 1.5–3 × 0.3–0.7 cm; petals: red, linear-oblong, 1–2 × 0.3–0.4 cm; corona: 1 series, cylindrical membrane 1–1.5 cm long, red or orange-red; ovary: ovoid, glabrous; fruit: globose, 1–1.5 cm diam., blackish; seeds: obovate, 0.2–0.3 × 0.1–0.2 cm, transversely 6- to 7-sulcate.

Passiflora oblongata Swartz

Prodr. Veg. Ind. Occ. 97 (1788)
oblongatus: elongated
Synonyms: *P. elongata* Poir., *P. lyraefolia* Tussac, *P. oblonga* DC.
Supersection *Decaloba*, Section *Decaloba*

As a result of Elma Kay's studies of passionflowers in the Greater Antilles, some species from this area have regained prominence. Surprisingly, many species endemic in this region are hummingbird-pollinated. Their flowers are purple or reddish, tubular, and odorless, with diurnal anthesis. Distinctions between these species are often based on the structure of the corona and other floral characteristics related to pollination. As a rule, the vegetative characteristics are typical of species from subgenus *Decaloba* and are conditional only for an unequivocal distinction.

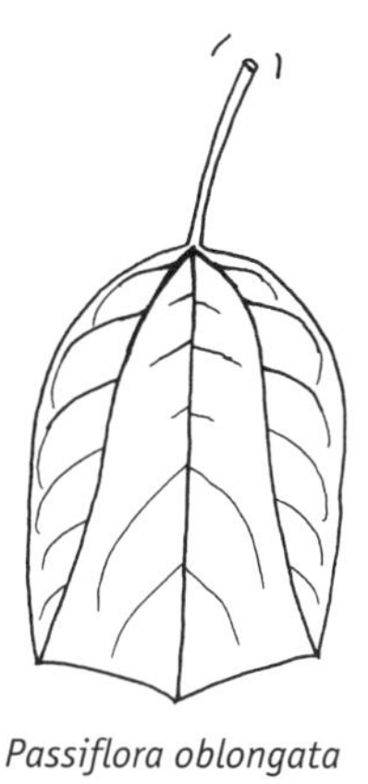

Passiflora oblongata

Passiflora oblongata (Plate 129) is hitherto a less well-known hummingbird-pollinated species. It is common throughout Jamaica, growing at elevations up to 900 m in tropical forest that is moist to wet. Its light purple to rose-pink flowers reach a diameter of about 5 cm. As is typical of the *Pseudomurucuja* group, the androgynophore juts far out, while the corona is very small. *Passiflora oblongata* is distinguished from the other Jamaican representatives of *Pseudomurucuja* (*P. perfoliata*, *P. tacsonioides*, *P. calcicola*) by its pseudoracemose inflorescence and by its leaves, which are usually longer than wide.

Though brought into cultivation as early as 1983, this rarity is only cultivated by a few enthusiasts in the United States. Its demands are similar to those of *Passiflora perfoliata*, which is at home in similar habitats.

Vine: glabrous; stem: angulate, striate; stipules: linear-subulate, 0.4–0.7 cm long; petioles: 1.5–4 cm long, without glands, often purplish; leaves: 2- or 3-lobed, variable, generally oblong, occasionally cuneate-obovate, bearing laminar nectaries, subcoriaceous, entire; peduncles: ca. 1 cm long, solitary or generally in pairs, borne on elongated branches with reduced leaves; bracts: setaceous, ca. 0.2 cm long, free; floral tube: cylindrical or tubular-campanulate, 1–2 cm long, purplish red outside; flowers: purplish red, crimson, scarlet, or rose, 3.5–5.5 cm diam.; sepals: purplish red, linear-oblong, 1.5–3 × 0.4–0.5 cm; petals: purplish red, linear, 0.8–2 × 0.3–0.4 cm; corona: 1 or 2 series, outer series 0.2–0.4 cm long, green with purplish apex, inner series minute (if present); ovary: ovoid, glabrous, green, 6-angled; fruit: globose, ca. 1 cm diam.; seeds: orbicular-ovate, 0.3–0.4 × 0.3 cm, transversely 6- to 8-sulcate.

Passiflora organensis Gardn.

London J. Bot. 4: 104 (1845)
organensis: from the Organ Mountains
Synonyms: *P. maculifolia* Mast., *P. porophylla* Vell.
Supersection *Decaloba*, Section *Decaloba*

Passiflora organensis (Plate 130) is a favorite among species with colorful leaves. Its decorative 2- or 3-lobed leaves are purple to light violet underneath, with upper sides that are dark green with a light

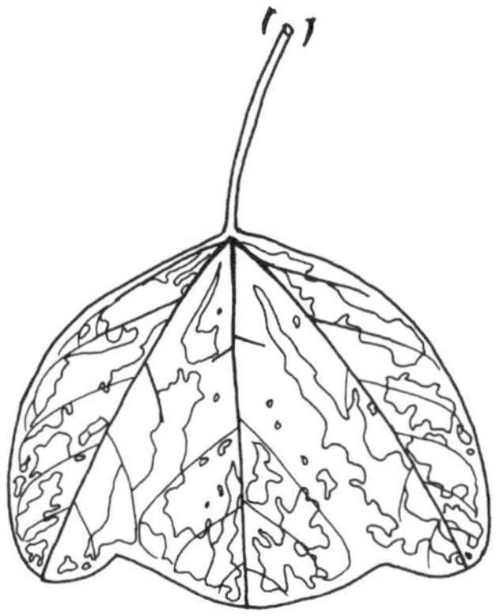

Passiflora organensis

green to cream, partially violet marbling. The intensity of this marbling is dependent on temperature and on the plant's supply of nutrients and trace elements.

Passiflora organensis is rather reluctant to produce its flowers, which are up to 5 cm diam. In comparison to other species with colorful leaves, especially those from *Decaloba*, it might even be regarded as unwilling to flower. However, the coloration of the leaves makes up for this. The sepals and petals vary in shade, from cream with a touch of purple to predominantly purple.

This species grows naturally in the Organ Mountains in southwestern Brazil. It dislikes a rooting medium containing too much peat, as this retains moisture too long. A stony potting mix, such as for cacti and succulents, should be added. Rather exacting in its demands, this species will suffer during spells of hot weather when it is cultivated under glass. Add a shading device to the greenhouse, or cultivate the plant outdoors during summer months. During overwintering, protect the plant from temperatures that are too low. *Passiflora organensis* will tolerate temperatures down to 5°C for a short while, but 12°C is more ideal.

Vine: glabrous; stem: subangular, compressed; stipules: linear-subulate, 0.2–0.4 cm long; petioles: 1.5–3 cm long, slender, without glands; leaves: 2-lobed, rarely 3-lobed, 3–7 × 2.5–10 cm, bearing laminar nectaries, entire; peduncles: up to 4 cm, generally in pairs; bracts: setaceous, 0.2–0.4 cm long, near middle of peduncle; flowers: cream-white with purple, 3–4.2 cm diam.; sepals: greenish white outside, cream or cream with purple inside, oblong-lanceolate, 1.4–2 × 0.5–0.6 cm; petals: cream to cream with purple, oblong, 0.7–1 × 0.3–0.4 cm; corona: 1 series, strongly dolabriform, ca. 0.5 cm long, cream in lower 1/2, deep purple above; ovary: ovoid, glabrous; fruit: globose, 1–1.5 cm diam., blackish; seeds: ovate, 0.4–0.5 × 0.2–0.3 cm, transversely 6- to 9-sulcate.

Passiflora penduliflora Bert.

Prodr. 3: 326 (1828)
penduliflorus: having pendent flowers
Synonyms: *P. rotundifolia* var. *swartzii* DC., *P. swartzii* (DC.) Mast.
Supersection *Decaloba*, Section *Decaloba*

Passiflora penduliflora (Plates 131 and 132) is among the most exceptional species of this subgenus. Its greenish campanulate flowers are pendent on long, sturdy peduncles. Though several other species from this subgenus have green or pendent flowers, none have a combination of the two. The sepals cannot enclose the growing flower buds completely, and so they always look a bit burst. The short, upright corona is also conspicuous—green with a rich yellow tip that contrasts with the rest of the flower. In addition, the flowers of *P. penduliflora* open early in the evening and remain open until early morning, effusing a musty smell. All these features are adaptations that suggest pollination by bats, a phenomenon otherwise unknown in *Decaloba* species. In fact, *P. penduliflora* turns out to be pollinated by the long-tongued glossophagine bat, *Monophyllus redmani*. Elma Kay (2001) furnished proof of this during her studies on the *Passiflora* species of the Greater Antilles. At a diameter of about 4 cm, the flowers of *P. penduliflora* are almost too small for their pollinators. Other bat-flowered species have larger flowers and belong to other subgenera.

All these peculiarities, in combination with other morphological characteristics, induced Killip (1938) to assign *Passiflora penduliflora* to the monotypic subgenus *Astephia*, though this name was never validly published. Quite recently, as a consequence of the new systematics, this species was moved into subgenus *Decaloba*, together with the closely related representatives of the former subgenera *Pseudomurucuja* and *Murucuia*.

The native range of *Passiflora penduliflora* is re-

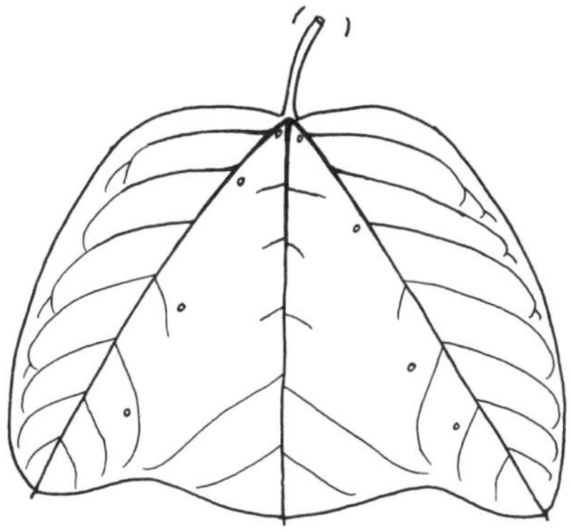

Passiflora penduliflora

stricted to Cuba (mainly the eastern part) and Jamaica. It has come into cultivation in the last few years and is now represented in many collections. In the 1990s, Lawrence E. Gilbert bred and flowered a hybrid of this species with its putative sister species, *P. perfoliata*. In 2002, Mark Cooper repeated the cross, and Myles Stewart Irvine named the beautiful hybrid *P.* 'Flying V' in reference to the form of the leaf.

Passiflora penduliflora is by no means difficult to cultivate. It prefers constant temperatures around 25°C. In regions with variable summers it should be grown under glass. In winter, temperatures below 10°C are to be avoided.

Vine: glabrous; stem: angular, striate; stipules: setaceous, 0.3–0.4 cm long; petioles: 0.5–2 cm long, slender, without glands; leaves: 3-lobed at apex, variable, occasionally nearly unlobed, 4–7.5 × 2.5–8 cm, bearing laminar nectaries, entire; peduncles: 2–10 cm, pendent, solitary or in pairs; bracts: subulate, 0.2–0.4 cm long, soon deciduous, free; floral tube: campanulate to cylindrical-campanulate, about 1 cm long, green outside; flowers: whitish green to green or greenish yellow, about 4 cm diam.; sepals: greenish white or green, oblong-lanceolate, 1.5–2 × 0.4–0.6 cm; petals: green, oblong-lanceolate, 1.7–2.4 × 0.5–0.7 cm; corona: 1 series, 0.2–0.5 cm long, erect, green with yellow apex, united at base forming a fleshy ring; ovary: ovoid, glabrous; fruit: globose or ovoid, 1–1.5 cm diam.; seeds: broadly ovate, transversely 8- to 10-sulcate.

Passiflora perfoliata L.

Sp. Pl. 956 (1753)
perfoliatus: having leaves pierced by the stem
Synonyms: *P. cephaleima* Bory, *P. normalis* L., *P. perfoliata* var. *normalis* (L.) Fawcett & Rendle
Supersection *Decaloba*, Section *Decaloba*

Passiflora perfoliata (Plate 133) is possibly the best-known hummingbird-flowered species indigenous to Jamaica, where it grows from sea level to elevations up to 1000 m. This interesting species has been cultivated for many years. It belongs to the group of easy-to-grow purple-red species that continue to grow happily in winter when provided temperatures above 12°C. Even in a 12-cm pot, it will set some flowers on short shoots, and so it is an ideal plant for a windowsill. This rather warmth-loving species does very well in a greenhouse; if planted out, it can reach an undreamed-of size. Under ideal conditions it may produce its purple-red, 5-cm-diam. flowers year-round with only a few short breaks. The remarkable structure of these flowers is typical of species in the *Pseudomurucuja* group. The androgynophore juts out far, the sepals and petals are long and narrow, and the corona is very short. Because of these features the flowers look a little scanty rather than compact. The petals are significantly larger than the sepals, so that the young flower buds have a campanulate form with a slightly open tip.

The equally conspicuous leaves are comparatively small and deeply 2-lobed, occasionally with an additional short lobe in the middle. The angle at which the 2 outer nerves meet is usually larger than 90°.

Vine: nearly glabrous; stem: angular, striate; stipules: linear-subulate, 0.2–0.4 cm long; petioles: about 0.5 cm long, without glands; leaves: deeply 2-lobed, occasionally with short 3rd lobe at middle, 1.5–3.5 × 4–12.5 cm, bearing laminar nectaries, entire; peduncles: 2–3 cm long, solitary or in pairs;

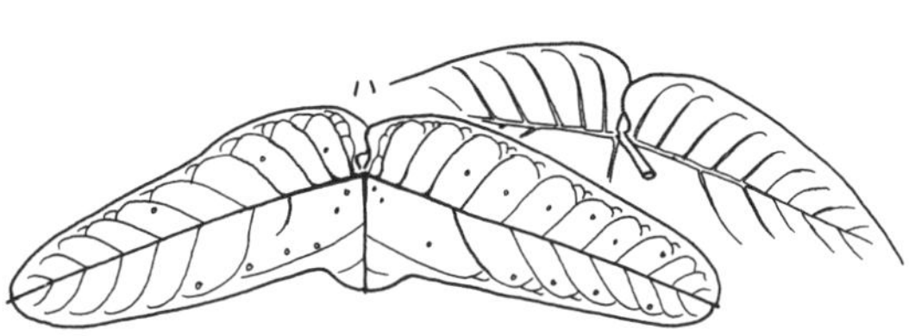

Passiflora perfoliata

bracts: setaceous, 0.2–0.3 cm long, free; floral tube: nearly cylindrical, 0.7–1.3 cm long, purple-red outside; flowers: purple-red, 3.5–5.2 cm diam.; sepals: purple-red, linear-subulate, 1–2 × 0.2–0.3 cm; petals: purple-red, oblong-lanceolate, 1.3–2.3 × 0.5–0.7 cm; corona: 1 series, 0.3–0.5 cm long, yellowish to green with orange apex; ovary: ovoid, subangulate, glabrous; fruit: globose, up to 1.5 cm diam.; seeds: ovate or slightly obovate, ca. 0.3 × 0.2 cm, transversely 5- or 6-sulcate.

Passiflora pilosa Ruíz & Pavón ex DC.

Prodr. 3: 330 (1828)
pilosus: covered in hair
Supersection *Bryonioides*

Many species in supersection *Bryonioides* have annual shoots from thick perennial roots, and *Passiflora pilosa* (Plate 134) is a prime example. This species is found in southern Mexico and Guatemala at 1250–2000 m altitude, growing in grassy oak thickets in areas with a strong dry season. It goes without rain for several months of the year, surviving by dying back to the ground. It is covered with short stiff hairs and has greenish fruits that drop to the ground, attracting mammals with their sweet scent. There are 2 subspecies. The typical one grows in southwestern Mexico and has very wide stipules that are foliose on both sides of the main vein. The other, *P. pilosa* subsp. *dimidiata*, has stipules that are foliose only on one side of the main vein, narrower bracts, no hooked trichomes, and a similar but smaller flower. This plant ranges from Guerrero, Mexico, to Guatemala. The flowers of both are typical of this group: white with a single-ranked white and violet corona. The species is not presently in cultivation.

Plant: hispid throughout with straight or hooked trichomes; stem: terete, annual; stipules: ovate to ovate-auriculate, 0.7–3 × 0.2–2.3 cm; petioles: 1–5 cm, with 2 glands usually near middle or above; leaves: 3- or nearly 5-lobed, 3–11.5 × 4–16 cm, laminar nectaries absent, denticulate to serrulate; peduncles: usually in pairs, 0.7–2.9 cm long, 1-flowered; bracts: linear, 0.4–1 × ca. 0.1–0.3 cm; flowers: white with white and purple corona, ca. 3–4 cm diam.; sepals: greenish outside, whitish inside, ovate-triangular, 0.9–2 × 0.4–0.9 cm; petals: white, ovate-triangular, 0.9–1.9 × 0.4–0.6 cm; corona: 1 series, 0.9–1.3 cm long, filiform, white with purple zones; ovary: glabrous, ellipsoid to ovoid; fruit: ovoid, 3–3.8 × 2.3–2.7 cm, pale green with whitish apex; seeds: ca. 0.5 × 0.3 cm, finely reticulate.

Passiflora punctata L.

Sp. Pl. 957 (1753)
punctatus: dotted with spots (in reference to the leaves)
Supersection *Decaloba*, Section *Decaloba*

The slightly mottled, purple-violet leaves of *Passiflora punctata* (Plate 135) bear many spotlike nectaries on their undersides, giving the plant a distinctive charm. Its flowers are less spectacular, about 4 cm diam. and borne on long, thin peduncles. However, during the morning hours they effuse a pleasant fruity scent that can be noticed when one comes near the plant. Purple shades can

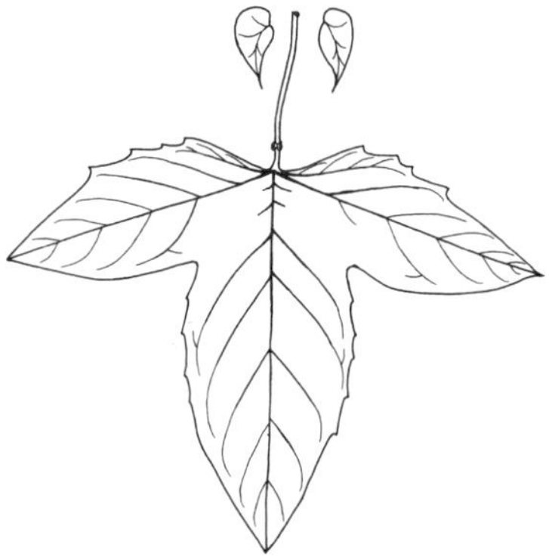

Passiflora pilosa

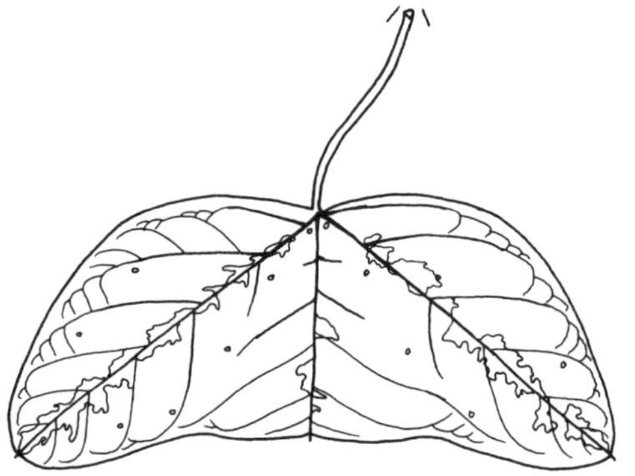

Passiflora punctata

be found on the cream-colored sepals and petals near the middle of the flower.

Passiflora punctata is indigenous to Panama, Colombia, Ecuador, Peru, and Bolivia, where it advances from sea level to altitudes up to 1500 m. An easy plant, it can spend the summer on a patio or balcony and may well flower even in a 20-cm pot. Its pretty foliage offers a welcome change. The minimum temperature is 7°C.

Sometimes *Passiflora punctata* is confused with *P. biflora*, *P. tuberosa*, and especially *P. misera*. All of these species share a similar leaf structure but all have completely different flowers. The closest relative is *P. colinvauxii*, endemic to the Galápagos Islands, which has nearly identical flowers and is barely distinct.

Vine: glabrous; stem: 3-angular, striate; stipules: linear-falcate, 0.3–0.5 cm long; petioles: 3–6(–8.7) cm long, without glands; leaves: shallowly 3-lobed at apex, rarely nearly 2-lobed, (0.7–)2.5–6.6 × 2.9–12.5(–17.5) cm, bearing laminar nectaries, membranous, entire; peduncles: (2–)5–8 cm long, slender, solitary or in pairs; bracts: setaceous, 0.1–0.2 cm long, scattered, soon deciduous; flowers: cream and purple, 2.5–4.5 cm diam.; sepals: greenish cream outside, cream and often with purplish basal part inside, oblong-lanceolate, 1.5–2.2 × 0.5–1 cm; petals: cream, often with purplish basal part, oblong-lanceolate, 0.8–1.2 × 0.2–0.6 cm; corona: 2 series, outer series 0.7–1 cm long, white at base, purplish in lower 1/2, cream-yellow above, dilated near middle, inner series 0.4–0.5 cm long, purplish; ovary: narrowly ovoid to ellipsoidal, puberulent to villosulous; fruit: subglobose to ellipsoid; seeds: ovate, ca. 0.3–0.4 × 0.2 cm, transversely 6- to 7-sulcate, the ridges rugulose.

Passiflora quinquangularis S. Calderón ex MacDougal

Novon 14 (2004)
quinquangularis: 5-angled (in reference to the stem)
Supersection *Decaloba*, Section *Xerogona*

Lumped by Ellsworth P. Killip into the related *Passiflora capsularis*, this wonderful species of *Xerogona* (Plate 136) was proved distinct when John MacDougal collected it alive in Guatemala in 1980 and

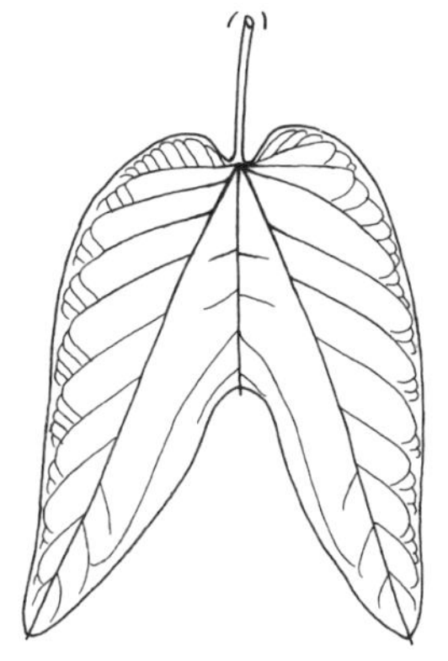

Passiflora quinquangularis

introduced it into horticulture. There is presently only this one clone in the trade. Although the plant is reminiscent of *P. rubra*, the pointy flower buds hang down on very slender peduncles, almost like spiders on their threads. The flowers turn to the side before opening horizontally and, still hanging on their threadlike peduncles, emit a sweet, very fruity fragrance.

Passiflora quinquangularis is native to thickets and brushy hillsides in Guatemala and El Salvador at 750–1600 m altitudes. It is easy to grow, quickly rewarding the gardener with abundant growth (up to a few meters) and many flowers. It will flower in a 12-cm pot. Min. temp.: 12°C.

Plant: lightly pubescent throughout; stem: 5-angulate; stipules: setaceous, falcate, 0.4–0.7 cm; petioles: 1–2 cm, without glands; leaves: 2-lobed to near middle, 4–14 × 4–11 cm, laminar nectaries absent, entire; peduncles: solitary, rarely in pairs, 1.5–5.5 cm long, very slender, 1-flowered; bracts: absent; flowers: whitish with white and purplish red corona, ca. 3 cm diam.; sepals: greenish yellow outside, often flushed with red, whitish inside, lanceolate-triangular, 1.5–3 × 0.6–0.9 cm; petals: white, lanceolate, 1.1–1.5 × 0.4–0.5 cm; corona: 2 series, outer series 1.1–1.8 cm long, filiform, white with basal 1/2 light to medium reddish purple or purplish red, inner series 0.3–0.5 cm long, capillary; ovary: puberulent, ellipsoid; fruit: ellipsoid to widely fusiform, 6-keeled, 3–6 × 1.5–3 cm, reportedly yellowish green, dehiscent; seeds: ca. 0.3–0.4 × 0.2 cm, transversely grooved.

Passiflora rovirosae Killip

J. Wash. Acad. Sci. 12: 259 (1922)
named after Mexican botanist J. N. Rovirosa, collector of the type plant
Supersection *Decaloba*, Section *Xerogona*

Passiflora rovirosae (Plate 137) has an unusually shaped 2-lobed leaf. It is basically ovate or widely lanceolate in outline but is cut off at the end to form a flat edge; or more typically, it has a 2-lobed, crescent-shaped apex. This warmth-loving species grows from Veracruz, Mexico, to northern Guatemala and Belize, from sea level to 350 m elevation, producing much attractive foliage before it blooms. It was introduced by Alfred Lau in 1984 and is not uncommon in collections.

Larger plants are cauliflorous, with clumps of flowers borne on short spurs or compact pads from old wood. The fruit is similar to that of its close relative, *Passiflora costaricensis*: long, ridged, wine-red fruits split open to reveal white insides. *Passiflora rovirosae* can be distinguished from this relative by its short pubescence and 5-sided, not 3-keeled, stem.

This species will flower in a 20-cm or larger pot, but patience will be required. Min. temp.: 15°C.

Plant: puberulent throughout; stem: 5-angulate, striate; stipules: setaceous, falcate, ca. 0.4–0.9 cm long; petioles: 1.5–2.5 cm, without glands; leaves: 2-lobed or truncate at apex, 7.5–12 × 5–9.5 cm, laminar nectaries absent, entire; peduncles: often cauliflorous, usually in pairs, or clumped in groups, 0.5–2.3 cm long, 1-flowered; bracts: absent; flowers: cream, pale yellow, or pale greenish yellow, with purplish coronal base, ca. 4 cm diam.; sepals: yellowish green outside, light greenish yellow inside, lanceolate-triangular, 2–2.8 × 0.5–0.7 cm; petals: cream, lanceolate, 1.8–2.6 × 0.4–0.5 cm; corona: 2 series, outer series 1.3–1.6 cm long, filiform, whitish, cream, or pale yellow tinged purplish red on basal 1/2, inner series ca. 0.3 cm long, capillary; ovary: densely puberulent, ellipsoid; fruit: fusiform, 6-ridged, 6–10.5 × 1.5–2.5 cm, dark reddish or purplish red, dehiscent; seeds: ca. 0.4 × 0.2 cm, transversely grooved.

Passiflora rubra L.

Sp. Pl. 956 (1753)
rubrus: red (in reference to the fruit)
Synonym: *P. cisnana* Harms
Supersection *Decaloba*, Section *Xerogona*

Passiflora rubra (Plate 138) is among the most rewarding species to grow within this subgenus. It flowers even on short shoots and can be cultivated in a small pot kept on a windowsill. In a 30-cm pot it will develop into an imposing specimen that can spend the summer outdoors, and in a greenhouse it can reach an enormous size when ideal conditions are provided. Its cream or white flowers with a pink or red corona base reach a diameter of 3–4 cm. This is a widespread and variable species; the color and size of the flowers and form of the fruit and leaves all depend on the origin of the clone. Various clones of *P. rubra* with such differences have come into cultivation. These differences become especially evident when the fruits are compared. Some plants produce fusiform, 6-angular fruits, typical of section *Xerogona*, while others produce subglobose fruits without angles.

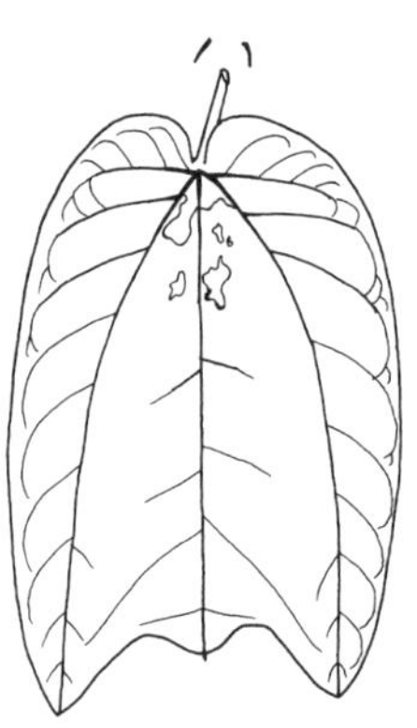

Passiflora rovirosae

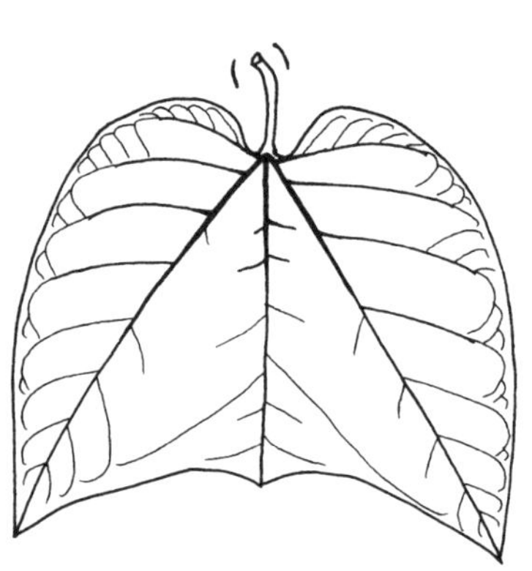

Passiflora rubra

This variability is not astonishing in view of the wide native range of *Passiflora rubra*. It is found from the Caribbean through French Guiana, Venezuela, and Colombia to Peru, Bolivia, and Brazil, occurring near sea level as well as at elevations up to 2200 m. A common feature, and a distinguishing criterion from *P. capsularis*, holds true for all plants belonging to this species: all have a densely white- or brown-hirsute ovary.

In winter *Passiflora rubra* is satisfied with temperatures above 7°C.

Vine: pubescent; stem: angulate, striate; stipules: subulate, 0.2–0.7 cm long; petioles: 0.4–3(–5) cm long, without glands; leaves: 2- or 3-lobed, 0.9–6(–8) × 1.5–10 cm, without laminar nectaries, entire; peduncles: 1.1–3.5 cm long, solitary or occasionally in pairs; bracts: absent; flowers: white to cream and pink to purplish, 2.8–4.5 cm diam.; sepals: green to reddish white outside, white to cream inside, oblong, 1.1–1.9(–2.5) × 0.3–0.7 cm; petals: white to cream, oblong, 0.7–1.5 × 0.1–0.3 cm; corona: (1–)2 series, outer series 0.5–1(–1.6) cm long, purplish or pink or red at base, cream to white above, inner series (if present) with few elements, 0.1–0.3 cm long; ovary: ellipsoid, densely white- or brown-hirsute; fruit: subglobose to ovoid or fusiform and 6-angular, 2.2–5 × 1.1–3.4 cm, reddish or red and white; seeds: ovate, ca. 0.3–0.4 × 0.1–0.2 cm, transversely 5- or 6-grooved.

Passiflora sagasteguii Skrabal & Weigend

Harvard Pap. Bot. 6: 321 (2001)
named after Abundio Sagástegui Alva, Peruvian botanist and collector
Supersection *Multiflora*

When the *Passiflora lobbii* group was revised (Skrabal et al. 2001), several new species were described, including *P. sagasteguii* (Plate 139). All of these species, growing at altitudes above 2500 m in the Andes, are characterized by dainty leaves with a 3-lobed tip. In addition, the petiole usually bears 2 nectaries. At a diameter of 2.5–3 cm, the flowers of the *P. lobbii* group are rather small. A very special case among these mostly yellowish-green-flowered species is *P. sagasteguii*, endemic in the departments of Cajamarca and La Libertad in northern Peru. At a diameter of about 4.5 cm, its flowers are the largest within the group and also stand out because of their rich purple or purplish red coloration.

This handsome species has been in cultivation for a few years and is enjoyed by more and more enthusiasts. It can be cultivated even in a 15-cm pot without problems and is also suited as a trailing plant.

In central Europe the main flowering periods are early spring and late fall. Although *Passiflora sagasteguii* tolerates heat spells better than many other species from high altitudes, it may show signs of distress. Cultivate it outdoors during summer months rather than in a greenhouse. Take it indoors in fall and provide it enough light in winter; the first flower buds will set after some weeks. Although older plants might tolerate a short spell of frost, temperatures above 5°C should be guaranteed.

Vine: finely pubescent; stem: subterete to angulate; stipules: setaceous, 0.3–0.5 cm long, soon deciduous; petioles: 1–3 cm long, with 2 glands below middle; leaves: 3-lobed, 1.5–4(–5.6) × (2.5–)4.5–9 cm, without laminar nectaries, subcoriaceous, entire; peduncles: 2.5–3.5(–5) cm long, solitary or generally in pairs; bracts: setaceous, 0.3–0.5 cm long, scattered; flowers: purple, 3.5–4.8 cm diam.; sepals: dark purple or whitish green outside, purple or reddish purple inside, oblong, 1.6–1.9 × 0.5–0.6 cm; petals: purple or reddish purple, oblong-ovate, 1.3–1.5 × 0.3–0.4 cm; corona: 2 series, outer series 0.4–0.7 cm long, dark purple in lower 1/2, whitish green or purple in upper 1/2, inner series 0.2–0.4 cm long, dark purple, generally with white apex; ovary: subglobose, glabrous; fruit: unknown.

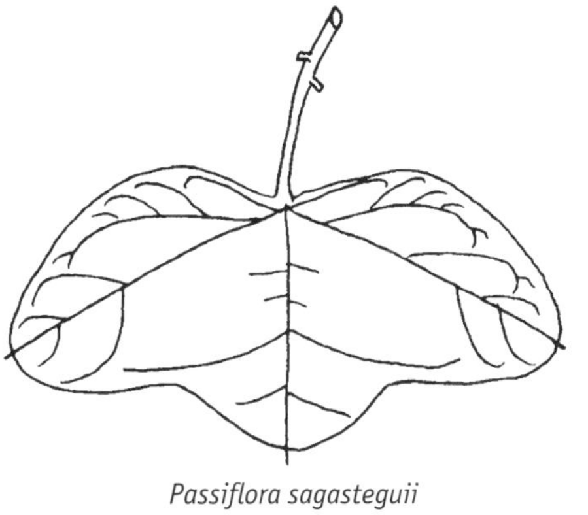

Passiflora sagasteguii

Passiflora sanguinolenta Mast.

Gard. Chron. 1868: 1162 (1868)
sanguinolentus: blood-red
Synonym: *P. mastersiana* Harms
Supersection *Decaloba*, Section *Xerogona*

One of the most beautiful small-flowered passionflowers, *Passiflora sanguinolenta* (Plate 140) is a very rewarding plant for beginners and advanced growers alike. This species comes from the mountainous regions of Ecuador at altitudes of 800–2800 m and is easy to cultivate. It will grow large enough to flower reliably even in a 12-cm pot on a windowsill and can also be grown in a much larger pot on a terrace or balcony. Under favorable conditions plants 2–3 m high may produce several hundred flowers in one season. *Passiflora sanguinolenta* flowers untiringly until fall. In winter, 7°C should be provided, although lower temperatures are tolerated for a short time.

The rose-pink flowers reach a diameter of about 5 cm. The long androgynophore and marked floral tube are unusual for *Decaloba* species. For a long time all species showing an adaptation to pollination by hummingbirds were banned from subgenus *Decaloba* and assigned to small subgenera of their own. This also applied to *Passiflora sanguinolenta*. Even Killip (1938) recognized that this species did not fit well into the old subgenus *Psilanthus*, together with *P. trinervia*, *P. bicuspidata*, and *P. hyacinthiflora*. He also pointed out a certain relationship with *P. rubra*. Since then subgenus *Psilanthus* has been dismissed, as have subgenera *Pseudomurucuja* and *Murucuia*, and all species have been assigned to subgenus *Decaloba*, in accordance with the new systematics by Feuillet and MacDougal. These pretty species belong to their own special groups, but not at the level of subgenus.

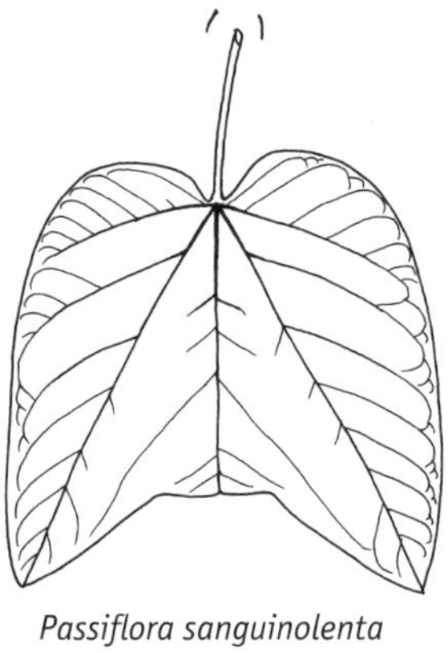

Passiflora sanguinolenta

Together with yellow-flowered *Passiflora citrina*, with which it has the most in common, and with *P. rubra* and *P. capsularis*, *P. sanguinolenta* was assigned to section *Xerogona*, in accordance with the real kinship relations. A cross of *P. sanguinolenta* with *P. citrina* brought about the excellent hybrid *P.* 'Adularia'.

Vine: densely hirsute; stem: angulate, striate; stipules: setaceous, 0.3–0.4 cm long; petioles: 0.6–2 cm long, without glands; leaves: 2- or 3-lobed, 1.5–5(–7) × 3–6.5(–11.5) cm, without laminar nectaries, entire; peduncles: 2.5–4.5 cm long, solitary; bracts: absent; floral tube: 1–2.5 cm long, reddish outside; flowers: light reddish, 3.5–5 cm diam.; sepals: reddish, oblong, (1–)1.5–2.2 × 0.4–0.5 cm; petals: reddish, oblong, 0.8–1.7 × 0.2–0.3 cm; corona: 2 series, outer series 0.4–0.7 cm long, red with white apex, inner series 0.1–0.2 cm inside floral tube; ovary: obovoid to ellipsoid, densely puberulent; fruit: ellipsoid, 2–4.5 × 0.7–2 cm, hexagonal, dark red; seeds: 0.2–0.3 × 0.1–0.2 cm, transversely grooved.

Passiflora sexflora Juss.

Ann. Mus. Natl. Hist. Nat. 6: 110, pl. 37, fig. 2 (1805)
sexflorus: having 6 flowers
Synonyms: *P. capsularis* var. *geminifolia* DC., *P. floribunda* Lemaire, *P. isotriloba* Cuf., *P. melahelos* Gilli, *P. miraflorensis* Killip, *P. obscura* Lindley, *P. pannosa* J. E. Sm.
Supersection *Decaloba*, Section *Decaloba*

The native range of *Passiflora sexflora* (Plate 141) extends from Florida through eastern Mexico and Mesoamerica, including the Greater Antilles, to Colombia and Ecuador in northeastern South America. It occurs from near sea level to elevations up to 2500 m. Because of its large native range, this frequently found species is quite variable, as can be seen by the number of its synonyms.

The first clones of *Passiflora sexflora* that circulated in Europe in the early 1990s turned out to be extremely reluctant to flower, but this fact was made up for by the plant's attractive, 2- or mostly 3-lobed, velvety leaves. Although these clones have almost disappeared from cultivation, *P. sexflora* has developed a reputation as a plant that is unwilling to flower and unworthy of cultivation. This demonstrates how a clone, having come into cultivation by chance, can influence judgment about a species. Quite recently, however, a clone was introduced that flowers quite willingly under cultivation, especially during winter.

The flowers of *Passiflora sexflora* are produced on paired inflorescences, appearing in the leaf axils, each with 1–3 or up to 5 individual flowers. They are greenish white to cream, with a little purple in the corona, and reach a diameter of only about 2.5 cm. This species is often confused with *P. capsularis* and *P. rubra* because of its similar leaves, which have no laminar nectaries, but its stem is subterete, not angulate, and it has bracts. It is a plant of humid areas and is best cultivated under glass, although it can be kept outdoors during summer months. The minimum temperature is 7°C.

Vine: densely hirsute; stem: subterete, slightly striate; stipules: linear-subulate, 0.4–0.6 cm long; petioles: 2.2–4.2 cm long, without glands; leaves: 2- or generally 3-lobed, 4–8 × 4.5–11 cm, subcoriaceous, entire; peduncles: in pairs, inflorescences with (1–)3–5 flowers; bracts: linear-lanceolate, 0.3–0.5 cm long, usually 3-cleft or 3-toothed; flowers: greenish white to cream and purplish violet, 1.8–2.6 cm diam.; sepals: greenish white to cream, oblong, 0.7–1.6 × 0.3–0.5 cm; petals: greenish white to cream, oblong, 0.6–1.1 × 0.1–0.2 cm; corona: 2 series, outer series 0.4–0.8 cm long, purplish violet at base, pale green to cream above, inner series 0.1–0.3 cm; ovary: globose, densely pubescent; fruit: ellipsoid to globose, 0.6–2.3 × 0.7–1.4 cm; seeds: obovate, ca. 0.3 × 0.2 cm, transversely 6- to 8-sulcate.

Passiflora standleyi Killip

J. Wash. Acad. Sci. 14: 110 (1924)
named after Paul Standley, collector of the type plant
Synonym: *P. macrostemma* Killip
Supersection *Decaloba*, Section *Decaloba*

Passiflora standleyi (Plate 142) is among the most beautiful *Decaloba* species, owing its attractiveness to a combination of decorative leaves and pretty, small lilac flowers. Its leaves, 2-lobed for about 1/2 or more of their length, are reminiscent in form of *P. ornithoura* and *P. tuberosa*. They are purple on their undersides and bear numerous nectaries along the main nerves. The flowers, 3–4 cm diam., are particularly eye-catching, their yellow coronal filaments forming a contrast to the rich purple-violet sepals and petals.

Passiflora standleyi is at home in the mountains from Chiapas, Mexico, to Costa Rica, where it generally grows at altitudes of 1000–2100 m. John MacDougal introduced it to the trade in the United States from Guatemala in 1980, and it has enriched the collections of many enthusiasts. By the 1990s at least one other clone was introduced.

This plant is excellently suited for cultivation in a container and proves to be very free-flowering when appropriate conditions are provided. In principle *P. standleyi* needs to reach a certain size to flower satisfactorily. Provide a pot of at least 20 cm diam. In nature the species is often found near cliffs in seasonally dry forests, so a potting mix that

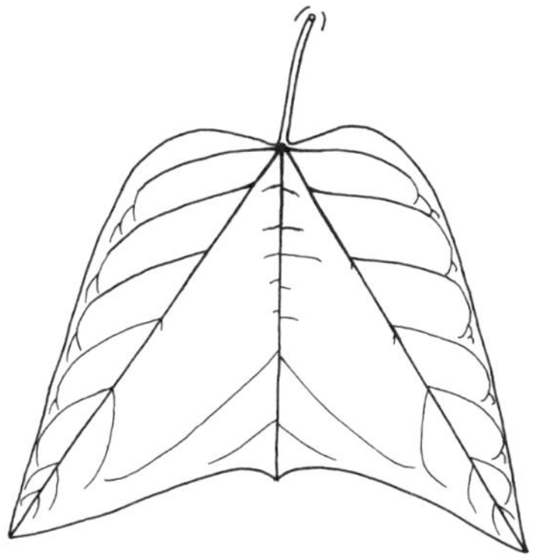

Passiflora sexflora

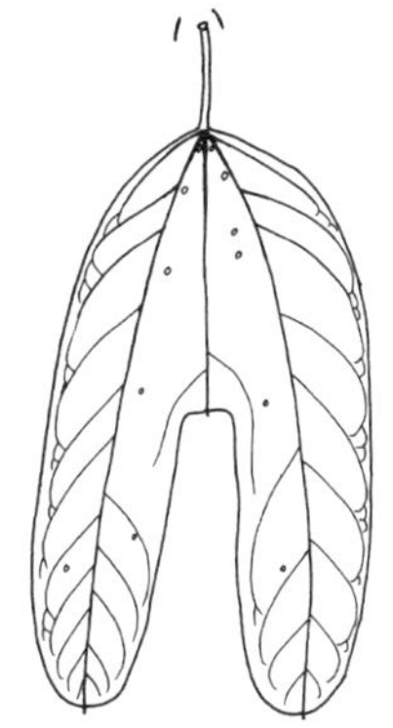

Passiflora standleyi

contains too much peat is disadvantageous, especially when it keeps moisture too long. Add gravel or coarse sand. Moreover, the long central European winter gives *P. standleyi* more trouble than most other *Decaloba* species. Generally speaking, this is a handsome, free-flowering species with occasional whims. Propagation is by cuttings; roots will only develop when optimum conditions are provided. Min. temp.: 10°C.

Vine: glabrous; stem: subterete, striate; stipules: narrowly linear-falcate, 0.2–0.6 cm long; petioles: 1.5–2.6 cm long, without glands; leaves: 2-lobed, 6–13 × 4–9 cm, bearing laminar nectaries, dark green to brownish green above, not variegated, sometimes purplish beneath, entire; peduncles: 2–5 cm long, slender, solitary or generally in pairs; bracts: setaceous, 0.3–0.6 cm long; flowers: bluish purple to violet, often pale or whitish toward tips of perianth, corona yellow, 3–4 cm diam.; sepals: whitish green outside, bluish purple to violet becoming increasingly white at apex, ovate-lanceolate, 1.5–2.3 × 0.4–0.5 cm; petals: bluish purple to violet becoming increasingly white at apex, oblong, 1.3–1.8 × 0.2–0.4 cm; corona: (2–)3–4 series, outer series 0.6–1.1 cm long, yellow distally, filiform to capillary, inner (1–)2–3 series very numerous, capillary, 0.3–0.7 cm; ovary: subglobose, glabrous; fruit: globose, 1.2–1.7 cm diam.; seeds: ovate or ovate-oblong, ca. 0.3 × 0.2 cm, transversely 5- to 7-sulcate.

Passiflora suberosa L.

Sp. Pl. 958 (1753)
suberosus: having corky bark
Synonyms: *P. angustifolia* Swartz, *P. calliaquatica* Krause, *P. flexuosa* Gardn., *P. glabra* Mill., *P. globosa* Vell., *P. hederacea* Cav., *P. hederaefolia* Lam., *P. heterophylla* Aiton non Lam., *P. hirsuta* L., *P. kohautiana* Presl., *P. limbata* Tenore, *P. longifolia* Lam., *P. nigra* Jacq., *P. olivaeformis* Mill., *P. oliviformis* Vell., *P. peltata* Cav., *P. pseudosuberosa* Fisch., *P. puberula* Hook., and more
Supersection *Cieca*

Although *Passiflora suberosa* (Plate 143) produces very small, unspectacular, apetalous flowers, it is valued and cultivated by many enthusiasts. It has shiny dark green leaves and is known for a rather uncommon phenomenon: the base of the stem becomes increasingly corky with age. It produces numerous green or whitish flowers that are about 1.3–2.5 cm diam., and even in winter some flowers may appear once in a while. Some clones are self-fertile so that fruit is regularly set, turning purplish black when ripe. The absence of petals and the marked petiolar glands are typical characteristics of species belonging to supersection *Cieca*.

Passiflora suberosa is widespread in the West Indies and in many parts of Mexico and Central and South America. It can be grown well on a windowsill, in the open, or under glass, and in winter it is satisfied with temperatures around 7°C.

The foliage of *Passiflora suberosa* is quite variable. There are 3-lobed leaves, which may be wider than long or longer than wide, and there are unlobed leaves. There are different flower types too, from the tiniest yellow-green flowers to the somewhat larger whitish flowers with a red operculum. The variability and frequent occurrence of this species has led to some confusion, and the *P. suberosa* complex is considered a taxonomic nightmare. Killip (1938) compiled a list of more than 60 synonyms for this single variable species.

Recent studies by Kristen Porter-Utley (2003) have shown that *Passiflora suberosa* is not as variable as it seems and should be subdivided into at least 3 completely distinct species, some with distinct subspecies. The tiny-flowered form found in the United States, Caribbean, and Yucatán should be called *P. pallida*, as originally suggested by Linnaeus. *Passiflora tridactylites* from the Galápagos is also distinct.

Vine: partly pubescent; stem: terete, striate, older stems with corky bark; stipules: subulate, 0.2–0.8 cm long; petioles: 0.5–4 cm long, with 2

Passiflora suberosa

stipitate or occasionally subsessile glands above middle; leaves: quite variable, 3-lobed, occasionally unlobed, 2.5–13 × 2.3–17 cm, laminar nectaries sometimes present, entire; peduncles: 1–4 cm long, solitary or generally in pairs; bracts: absent or 1 or 2, setaceous; flowers: greenish or whitish, 1.3–2.5 cm diam.; sepals: green to white, oblong, 0.5–1.5 × 0.2–0.5 cm; petals: absent; corona: 2 series, outer series 0.3–0.7 cm long, brown, reddish, or violet at base, green above and greenish yellow to yellow at apex, inner series 0.1–0.3 cm long, generally purplish brown; ovary: ovoid, glabrous; fruit: globose to ovoid, 0.6–1.5 × 0.7–1.6 cm, black; seeds: ovate, ca. 0.3 × 0.2 cm, slightly curved, reticulate.

Passiflora tatei Killip & Rusby

Phytologia 1: 66 (1934)
named after George Henry Hamilton Tate, collector of the type
Supersection *Decaloba*, Section *Decaloba*

This species (Plate 144) is not yet well known among enthusiasts. *Passiflora tatei* is at home at altitudes around 2500 m in the Andes of western Bolivia. It is closely related to *Passiflora alnifolia*, growing at similar altitudes in the Andes of Ecuador and Colombia, though the two can be distinguished by their bracts and floral cup. In contrast to *P. alnifolia*, which has bracts with an entire margin, the bracts in *P. tatei* are deeply cleft, which is unusual in *Decaloba* species. In addition, the floral cup in *P. tatei* is often wider than in *P. alnifolia*. The flowers of *P. tatei*, 3–4 cm diam., are whitish with purple and have a white-banded corona.

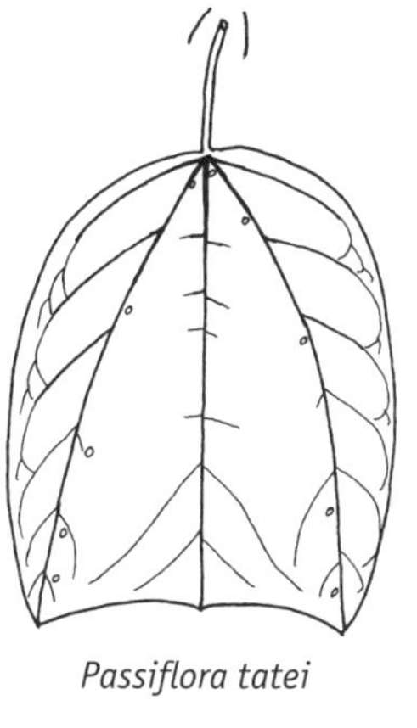

Passiflora tatei

This species has come into cultivation during the last few years and is not represented in many collections. As a species from high altitudes it dislikes heat spells and is recommended for cultivation outdoors, where it should be grown in a pot of at least 20 cm diam. As soon as temperatures drop to near freezing, *Passiflora tatei* should be taken to its winter quarters and guaranteed a temperature of 7°C.

Vine: partly pilosulous; stem: trigonous; stipules: narrowly linear, 0.7–1 cm long, purplish; petioles: 1–2.5 cm long, without glands; leaves: 2- or 3-lobed at apex, 3.5–10 × 3–7 cm, subcoriaceous, entire; peduncles: 3.5–4 cm long, slender, solitary or in pairs; bracts: setaceous, deeply 2- to 3-cleft, purplish; flowers: greenish white to white and purple, 3–4 cm diam.; sepals: green outside, greenish white or white inside, oblong, 1.4–1.7 × 0.4–0.5 cm; petals: white, narrowly linear, 0.7–0.9 × 0.2 cm; corona: 2 series, outer series 0.7–0.9 cm long, white at base, banded white and purple above, white in upper 1/2, inner series 0.2–0.3 cm long, greenish white; ovary: globose, densely lanate; fruit: globose; seeds: ovate-orbicular, transversely sulcate.

Passiflora telesiphe Knapp & Mallet

Novon 8: 162 (1998)
named after the butterfly *Heliconius telesiphe*
Supersection *Decaloba*, Section *Decaloba*

Once again, a butterfly helped in discovering a *Passiflora* species. In the 1990s, when Sandra Knapp and James Mallet did field studies on the natural history and foodplants of *Heliconius telesiphe*, they discovered this new species. *Passiflora telesiphe* (Plates 145 and 146) is endemic to the province of Zamora-Chinchipe, southern Ecuador, where it is found at an altitude of 1700 m. It belongs to a special group of species endemic in Ecuador at altitudes around 1200–2000 m. Among these are *P. hirtiflora*, *P. indecora*, and *P. subpurpurea*, distinguished by colorful leaves and exceptionally large bracts, *P. subpurpurea* being very untypical of *Decaloba* species. Before the new systematics, these species had been united in section *Pseudogranadilla* by Killip.

Passiflora telesiphe is most closely related to *P. hirtiflora*, which is also in cultivation. In both spe-

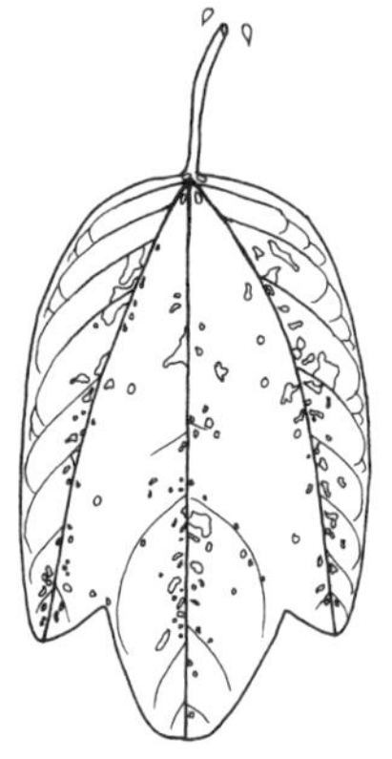

Passiflora telesiphe

cies the young leaves are purple with a paler mottling, changing to green when the plants get older. Moreover, their leaves hang down laxly, giving the impression of being wilted from lack of water. *Passiflora telesiphe* can be distinguished from *P. hirtiflora* by its lesser pubescence and by its flowers. While the outermost coronal series is significantly shorter in *P. hirtiflora*, reaching about 1/2 the length of the sepals, the filaments in *P. telesiphe* are distinctly longer than the sepals. Its pendent cream flowers can reach a diameter of up to 4.5 cm. However, the sepals and petals reflex widely.

Fortunately, Mallet succeeded in bringing this pretty species into cultivation at the time of its discovery. Now many collections cannot be imagined without *Passiflora telesiphe*. Even without flowers, it is always an attractive plant. It can be induced to flower reliably even in a 12-cm pot and may be kept in a 20-cm pot for a longer time. However, a potting mix containing too much peat may cause problems, as it can keep moisture too long. Therefore, gravel or coarse sand should be added to the rooting medium. In midsummer, temperatures in a greenhouse are too high for this species, and so cultivation in the open is preferred. The flowers of *P. telesiphe* appear mostly in fall. Min. temp.: 5°C.

Vine: pubescent; stem: angulate; stipules: linear, 0.2–0.3 cm long, purplish, deciduous; petioles: 1–1.5 cm long, without glands; leaves: 3-lobed at upper 1/3, 5–11 × 3–8 cm, upper surface dark green, occasionally variegated with cream-white or pale purple, dark purple beneath, with 14–19 laminar nectaries, entire; peduncles: 2.5–3 cm long, solitary or generally in pairs; bracts: elliptic to ovate, 7–13 × 0.2–0.5 cm, greenish tinged purple; flowers: cream-white, 3.5–4.5 cm diam.; sepals: cream-white, oblong, 2–2.5 × 0.5–0.6 cm; petals: cream-white, oblong, 1–1.4 × 0.3–0.4 cm; corona: 2 series, outer series 2.2–2.6 cm long, white flushed with purple in lower 1/2, white above, slightly wavy in upper 1/3, inner series 0.3–0.5 cm long, mottled with purple speckles toward apex; ovary: ovoid, densely pubescent, purplish; fruit: unknown.

Passiflora trifasciata Lemaire

Ill. Hort. 15: t. 544 (1868)
trifasciatus: having 3 bands
Supersection *Decaloba*, Section *Decaloba*

Passiflora trifasciata (Plate 147) was among the first passionflowers cultivated for its attractive foliage. On the upper side, its 3-lobed leaves are dark green and light green or dark pink along the main nerves. Underneath, they have a more or less intense purple touch. Because of these features this species is well known and widespread among enthusiasts. Although *P. hirtiflora* and *P. telesiphe* have more intensely colored leaves, they cannot compete with the unpretentiousness of *P. trifasciata* and with its willingness to flower. It can be cultivated indoors as well as on a balcony, on a terrace, or under glass. It is an easy-to-grow species and tolerates temperatures down to 5°C for a short time in winter.

At home in Brazil, Ecuador, and Peru, *Passiflora trifasciata* occurs in the Amazonian lowlands as well as on the eastern foothills of the Andes at altitudes up to 1500 m. The flowers, abundantly produced during summer, are about 3 cm diam. At times the pink operculum gives the fragrant flow-

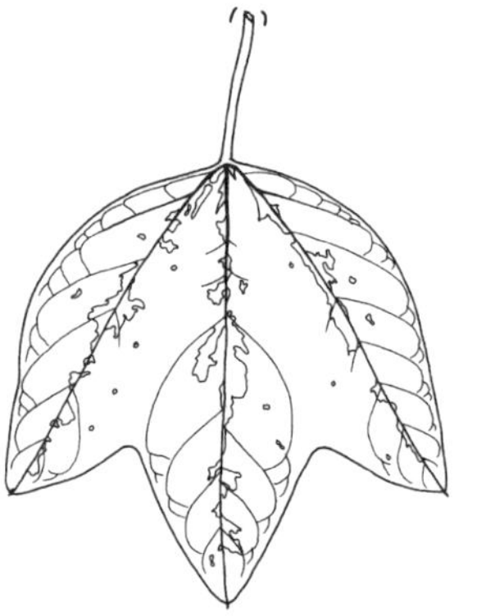

Passiflora trifasciata

ers some color. The sepals and petals reflex very soon when the flower has opened. The innermost coronal series is conspicuous, the tips of the filaments markedly broadened. This characteristic is found in a group of species that grow mainly in Peru. Before the new systematics, these species had been united by Killip in the invalidly published series *Miserae*.

Occasionally one comes across a plant cultivated under the name *Passiflora tricuspis* that is actually *P. trifasciata*. These sister species can be distinguished by their leaves, though some specimens are more difficult to classify than others. In *P. tricuspis* the 2 outer leaf lobes are usually noticeably longer than the middle one.

Vine: glabrous; stem: angulate, striate; stipules: subulate, 0.2–0.5 cm long; petioles: 1.5–5 cm long, without glands; leaves: shallowly 3-lobed, 3.3–13 × 4–11.8 cm, upper surface dark green, variegated with white or yellowish green, purple beneath, bearing laminar nectaries, membranous, entire; peduncles: 2.1–3.3 cm long, slender, solitary or generally in pairs; bracts: setaceous, about 0.2 cm long; flowers: white, 2.5–3.5 cm diam.; sepals: greenish white, oblong, 1.4–1.5 × 0.6–0.7 cm; petals: white, oblong, 0.6–1 × 0.3 cm; corona: 2 series, outer series 0.8–1 cm long, white, inner series 0.2–0.4 cm long, white, broadly lobulate at apex; ovary: subglobose to ovoid, glabrous; fruit: globose, 1.5–2.5 cm diam., dark purple; seeds: oblong-ovoid, 0.4 × 0.2 cm, transversely 6- to 8-sulcate.

Passiflora trinervia (Juss.) Poir.

Encycl., Suppl. 2: 843 (1811)
trinervius: having 3 nerves or veins
Supersection *Decaloba*, Section *Decaloba*

Passiflora trinervia (Plate 148) is an exceptional and extremely beautiful representative of this subgenus. At first sight its pendent blossom seems more like that of a fuchsia than a passionflower. As with *Tacsonia* species, the flower is dominated by the extended floral tube, which is by far the largest within the whole subgenus. It is a deep rose-pink, as are the sepals and petals. At the tip of the narrow floral tube is a cream-colored corona in one series, which reaches a length of only 0.3 cm. Flowers are about 5 cm diam.

Like all hummingbird-pollinated species, *Passiflora trinervia* was at first not assigned to subgenus *Decaloba*. It was instead placed into the then independent genus *Tacsonia* by Antoine Laurent de Jussieu (1805). Later on it was incorporated into the small subgenus *Psilanthus*. It should be noted that its lanceolate leaves with a 3-lobed tip and almost parallel main nerves are not very much in line with the standard features of species from subgenus *Decaloba*. Only about 20 species with such leaves are known. Nevertheless, *P. trinervia* fits well into this subgenus because its morphology is otherwise characteristic.

Its native range is in the central part of the Central Cordillera of Colombia at altitudes around 3000 m. Dreams of growing this species cause much heartache, as it has not yet been brought into cultivation.

Vine: tomentose; stem: angulate; stipules: setaceous, up to 1 cm long; petioles: up to 1.5 cm long, without glands; leaves: 3-lobed at apex, 6–10 × 3.5–5 cm, glabrescent or pilosulous on upper surface, densely tomentose beneath, entire; peduncles: up to 10 cm long, slender, pendent; bracts: setaceous, ca. 1 cm long, slightly below middle of peduncle; floral tube: cylindrical, up to 12 cm long, dilated just above base, rose-red outside; flowers: rose red, 4.5–5.5 cm diam.; sepals: rose-red, with white parts inside, oblong-lanceolate, ca. 3 × 0.6 cm; petals: rose-red, linear-oblong, ca. 1.5 × 0.3 cm; corona: 1 series, 0.2–0.3 cm long, cream; ovary: narrowly ovoid, densely hirsute; fruit: ovoid, 3.5–4 × 2 cm, densely hirsute; seeds: obovate or obcordate, 0.4–0.6 × 0.2–0.3 cm, transversely 8- to 9-sulcate.

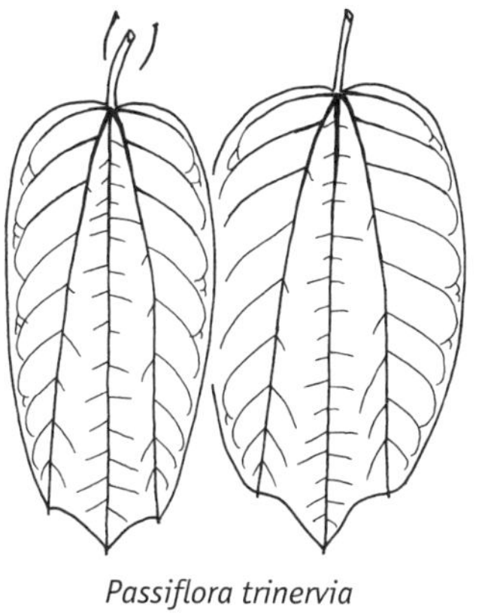

Passiflora trinervia

Passiflora tulae Urban

Symb. Antill. 1: 374 (1899)
Supersection *Decaloba*, Section *Decaloba*

This recommendable passionflower (Plate 149) stands out for its unusual combination of flower colors. The pink sepals and petals reflex immediately upon opening and contrast with the rich orange corona that encloses the androgynophore in a membranous tubular form. When the flower has just opened, its diameter is 5.5–7 cm. Elma Kay has found that in its natural surroundings, the mountains of Puerto Rico, this species is pollinated by hummingbirds.

Many enthusiasts cultivate *Passiflora tulae* and its red-flowered sister species, *P. murucuja*, and in central Europe it is offered by some flower shops and garden centers. It is suited for cultivation on a windowsill and is particularly recommended for a balcony or terrace during summer months. In a protected location it will develop an abundance of flowers, although flowering is occasionally interrupted for a short time. Because *P. tulae* sets flowers even on short shoots, it can be grown in a 12-cm pot, although a container of at least 20 cm diam. is needed to avoid changing pots too frequently. Min. temp.: 10°C.

Vine: glabrous; stem: angulate, striate; stipules: linear-subulate, 0.1–0.2 cm long, soon deciduous; petioles: 1–3 cm long, without glands; leaves: 2- or 3-lobed, 1.5–7 × 5–10.5 cm, entire; peduncles: 2–6 cm long, solitary or in pairs; bracts: setaceous, 0.2–0.3 cm long; floral tube: bowl-shaped, 0.4–0.5 cm long, 10-sulcate, pink outside; flowers: pink and orange, 5.5–7 cm diam.; sepals: pink, linear-oblong, 3–4 × 0.5–0.6 cm; petals: pink, linear-oblong, 2.5–3 × 0.3–0.4 cm; corona: 1 series, erect, cylindrical membrane, 1.5–2 cm long, orange; ovary: subglobose, glabrous; fruit: globose, 1–1.5 cm diam., blackish; seeds: ovate, 0.3–0.4 × 0.2 cm, transversely 10- to 12-sulcate.

Passiflora vespertilio L.

Sp. Pl. 957 (1753
vespertilius: resembling a bat (in reference to the leaf silhouette)
Synonyms: *P. europhylla* Mast., *P. geminiflora* DC., *P. hemicycla* G. F. W. Mey., *P. surinamensis* Miq.
Supersection *Decaloba*, Section *Decaloba*

Passiflora vespertilio (Plate 150) is widespread in South America. It is especially common in the Guianas but is also at home in Colombia, Ecuador, Trinidad, Peru, Brazil, and Bolivia. It occurs in the lowlands from near sea level up to elevations of only 500 m. It is known that at least the collections from eastern Ecuador are confined to black water areas (Holm-Nielsen et al. 1988).

Although this species has been in cultivation for some years, it is not represented in many collections. This may be due to the fact that its overwintering is quite problematic. It is valued, however, among lepidopterists as a foodplant for the caterpillars of their butterflies. The white flowers reach a diameter of 4–5 cm. The wavy white corona is remarkable, in which the filaments are connate at their base and form a membrane. Occasionally the flowers do not open completely in cool weather or fall off prematurely. The globular, dark purple fruits are said to be edible. This rather warmth-loving species will flourish in a greenhouse, but flowers can

Passiflora tulae

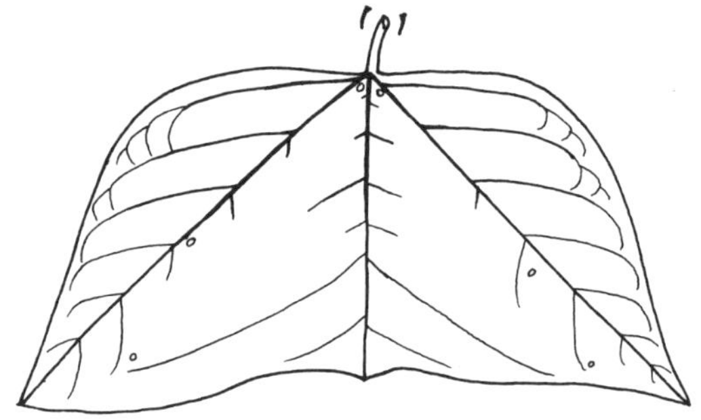

Passiflora vespertilio

be expected during summer months even when it is grown on a balcony or terrace. Min. temp.: 10°C.

Vine: glabrous; stem: terete or subangular, striate; stipules: subulate, 0.3–0.6 cm long; petioles: 0.5–1.5 cm long, without glands; leaves: 2- or generally 3-lobed, 2–5.5(–6.5) × 4–10.5 cm, bearing laminar nectaries, subcoriaceous, entire; peduncles: 0.5–1.5 cm long, solitary or in pairs; bracts: linear, 0.2–0.4 cm long, entire; flowers: white, 4–5 cm diam.; sepals: greenish outside, white inside, ovate-oblong, 1.5–2.1 × 0.7–1.2 cm; petals: white, ovate-oblong, 1–1.2 × 0.4–0.5 cm; corona: 2 series, outer series 1.3–1.7 cm long, white, united into a membrane at base, wavy at middle, inner series 0.3–0.4 cm long, white; ovary: ovoid, glabrous to minutely puberulent; fruit: subglobose, 1.3–1.9 × 1–1.6 cm, glabrous, purple; seeds: obovate, 0.4 × 0.2 cm, transversely sulcate.

Passiflora viridescens Escobar

Phytologia 66: 81 (1989)
viridescens: becoming green
Synonym: *P. ulmeri* Schwerdtfeger
Supersection *Decaloba*, Section *Decaloba*

Passiflora viridescens (Plate 151) is a truly strange and interesting species. Its green, pendent flowers could be compared to those of *P. penduliflora*. While proof could be established that the latter is pollinated by bats, there is still speculation about the pollinators of *P. viridescens*. Of interest in this context is the corona. It consists of a single series of 10–15 club-shaped bodies that are often absent in the open flowers, although they can be found in the flower buds (Schwerdtfeger 1997). Peter Møller Jørgensen and John MacDougal (1997) surmise that the flowers of this species open early in the morning and that the pollinators eat a large part of the corona, perhaps as the reward for visiting the flower. Nectar, which in other species attracts pollinators, seems to be absent in *P. viridescens*. The only other passionflower without nectar is *P. xiikzodz*, with a flower of entirely different form. *Passiflora viridescens* seems to be the only species in the entire genus in which something other than nectar or pollen is food for pollinators. Nevertheless, the question remains as to what animals the pollinators may be. The diameter of a flower is 6–7 cm and can even reach 10 cm, which is quite spectacular for a *Decaloba* species. And this species is imposing not only with respect to its flowers: the general impression is that of a large, woody liana.

The native range of *Passiflora viridescens* reaches from the province of Pichincha in north-central Ecuador to the department of Cajamarca in northern Peru. It occurs exclusively along the chain of the Andes at altitudes of 2300–3200 m. In *Flora of Ecuador* (Holm-Nielsen et al. 1988), collections of this species are erroneously regarded as *P. cuspidifolia*, a species with similar leaves. The same mistake appears in *Passionsblumen* (Ulmer and Ulmer 1997), in which a photo of *P. viridescens* is erroneously presented as *P. cuspidifolia*.

A clone introduced by Michael Schwerdtfeger in the late 1990s is now in cultivation in Europe. Alas, this clone is reluctant to flower. Plants that are 2–3 m tall have not been induced to flower even after several years. Perhaps they should be planted out in a cool greenhouse to achieve longer shoots. As a typical species from high altitudes, it does well outdoors during summer months. Older specimens will even survive a slight frost. Nevertheless, *Passiflora viridescens* should be overwintered at temperatures above 5°C.

Vine: minutely pubescent to glabrescent; stem: subterete, striate; stipules: linear-falcate, 0.2–0.6 × 0.1 cm; petioles: 0.8–3.2 cm long, without glands; leaves: 2- or 3-lobed at very apex, occasionally unlobed, 6.2–14 × 2.5–8.3 cm, bearing 4–18 laminar nectaries, subcoriaceous, entire; peduncles: 4–9.5 cm long, solitary or generally in pairs, pendent; bracts: setaceous, 0.3–0.6 cm long, entire; flowers:

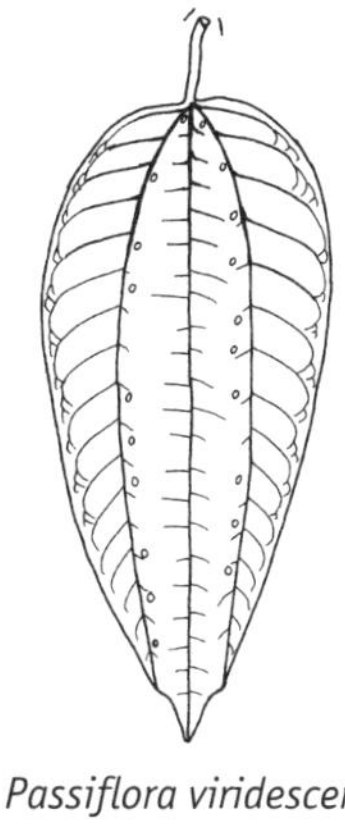

Passiflora viridescens

light green, 6–7(–10) cm diam.; sepals: green, oblong-lanceolate, 2.4–3.7(–5.5) × 0.6–0.8(–1.1) cm; petals: green, oblong to lanceolate, 1.5–2.8 × 0.4–0.7 cm; corona: 1 series, 0.7–1.4 × 0.2 cm, white to whitish green, ligulate, thick-fleshy, often removed in open flowers; ovary: ellipsoid, densely pubescent; fruit: ellipsoid to slightly obovoid, 5–6 × 2.5 cm, green; seeds: obovate, transversely 7- to 8-sulcate.

Passiflora viridiflora Cav.

Icon. Pl. 5: 15, pl. 424 (1799)
viridiflorus: having green flowers
Synonym: *P. tubiflora* Kunth
Supersection *Cieca*

Passiflora viridiflora (Plate 152) is an unusual species that is occasionally cultivated. Its unique flowers reach a diameter of 3–3.5 cm. The androgynophore, which protrudes far out of the flower, and the floral tube, about 1.5 cm long, are both remarkable adaptations to pollination by hummingbirds. The bright yellow-green curved sepals adjoin the floral tube, while the petals are missing. The corona is reduced, and there is no floral odor. This species typically has bright red stems and pale, rather whitish seeds at maturity.

Its native range is southern Mexico, where it grows in the lower coastal regions in deciduous forest or dune forest. A greenhouse is the ideal location to cultivate such a warmth-loving species.

Passiflora viridiflora is the sister species to *P. juliana*, described in 1992. *Passiflora juliana* is very similar vegetatively except that it usually has very wide stipules. Some forms have narrow stipules, however, and in these cases the 2 species are difficult to distinguish without the flowers. In winter, temperatures should not drop below 15°C, although short periods to 12°C may be tolerated if the plant is kept on the dry side.

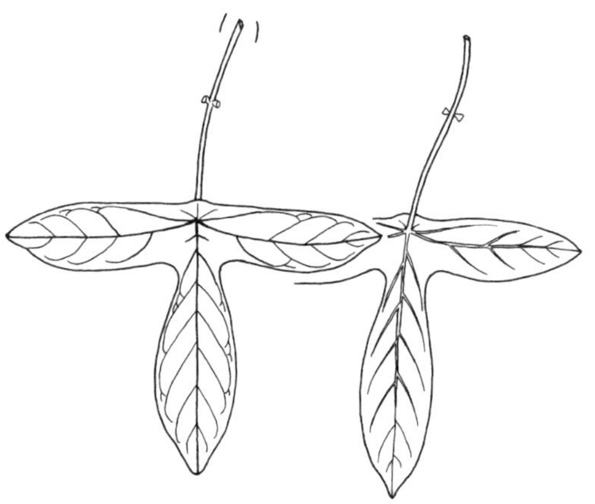

Passiflora viridiflora

Plant: minutely puberulent to glabrescent throughout; stem: almost terete to almost angulate, red, corky when old; stipules: linear-lanceolate, 0.3–0.6 cm long, red; petioles: 1.5–3.5 cm, with 2 glands usually in upper 1/3; leaves: 3-lobed with middle lobe narrowed at base, 4–16 × 6–25 cm, bearing laminar nectaries near margin, peltate, entire; peduncles: solitary or in pairs, 1.5–3.3 cm long, 1-flowered, often in condensed terminal inflorescences to 15 cm long with aborted leaf blades; bracts: usually absent, setaceous (if present); flowers: yellow-green, 3–3.5 cm diam., cylindrical-tubular in lower 1/2; floral tube: 0.8–1.7 cm long, ca. 0.5 cm diam., swollen at base; sepals: yellow-green, narrowly oblong-lanceolate, 1.5–2.2 × 0.3–0.4 cm; petals: absent; corona: 1 series, ca. 0.3 cm long, borne at opening of floral tube, rarely a trace of inner series present inside at base of tube, capillary; ovary: glabrous, narrowly ellipsoid; fruit: subglobose to prolate, 1.5–2 cm diam., blackish purple; seeds: ca. 0.4–0.5 × 0.3 cm, reticulate.

Passiflora xiikzodz subsp. *itzensis* MacDougal

Novon 2: 363 (1992)
xiikzodz: native Mayan name from *xi'ik* (wing), *zodz* (bat); Chichen Itza: Mayan pyramid
Supersection *Cieca*

Passiflora xiikzodz (Plate 153) is among the most bizarre passionflowers. It has no nectary and produces no nectar; the pollinators of the plant are unknown, and their reward is unknown. The dark, shiny convex center of the flower is conspicuous and may play a role. Perhaps this flower mimics that of another family, or perhaps insects are attracted by deceit. It is a mystery.

The typical subspecies has an androgynophore ca. 0.2–0.5 cm long, and styles and stamens typical of subgenus *Decaloba*. *Passiflora xiikzodz* subsp. *itzensis* has almost no androgynophore, short nonmoving styles, and pollen that faces upward. Both subspecies have the bizarre dark corona, and the peduncles are articulated 1.2–3 cm below the flower.

In nature *Passiflora xiikzodz* is found in the Yucatán Peninsula of Mexico, through the Petén of Guatemala, and into adjacent Belize. Found on limestone outcrops, walls of limestone sinkholes, and at edges of forests with exposed limestone, almost all collections of *P. xiikzodz* have been made from Mayan ruins. The peltate leaf is nearly identical to *P. coriacea*, which often grows with it, but the position of the petiolar nectary glands offers an immediate distinction: at or near the apex in *P. xiikzodz*, near the middle in *P. coriacea*. Recent studies by Kristen Porter-Utley (2003) suggest that the 2 subspecies should be recognized as distinct species.

This is not a difficult species for small pot culture, so long as the growing medium is very well drained and provided with limestone or marble chips. The plant should be restarted every 2 years from cuttings. Min. temp.: 15°C.

Plant: minutely, lightly puberulent throughout, leaves almost glabrous; stem: almost terete, purplish black; stipules: linear-lanceolate, 0.3–0.7 cm long; petioles: 1–3 cm, with 2 glands near or at the apex; leaves: transversely 2-lobed or nearly 3-lobed with truncate and obscure middle lobe, 1.5–7.5 × 3.5–19 cm, bearing laminar nectaries near margin, peltate, entire; peduncles: solitary or usually in pairs, 0.2–3.5 cm long, 1-flowered, often in condensed terminal inflorescences to 20 cm long with aborted leaf blades; bracts: absent; flowers: light yellow-green with lustrous dark purple or brown center, the corona dark violet with yellow tips, ca. 2 cm diam.; sepals: light yellow-green, oblong-lanceolate, 0.8–1.5 × 0.4–0.6 cm; petals: absent, but rare sports reported with 1 or 2 pale green petals up to 0.7 mm long; corona: 5–8 series, outer series 0.8–1 cm long, slender filiform, very dark purple with light yellow tips, inner 3–7 series less than 0.3 cm long, operculum absent; ovary: glabrous, ellipsoid; fruit: ovoid, 2–3 × 1.5–2 cm, blackish purple; seeds: ca. 0.5 × 0.2 cm, reticulate.

Passiflora yucatanensis Killip

Publ. Field Columbian Mus., Bot. Ser. 8: 26 (1930)
yucatanensis: from the Yucatán Peninsula
Supersection *Decaloba*, Section *Decaloba*

Passiflora yucatanensis (Plate 154) is a rare species only recently introduced to horticulture. Contrary to its name, this species is endemic to the island of Cozumel off the coast of the Yucatán Peninsula and has not been found on the mainland. Known for many years from only the type collection, *P. yucatanensis* was finally collected a 2nd time by Dottie Koch for John MacDougal, who introduced it around 1992. In 2003 Sally Ransom collected and introduced additional clones to the trade.

The foliage is indistinguishable from *Passiflora biflora*, its probable sister species. The flowers are also similar at first sight, but the androgynophore is much longer in *P. yucatanensis*, 1.1–1.6 cm versus 0.7–0.9 cm. The corona affords another distinction, being shorter, at first erect, and then abruptly flaring outward 90°.

Free-flowering and very easy to grow, this species can be treated like *Passiflora biflora* or other easy decalobas, producing its attractive white, yellow, and red flowers early in the morning. Its foliage is fairly dense and is nicely displayed in a hang-

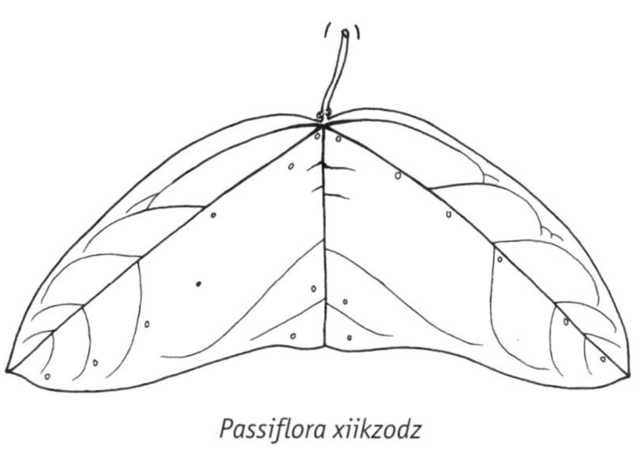

Passiflora xiikzodz

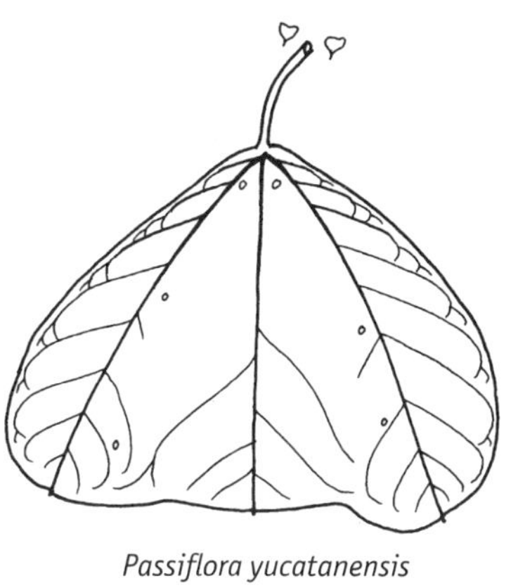

Passiflora yucatanensis

ing basket. Overwintering temperatures should not drop below 10°C.

Plant: minutely puberulent to glabrescent throughout; stem: angulate, grooved-striate; stipules: setaceous, falcate, 0.3–0.4 cm long; petioles: 1.5–2 cm, without glands; leaves: very shallowly 2- or 3-lobed with short middle lobe, or truncate, 2.3–5 × 4.8–8 cm, bearing laminar nectaries, entire; peduncles: in pairs, 2–4 cm long, 1-flowered; bracts: setaceous, 0.2–0.3 cm; flowers: white with short yellow and reddish corona, ca. 3 cm diam.; sepals: pale yellow-green outside, white inside, lanceolate-oblong, 1.8–2.5 × 0.6–0.8 cm; petals: white, narrowly oblong, 1.3–2 × 0.5–0.7 cm; corona: 2 series, outer series 0.7–1 cm long, erect in lower 1/2 and red or purplish red, curving out and bright yellow apically, inner series ca. 0.3–0.5 cm, reddish purple, capillary; ovary: glabrous to slightly puberulent, ellipsoid; fruit: purple.

Plate 1. An array of passionflowers.

Plate 2. *Passiflora insignis* in Bolivia.
Photo by Linda K. Escobar.

Plate 3. Various leaf forms found in the genus *Passiflora*.

Plate 4. The leaves of many species of subgenus *Decaloba* are covered with extrafloral glands that secrete nectar on their undersides. When looked at from above, these glands appear as pale, mostly yellowish dots, imitating the eggs of butterflies. Photo by Lawrence E. Gilbert.

Plate 5. *Passiflora alata* is among the most beautiful passionflowers.

Plate 6. *Passiflora mixta*, a typical long-tubed representative of supersection *Tacsonia*, pollinated by a sword-billed hummingbird (*Ensifera ensifera*). Photo by Luis A. Mazariegos.

Plate 7. *Passiflora coccinea*, pollinated by a hummingbird. Photo by Annika Lindberg.

Plate 8. Hummingbird-flowered species of supersections *Distephana* and *Coccinea* commonly attract ants by means of extrafloral nectaries on the bracts (and occasionally on the back of the sepals) as a protection against nectar robbers. Pictured here is *Passiflora aimae* in French Guiana. Photo by Helga Gelewsky.

Plate 10. *Passiflora* species from Ecuador. Photo by Ron Boender.

Plate 9. The Brazilian *Passiflora ovalis* is a typical bat-pollinated species. Its racemose inflorescences are pollinated by the nocturnal bat *Glossophaga soricina*. Photo by João Vasconcellos-Neto.

Plate 11. *Passiflora* species from Peru. Photo by Ron Boender.

Plate 13. A fruit of *Passiflora quadrangularis* at an exhibition of exotic fruits in the Dominican Republic. Photo by Brigitte Mußgay.

Plate 12. Passionfruits can be many sizes and colors. Clockwise from top left: *Passiflora caerulea, P. ligularis, P. rubra, P. edulis* f. *edulis, P. suberosa,* and *P. tarminiana.*

Plate 14. Fruit of *Passiflora quadrangularis*. Photo by Brigitte Mußgay.

Plate 15. *Passiflora tarminiana*, a popular fruit plant in native gardens in Colombia. Photo by Guillermo Guzmán (CIRAD/IPGRI).

Plate 16. Such fruiting of *Passiflora nitida* can only be expected in its natural surroundings (in this case, French Guiana) or when it is planted in a spacious, warm greenhouse. Photo by Hilaire Annonay.

Plate 18. *Passiflora edulis* f. *flavicarpa*—a flowering twig with fruit. Photo by Heinrich Drexler.

Plate 17. A plantation of *Passiflora edulis* f. *edulis* in Uganda. Photo by Anette Schumacher.

Plate 19. A plantation of *Passiflora edulis* f. *flavicarpa* in Brazil. Photo by Heinrich Drexler.

Plate 20. Delivering and sorting the fruit. Photo by Heinrich Drexler.

Plate 21. Bottling the juice of *Passiflora edulis* f. *flavicarpa*, or maracuja juice. Photo by Heinrich Drexler.

Plate 22. Fruit of *Passiflora laurifolia*. Photo by Hilaire Annonay.

Plate 23. Fruit of *Passiflora ligularis*.

Plate 24. *Passiflora cumbalensis* is one of many *Tacsonia* species with edible fruit. Photo by Gwilym Lewis.

Plate 25. *Passiflora tarminiana*—a flower and a ripening fruit. Photo by Hilaire Annonay.

Plate 26. *Passiflora mayarum* in Belize. Photo by Jan Meerman.

Plate 28. *Passiflora rovirosae* in Belize. Photo by Jan Meerman.

Plate 27. *Passiflora pittieri* in Belize. Photo by Jan Meerman.

Plate 29. *Passiflora foetida* var. *nicaraguensis* in Belize. Photo by Jan Meerman.

Plate 30. Michael Schwerdtfeger with *Passiflora harlingii* in the Andes of Ecuador. Photo by Gaba Weis.

Plate 31. A comparison of flower size and color in various species of supersection *Tacsonia*, including (from left to right) *Passiflora tarminiana*, a white-flowered form of *P. mathewsii*, *P. mixta*, and *P. luzmarina*.

Plate 32. *Ensifera ensifera* in Ecuador. Photo by Thomas Fechtler.

Plate 33. *Passiflora sanctae-barbarae* in Ecuador. Photo by Michael Schwerdtfeger.

Plate 34. *Passiflora manicata* in Ecuador. Photo by Gwilym Lewis.

Plate 35. *Passiflora eichleriana* in São Paulo, Brazil. Photo by Reto Gamma.

Plate 36. The reddish form of *Passiflora speciosa*, pollinated by *Phaethornis idaliae*. Photo by João Vasconcellos-Neto.

Plate 37. Flower buds of the rare *Passiflora ischnoclada* in Brazil. Photo by Alan Jouy.

Plate 38. *Passiflora deidamioides*. Photo by Lawrence E. Gilbert.

Plate 39. The habit of *Passiflora sphaerocarpa* in cultivation. Photo by Ron Boender.

Plate 40. Most *Astrophea* species have a yellow corona. However, in *Passiflora sclerophylla* the corona and parts of the sepals and petals are a mottled purple. Photo by Wilhelm Barthlott.

Plate 41. A typical flower of genus *Astrophea*. This specimen from French Guiana may be a new species. Photo by Helga Gelewsky.

Plate 42. *Passiflora amoena* fruits. Photo by Pierre Pomié.

Plate 43. *Passiflora amoena* flowers. Photo by Pierre Pomié.

Plate 44. *Passiflora arborea* in Colombia. Photo by Linda K. Escobar.

Plate 45. *Passiflora candida* in French Guiana. Photo by Hilaire Annonay.

Plate 46. *Passiflora citrifolia*. Photo by Ron Boender.

Plate 47. *Passiflora kawensis* in French Guiana. Photo by Hilaire Annonay.

Plate 49. *Passiflora longiracemosa* in the state of Amazonas, Venezuela, showing the typical long inflorescence with flowers that abound at its base. Photo by Paul E. Berry.

Plate 48. *Passiflora lindeniana*. Photo by Linda K. Escobar.

Plate 50. This *Passiflora longiracemosa* from the state of Roraima in the northernmost part of Brazil has significantly shorter inflorescences. Photo by Gwilym Lewis.

Plate 51. *Passiflora macrophylla* in Ecuador. Photo by Gwilym Lewis.

Plate 52. This plant growing in Venezuela is thought to be *Passiflora maguirei*. There are slight doubts about the correct identification, because the operculum does not seem to be 1 cm above the edge of the floral tube as is usual in *P. maguirei*. Photo by Günter Gerlach.

Plate 53. *Passiflora mansoi* in Brazil. Photo by Armando Carlos Cervi.

Plate 54. *Passiflora pittieri*. Photo by Lawrence E. Gilbert.

Plate 55. *Passiflora putumayensis*. Photo by Ron Boender.

Plate 56. *Passiflora pyrrhantha* in Ecuador. Photo by John M. MacDougal.

Plate 57. *Passiflora quelchii* in Guyana. Photo by Sandra Knapp and James Mallet.

Plate 58. *Passiflora rhamnifolia* in Brazil. Photo by Alan Jouy.

Plate 59. *Passiflora sclerophylla* in Venezuela. Photo by Tassilo Feuerer.

Plate 60. *Passiflora securiclata* in Venezuela. Photo by Appel.

Plate 61. *Passiflora sphaerocarpa*. Photo by Ron Boender.

Plate 62. *Passiflora spinosa*. Photo by Sandra Knapp and James Mallet.

Plate 63. *Passiflora tica* in Costa Rica. Photo by Stefan Vogel.

Plate 64. *Passiflora tina*. Photo by Ron Boender.

Plate 65. *Passiflora cirrhiflora*, flowering in pairs midway along the tendrils, may be the most beautiful species within subgenus *Deidamioides*. In French Guiana it is often found along roadsides. Note the color of lateritic soil. Photo by Pierre Pomié.

Plate 66. *Passiflora arbelaezii*.

Plate 67. *Passiflora cirrhiflora* in French Guiana. Photo by Pierre Pomié.

Plate 68. *Passiflora deidamioides*. Photo by Lawrence E. Gilbert.

Plate 69. *Passiflora discophora*. Photo by Michael Schwerdtfeger.

Plate 70. *Passiflora macdougaliana* in Panama. Photo by Sandra Knapp and James Mallet.

Plate 71. *Passiflora obovata* in Costa Rica. Photo by Sandra Knapp and James Mallet.

Plate 72. *Passiflora ovalis* has all the features characteristic of a bat-flowered species. Photo by João Vasconcellos-Neto.

Plate 73. An atypical form of *Passiflora ovalis* that may be assigned to a new species. Photo by Lawrence E. Gilbert.

Plate 74. *Passiflora ornithoura* combines attractive foliage with handsome flowers that are 1.5–2 cm wide.

Plate 75. *Passiflora citrina*, a representative of section *Xerogona*, has the most intense yellow flowers of any species in the genus.

Plate 76. *Passiflora adenopoda*. Photo by Ron Boender.

Plate 77. *Passiflora affinis*. Photo by Lawrence E. Gilbert.

Plate 78. *Passiflora allantophylla*.

Plate 79. *Passiflora alnifolia*.

Plate 80. *Passiflora anfracta*. Photo by Ron Boender.

Plate 81. *Passiflora apetala*.

Plate 82. *Passiflora aurantia*.

Plate 83. *Passiflora auriculata* in Bolivia. Photo by Günter Gerlach.

Plate 84. *Passiflora bicornis*. Photo by Ron Boender.

Plate 85. *Passiflora biflora*.

Plate 86. *Passiflora boenderi*.

Plate 87. *Passiflora bryonioides* in Mexico. Photo by Alan Jouy.

Plate 88. *Passiflora capsularis*.

Plate 89. *Passiflora chrysosepala* in Ecuador. Photo by Michael Schwerdtfeger.

Plate 90. *Passiflora cinnabarina*.

Plate 91. *Passiflora citrina*.

Plate 92. *Passiflora cobanensis*. Photo by Ron Boender.

Plate 93. *Passiflora colimensis*. Photo by John M. MacDougal.

Plate 94. *Passiflora conzattiana*.

Plate 95. *Passiflora coriacea* (or *P. sexocellata*).

Plate 96. *Passiflora costaricensis*. Photo by Lawrence E. Gilbert.

Plate 97. *Passiflora cuneata*.

Plate 98. *Passiflora cupraea*.

Plate 99. *Passiflora dioscoreifolia* exhibits the characteristic arrangement of the anthers and the 4 stigmas, as seen in this longitudinal section through the flower. Photo by Lawrence E. Gilbert.

Plate 100. A form of *Passiflora dioscoreifolia* with a broad purple corona band. Photo by John M. MacDougal.

Plate 101. *Passiflora escobariana* with pale pink flowers. Photo by Lawrence E. Gilbert.

Plate 102. *Passiflora escobariana* with nearly white flowers. Photo by John M. MacDougal.

Plate 103. *Passiflora eueidipabulum* in Panama. Photo by Sandra Knapp and James Mallet.

Plate 104. *Passiflora exoperculata*.

Plate 105. *Passiflora fanchonae* in French Guiana. Photo by Hilaire Annonay.

Plate 106. *Passiflora gilbertiana*. Photo by John M. MacDougal.

Plate 107. *Passiflora gracilis*.

Plate 108. *Passiflora guatemalensis*. Photo by Lawrence E. Gilbert.

Plate 109. *Passiflora hahnii*.

Plate 110. *Passiflora helleri*.

Plate 111. *Passiflora herbertiana*.

Plate 112. *Passiflora hirtiflora*.

Plate 113. *Passiflora holosericea*.

Plate 114. *Passiflora indecora*.

Plate 116. *Passiflora jorullensis* in Mexico. Photo by Alan Jouy.

Plate 115. *Passiflora jatunsachensis*. Photo by Michael Schwerdtfeger.

Plate 117. *Passiflora juliana*. Photo by John M. MacDougal.

Plate 118. *Passiflora karwinskii*.

Plate 119. *Passiflora lancearia* in Costa Rica. Photo by Stefan Vogel.

Plate 120. *Passiflora lancetillensis*.

Plate 121. *Passiflora lobata* is characterized by stamens held to the lower side and by the position of the 4 stigmas. Photo by John M. MacDougal.

Plate 122. The conspicuous fruit of *Passiflora lobata* contains brilliant orange-red arils. Photo by Lawrence E. Gilbert.

Plate 123. *Passiflora lobbii* in Peru. Photo by Maximilian Weigend.

Plate 124. *Passiflora lutea*. Photo by Karl-Heinz Kahlert.

Plate 125. *Passiflora membranacea* in Guatemala. Photo by Jürgen Lautner.

Plate 126. *Passiflora microstipula*.

Plate 127. *Passiflora morifolia*.

Plate 128. *Passiflora murucuja*.

Plate 129. *Passiflora oblongata* in Jamaica. Photo by Elma Kay.

Plate 130. *Passiflora organensis*.

Plate 131. The flowers of *Passiflora penduliflora* are among the most exceptional within subgenus *Decaloba*. Photo by Lawrence E. Gilbert.

Plate 132. *Passiflora penduliflora*. Photo by Lawrence E. Gilbert.

Plate 133. *Passiflora perfoliata*.

Plate 134. *Passiflora pilosa*. Photo by John M. MacDougal.

Plate 135. *Passiflora punctata*.

Plate 136. *Passiflora quinquangularis*.

Plate 137. *Passiflora rovirosae*. Photo by Ron Boender.

Plate 138. *Passiflora rubra*.

Plate 140. *Passiflora sanguinolenta*.

Plate 139. *Passiflora sagasteguii*.

Plate 141. *Passiflora sexflora* in Guatemala. Photo by Harald Förther.

Plate 142. *Passiflora standleyi*.

Plate 143. *Passiflora suberosa*. Photo by Karl-Heinz Kahlert.

Plate 144. *Passiflora tatei* in Bolivia. Photo by Reto Gamma.

Plate 145. *Passiflora telesiphe* produces cream-white pendent flowers. Photo by Sandra Knapp and James Mallet.

Plate 146. The foliage of *Passiflora telesiphe* is particularly attractive, variegated above and purple beneath.

Plate 147. *Passiflora trifasciata*.

Plate 148. *Passiflora trinervia* in Colombia. Photo by Geo Coppens (CIRAD/IPGRI).

Plate 149. *Passiflora tulae*. Photo by Lawrence E. Gilbert.

Plate 150. *Passiflora vespertilio*.

Plate 151. *Passiflora viridescens* in Peru. Photo by Jürgen Skrabal.

Plate 152. *Passiflora viridiflora*. Photo by John M. MacDougal.

Plate 153. *Passiflora xiikzodz* subsp. *itzensis*. Photo by Ron Boender.

Plate 154. *Passiflora yucatanensis*.

Plate 155. *Passiflora incarnata* is the type species of subgenus *Passiflora*. This form found in southern Missouri has particularly sumptuous flowers. Photo by Eldon Leiter.

Plate 156. The large red flowers of *Passiflora coccinea* are pollinated by hummingbirds. Photo by Annika Lindberg.

Plate 157. *Passiflora setulosa*, a typical representative of section *Granadillastrum* from Brazil with a large, conspicuous corona. Photo by Mauro Peixoto.

Plate 158. *Passiflora subrotunda*, another representative of section *Granadillastrum* from Brazil. Photo by Mauro Peixoto.

Plate 160. *Passiflora acuminata* in Brazil. Photo by Lawrence E. Gilbert.

Plate 161. *Passiflora aimae* in French Guiana. Photo by Hilaire Annonay.

Plate 159. *Passiflora actinia*.

Plate 162. *Passiflora alata*.

Plate 163. *Passiflora ambigua* in Belize. Photo by Jan Meerman.

Plate 164. The short-peduncle form of *Passiflora amethystina*.

Plate 165. The long-peduncle form of *Passiflora amethystina*.

Plate 166. A clone of *Passiflora amethystina* from Minas Gerais that has come into cultivation quite recently.

Plate 167. *Passiflora arida*.

Plate 168. The rare *Passiflora bomareifolia*, found near the Río Surucun, Venezuela. Photo by Günter Gerlach.

Plate 169. *Passiflora caerulea*.

Plate 170. *Passiflora caerulea* 'Constance Eliott'.

Plate 171. *Passiflora cerasina* in French Guiana. Photo by Hilaire Annonay.

Plate 172. *Passiflora chocoensis*. Photo by Günter Gerlach.

Plate 173. *Passiflora chrysophylla* in Brazil. Photo by Armando Carlos Cervi.

Plate 175. Different forms of *Passiflora coccinea* seem to exceed the limits of normal variability. This specimen was found in Brazil. Photo by Helga Gelewsky.

Plate 174. *Passiflora cincinnata*.

Plate 176. *Passiflora coccinea* in French Guiana. Photo by Hilaire Annonay.

Plate 177. The form of *Passiflora coccinea* cultivated in Europe.

Plate 178. *Passiflora crenata* in French Guiana. Photo by Hilaire Annonay.

Plate 179. *Passiflora deltoifolia* in Ecuador. Photo by Michael Schwerdtfeger.

Plate 180. *Passiflora edmundoi* in Brazil, exhibiting the characteristic petiolate, purple bracts. Photo by Wilhelm Barthlott.

Plate 182. *Passiflora edulis* f. *edulis*.

Plate 181. *Passiflora edulis* f. *flavicarpa*.

Plate 183. *Passiflora edulis* 'Norfolk'.

Plate 184. *Passiflora exura*. Photo by Lawrence E. Gilbert.

Plate 185. *Passiflora foetida* var. *ciliata* from Honduras. Photo by John M. MacDougal.

Plate 186. The Mexican *Passiflora foetida* var. *lanuginosa*. Photo by John M. MacDougal.

Plate 187. *Passiflora foetida* var. *gossypiifolia*.

Plate 188. *Passiflora foetida* var. *hirsuta*.

Plate 189. *Passiflora foetida* var. *hirsutissima*.

Plate 190. *Passiflora garckei*.

Plate 191. *Passiflora gibertii*.

Plate 192. Red-flowered *Passiflora glandulosa* from French Guiana. Photo by Helga Gelewsky.

Plate 193. Rose-pink-flowered *Passiflora glandulosa* from French Guiana. Photo by Pierre Pomié.

Plate 194. *Passiflora gritensis*.

Plate 195. *Passiflora incarnata* is quite variable with regard to the color of its flowers.

Plate 196. *Passiflora incarnata*.

Plate 197. *Passiflora incarnata*.

Plate 198. *Passiflora kermesina*.

Plate 199. *Passiflora laurifolia*. Photo by Karl-Heinz Kahlert.

Plate 201. *Passiflora loefgrenii*.

Plate 200. *Passiflora ligularis* in Ecuador. Photo by Michael Schwerdtfeger.

Plate 202. *Passiflora luetzelburgii* in Brazil. Photo by Stefan Vogel.

Plate 203. *Passiflora magnifica*. Photo by Lawrence E. Gilbert.

Plate 204. *Passiflora maliformis*. Photo by Karl-Heinz Kahlert.

Plate 205. *Passiflora mayarum*. Photo by Ron Boender.

Plate 206. *Passiflora mendoncae* in Brazil. Photo by Alan Jouy.

Plate 207. *Passiflora menispermifolia*.

Plate 208. *Passiflora miersii*. Photo by Ron Boender.

Plate 209. *Passiflora mooreana*.

Plate 210. *Passiflora mucronata*. Photo by Stefan Vogel.

Plate 211. *Passiflora nitida* in Bolivia. Photo by Günter Gerlach.

Plate 212. *Passiflora oerstedii*.

Plate 213. A clone resembling *Passiflora oerstedii* var. *choconiana*, widely cultivated under the erroneous name of *P. porphyretica*.

Plate 214. *Passiflora dispar*, erroneously cultivated as a variety, "alba," of *P. oerstedii*.

Plate 216. *Passiflora pergrandis*. Photo by Günter Gerlach.

Plate 215. *Passiflora pectinata*. Photo by John M. MacDougal.

Plate 217. *Passiflora phoenicea*. Photo by Maurizio Vecchia.

Plate 218. *Passiflora platyloba*.

Plate 219. *Passiflora popenovii*. Photo by Geo Coppens (CIRAD/IPGRI).

Plate 220. *Passiflora quadrangularis*. Photo by Hilaire Annonay.

Plate 221. *Passiflora quadriglandulosa* in Venezuela. Photo by Günter Gerlach.

Plate 223. *Passiflora recurva* in Brazil. Photo by Stefan Vogel.

Plate 224. *Passiflora reflexiflora*.

Plate 222. *Passiflora racemosa*.

Plate 225. *Passiflora resticulata*.

Plate 226. *Passiflora retipetala*.

Plate 227. *Passiflora seemannii* in Belize. Photo by Jan Meerman.

Plate 228. *Passiflora serratifolia*.

Plate 229. *Passiflora serratodigitata* in Bolivia. Photo by Axel Frank.

Plate 230. *Passiflora setacea* in Brazil. Photo by Stefan Vogel.

Plate 231. The variability of the Brazilian *Passiflora speciosa* is astonishing. This pink form from the state of Minas Gerais has a purple corona. Photo by Armando Carlos Cervi.

Plate 232. The red-flowering form of *Passiflora speciosa*, reminiscent of *P. coccinea*, stems from the Reserva Florestal de Linhares in the state of Espirito Santo. Photo by João Vasconcellos-Neto.

Plate 233. The pendent flower of the *Passiflora speciosa* from Santa Teresa reaches an astounding diameter of 22 cm. Photo by Reto Gamma.

Plate 234. This form of *Passiflora speciosa* with a banded corona stems from the Pico de Frade. Photo by Reto Gamma.

Plate 236. *Passiflora sublanceolata*. Photo by Ron Boender.

Plate 235. *Passiflora sprucei*.

Plate 237. *Passiflora subpeltata.*

Plate 238. The long peduncles of the Peruvian *Passiflora tarapotina* are very conspicuous. Photo by Sandra Knapp and James Mallet.

Plate 239. *Passiflora tenuifila.*

Plate 240. *Passiflora trialata*. Photo by Christian Houel.

Plate 241. *Passiflora triloba* is unique within genus *Passiflora* for its 6 sepals and petals. Photo by Roberto Vásquez Chavez.

Plate 242. *Passiflora trisulca*. Photo by Ron Boender.

Plate 243. *Passiflora tucumanensis*.

Plate 244. *Passiflora umbilicata*.

Plate 245. *Passiflora urbaniana*. Photo by Ron Boender.

Plate 246. *Passiflora variolata* in French Guiana. Photo by Pierre Pomié.

Plate 247. *Passiflora vitifolia.*

Plate 248. *Passiflora mixta*, a typical hummingbird-flowered representative of supersection *Tacsonia*, produces large pink flowers. The corona is merely a purplish band with white teeth, at the tip of the long floral tube.

Plate 249. The type specimen of *Passiflora rugosa* var. *venezolana*, a typical representative of section *Colombiana*, series *Colombianae*, with lanceolate stipules and unlobed, rugose leaves. Photo by Paul E. Berry.

Plate 250. Flower coloration is a special feature of the Colombian *Passiflora fimbriatistipula*. The inside is whitish with a touch of pinkish salmon, while the outside is intensely pinkish salmon to attract hummingbirds. Photo by Linda K. Escobar.

Plate 251. *Passiflora amazonica* in Peru—flower and characteristic longitudinally grooved fruit. Photo by Maximilian Weigend.

Plate 252. *Passiflora ampullacea* is characterized by horizontal green flowers. Photo by Giselher Behnken.

Plate 253. A closer inspection of *Passiflora ampullacea* reveals the zygomorphic androecium and a coronal series that occasionally reaches a length of 0.5 cm. Photo by Giselher Behnken.

Plate 254. *Passiflora anastomosans* in Peru. Photo by Jürgen Skrabal.

Plate 255. *Passiflora andina* in Ecuador. Photo by Thomas Fechtler.

Plate 256. The purplish red flower of *Passiflora antioquiensis*. Photo by Henk Wouters.

Plate 257. *Passiflora bracteosa* in Colombia. Photo by Sergio Segura (CIRAD-IPGRI).

Plate 258. The gigantic flowers of the Ecuadorian *Passiflora coactilis* are pendent on long peduncles. Photo by Michael Schwerdtfeger.

Plate 259. *Passiflora crispolanata* in Colombia. Photo by Sergio Segura (CIRAD-IPGRI).

Plate 260. *Passiflora cumbalensis* var. *pilosa*. Photo by Michael Schwerdtfeger.

Plate 261. *Passiflora cumbalensis* var. *cuencana*. Photo by Michael Schwerdtfeger.

Plate 262. *Passiflora cumbalensis* produces unusual masses of flowers. Photo by Peter Møller Jørgensen.

Plate 263. *Passiflora gracilens* in Peru. This species has the smallest flowers within supersection *Tacsonia*. Photo by Maximilian Weigend.

Plate 264. *Passiflora harlingii* in Ecuador. Photo by Michael Schwerdtfeger.

Plate 265. *Passiflora insignis* in Bolivia. Photo by Roberto Vásquez Chavez.

Plate 266. *Passiflora jamesonii* in Ecuador. Photo by John M. MacDougal.

Plate 267. *Passiflora jardinensis* in Colombia. Photo by John M. MacDougal.

Plate 268. *Passiflora lanata* in Colombia. Photo by Sergio Segura (CIRAD-IPGRI).

Plate 269. *Passiflora leptomischa* is characterized by a long peduncle. Photo by Linda K. Escobar.

Plate 270. A look into the flower of *Passiflora leptomischa* reveals 2 contrasting rings in the center of the flower. Photo by Linda K. Escobar.

Plate 271. *Passiflora loxensis* in Ecuador. Photo by Carmen Ulloa.

Plate 272. *Passiflora luzmarina* in Ecuador. Photo by Gwilym Lewis.

Plate 273. *Passiflora macropoda* in Bolivia. Photo by Helga Gelewsky.

Plate 274. *Passiflora mandonii* in Bolivia. Photo by Reto Gamma.

Plate 275. *Passiflora manicata* in Ecuador. Photo by Michael Schwerdtfeger.

Plate 276. *Passiflora mathewsii* in Ecuador. Photo by Gwilym Lewis.

Plate 277. A cultivar of *Passiflora mathewsii* with white flowers.

Plate 278. *Passiflora mixta* var. *eriantha* in its habitat, exhibiting the characteristic whitish pubescence of the hypanthium. Photo by Michael Schwerdtfeger.

Plate 279. A cultivated clone of *Passiflora mixta*.

Plate 280. *Passiflora parritae*. Photo by Ron Boender.

Plate 282. *Passiflora pilosicorona* in Bolivia. Photo by Roberto Vásquez Chavez.

Plate 281. *Passiflora peduncularis* in Peru. Photo by Maximilian Weigend.

Plate 283. *Passiflora pinnatistipula* in Peru. Photo by Paul E. Berry.

Plate 285. *Passiflora roseorum* in Ecuador. Photo by Gwilym Lewis.

Plate 284. Under cultivation, a white-flowered variety of *Passiflora pinnatistipula* proves to be very free-flowering.

Plate 286. *Passiflora sanctae-barbarae* in Ecuador. Photo by Michael Schwerdtfeger.

Plate 287. The frequently cultivated *Passiflora tarminiana*, with its typical reflexed sepals and petals. Photo by Giselher Behnken.

Plate 288. *Passiflora tenerifensis* in Colombia. Photo by Guillermo Guzmán and Geo Coppens (CIRAD/IPGRI).

Plate 289. The paler form of *Passiflora trifoliata* stems from the permanently dry habitats in southern and western Peru. The coloration is constant within populations and regions. Photo by Maximilian Weigend.

Plate 290. The more typical, more brightly colored form of *Passiflora trifoliata* occurs on the Cordillera Blanca, a considerably more humid region. Photo by Maximilian Weigend.

Plate 291. *Passiflora tripartita* var. *mollissima*, with a campanulate perianth. Photo by Gwilym Lewis.

Plate 292. *Passiflora tripartita* var. *azuayensis* has pink styles and coronal filaments. Photo by Michael Schwerdtfeger.

Plate 293. *Passiflora trisecta*, flowering during the night, has several characteristics that suggest pollination by bats.

Plate 294. A comparison of flower size in *Passiflora caerulea*, left, and the tetraploid *P.* 'Clear Sky', right.

Plate 295. *Passiflora* 'Adularia'.

Plate 296. *Passiflora* ×*allardii*.

Plate 297. *Passiflora* 'Amethyst'.

Plate 298. *Passiflora* 'Anastasia'.

Plate 299. *Passiflora* 'Aurora'. Photo by Maurizio Vecchia.

Plate 300. *Passiflora* ×*belotii*.

Plate 301. *Passiflora* 'Byron Beauty'.

Plate 302. *Passiflora* 'Cary'.

Plate 303. *Passiflora* ×*colvillii*.

Plate 304. *Passiflora* 'Coral Glow'.

Plate 305. *Passiflora* 'Corry Rooymans'.

Plate 306. *Passiflora* 'Crimson Tears'.

Plate 307. *Passiflora* ×*decaisneana*.

Plate 308. *Passiflora* 'Elizabeth'.

Plate 309. *Passiflora* ×*exoniensis*.

Plate 310. *Passiflora* 'Incense'.

Plate 311. *Passiflora* 'Inspiration'.

Plate 312. *Passiflora* 'Jara'.

Plate 313. *Passiflora* 'Jutta'. Photo by Henk Wouters.

Plate 314. *Passiflora* 'Lady Margaret'.

Plate 315. *Passiflora* 'Livie'.

Plate 316. *Passiflora* 'Marijke'.

Plate 317. *Passiflora* 'Minai'.

Plate 318. *Passiflora* ×*piresea*.

Plate 319. The deep purple form of *Passiflora* 'Pura Vida'.

Plate 320. *Passiflora* 'Purple Haze'.

Plate 321. *Passiflora* 'Roedie'.

Plate 322. *Passiflora* 'Sandra'.

Plate 323. *Passiflora* 'Stradivarius'. Photo by Maurizio Vecchia.

Plate 324. *Passiflora* 'Sunburst'. Photo by Karl-Heinz Kahlert.

Plate 325. *Passiflora* ×*violacea* 'Atropurpurea'.

Plate 326. *Heliconius hewitsoni*. Photo by Ron Boender.

Plate 327. Eggs of *Heliconius hewitsoni* on *Passiflora pittieri*. Photo by Ron Boender.

Plate 328. Two *Heliconius doris* ovipositing on *Passiflora ambigua*. Photo by Ron Boender.

Plate 329. *Heliconius doris* larvae on *Passiflora ambigua*. Photo by Ron Boender.

Plate 330. *Heliconius ismenius* larva. Photo by Ron Boender.

Plate 331. *Heliconius eleuchia* larvae on *Passiflora macrophylla*. Photo by Ron Boender.

Plate 332. *Heliconius doris* pupae. Photo by Ron Boender.

Plate 333. *Heliconius sapho* pupae. Photo by Ron Boender.

Plate 334. *Heliconius sapho* male piercing the pupa cuticle prior to mating. Photo by Ron Boender.

Plate 335. *Heliconius doris* with pollen. Photo by Ron Boender.

Plate 336. *Psiguria triphylla*, a foodplant for *Heliconius* butterflies. Photo by Ron Boender.

Plate 337. *Heliconius erato* nectaring on *Psiguria*. Photo by Ron Boender.

Plate 338. *Gurania* species, another foodplant for *Heliconius* butterflies. Photo by Ron Boender.

Plate 339. Spectacular "piano key" hybrids from Butterfly World. Photos by Ron Boender.

Plate 340. A survey of *Heliconius* species, part I. Clockwise from top left: *H. atthis*, *H. charitonia*, *H. doris*, *H. erato cyrbia*, *H. eleuchia*, and *H. cydno galanthus*. Photos by Ron Boender.

Plate 341. Rearing-cubicles at Butterfly World. Photo by Ron Boender.

Plate 342. Nectaring platform. Photo by Ron Boender.

Plate 343. A survey of *Heliconius* species, part II. Clockwise from top left: *H. erato venustus*, *H. ethilla*, *H. numata*, *H. zuleika*, *H. sapho candida*, and *H. hecalesia eximius*. Photos by Ron Boender.

15
Subgenus *Passiflora*

SUBGENUS *Passiflora* comprises about 240 species, including those generally regarded as typical passionflowers. Among these are numerous species that have been cultivated for many years for their sumptuous flowers. One example is *P. caerulea*, possibly the best-known species and a standard offering in flower shops and garden centers. Another favorite is *P. incarnata* (Plate 155), which grows in the wild in many parts of the United States and as a perennial in climatically favorable regions of Europe.

Species of subgenus *Passiflora* are characterized by handsome flowers that are usually dominated by a corona that is zoned or banded in several colors, less frequently in one color. The corona is the seat of scent production, and this subgenus contains the majority of intensely fragrant species. The corona is also the landing platform for bees and other insects attracted by the odor. Subgenus *Passiflora* contains many hummingbird-pollinated species as well, often with a reduced corona, notably the representatives of supersection *Tacsonia* (see chapter 16). Supersections *Distephana* and *Coccinea* contain hummingbird-pollinated species that are favored in cultivation for their large, brilliant red flowers, especially *P. coccinea* (Plates 7, 156, and 175–177) and *P. vitifolia*.

Time and again the species from this subgenus have been used for crossbreeding experiments, spawning well-known and traditional hybrids such as *Passiflora* ×*belotii* 'Empress Eugenie', *P.* ×*decaisneana*, and *P.* ×*violacea*. Recent years have brought a real flood of beautiful hybrids in this subgenus.

The native range of the species from subgenus *Passiflora* reaches from latitude 42° north to latitude 40° south. Thus it covers the southern half of the United States, Central America, and all of South America except its southern tip. The abundance in species is largest in north-central South America and gradually decreases in both directions. Representatives of subgenus *Passiflora* grow in tropical lowlands such as the Amazon basin as well as at the moderately high altitudes of the Andes. There is hardly a bee-pollinated species from subgenus *Passiflora* that grows at altitudes above 2500 m. This is not surprising, because the activity of poikilothermic insects (insects with a variable body temperature) decreases with increasing altitude, and pollination can only be guaranteed by warm-blooded hummingbirds and bats.

Quite a number of species in this group produce tasty fruit, with *Passiflora edulis*, *P. ligularis*, and *P. tarminiana* among the most economically important. Besides these, several others are grown on plantations or, more often, cultivated locally in gardens or exploited as a natural resource where they grow in the wild. With the exception of several well-known species, including *P. edulis* f. *edulis*, *P. tarmi-*

niana, and *P. foetida*, nearly all species from subgenus *Passiflora* are self-incompatible and require cross-pollination.

Systematics—Subdivision in Supersections

With *Passiflora incarnata* as the type species, subgenus *Passiflora* can be regarded as the heart of the genus. In 1753, Linnaeus described 24 species of *Passiflora*. As an increasing number of species became known, de Candolle (1822) subdivided the genus into 7 sections. Section *Granadilla* comprised those species that now form subgenus *Passiflora*. *Tacsonia* was considered a separate genus. These subdivisions were partly revised in the years to follow, and sections were raised to the rank of subgenus. Masters (1871) subdivided the genus *Passiflora* into 4 subgenera: *Astrophea*, *Plectostemma*, *Murucuia*, and *Granadilla*, the latter being of importance in this context. This classification was taken over by Triana and Planchon (1873) in their work on Colombian Passifloraceae and was enlarged by moving the long-tubed species comprising genus *Tacsonia* into a new, 5th subgenus: *Passiflora* subgenus *Tacsonia*. Even when Harms used the term "section *Granadilla*" again in 1925, little had changed regarding its composition, but he added many new sections to hold the many other species found by that time. The comprehensive revision of American *Passiflora* species by Killip (1938) was strongly influenced by the 21 sections of the genus *Passiflora* described by Harms. Killip raised most of Harms's sections to subgenera. Many of the names are the same, but the ranks differ.

Since then, new, binding versions of the *International Code of Botanical Nomenclature* have been established. In accordance with this code, the name of every subgenus that includes the type species is formed by the unchanged repetition of the genus' name as an epithet. Consequently, subgenus *Granadilla*, containing the type species of the genus *Passiflora*, *P. incarnata*, must be renamed subgenus *Passiflora*.

In the 1980s and 1990s, Linda Escobar presented outstanding studies on systematics that dealt mostly with the subgenera *Tacsonia* and *Astrophea*. In 1997, Armando Carlos Cervi revised the Brazilian species of subgenus *Passiflora* and presented a short review of the history of its classification.

The most considerable changes are due to the recently revised systematics set forth by Feuillet and MacDougal (see chapter 2). In essence, they reduced the number of subgenera previously recognized by Killip and others by incorporating several closely related subgenera into subgenus *Passiflora*: *Adenosepala* (invalid), *Calopathanthus*, *Distephana*, *Dysosmia*, *Dysosmioides* (invalid), *Granadilla*, *Granadillastrum*, *Manicata*, *Rathea*, *Tacsonia*, *Tacsonioides*, and *Tacsoniopsis*.

Altogether, Feuillet and MacDougal subdivide the representatives of subgenus *Passiflora* into 6 supersections. Supersection *Passiflora*, containing the type of the genus, comprises those species that have linear to lanceolate stipules, serrate leaves, free serrate bracts, and upright flowers with a dominating corona (for example, *P. incarnata*, *P. cincinnata*, *P. mayarum*, and *P. serratifolia*).

The representatives of supersection *Stipulata* are divided into 3 sections, with section *Granadillastrum* the most rich in species. This supersection contains plants with conspicuous, mostly subreniform or oblong-ovate stipules, free bracts, mainly 3-lobed or entire leaves (except section *Dysosmia*), and upright or rarely pendent flowers with a large, conspicuous corona (except the hummingbird-pollinated species from sections *Tacsonioides* and *Kermesinae*). Examples are *Passiflora caerulea*, *P. foetida*, *P. reflexiflora*, *P. setulosa* (Plate 157), *P. subpeltata*, and *P. subrotunda* (Plate 158).

An essential characteristic of the species from supersection *Laurifolia* are large pendent flowers with a predominant corona that surrounds the ovary in a campanulate fashion. This group is divided into 3 series, with the series *Laurifoliae* having stipules that are linear or linear-lanceolate, bracts that are free and usually glandular, and leaves that are lanceolate to oblong-ovate and entire. Examples are *Passiflora laurifolia* and *P. nitida*. When the bracts are connate, at least at base, this characteristic is sufficient to assign the species to series *Tiliifolia*, even when the leaves and stipules have different forms. Examples are *P. ligularis*, *P. maliformis*, and *P. triloba*. The series *Quadrangulares* comprises the species with winged, angled stems. Moreover, these have large, unlobed leaves with entire margins, free bracts, and ovate-lanceolate stipules. This series includes such well-known and popular species as *P. alata* and *P. quadrangularis*. In contrast to the other species from supersection *Laurifolia*, the monotypic series *Marginatae* is characterized by small leaves and flowers with very narrow bracts.

The few species of supersection *Distephana*, notably *Passiflora glandulosa* and *P. variolata*, are characterized by only a slight pubescence or no pubescence at all, linear-lanceolate stipules, unlobed entire leaves, very narrow free bracts, and upright hummingbird-pollinated flowers. This makes them easily distinguishable from species of the related supersection *Coccinea*, the representatives of which (*P. coccinea* and *P. speciosa*, for example) have linear-lanceolate stipules and sumptuous red flowers with a short, upright corona but also feature a fine pubescence, serrate leaves, and, as a rule, significantly larger bracts.

The representatives of supersection *Tacsonia* are treated in chapter 16.

Cultivation

The horticultural needs of this large subgenus can vary considerably depending on the origins of various representatives. The whole range of conditions becomes evident when one regards the minimum temperatures that different species will tolerate. The 2 species with the greatest resistance to frost, *Passiflora incarnata* and *P. caerulea*, belong to this subgenus, as do numerous species from the Amazonian lowlands that need year-round temperatures around 25°C. General rules cannot be given until the subgenus is split into several groups.

Species from supersection *Laurifolia* are mostly at home in the tropical lowlands of Central and South America, including the Amazon basin. Only a few species, such as *Passiflora ligularis* or *P. popenovii*, proceed westward to the Andes. Lowland species from supersection *Laurifolia*, such as *P. quadrangularis*, *P. laurifolia*, *P. platyloba*, and *P. serratodigitata*, should preferably be cultivated in a warm greenhouse. It is advisable to plant these species into an earth bed, because only in this way will they reach the necessary size to set their gorgeous, pendent flowers. Some of these species tend to produce chlorotic leaves under cultivation, so it is necessary to administer an adequate supply of nutrients and trace elements and to provide an adequate soil mix. Soil for susceptible species should not contain too much peat. Mixing in some lateritic soil with a high content of iron and aluminum will be a great help. During winter a minimum temperature of 15°C should be guaranteed, except for the few species from the Andes. It may be helpful to use heating cables in the soil or heating mats under the pots. Sometimes additional lighting is needed as well.

Similar conditions are required by representatives from supersections *Distephana* and *Coccinea*, for the most part also at home in the tropical lowlands of South America. In a sunny, protected location, *Passiflora vitifolia*, for example, can be cultivated in a 40-cm pot during summer months. These species will do much better, however, in a warm greenhouse. Although they may flower reliably when they have reached a height of about 1.5–2 m, they will display all their splendor only when planted directly in the soil. In their natural habitats, they are sometimes found creeping over the ground.

Species from section *Dysosmia* can be described as very free-flowering, even in a small pot, provided they are grown in a warm location, either in a protected, sunny place outdoors or under glass or near a window. Standard potting mix is sufficient. Well-known species such as *Passiflora foetida*, *P. sublanceolata*, and *P. urbaniana* are representatives of this group.

Most species that have been cultivated in enthusiasts' collections for many years are distributed in the other groups of subgenus *Passiflora*. Often these species are unpretentious: they are satisfied with a standard potting mix and will flower without problems under cultivation. One only has to think of *P. amethystina*, *P. caerulea*, *P. edulis*, *P. subpeltata*, and *P. tucumanensis*, all easy and beautiful species. In addition, a great many of them will withstand rather low temperatures in winter.

In most cases, the propagation of species from subgenus *Passiflora* is quite easy. At a soil temperature around 25°C, cuttings will root in a propagating chamber within 10–25 days. The same holds true for sowing seed. Fresh seeds may well germinate after just a few days.

Passiflora actinia Hook.

Bot. Mag. 69: pl. 4009 (1843)
actinia: sea anemone (in reference to the appearance of the corona)
Synonym: *P. paulensis* Killip
Supersection *Stipulata*, Section *Granadillastrum*

Passiflora actinia (Plate 159), indigenous to the Organ Mountains in southern Brazil, produces beautiful flowers 6–9 cm diam. with an intense scent of lilac. This plant's main flowering period in its natural surroundings is September to December, but it flowers February to April in central Europe. For the rest of the year only a few single flowers can be expected. As conditions are too cold outdoors in February to April, *P. actinia* is recommended for cultivation in a greenhouse or conservatory. Only in such a location are conditions ideal for flower bud formation. In frost-free climatic zones the plant might be cultivated outdoors year-round.

The impressive corona of the pendent flowers is for the most part violet-blue and white; at the very base, it is violet-red and white. When the flower is open, the white sepals and petals are bent back widely. Occasionally the plant produces fruit with its own pollen. These, however, only reach the size of a hazelnut and fall off while still unripe. In order to produce ripe fruit, the flowers must be cross-pollinated.

Passiflora actinia belongs to the group of slightly frost-resistant passionflowers. Plants that are several years old can withstand temperatures down to −5°C for a short time, though the upper parts of the shoots die off. Overall it prefers more moderate temperatures and should be overwintered at temperatures above 7°C.

Vine: glabrous; stem: subterete, slender, wiry; stipules: semiovate to subreniform, 1–4 × 0.5–2 cm, entire to slightly crenulate; petioles: slender, (0.5–)1.5–3(–5) cm, with 4 or occasionally 6 glands, mostly paired above and below middle; leaves: unlobed, ovate, 3–11 × 2–8 cm, membranous to subcoriaceous, entire; peduncles: slender, 1.5–3.5 cm, solitary; bracts: cordate-ovate, 1.5–2.5 × 1–2 cm, entire, free; flowers: white and violet, 6–8(–9) cm diam.; sepals: greenish white inside, green outside, oblong-lanceolate, 2–3.5 × 1–1.5 cm, slightly aristate; petals: white, lanceolate, 2.5–4 × 1 cm; corona: 4 or 5 series, violet banded with white, outer 2 series 3–4 cm long, wavy in upper 1/3, innermost series 0.1 cm long; ovary: ovoid, glabrous; fruit: ovoid to globose, 2.5–4 × 2–3 cm, yellow, edible; seeds: ovoid, 0.4–0.5 × 0.3–0.4 cm, reticulate.

Passiflora acuminata DC.

Prodr. 3: 328 (1828)
acuminatus: tapering to a slender point (in reference to the leaves)
Supersection *Laurifolia*, Series *Laurifoliae*

This is a previously little-known species from series *Laurifoliae*, a series characterized by pendent flowers with a conspicuous, campanulate corona in combination with linear-lanceolate stipules and unlobed, entire leaves. In *Passiflora acuminata* (Plate 160), at home along the Amazon River up to

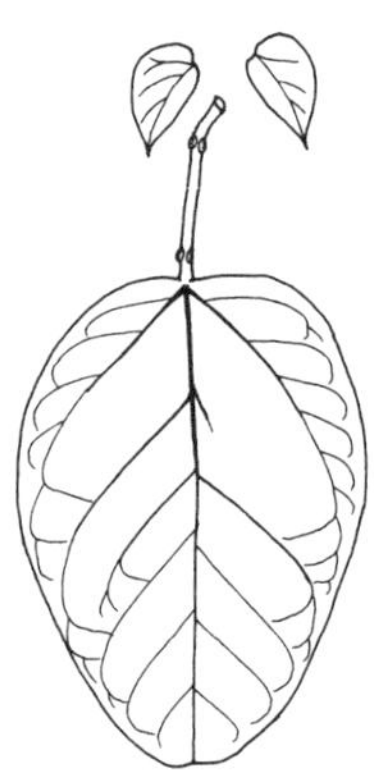

Passiflora actinia

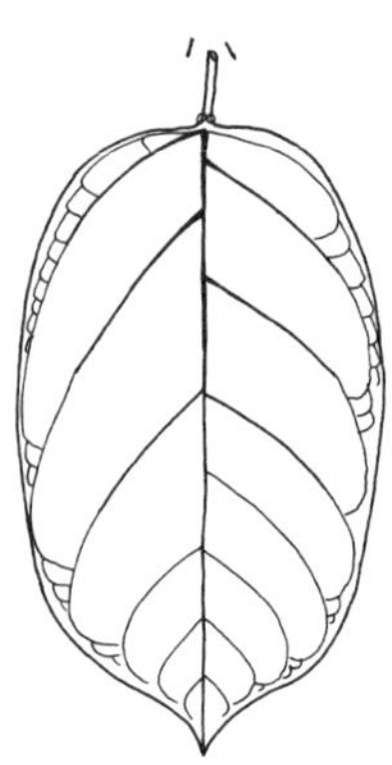

Passiflora acuminata

its delta in northeastern Brazil, the filaments of the outer 2 coronal series are particularly slender, making it easily distinguishable from the closely related, widely distributed *P. laurifolia*. Similar delicate coronal filaments can otherwise be found in only one other species from this series: *P. gleasonii*. This species differs from *P. acuminata* in having much smaller, reddish bracts, a different operculum, and much shorter inner coronal series.

On the whole, the scented flowers of *Passiflora acuminata* are particularly beautiful among passionflowers, though they only reach a diameter of about 6 cm. In its habitat, it flowers from November to February (Cervi 1997). It is not known whether this rare species is under cultivation. The tropical conditions in its home range—the high atmospheric humidity of the banks of the Amazon—certainly suggest that it should be cultivated in a warm greenhouse.

Vine: glabrous except ovary; stem: terete or subangular; stipules: linear, ca. 0.4 cm long, soon deciduous; petioles: ca. 1 cm, with 2 glands at apex; leaves: unlobed, lanceolate or oblong-lanceolate, 7–14 × 2–5 cm, membranous to subcoriaceous, entire; peduncles: 3–4 cm, solitary; bracts: oblong, 2.5–4 × 1–1.5 cm, entire, free; flowers: white and violet, 5–7 cm diam.; sepals: white inside, greenish outside, narrowly lanceolate, 2–2.5 × 0.6–0.7 cm, aristate, awn 0.3 cm long; petals: white, mottled with purple spots, narrowly lanceolate, 1.5–2 × 0.3–0.4 cm; corona: 5 series, banded with violet and white, outer 2 series linear, 3–4 cm long, wavy in upper 1/3, innermost series 0.5–0.6 cm long; ovary: ovoid, minutely puberulent; fruit: not seen.

Passiflora aimae Annonay & Feuillet

Adansonia 20: 295 (1998)
named after Aimee Govindoorazoo, mother of Hilaire Annonay
Supersection *Coccinea*

This extremely rare passionflower (Plate 161) is at home in French Guiana, where it has been found only in the region of Sinnamary and in the Bassin de l'Arataye at altitudes of 10–400 m. It combines the characteristics of the best-known and most widely distributed red-flowering species from French Guiana: *Passiflora glandulosa* from supersection *Distephana*, and *P. coccinea* from supersection *Coccinea*. It differs from these 2 species, and from the other representatives with unlobed leaves from these 2 supersections, by its combination of crenate leaves and small bracts. It is also remarkable for its leaf margins and for the reddish violet venation on the underside of its leaves.

The brilliant red flowers of *Passiflora aimae* reach a maximum diameter of 9 cm. A contrast is formed by the short, white corona that stands out in the middle of the flower with its tip inclined toward the androgynophore. As with most hummingbird-pollinated species, the flowers are scentless.

Passiflora aimae has not yet come into cultivation. Considering the conditions of its natural surroundings, and considering growers' experience with other species from similar environments in French Guiana, this is likely to be a very warmth-loving, picky plant that should be cultivated in a warm greenhouse year-round.

Vine: partly pubescent; stem: terete, brownish; stipules: triangular to linear, ca. 0.5 cm long, with 2 glands; petioles: up to 2 cm long, with 2 paired glands near base, occasionally 2 additional glands above middle; leaves: unlobed, ellipsoid to oblong-ovate, 8–15 × 3–8 cm, glabrous above, pubescent beneath, crenate; peduncles: 4–5 cm, solitary; bracts: oblong-ellipsoid, ca. 1.3 × 0.4 cm, with 6 paired glands at margin, free; flowers: bright red and whitish, 8–9 cm diam.; sepals: red, oblong, 4–4.2 × 1–1.5 cm, aristate, awn 0.3–0.5 cm long; petals: red, subequal to sepals; corona: 2 series, outer series ca. 0.7 cm long, whitish, inner series ca. 0.5 cm long, reddish; ovary: ovoid, pubescent; fruit: unknown.

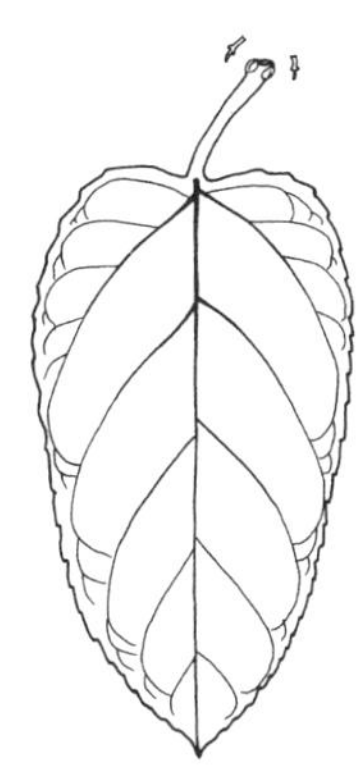

Passiflora aimae

Passiflora alata Curtis

Bot. Mag. 2: t. 66 (1788)
alatus: winged (in reference to the stem)
Synonyms: *P. brasiliana* Desf., *P. mascarensis* Presl., *P. mauritiana* Du Pet. Thouars, *P. oviformis* Roemer, *P. sarcosepala* Barb. Rodr.
Supersection *Laurifolia*, Series *Quadrangulares*

Passiflora alata (Plate 162), with its beautiful, fragrant flowers, is a definite favorite among passionflowers. At a diameter of about 10 cm, rarely a little smaller, the pendent flowers are quite imposing. The particular charm of the flower results from the deep red sepals and from the petals with a corona banded violet and white. The 4-angled stem is conspicuous for its more or less distinct wings.

In many regions of South America, *Passiflora alata* is cultivated because of its edible fruit. However, to set fruit under cultivation the plants must be hand-pollinated with pollen from another clone. Its original native range is unclear but appears to be eastern Brazil.

Passiflora alata is closely related to *P. quadrangularis* and *P. phoenicea*. *Passiflora phoenicea* differs from the 2 others by the bright yellow nectar glands on the upper 1/3 of the petiole. *Passiflora alata* can be distinguished from *P. quadrangularis* and the numerous other common crossbreedings, especially *P.* ×*decaisneana*, by its corona. While the corona opens wide in *P. quadrangularis* and *P.* ×*decaisneana*, it keeps standing like a lampshade around the column and ovary of *P. alata*. Moreover, the innermost coronal filaments (not counting the membranous operculum) are short and separate in *P. alata*, while in *P. quadrangularis* they are more or less merged together into a flaplike 2nd operculum (Sowerby 1794). A classification based on the number of glands on the petioles can lead to false results. Although *P. alata* usually has 2–4 petiolar nectar glands and *P. quadrangularis* 4–6, both can have 2–6 nectar glands, as can their crossbred offspring. The genuine *P. quadrangularis* can easily be distinguished from *P. alata* by its stipules, which are 2–4.5 cm long, and by the color of the sepals. *Passiflora quadrangularis*, however, is rarely represented in European collections. Instead, *P.* ×*decaisneana* is mistakenly cultivated as *P. quadrangularis*.

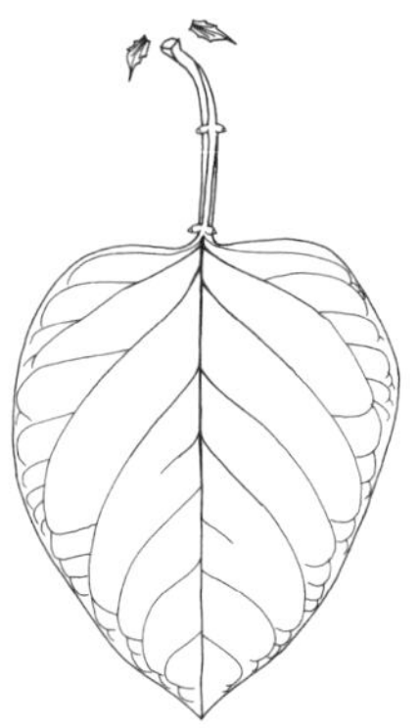

Passiflora alata

In principle, cultivation of *Passiflora alata* is not very complicated. However, 2 or 3 years may go by before young plants that were planted directly in the ground or in large containers start to flower. Often the plants use all their power to produce leaves at first, beginning to set flowers only when the roots have completely permeated the potting mix.

Vine: glabrous; stem: 4-angular, the angles winged, striate; stipules: linear-lanceolate to ovate-lanceolate, 0.7–2 × 0.2–1 cm, entire or slightly serrate; petioles: 2.5–5 cm, with 2–4 or occasionally 6 mostly paired glands at and above middle; leaves: unlobed, ovate, 8–15(–22) × 7–10(–13) cm, entire to slightly serrate; peduncles: 2–4(–6) cm, solitary; bracts: ovate, 0.4–2 × 0.3–1.5 cm, entire, free; flowers: red, violet and white, 7–10(–12) cm diam.; sepals: red inside, green outside, oblong-ovate, 3–5 × 1–2 cm, aristate; petals: red inside, whitish red outside, oblong, slightly smaller than sepals; corona: 4–6 series, banded violet and white, outer 2 series (2–)3–4 cm long, wavy in upper 1/3, inner series 0.1–0.3 cm long; ovary: oblong to obovoid, glabrous; fruit: obovoid or pyriform, 8–11 × 5–8 cm, yellow, edible; seeds: cordate or cordate-oblong, 0.7–0.9 × 0.5–0.6 cm, reticulate.

Passiflora ambigua Hemsl.

Bot. Mag. 128: t. 7822 (1902)
ambiguus: doubtful, ambiguous
Supersection *Laurifolia*, Series *Laurifoliae*

This species (Plate 163), known since the early 1900s, is a fastidious one that should be cultivated exclusively in the warm greenhouse. It is also advisable to plant it directly into an earth bed, as flowers will not form when it is grown in a pot.

In comparison with the pink- to violet-tinged sepals, the smaller petals are more strongly colored—an intense pink-violet. The coronal filaments, up to 5 cm long, are not undulated but banded white and deep violet. *Passiflora ambigua* differs from *P. laurifolia*, to which it is closely related, by its significantly larger flowers, 8–13 cm diam. Moreover, both nectar glands in *P. ambigua* are located in the middle of the petiole or slightly lower, while in *P. laurifolia* they occur at the tip of the petiole. The specific epithet refers to the frequent confusion of these species. *Passiflora ambigua* is also closely related to *P. pergrandis*, which was described in the 1980s and which has been cultivated for only a short time.

In its homelands, from southern Mexico to Ecuador and Venezuela, *Passiflora ambigua* is renowned for its thick-rinded edible fruit. Propagation by cuttings is rarely successful, which has undoubtedly contributed to this magnificent species being so rarely cultivated. The clone from Venezuela that is cultivated in Europe is prone to chlorosis.

Vine: glabrous except ovary and inflorescence; stem: terete, striate; stipules: linear, 0.7–1.5 cm long, soon deciduous; petioles: 2–4 cm, with 2 ovate, sessile glands near middle; leaves: unlobed, oblong-ovate, 8–20 × 4.5–11 cm, coriaceous, entire; peduncles: 4–10 cm, solitary or inflorescence with 1–2 (occasionally more) flowers; bracts: ovate, 3.5–7 × 3–5 cm, entire with several glands; flowers: purplish, violet, and white, 9–14 cm diam.; sepals: more or less intensely mottled with purple inside, greenish white outside, oblong-ovate, 3–5 × 1–2 cm, aristate, awn 0.3 cm long; petals: purple, oblong, 3–4 × 0.5–1 cm; corona: 3 or more series, outer series 0.7–1.5 cm, 2nd series 3–5 cm long, banded violet and white, inner series 0.1–0.3 cm; ovary: ovoid, densely puberulent; fruit: orange to dark yellow, ovoid, 4–14 × 3–12 cm, edible; seeds: ovate, 0.6–1 × 0.5–0.6 cm, strongly flattened, closely reticulate, dark brown.

Passiflora amethystina J. C. Mikan

Del. Fl. and Faun. Bras. Fasc. 4: t. 20 (1825)
amethystinus: amethyst-colored, purple
Synonyms: *P. bangii* Mast., *P. cornuta* Mast., *P. laminensis* Barb. Rodr., *P. lilacina* Roemer, *P. onychina* Lindley, *P. violacea* Vell.
Supersection *Stipulata*, Section *Granadillastrum*

Three clones of this pretty species (Plates 164–166) have been established in enthusiasts' collections for quite some time: one from Paraná, Brazil, with a short peduncle; one from Espirito Santo, Brazil, with a long peduncle; and one from the Yungas near La Paz, Bolivia. And an increasing number of clones, from various habitats, have come into cultivation during the last few years.

The palette of *Passiflora amethystina* was enriched quite recently in the wake of Mauro Peixoto's field studies in Brazil; the clones he found show an intense blue coloration. Some of these new clones have come into cultivation. Among enthusiasts, the question is often raised as to whether the variability of the individual clones exceeds the

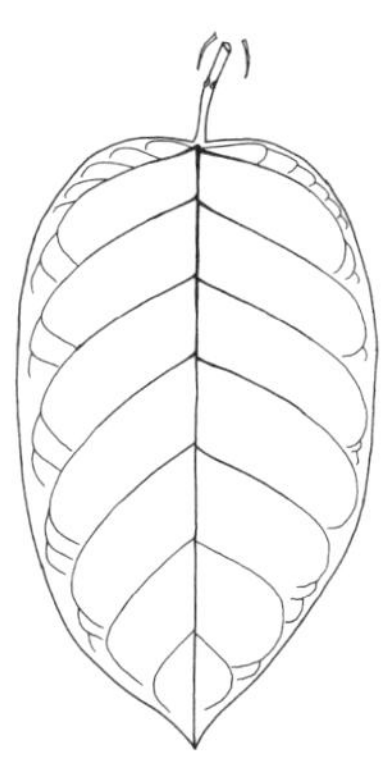

Passiflora ambigua

Passiflora amethystina

normal limits. A number of collectors think that the form with a long peduncle is an independent species—perhaps *P. violacea* (a later homonym of *P. ×violacea*)—and should be regarded as different from *P. amethystina*. This contradicts the conclusions drawn by José da Costa Sacco in 1980 and by Armando Carlos Cervi somewhat later. According to their studies, *P. cornuta*, *P. violacea*, and *P. amethystina* are not, as Killip surmised, 3 closely related species but one and the same species. The length of the peduncles, the size of the stipules, and the number of coronal series differ only gradually between the 3. As a consequence, *P. cornuta* and *P. violacea* were declared synonyms of *P. amethystina*. Cervi (1997) has recognized 2 varieties: *P. amethystina* var. *bolosii* and the typical variety, *P. amethystina* var. *amethystina*. These differ with regard to the pilosity of the ovary. While the ovary is glabrous in variety *bolosii*, it is pilose in variety *amethystina*.

It should be pointed out that *Passiflora violacea* is not to be confused with the well-known hybrid *P. ×violacea*.

The native range of *Passiflora amethystina* covers Argentina, Brazil, Bolivia, Guyana, Paraguay, Peru, and Uruguay. Its habitats are the tropical lowlands as well as the elevations up to 2600 m.

This charming passionflower produces violet flowers that reach 6–10 cm diam. even in a small pot. It can thus be cultivated on a windowsill, though it produces more flowers when grown in a greenhouse or conservatory. The coloration of the corona is variable, but as a rule it is deep purple at the base, white in the middle (partially with some blue-violet), and purple at the tip. *Passiflora amethystina* can easily be distinguished from the far more common *P.* 'Amethyst' by the colored zones on its coronal filaments and by its awns or cornicles, often about 2 cm long at the tip of the sepals. Although some flowers may appear in summer, the main flowering period is late summer and fall, so that *P. amethystina* might be called a short-day plant. The overwintering temperature varies with regard to the origin of the clones; some are from tropical lowlands, others from high altitudes. Both the short- and long-peduncle forms, as well as the Yungas clone, tolerate temperatures around 5°C for a short time, though a temperature around 12°C is recommended. On the whole, the clone with short peduncles seems to be fussier about its conditions.

The 3-lobed deep green foliage of *Passiflora amethystina* has a characteristic smell that is even more intense when the leaves are crushed. Another remarkable feature is that dried-up parts of the plant turn black.

Vine: generally glabrous, except the ovary; stem: terete, wiry, striate; stipules: ovate-lanceolate to ovate-oblong, 0.5–3.7(–5) × 0.2–2.5 cm, entire; petioles: 1.5–7(–12) cm, with 3–8 short stipitate glands; leaves: 3-lobed, 4–12.5 × 5–20 cm, membranous or subcoriaceous, entire; peduncles: 2.5–12(–21) cm, solitary; bracts: oblong, 0.8–3 × 0.5–1.3 cm, entire, free; flowers: violet to purple-violet or bluish, 6–10 cm diam.; sepals: violet inside, green outside, oblong to oblong-lanceolate, 2.5–4 × 0.5–1 cm, aristate, awn 0.5–2.5 cm; petals: violet, oblong, 2.7–4.3 × 0.5–0.9 cm; corona: 4–7 series, purple-violet at upper 1/3, white with violet at middle and dark purple at base, outer 2 series 1.7–2.6 cm long, inner series 0.4–0.7 cm long; ovary: ellipsoid to ovoid, densely pilose or occasionally glabrous; fruit: ellipsoid, 5–8 × 2–2.5 cm, orange; seeds: ovoid, 0.3–0.5 × 0.2–0.4 cm, reticulate.

Passiflora arida (Mast. & Rose) Killip

J. Wash. Acad. Sci. 12: 256 (1922)
aridus: at home in arid areas
Supersection *Stipulata*, Section *Dysosmia*

It is astonishing how different the habitats colonized by passionflowers can be. On the one hand there are the high-altitude species from the Andes, which are adapted to nightly drops in temperature down to the freezing point, and on the other hand there are the numerous species from the tropical lowlands, which need constant high temperatures and do not tolerate frequent variations. Although there are many forms between these 2 extremes, all *Passiflora* species have one characteristic in common: they cannot survive without a generous water supply. An exception to this is *P. arida* (Plate 167). It is native to northwestern Mexico but has recently been introduced to southern Arizona. As a protection against evaporation and sunburn, it is usually covered with a fuzz of dense, soft white hair. However, Killip (1938) speaks of 2 varieties, one characterized by a less dense pubescence and the other by a different form of the leaf lobes. Variation is continuous, though, and these varieties are not generally used now.

With its 2-pinnatifid bracts, *Passiflora arida* belongs to section *Dysosmia*, which also contains such well-known species as *P. foetida*, *P. sublanceolata*, and *P. urbaniana*. In fact, the purple, blue-violet, and white flowers are very similar to those of several varieties of *P. foetida*. They can reach a diameter of up to 4 cm. A unique characteristic of this species is the absence of oily, sticky secretions typical of other representatives from section *Dysosmia*, a difference that can be felt when one touches the leaves and particularly the bracts. The seeds are 2.5 mm wide or less, another useful feature for identification purposes. At times seeds from this species are available, and it appears in collections now and then. At least the clones from Arizona are known to be self-compatible, and this species has the potential to be a weed if introduced outside its native habitat. The small, unusually sweet, fragrant fruit is among the most delicious. Plants sold as *P. foetida* var. *arizonica* are actually this species.

Vine: densely lanate with soft indument; stem: profusely branched at base; stipules: semiannular, deeply cleft in subulate divisions, one segment much longer than the rest; petioles: 0.5–1(–3.5) cm without glands; leaves: 3-lobed, occasionally deeply 5-lobed, 3–7 × 4–8 cm; peduncles: 1.5–3 cm, solitary; bracts: 2-pinnatifid, into linear-subulate segments ca. 3 × 2 cm, free; flowers: white, purple and blue-violet, 3–4 cm diam.; sepals: white inside, whitish green or white outside, ovate-lanceolate, ca. 1.5 × 0.7 cm, slightly keeled outside; petals: white, oblong, ca. 1.5 × 0.4 cm; corona: ca. 6 series, purple or violet at base, white at middle and blue-violet at apex, outer 2 series 1–1.5 cm long, inner series 0.1–0.2 cm long; ovary: subglobose, densely sericeous; fruit: ovoid, ca. 3.5–4 × 3 cm, pubescent; seeds: oblong, 0.3–0.5 × 0.2–0.25 cm, 3-toothed at apex, truncate at base, reticulate.

Passiflora bomareifolia Steyermark & Maguire

Mem. New York Bot. Gard. 17: 455 (1967)
bomareifolius: having leaves like *Bomarea*
Supersection *Distephana*

This rare species (Plate 168) has been found only in a few places in the department of Bolívar, southeastern Venezuela, in the region of the Chimanta Massif at altitudes of 470–1275 m. Unfortunately, it has yet to come into cultivation.

Passiflora bomareifolia and the closely related *Passiflora amicorum*, *P. glandulosa*, *P. ernestii*, and *P. variolata* constitute the small supersection *Distephana*. These species are characterized by linear-lanceolate to setaceous stipules, unlobed, entire, glabrous or nearly glabrous leaves, and hummingbird-pollinated, mostly reddish flowers with a short corona. *Passiflora bomareifolia* can be distinguished at first sight from the first 2 species by having leaves that are often up to 10 cm long. The inside of the sepals and petals in *P. amicorum* is whitish, while the outer coronal series in *P. glandulosa* consists of clearly visible single filaments that do not adjoin to the androgynophore. In *P. bomareifolia*, as well as in *P. ernestii* and *P. variolata*, the corona forms a ring that is connate except for a few millimeters of

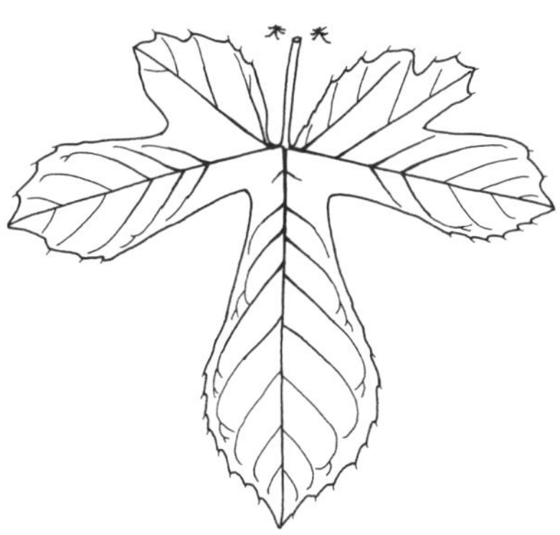

Passiflora arida

Passiflora bomareifolia

its tip, and it is inclined toward the androgynophore, thus preventing nectar robbers from reaching the innermost part of the flower. By means of its narrower leaves, fimbriate operculum, and above all its coral-rose flowers, *P. bomareifolia* can be distinguished from the red-flowering *P. ernestii* and *P. variolata*. Stephen S. Tillett (2003) believes *P. bomareifolia* to be a synonym of *P. variolata*.

Vine: glabrous; stem: subangular; stipules: setaceous, soon deciduous; petioles: 0.7–1 cm, with 2 glands near base; leaves: unlobed, narrowly oblong-lanceolate, 9.5–12.5 × 1.7–2.3 cm, dark green above, pale green beneath, membranous, entire; peduncles: 5–7 cm, solitary; bracts: linear-lanceolate, free; flowers: pink to coral-rose, 10–12 cm diam.; sepals: pink to coral-rose, narrowly oblong, 5–6 × 0.6–0.7 cm, slightly keeled outside, keel terminating in an awn 0.2–0.3 cm long; petals: pink to coral-rose, ca. 5.5 × 0.5 cm; corona: 2 series, pink, paler in upper 1/3, pale green or white at tips, outer series ca. 1.4 cm long, leaned against androgynophore, inner series 0.4–0.5 cm long; ovary: ovoid, glabrous; fruit: unknown.

Passiflora caerulea L.

Sp. Pl. 959 (1753)
caeruleus: sky-blue (in reference to the corona)
Supersection *Stipulata*, Section *Granadillastrum*

Passiflora caerulea (Plate 169) is possibly the best-known representative of the genus *Passiflora*. It has been offered by nurseries and flower shops for many years, typically tied to a stake or wrapped around a bowed wire. Since its demands are rather modest, it is suited for any location. It always develops into a decorative eye-catcher, whether as a houseplant on a windowsill, as a robust container plant, planted out directly in the garden, or cultivated in a greenhouse or conservatory. Moreover, *P. caerulea* tolerates temperatures down to −15°C when young and even lower temperatures when several years old. It can be overwintered outdoors in many parts of the United States and Europe. Accordingly, the wearisome removal to winter quarters can be avoided. In May, or at that month's end, the vine is planted in the garden directly in a hole with good drainage, preferably at a location facing south. Normally it develops so well and develops such vigorous roots that it is prepared for the next winter. To protect it from severe frost, cover it with a layer of fallen leaves under Bubble Wrap, or in milder regions, with fir twigs. If a plant perishes during winter, this is generally due not to the cold but to the rotting of the roots, caused by continuous rain. The drier the location in winter, the better the plant will withstand frosty spells.

This species is indigenous to Argentina, Brazil, and Paraguay. Flowers are about 7–9 cm diam. White petals form the background for the straight corona, which is colored in bands. Toward the end of summer the plant bears edible orange fruits that are 5–7 cm long. In combination with the flowers, these give the plant a decorative effect.

There are several varieties of *Passiflora caerulea*. One is the popular, completely white *P. caerulea* 'Constance Eliott' (Plate 170), which is also resistant to frost down to −15°C. Varieties such as such as *P. caerulea* 'China Blue' have a light blue corona, and the rather recent tetraploid cultivar *P.* 'Clear Sky' has larger flowers. 'Clear Sky' is a complex crossbreed between different *P. caerulea* varieties and hybrids that have *P. caerulea* as parent plants. All these cultivars tolerate temperatures below the freezing point very well.

Vine: glabrous; stem: subangular, striate, grooved; stipules: semiovate to subreniform, 1–2 × 0.5–1 cm, slightly dentate to subentire; petioles: 1.5–5 cm, with 2–4 or occasionally 6 short stipitate glands; leaves: 5- or 7-lobed or occasionally 3- or 9-lobed, 5–16 × 6–18 cm, entire; peduncles: 3–7 cm, solitary; bracts: ovate to ovate-oblong, 1.5–2.7 × 1–2.3 cm, pale green, entire, free; flowers: white, blue,

Passiflora caerulea

and blackish purple, (6–)7–9(–10) cm diam.; sepals: white inside, green outside, oblong, 2–3.5 × 1.3–1.8 cm, slightly keeled outside, keel terminating in a short awn; petals: white, oblong, 2–3.8 × 0.8–1.4 cm; corona: 4 series, blue at upper 1/3, white at middle and blackish purple at base, outer 2 series (0.8–)1.5–2.5 cm long, inner series 0.1–0.2 cm long; ovary: ovoid or subglobose, glabrous; fruit: ovoid, (4–)5–7 × 3–4 cm, orange, edible; seeds: obcordate or cuneate, 0.5–0.6 × 0.4 cm, coarsely reticulate.

Passiflora cerasina Annonay & Feuillet

Sida 17: 551 (1997)
cerasinus: cherry-red (in reference to the bracts, the sepals and petals, and the endocarp of the fruit)
Supersection *Laurifolia*, Series *Laurifoliae*

Passiflora cerasina (Plate 171), another species from the series *Laurifoliae*, was described only a few years ago, from a location in northeastern French Guiana. This relatively small area around the Montagne de Kaw is a real treasure trove for passionflower enthusiasts: about 20 species can be found there. It is not astonishing, then, that this region is a favorite destination for passionflower hunters. Consequently, seeds of various species, among them *P. cerasina*, have been brought to Europe, mainly by Pierre Pomié.

The pendent, fragrant flowers of *Passiflora cerasina* reach a diameter of 7–8 cm. The cherry-red sepals and petals differ from those of the closely related *P. laurifolia*, which can also be found in French Guiana. As is the case with all representatives from the series *Laurifoliae*, *P. cerasina* can produce pseudoracemose inflorescences at the ends of lateral shoots by reduction of the leaves and the internodes. This species has beautiful flowers but is also finicky and difficult to cultivate. There is hardly any problem during summer, but as soon as days become shorter and temperatures begin to drop, the plant stops growing and the already essentially chlorotic leaves fall off. Most plants do not recover from this setback. Even if a grower succeeds in getting a plant through winter with the help of soil heating, additional lighting, and a lot of instinct, it takes the plant some years to develop into a size at which it can produce flowers. Only botanical gardens or similar institutions are likely to be capable of providing the conditions in which *P. cerasina* can be cultivated. For most enthusiasts, the cultivation of this species will result in many disappointments.

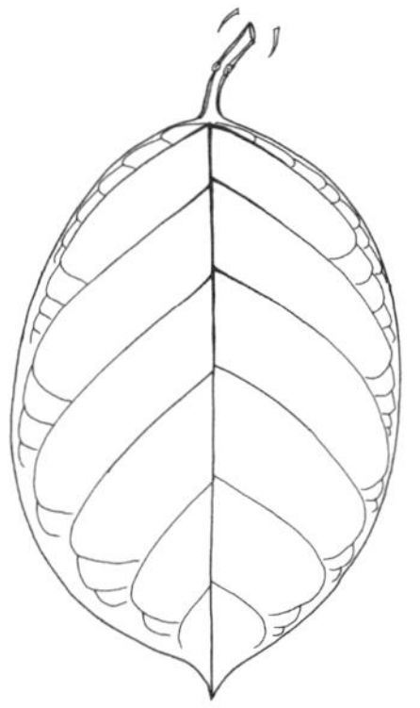

Passiflora cerasina

Vine: glabrous; stem: terete; stipules: linear, 0.3–0.5 cm long, pale green, soon deciduous; petioles: 2–2.2 cm, with 2 large paired glands near middle; leaves: unlobed, ovate, 13–15 × 8–8.5 cm, entire; peduncles: ca. 2.2 cm long, solitary; bracts: ovate, 4.2–4.8 × 2.2–2.9 cm, cherry-red in flower, entire with yellow glands, free; flowers: cherry-red, violet, and white, 7–8 cm diam.; sepals: cherry-red inside, whitish outside, oblong-ovate, ca. 4 × 0.8–1.2 cm, aristate, yellowish awn ca. 0.7 cm long; petals: cherry-red, narrowly oblong, 4 × 0.4–0.6 cm; corona: several series, banded violet and white, violet in upper 1/3, outermost series 2–2.5 cm long, 2nd series 4.7–5.2 cm long, wavy at apex; ovary: obovoid, glabrous; fruit: obovoid, 4.5–5 × 3.9–4.2 cm, dark purple, mottled with pale green; seeds: elliptic, ca. 0.6 × 0.5 cm, flattened, coarsely reticulate at middle, radiantly striate at margin, dark brown.

Passiflora chocoensis Gerlach & T. Ulmer

Caldasia 22: 231, figs. 1, 2 (2000)
named after the department of Chocó, Colombia, provenance of the type plant
Supersection *Stipulata*, Section *Granadillastrum*

Passiflora chocoensis (Plate 172) is a new species first found in the extremely rainy department of Chocó, western Colombia, on the Pacific Coast. It ranges from eastern Panama to the department of Valle in Colombia. It is closely related to *P. menispermifolia* and *P. oerstedii*, both of which are also found in those countries. All these species are capable of saving up large buds for several weeks and opening them all together within a few days. By its entire, larger bracts, great number of petiole glands, and nearly entire stipules, *P. chocoensis* can be distinguished from *P. menispermifolia*. Further criteria, like a comparatively weak pubescence and thinner leaves, are of conditional usefulness, because extreme variations of *P. menispermifolia* are known in which the pubescence is missing and the leaves vary in thickness. *Passiflora chocoensis* may be distinguished from *P. oerstedii* by its only obscurely 3-lobed leaves, larger bracts, and additional petiolar nectaries.

A specimen of *Passiflora chocoensis* has been cultivated in the Munich Botanical Garden since the mid-1990s, setting its first flowers in 1998. It is a rather warmth-loving species and should be cultivated in a greenhouse where temperatures do not fall below 12°C in winter. This species can be called free-flowering when it is grown under favorable conditions. Even small plants can produce a couple of flowers. Its violet flowers reach a diameter of about 7 cm and have a sweetish scent reminiscent of honey. In Panama the flowers are much paler, even nearly white. Under cultivation, *P. chocoensis* often shows symptoms of a deficiency of trace elements, older leaves turning chlorotic from the margin inward.

Vine: pubescent to subglabrous; stem: terete, striate; stipules: subreniform, 1.4–1.8 × 0.6–0.8 cm, entire or slightly serrate near base; petioles: 3–4 cm, pubescent or only slightly so, with 8–12 elongated glands, mostly in pairs; leaves: obscurely 3-lobed, sometimes appearing unlobed, 9–13 × 7–9.2 cm, membranous to subcoriaceous, glabrous above, pubescent beneath, entire; peduncles: 2.5–3 cm long, solitary; bracts: cordate to ovate-oblong, 2–3 × 1.3–1.5 cm, entire, free; flowers: violet to purplish violet, or white in Panama, 6.5–7.5 cm diam.; sepals: pale violet to whitish inside, greenish outside, lanceolate, 3.1–3.3 × 1.1 cm, slightly keeled, keel terminating in an awn ca. 0.3 cm long; petals: purplish violet or pale purple, lanceolate, 2.9–3 × 0.8–0.9 cm; corona: several series, violet with paler bands, outer series ca. 1.7 cm long, inner series up to 0.3 long; ovary: ovate, glabrous; fruit: ellipsoid-ovoid, 8–9.5 × 4.3–5.7 cm, greenish yellow to yellow; seeds: oblong-ovoid, 0.7–0.8 × 0.4 cm, blackish brown, reticulate.

Passiflora chrysophylla Chod.

Bull. Herb. Boissier 7 (1): 75 (1899)
chrysophyllus: having golden yellow leaves
Synonyms: *P. australis* Chod., *P. foetida* var. *vitacea* Mast.
Supersection *Stipulata*, Section *Dysosmia*

This rare species from section *Dysosmia* (Plate 173) is not under cultivation. Its native range extends from southern Bolivia through northern Argentina, Paraguay, Uruguay, and southern Brazil, from sea level to 1100 m.

The name of the species is derived from the conspicuous greenish yellow hair that covers the leaves and peduncles. At a diameter of 3–6 cm, the pink and white flowers are similar to or larger than those of typical *Passiflora foetida*. One distinguishing characteristic that separates this species from *P.*

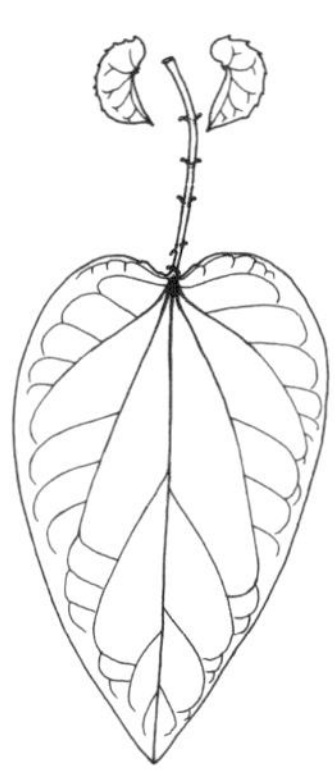

Passiflora chocoensis

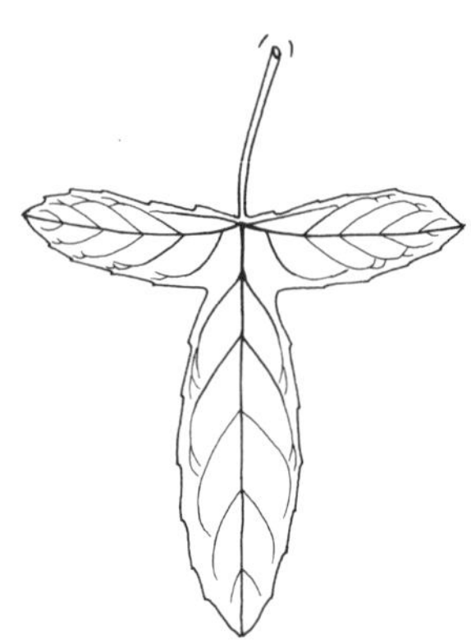

Passiflora chrysophylla

foetida is the reticulate sculpturing on its seeds, which is much finer in *P. chrysophylla*, with more than 35 pits per seed face (Deginani 2001). Furthermore, *P. chrysophylla* often differs in having 1-pinnatifid bracts, although this is variable.

Only 4 of the total 20 species within section *Dysosmia* are under cultivation: *Passiflora arida*, *P. foetida*, *P. sublanceolata*, and *P. urbaniana*. It is hoped that *P. chrysophylla* will be cultivated in the future, especially since the *Dysosmia* species under cultivation have turned out to be rather free-flowering even as small plants.

Vine: densely yellowish tomentose; stem: terete, sulcate, basal part woody; stipules: minute, semi-annular, 0.4–0.9 × 0.3–0.5 cm, deeply cut; petioles: 1–4 cm, without glands; leaves: 3-lobed, 3.5–10 × 4–10 cm, serrate to subentire; peduncles: 2–3.5 cm long, solitary; bracts: 1- to 2-pinnatifid, the middle part linear, 1.2–1.8 cm long with few segments up to 0.4 cm long, free; flowers: whitish pink, 3–6 cm diam.; sepals: whitish inside, greenish yellow and glandular outside, ovate-lanceolate, 1.5–2.5 × 0.5–1 cm, aristate; petals: white, lanceolate, 1.5–1.7 cm; corona: 5–6 series, white with purple or pink, outer 2 series 0.6–1.8 cm long, inner series 0.1–0.2 cm long; ovary: subglobose, pilose; fruit: subglobose, ca. 2–3.5 cm diam., yellow; seeds: ca. 0.6–0.7 × 0.3 cm, finely reticulate.

Passiflora cincinnata Mast.

Gard. Chron. 26: 966 (1868)
cincinnatus: curled (in reference to the corona)
Synonyms: *P. corumbaensis* Barb. Rodr., *P. perlobata* Killip
Supersection *Passiflora*, Section *Passiflora*, Series *Passiflora*

This species (Plate 174), whose large native range covers Argentina, Brazil, Bolivia, Colombia, Paraguay, and Venezuela, generally develops its sumptuous flowers only in a greenhouse. Under glass the large, fragrant flowers, 7–9 cm diam., appear from June to September. *Passiflora cincinnata* is conspicuous for its extremely undulate corona, which is deep purple at the base, banded lighter in the middle, and violet-blue in the upper part. The corona reflexes slightly (as do the bluish violet sepals and petals) immediately upon the opening of the flower and covers the sepals and petals almost completely. Depending on the area of origin, the intensity of the colors and zoning of the corona vary greatly within this species. *Passiflora cincinnata* 'Dark Pollen' possesses very intense colors.

A crossing of *Passiflora cincinnata* with the related *P. incarnata* resulted in the beautiful *P.* 'Incense', which has been cultivated in the United States and Europe for many years. This hybrid is often confused with *P. cincinnata*, but the flowers of *P.* 'Incense' are usually larger and a little darker in color than those of its parent. The warmth-loving *P. cincinnata* is suited for cultivation only under glass or in a container outdoors in sunny regions, while the robust and free-flowering *P.* 'Incense' is cultivated outdoors nearly everywhere passifloras can be grown. Even as a small plant it can produce some flowers, but when planted directly into an earth bed in a greenhouse it develops into an imposing plant within a year and produces countless flowers.

The minimum overwintering temperature for *Passiflora cincinnata* is 12°C, although it will tolerate a few slightly colder nights.

Vine: glabrous or rarely pilose; stem: terete to subangular; stipules: linear-subulate, 0.6–1.5 × 0.1 cm, entire; petioles: 1.5–5 cm, with 2 or occasionally 4 large glands in lower 1/3; leaves: deeply 5-lobed or occasionally 3-lobed, 4–11 × 5–12 cm, serrate; peduncles: 2–8.5 cm long, stout, solitary; bracts: ovoid, 2–4 × 1.5–2.5 cm, entire, glandular at base, free; flowers: bluish violet to purplish violet, 7–9(–12) cm diam.; sepals: violet inside, green outside, oblong-lanceolate, 3–5 × 1–2 cm, slightly keeled outside, keel terminating in a short awn;

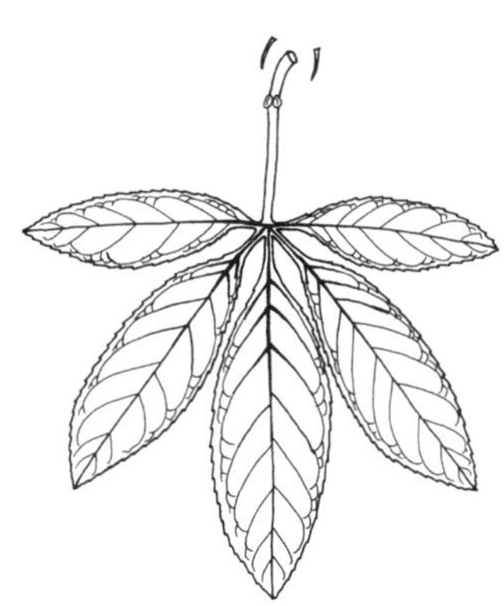

Passiflora cincinnata

petals: violet, lanceolate, 2.5–3.5 × 0.8–1.3 cm; corona: several series, basal part deep purple, banded purple-violet and whitish blue at middle, violet at apex, wavy in upper 1/2, outer series 2–4 cm long, innermost series 0.1–0.2 cm long; ovary: globose, glabrous; fruit: ovoid, 5–6 × 3–4 cm, greenish with white speckles; seeds: ovate, 0.5–0.6 × 0.4 cm, coarsely reticulate.

Passiflora coccinea Aubl.

Hist. Pl. Guiane 2: 828, t. 324 (1775)
coccineus: scarlet-red
Synonyms: *P. fulgens* Wallis ex Morren, *P. toxicaria* Barb. Rodr., *P. velutina* DC.
Supersection *Coccinea*

The native range of *Passiflora coccinea* (Plates 7, 156, and 175–177) is Bolivia, Brazil, Guyana, Peru, and southern Venezuela. It mainly grows in the tropical Amazonian lowlands, but in Bolivia it proceeds to altitudes of about 2000 m in the western foothills of the Andes. It occurs on noninundated lateritic soil (Gentry 1981). Sometimes its shoots simply creep along the ground, and it can sucker many meters from the parent. It has unlobed double-serrate leaves and large bracts, distinguishing it from other species of supersection *Coccinea*, especially *P. vitifolia*.

A discussion about *Passiflora coccinea* was raised a few years ago but has not yet been settled. The specimens of *P. coccinea* from French Guiana, the type location, differ from those of other origins in several respects. The plants from French Guiana have very large, deep red bracts and a mostly pure white corona that is closely inclined toward the androgynophore. In specimens of another origin, the outermost series is dark purple, sometimes with a partly white tip, and is not closely inclined. Moreover, the bracts are smaller and without an intense red coloration. It is possible there are 2 different species going by the name *P. coccinea*. This uncertainty demonstrates the importance of conducting a thorough revision of the species from this supersection. Besides the problems around *P. coccinea*, great variation in related species suggests that more species from this complex might be described in times to come.

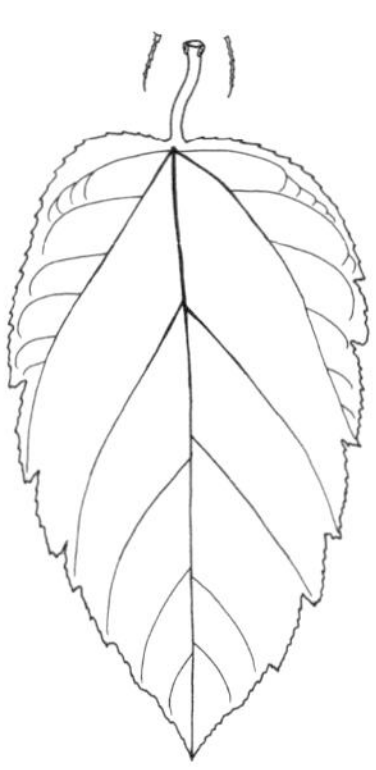

Passiflora coccinea

In Europe the variant with upright, purple outer filaments and small, brownish red bracts has always been cultivated as *Passiflora coccinea*. It is a favorite species, with scarlet-red flowers that reach a diameter of 8–12 cm. The greenhouses of some European botanical gardens display gorgeous specimens that remain in full bloom from January to April. As *P. coccinea* needs much warmth and a high atmospheric humidity, it is not suited for outdoor cultivation in a container in most regions; if a red-flowering species is wanted for a balcony or terrace, consider the more robust *P. vitifolia*. In winter *P. coccinea* does not tolerate temperatures of less than 12°C.

Vine: densely pubescent; stem: terete to subangular, striate, older stems brownish; stipules: linear to linear-lanceolate, 0.4–1.5 cm long, entire or minutely glandular-serrate; petioles: 1.8–3.5 cm, with 2 glands at base, rarely missing; leaves: unlobed, oblong, 6–14 × 3–7(–9.8) cm, double-serrate or -crenate; peduncles: (2.5–)5–8(–9.7) cm long, stout, solitary; bracts: ovate or oblong, (2.3–)4–6 × (0.5–)1.5–3.5 cm, serrate, glandular at margin, free; flowers: red, 8–12 cm diam.; sepals: red, linear-lanceolate, 3–5 × 0.8–1.4 cm, slightly keeled, keel terminating in an awn up to 1.3 cm long; petals: red, linear, 3.5–4 × 0.7–0.9 cm; corona: 3 series, outer series 1–1.5 cm long, white or purplish, inner series 0.6–0.8 cm long; ovary: ovoid, densely tomentose; fruit: ovoid or subglobose, 5–6 × 4–5 cm, greenish or greenish yellow, 6-striped with green, edible; seeds: narrowly obovate, ca. 0.6 × 0.4 cm, dark brown, minutely reticulate.

Passiflora crenata Feuillet & Cremers

Proc. Kon. Ned. Akad. Wetensch., C 87: 378 (1984)
crenatus: crenate (in reference to the seeds)
Supersection *Laurifolia*, Series *Laurifoliae*

Seeds and cuttings from this species (Plate 178), which is indigenous to French Guiana, have occasionally come to Europe. Although the seeds have germinated quite well, this species has remained a great rarity. Meanwhile, many *Passiflora* species from French Guiana have turned out to be difficult to cultivate. *Passiflora crenata*, which must be grown in a warm greenhouse, is certainly one of them, although it develops satisfactorily during summer months. But as soon as days become shorter and temperatures drop, its tropical origins become apparent: the plant stops growing and sheds some of its leaves. Only in greenhouse with additional lighting and a temperature above 15°C will overwintering be successful. The soil mix should not contain too much peat. It is not known whether plants under cultivation have reached the necessary height to set flowers.

The white sepals and petals form a contrast to the corona, which is banded dark violet and white. The flowers, up to 10 cm diam., can appear singly or in groups. This grouping of flowers on axillary short shoots is a phenomenon that can be observed now and then in many representatives of section *Laurifolia*. Other representatives of this closely related group are *Passiflora ambigua*, *P. pergrandis*, and *P. riparia*, which are also rather reluctant to flower under cultivation.

Vine: glabrous in most parts; stem: terete to subangular; stipules: linear, 1.5–2 × 0.1–0.2 cm, entire; petioles: 1.5–2 cm, with 2 glands, paired in lower 1/3; leaves: unlobed, ellipsoid to oblong, 6–13 × 2.5–7 cm, entire; peduncles: ca. 2.5 cm long, solitary or paired in fascicles; bracts: oblong, 4–5 × 2–3 cm, more or less 2-lobed at apex, whitish, entire with glands at margin, free; flowers: white and dark violet, 9–11 cm diam.; sepals: white inside and out, oblong-ovate, ca. 6 × 3 cm, aristate, awn up to 0.2 cm; petals: white, oblong-lanceolate, 5 × 1–1.5 cm; corona: several series, banded dark violet and white, outer series ca. 5 cm long, innermost series up to 0.3 cm long; ovary: ovoid, densely pubescent; fruit: ovoid, 6 × 4 cm; seeds: ovate, ca. 0.7 × 0.5 cm, reticulate at middle, radiately striate at margin.

Passiflora deltoifolia Holm-Nielsen & Lawesson

Ann. Missouri Bot. Gard. 74: 501, fig. 3 (1987)
deltoifolius: delta-shaped or triangular (in reference to the leaves)
Supersection *Stipulata*, Section *Granadillastrum*

This species (Plate 179), endemic in the province of Napo, northeastern Ecuador, was brought into cultivation in Germany by Michael Schwerdtfeger in 1995. Unfortunately, the imported clone turned out to be extremely difficult and does not seem to be represented in collections anymore, or is at least extremely rare. This is all the more regrettable since it produces beautiful flowers, 7 cm diam., that appear singly in the leaf axils. The large corona of many series is a deeper violet than the sepals and petals and shows a narrow white band only in its middle.

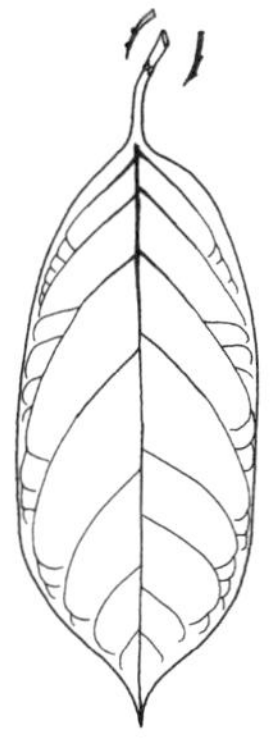

Passiflora crenata

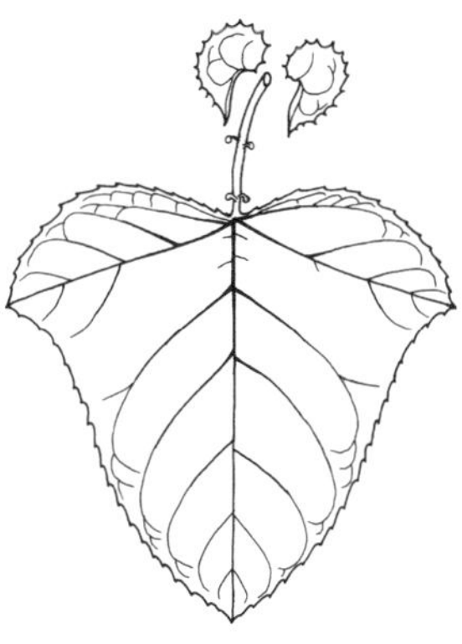

Passiflora deltoifolia

Its delta-shaped leaves are remarkable, almost smooth on the upper side, while the stem and stipules are pubescent. The flowers and leaves, the way in which dried-up parts of the plant turn black, and other details suggest a close relationship to *Passiflora menispermifolia* and *P. crassifolia*. By means of the missing awns at the tips of the sepals and a more plicate operculum, *P. deltoifolia* can be distinguished from these 2. Nevertheless, there is some doubt as to whether a separation from *P. menispermifolia* is justified, as recent findings indicate that the pubescence of that species is much more variable than expected.

Vine: weakly puberulent, except upper leaf surface and flower; stem: terete, striate; stipules: oblong-ovate to subreniform, ca. 2 × 1 cm long, glandular-serrate; petioles: ca. 3 cm, with 4 alternate, short stipitate glands; leaves: shallowly 3-lobed, deltoid, 8 × 9–10 cm, serrate; peduncles: 5–6 cm long, solitary; bracts: oblong-ovate, 0.7 × 0.3 cm, inserted 1 cm below base of flower, glandular-serrate, free; flowers: violet, 6–7 cm diam.; sepals: violet, oblong-ovate, 3–4 × 1–2 cm, without awn; petals: violet, oblong-lanceolate, 3–4 × 1–2 cm; corona: 5–8 series, violet with white band near middle, outer 2 series 2.5–3 cm long, ligular, inner series 0.1–0.5 cm long; ovary: ovoid, glabrous; fruit: ovoid to ellipsoid, 5.5 × 2 cm.

Passiflora edmundoi Sacco

Sellowia 18: 44 (1966)
named after Edmundo Pereira, collector of the type
Supersection *Stipulata*, Section *Kermesinae*

The first collections of *Passiflora edmundoi* (Plate 180) date from the beginning of the 20th century. However, these were thought to be *P. kermesina* until the first description was published in 1966. Even Killip (1938) did not notice the difference and presented a mixed description of both species.

Admittedly, their flowers look very similar at first glance. They are about the same size, very similar in color, and both species are pollinated by hummingbirds. There are decisive differences, however, with regard to the bracts, corona, and fruit. *Passiflora edmundoi* has petiolate, purple bracts that become wider in an ovate form above the peduncle and are not soon deciduous. In contrast to

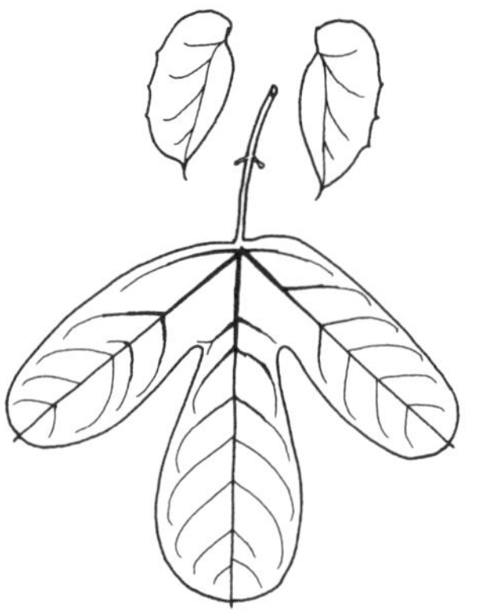

Passiflora edmundoi

P. kermesina, the corona consists of 2 or 3 very short series in which the filaments of the 2 innermost series are connate into a membrane each. Finally, the fruit of *P. edmundoi* is oblong and ellipsoid.

Passiflora edmundoi, at home in western to southwestern Brazil at altitudes around 1000 m, has not yet come into cultivation. Concerted efforts in recent years, however, have been made to bring *Passiflora* species from Brazil into cultivation. One day *P. edmundoi*, desired for its reddish flowers, may be among them.

Vine: glabrous; stem: terete, striate; stipules: reniform, 2–3 × 1–1.5 cm, entire; petioles: 2.5–3.5 cm, with 2–6 mostly paired, elongated glands, glands 0.3–0.5 cm long; leaves: 3-lobed, ovate, 5–9 × 5.5–8 cm, entire; peduncles: 6.5–17 cm, solitary or in pairs; bracts: petiolate, petioles 0.2–0.3 cm long, ovate above, 0.4–0.6 × 0.2–0.3 cm, purplish, scattered along peduncle, entire; floral tube: cylindrical, 1–1.5 cm long, green outside; flowers: purplish red, 6–7 cm diam.; sepals: purplish red, oblong-lanceolate, 2.5–3 × 0.8 cm, keeled outside, keel terminating in an awn 0.4–0.5 cm; petals: purplish red, lanceolate, 2.2–2.5 × 0.5 cm; corona: 2 or 3 series, outer series up to 0.6 cm long, bluish violet, inner series a membrane, 0.2–0.4 cm long; ovary: ellipsoid, glabrous; fruit: ellipsoid, 6–9 × 2–3 cm.

Passiflora edulis Sims

Bot. Mag. 45: pl. 1989 (1818)
edulis: edible (in reference to the fruit)
Synonyms: *P. diaden* Vell., *P. gratissima* St. Hil., *P. iodocarpa* Barb. Rodr., *P. middletoniana* Paxton, *P. pallidiflora* Bert., *P. picroderma* Barb. Rodr.,

P. pomifera Roemer, *P. rigidula* Jacq., *P. rubricaulis* Jacq., *P. vernicosa* Barb. Rodr., *P. verrucifera* Lindley

Supersection *Passiflora*, Section *Passiflora*, Series *Passiflora*

Because of its tasty fruit, *Passiflora edulis* is among the best-known passionflowers worldwide. Originally indigenous to Brazil, Paraguay, and northern Argentina, it is now grown in almost all countries of the tropics and subtropics. Much cultivating and crossbreeding has been done in the past several decades to optimize the succulence, size, and taste of the fruit, as well as the resistance and yield of the plants. Presently about a hundred varieties are recognized. The best-known forms are *P. edulis* f. *flavicarpa* (Plate 181), which must be cross-pollinated, and *P. edulis* f. *edulis* (Plate 182), which is self-fertile. Several other varieties (or hybrids between these 2) are of great economic importance: 'Australian Purple', 'Common Purple', 'Norfolk' (Plate 183), 'Ouropretana', and 'Perfecta'. Plants of these and other varieties are occasionally available in the United States and Europe.

The purple to brown-violet fruit of *Passiflora edulis* f. *edulis*, approximately the size of a hen's egg, is available throughout the year on the market. It is the best-known fruit of the genus. Its coriaceous, smooth skin begins to shrivel after just a few days of storage, marking it as ripe and ready to eat. *Passiflora edulis* f. *edulis* is somewhat resistant to cold and is best suited for cultivation at altitudes of 1000–2300 m, while its sister, *P. edulis* f. *flavicarpa*, is recommended for warmer, lower regions. The lemon-yellow, ellipsoid to ovoid fruit of *P. edulis* f. *flavicarpa* is about 8 cm long. It is significantly larger than the purple fruit of forma *edulis* but does not come close in intensity of flavor. In addition to the pure forms, crossbreedings of both are available in many countries where passionflowers are grown commercially or in private gardens. These forms combine the properties of the purple and yellow fruit in varying proportions.

New plants can easily be raised from the numerous seeds contained in the fruit. As a rule, flowers appear in the 2nd or 3rd year after sowing. When plants are available, propagation by cuttings is the preferred method for most species. *Passiflora*

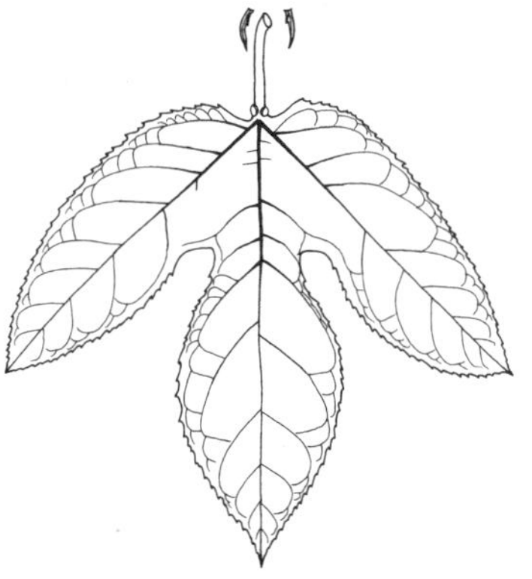

Passiflora edulis

edulis is among the least finicky species. In a conservatory or greenhouse it rapidly develops into a magnificent plant when set directly in the ground. A sunny location in a container outdoors also provides it with flowers and fruits from May to October. Its flowers reach a diameter of 7 cm, and those of *P. edulis* f. *flavicarpa* are even a little larger. The white sepals and petals are almost completely covered by the extremely undulating, tortuous corona. Up to the middle, occasionally a little further, the corona is purple to dark purple; in the upper 1/2 it is white. Its serrate leaves are 5–20 cm long, 3-lobed, and often have an undulate margin.

Both forms are overwintered at 10°C, although *Passiflora edulis* f. *edulis* tolerates temperatures down to 5°C and can even withstand a slight frost for a short time. In view of the apparent differences between forma *edulis* and forma *flavicarpa*, the question has been raised as to whether they might be 2 different species. They certainly deserve a greater separation than as mere forms. Results of genetic investigations are still pending.

Vine: in general glabrous; stem: terete or subangular, striate, stout; stipules: linear-subulate, 1–1.3 × 0.1–0.2 cm long, entire; petioles: 3–4 cm, with 2 glands in upper 1/3; leaves: 3-lobed, 5–25 × 5–25 cm, lustrous above, serrate; peduncles: 2–6 cm long, solitary, stout; bracts: ovate to oblong-ovate, 1.7–2.5 × 0.8–1.5 cm, serrate, free; flowers: white and purple, (5–)6–8 cm diam.; sepals: white inside, green outside, oblong, 2–3.3 × 0.7–1 cm, slightly keeled outside, keel terminating in an awn 0.3–0.6 cm long, in forma *flavicarpa* some wartlike nectary glands on the margins of the outer sepals; petals: white, oblong, 1.8–2.9 × 0.5–0.8 cm; corona: 4 or 5 series, purple or dark purple in basal 1/2, white and wavy in upper 1/2, outer 2 series 1–2.3 cm long, inner series 0.1–0.3 cm long; ovary: globose or ovoid, densely tomentose; fruit: in forma *edulis* globose, 4–6 × 4–5 cm, purple to brownish purple, in forma *flavicarpa* ellipsoid to ovoid, 5–10 × 4–8 cm, greenish yellow to yellow; seeds: ovoid, ca. 0.6 × 0.4 cm flattened, minutely reticulate, brown.

Passiflora exura Feuillet

Novon 4: 238, fig. 2 (1994)
exura: ending in a tail
Supersection *Stipulata*, Section *Granadillastrum*

Passiflora exura (Plate 184) hails from northeastern French Guiana, where it mainly occurs in disturbed and cut forests near roadsides or farms. In the 1990s this and many other species from French Guiana where described by Christian Feuillet, who also brought *P. exura* into cultivation in the United States. The species was introduced to Europe by Pierre Pomié some years later.

Passiflora exura has turned out to be rather hard to grow. This warmth-loving species develops quite well during summer months but does not flower reliably every year. Flowers appear only occasionally even when it is cultivated under glass. This is all the more regrettable as its purple-violet flowers are imposing at a diameter of 9 cm. It is distinguished by its sepalar awns of up to 2.5 cm long on the sepals, and by its innermost coronal series, consisting of triangular processes, that is inclined toward the androgynophore. Problems can arise during overwintering when the plant suffers due to insufficient light and temperatures below 12°C.

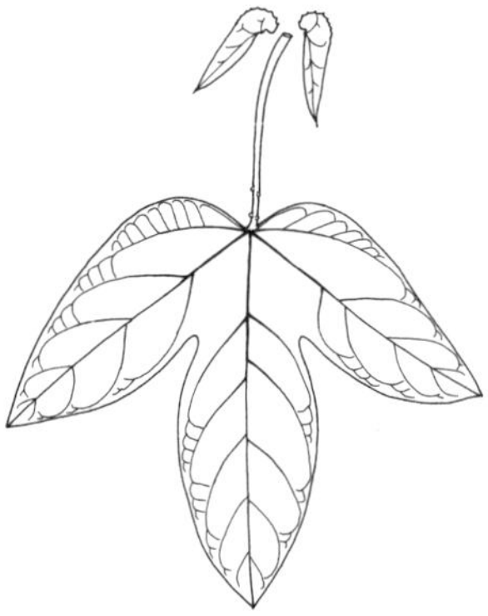

Passiflora exura

Passiflora exura has one characteristic in common with *P. amethystina* and *P. menispermifolia*: the dried-up parts of the plant turn black. It is distinguished from the closely related *P. rubrotincta* by its larger leaves, petiole glands, and in particular by the structure of the innermost coronal series.

Vine: glabrous; stem: terete; stipules: semicordate-oblong, 2.2–3 × 0.5 cm long, soon deciduous; petioles: (1.5–)8–11 cm, with 2 or 4 short stipitate glands near apex; leaves: 3-lobed, 5–11 × 6.5–13 cm, entire; peduncles: ca. 3.5 cm, solitary or in pairs; bracts: ovate, 2.5–2.7 × 1.5–2 cm, entire; flowers: purplish violet, 8–9 cm diam.; sepals: purplish inside, greenish outside, oblong, ca. 4 × 1.2 cm, thick, aristate, awn 2.5 cm long; petals: purplish, ca. 4 × 1.2 cm, thin; corona: several series, more or less intense purplish violet to pale lilac, outer series ca. 3 cm long, innermost series white, flat, triangular, ca. 1 cm long; ovary: obscurely trigonous, glabrous; fruit: ovoid, 10 × 7.5 cm, green; seeds: oblong, ca. 0.7 × 0.3 cm, flattened, coarsely reticulate.

Passiflora foetida L.

Sp. Pl. 959 (1753)
foetidus: stinking
Synonyms: *P. gossypiifolia* Hamilton, *P. hibiscifolia* Lam., *P. hirsuta* Lodd., non L., *P. hispida* DC., *P. liebmannii* Mast., *P. muralis* Barb. Rodr., *P. nigelliflora* Hook., *P. polyadena* Vell., *P. pseudociliata* Britton, and more
Supersection *Stipulata*, Section *Dysosmia*

Passiflora foetida is possibly the most variable species of the whole genus, particularly with respect to flowers and fruits. Almost 50 varieties have been suggested (see Plates 185–189), making it almost impossible to identify one's own plant. It is beyond the limits of this book to list the common synonyms for all varieties or to segregate all that might really be distinct species. Some of the red-fruited variants, in particular, appear to be "good" species (for example, variety *ciliata*, variety *hibiscifolia*, and variety *liebmannii* could justifiably be called *P. ciliata*), but the situation is very complex. It is time for someone to study these critically and give a detailed account of the all the varieties and segregate species of *P. foetida*.

About 15 varieties are under cultivation, but these do not flower equally well. Some produce flowers abundantly and bear many fruits (for example, variety *galapagensis*, variety *gossypiifolia*, variety *hirsuta*, and variety *maxonii*), while others can hardly be induced to flower. In most cases the flowers, 2.5–6 cm diam., are white with pink, lavender, or violet, but there are forms with pure white, pure pink, and pale bluish violet flowers. The sweetly scented flowers open in the very early morning hours and close around midday. *Passiflora foetida* is characterized by its feathery bracts that cover the flower buds and fruit. They look like venation that has been separated out of the leaf, and they are covered with numerous tiny sticky glands and sticky hairs. The leaves are normally more or less 3- or 5-lobed, often with the center lobe distinctly longer. Depending on the variety, the pubescence can vary from completely absent to densely woolly. Most varieties give off an unpleasant odor when the hairs are touched, particularly on the bracts.

The color of the thin-skinned fruits ranges from greenish yellow to yellow to brilliant red. They are eaten in many countries, for the whitish pulp is pleasantly sweet and juicy when the fruit is ripe. Due to their content of cyanogenic glycosides, however, unripe fruits are poisonous, as with all species of *Passiflora*. In India, unripe fruits are used in folk medicine to induce vomiting. At a diameter of 1.5–3.5 cm, the globular to ovate fruits are quite small. In some regions of the tropics and subtropics, *P. foetida* is used as a groundcover in various plantations and is also used to fight erosion. This species has escaped to the wild not only in its area of origin in tropical America but also in Africa, Southeast Asia, and Australia, and has become a noxious weed. Some varieties need cross-pollination, but others are more weedy, being self-compatible and even self-pollinating.

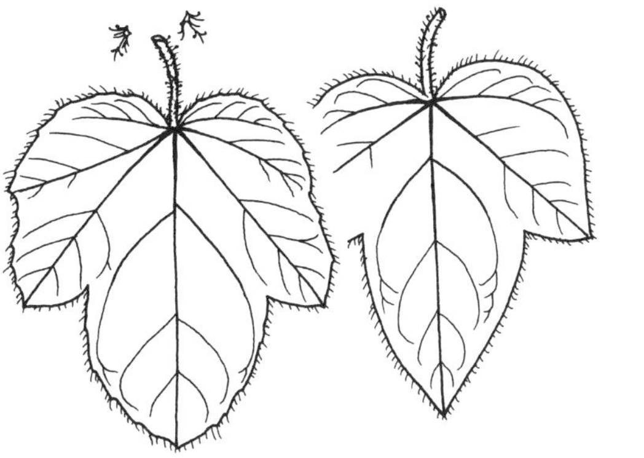

Passiflora foetida

The easy-to-grow but warmth-loving *Passiflora foetida* is suited for a greenhouse or for outdoor cultivation in summer but can also be grown on a windowsill. In winter it needs temperatures above 10°C.

Vine: herbaceous, quite variable indument; stem: terete, slender or stout; stipules: semiannular, deeply cleft into filiform, occasionally pinnatisect, gland-tipped divisions; petioles: 0.7–8.3 cm long with gland-tipped hairs; leaves: variable in shape and size, 3- or 5-lobed, 2–15 × 3–16 cm, membranous, serrate; peduncles: 2–7.5 cm long, solitary, occasionally in pairs; bracts: (1–)2- to 4-pinnatifid or pinnatisect, 1–4.5(–6.5) × 0.8–4(–5.5) cm, gland-tipped, free; flowers: variable in color, white, pink, violet, or purple, (2–)3.5–5(–6.2) cm diam.; sepals: mostly white or pink inside, green outside, ovate-oblong to ovate lanceolate, up to 2.5 cm long, slightly keeled outside, keel terminating in a short awn; petals: mostly white or pink, slightly shorter than or longer than the sepals; corona: ca. 4–6 series, white, pink, purple, or violet, the 2 outer series ca. 1.5 cm long; ovary: ovoid or subglobose, glabrous or with variable indument; fruit: ovoid to subglobose, red, orange, yellow, or green, 2.5–4 cm long, edible; seeds: oblong to obovoid, 0.5 × 0.2–0.3 cm, tridentate at apex, coarsely reticulate, dark brown.

Passiflora garckei Mast.

Trans. Linn. Soc. London 27: 639 (1871)
named after Friedrich August Garcke, author of the first description
Synonyms: *P. glaucophylla* Pulle, *P. pruinosa* Mast.
Supersection *Stipulata*, Section *Granadillastrum*

The native range of this species (Plate 190) extends from Guyana over Suriname to French Guiana in northeastern South America. Although many species from French Guiana are difficult to cultivate, *Passiflora garckei* is a pleasing exception. It is warmth-loving but also flowers reliably, producing flowers even in a 15-cm pot, though more blooms can be expected when the plant has grown larger.

Passiflora garckei

It will do well in a 30-cm container that can be placed outdoors during summer; otherwise it should be cultivated under glass. In winter, temperatures around 15°C are desirable, although nightly drops down to 7°C are tolerated. The relatively short internodes grow in a zigzag manner.

Its bluish purple and white flowers, which often open in batches, have a faint scent. They reach a diameter of about 8 cm and are quite variable with regard to intensity of color. Some cultivars have whitish petals and a relatively pale corona, while others have pale blue or purple petals and an intensely colored corona. The operculum is remarkable, its filamentous tips surrounding the androgynophore, similar to an additional coronal series. It can be distinguished from the related *Passiflora cyanea* by means of its larger, thicker leaves.

Vine: glabrous; stem: terete or slightly angulate, striate; stipules: semiovate or subreniform, 3–5 × 1.5–2, coriaceous, remotely glandular-serrate; petioles: 3–10 cm long, with 4–6 sessile glands; leaves: 3-lobed, 8–15 × 10–25 cm, coriaceous, entire or minutely serrulate; peduncles: up to 6 cm long, solitary; bracts: oblong-lanceolate, 0.6–1 × 0.4–0.6 cm, serrate, free; flowers: blue or purplish and white, 7.5–8.5 cm diam.; sepals: white or pale blue inside, green outside, oblong, 3.5–4 × 0.8–1 cm, aristate, awn 0.2–0.4 cm long; petals: whitish or pale blue, oblong, 3–3.5 × 1–1.2 cm; corona: several series, white in lower 1/3, bluish purple above, outer 2 series 3–3.5 cm long, wavy above middle, inner series 0.5–0.8 cm long, white; ovary: ovoid, glabrous; fruit: almost ellipsoid.

Passiflora gibertii N. E. Brown

Trans. and Proc. Bot. Soc. Edinburgh 20: 58 (1896)
named after José Ernesto Gibert, collector of the type plant
Supersection *Stipulata*, Section *Granadillastrum*

Though absolutely recommendable, *Passiflora gibertii* (Plate 191) is rarely cultivated except in larger collections. Perhaps its flowers are considered less colorful than those of other species. Regardless, this plant produces numerous fragrant flowers even in a small pot and is very suitable for cultivation indoors or outdoors in a container. A plant that is several years old and has reached 2–3 m in height is capable of producing up to 10 flowers a day during the flowering period, June to September. Its white flowers, with a corona banded purple and white, reach a diameter of 7 cm. The awns are interesting, 1–1.5 cm long and on the back of the sepals. Occasionally some fruits are set, and these are said to be edible when ripe. This species resembles *P. eichleriana*, *P. elegans*, and *P. subpeltata*, to which it is closely related, for its habit and its 3-lobed leaves.

Older specimens of this species, indigenous to Argentina, Brazil, and Paraguay, are said to have survived temperatures down to 2°C; temperatures above 10°C, however, are more ideal. Plants that go by the name *Passiflora* 'St. Rule' are almost always specimens of *P. gibertii*.

Vine: glabrous; stem: terete; stipules: semiovate, 2–5 × 0.7–3, subentire; petioles: slender, 1–5 cm, with (2–)4–5(–6), subclavate, paired glands; leaves: 3-lobed, rarely 5-lobed, 5–9(–18) × 5–12(–22) cm, membranous, entire; peduncles: 4–8.5 cm long,

Passiflora gibertii

solitary; bracts: ovate, 1.5–3.3 × 1–2 cm, entire, free; flowers: white and purple, 6–9 cm diam.; sepals: whitish inside, green outside, oblong-lanceolate, 2.5–4 × 0.8–1.1 cm, aristate, awn 0.7–1.5 cm long; petals: white, oblong, 2–2.9 × 0.6–1.1 cm; corona: 5–6 series, white banded with purple, outer 2 series 1–2.8 cm long, inner series increasingly short, 0.1–0.7 cm, greenish with purple; ovary: ellipsoid, glabrous; fruit: subglobose to ovoid, 4–6 × 3–6 cm, yellow to orange; seeds: ca. 0.5 × 0.2–0.4 cm, reticulate.

Passiflora glandulosa Cav.

Diss. 10: 453, t. 281 (1790)
glandulosus: having glands
Synonyms: *P. imthurnii* Mast., *P. silvicola* Barb. Rodr.
Supersection *Distephana*

Passiflora glandulosa (Plates 192 and 193) is probably the best-known species from the small supersection *Distephana*. It is indigenous to the tropical lowlands of Brazil, the Guianas, and Venezuela. Characteristics of *Distephana* species are the missing or weak pubescence, small bracts, and large, reddish, hummingbird-pollinated flowers. The flowers are comparable with those from supersection *Coccinea*, which, however, are distinguished by their fine pubescence and serrate leaves.

With its large rose-pink flowers, *Passiflora glandulosa* is regarded as one of the most beautiful species of the genus. It bears the greatest resemblance to *P. bomareifolia* and *P. variolata*. However, these 2 have an outermost coronal series that adjoins the androgynophore, while in *P. glandulosa* it is erect.

Because French Guiana is a favorite destination among passionflower enthusiasts, seeds of *Passiflora glandulosa* from that country have often come to Europe. As is the case with many other tropical species, *P. glandulosa* asks too much from the average enthusiast. It needs high temperatures year-round and plenty of light, and can only be cultivated appropriately in a warm greenhouse. Additional lighting is recommended in higher latitudes. The composition of the soil mix plays an important role in cultivating tropical species. A standard potting mix containing peat does not seem to be a good choice.

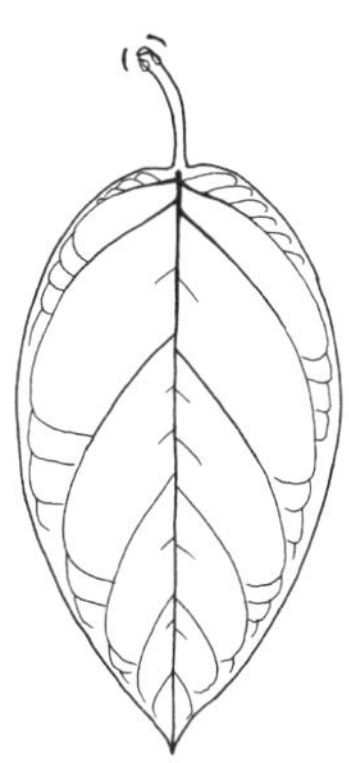

Passiflora glandulosa

Vine: glabrous or rarely minutely pubescent; stem: terete to subangular; stipules: linear-subulate, soon deciduous; petioles: up to 2.5 cm, with 2 sessile glands at or below middle; leaves: unlobed, ovate-oblong to oblong-lanceolate, 6–15 × 4–10 cm, coriaceous, entire or slightly undulate; peduncles: up to 8 cm long, solitary; bracts: linear-lanceolate, 0.5–1 × 0.1–0.2 cm, glandular at margin, free; flowers: red, 8–12 cm diam.; sepals: red, oblong or linear-oblong (2–)4–6 × 0.6–1.5 cm, aristate, awn 0.2–0.3 cm long; petals: red, slightly shorter than sepals; corona: 2 series, outer series ca. 1 cm long, white or pale pink, inner series a cylindrical membrane, filamentous at apex; ovary: ellipsoid, glabrous or white puberulent; fruit: ovoid, 5–6 × 2.5–3 cm; seeds: oblong-ovoid, ca. 0.8 × 0.4 cm, minutely reticulate.

Passiflora gritensis Karst.

Linnaea 30: 163 (1859)
named after La Grita, Venezuela, the type location
Supersection *Stipulata*, Section *Granadillastrum*

Passiflora gritensis (Plate 194), exceptional in several respects, has one of the most beautiful flowers of the whole genus. Endemic in Venezuela at altitudes around 2500 m, it was first brought into cultivation by Timothy Skimina in the early 1990s, but it did not flourish. Shortly thereafter, Gerrit Graat and Peter Ploeger succeeded in introducing, propagating, and distributing it. Thus a long-cherished wish of enthusiasts had come true. As *P. gritensis* can be propagated from cuttings without problems, it is represented in many collections. Although the flowers are pendent on peduncles more than 25 cm long, they are nevertheless turned upright and permit the beholder to marvel at the bushy corona. Flowers can reach a diameter of up to 10 cm. Confusion with other *Passiflora* species is does not occur because of this species' long peduncle and unique floral morphology.

Unfortunately its cultivation is not easy, and 2 conditions should be avoided: a sunny location and temperatures beyond 30°C. In contrast to most other passionflowers under cultivation, *Passiflora gritensis* can be called a shade plant. It reacts to excessive sunshine by developing yellow leaves and stopping growth. On the other hand, mild day temperatures around 20°C and nightly drops to 10°C have a positive effect. Consequently, *P. gritensis* is recommended for cultivation outdoors in summer.

Passiflora gritensis will reliably set flowers in central Europe when conditions are appropriate, especially between September and May. Flower buds are especially prone to be set in the period after winter when days are growing longer. In a 25-cm pot and at a height of about 90 cm, *P. gritensis* can produce its first flowers. In winter, temperatures above 7°C are recommended, but even young plants are capable of surviving a light frost.

Vine: glabrous; stem: terete, slender; stipules: semioblong, 1.3–2.5 × 0.5–1 cm, serrate to dentate; petioles: 1.8–3 cm, with 4 paired, stipitate glands, 0.3 cm long; leaves: 3-lobed, 2.5–10 × 5–12.5 cm, membranous, entire; peduncles: 25–35 cm, usually solitary, stout; bracts: ovate, 1–1.2 × 0.6–0.7 cm, serrate, free; flowers: rose to purplish red, 8–10 (–12) cm diam.; sepals: rose to purplish red inside, greenish outside, oblong-lanceolate, 3–4.5 × 1–1.3 cm, keeled outside, keel terminating in an awn 1–1.3 cm long; petals: rose to purplish red, oblong, slightly shorter than sepals; corona: 2 series, pinkish orange with whitish apex, outer series 2–2.5 cm long, inner series 0.5–0.7 cm long; ovary: ellipsoidal, glabrous; fruit: ellipsoid, 5–6 × 3 cm, yellow, edible.

Passiflora incarnata L.

Sp. Pl. 959 (1753)
incarnatus: flesh-colored (in reference to the flower)
Synonyms: *P. edulis* var. *kerii* Mast., *P. kerii* Spreng.
Supersection *Passiflora*, Section *Passiflora*, Series *Passiflora*

Passiflora incarnata (Plates 155 and 195–197) is much sought after for its hardiness and for its attractive, fragrant flowers, which reach a diameter of 6–8 cm. The color of the sepals and petals is variable. In some specimens the flowers are pale pinkish purple, in others they range from delicate lavender-pink to reddish violet, and in still others they are bright purple or even violet. There is also a variety with white flowers: *P. incarnata* f. *alba*. On rare occasions the color of the flowers differs on one and the same plant.

This species is native to the southeastern United States. The northern border of its native range runs through Missouri, Illinois, Indiana, Ohio, and Pennsylvania. In the west it has advanced to Texas, Oklahoma, and Kansas. It was also introduced to the Bermudas, where it has escaped into the wild. In many parts of the United States it is known as maypop. Generally, the flowers from the eastern seaboard are not as large or violet as those from the extreme south or southwest of its range.

Since *Passiflora incarnata* withstands tempera-

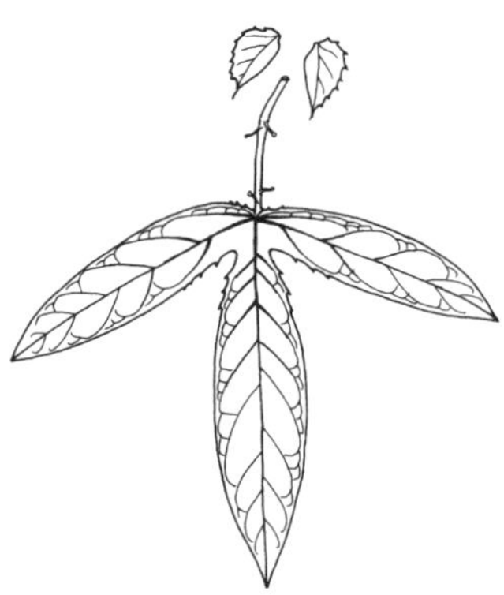

Passiflora gritensis

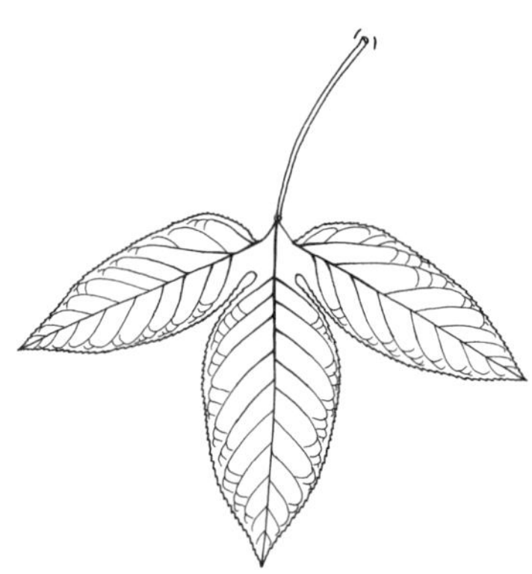

Passiflora incarnata

tures down to –15°C, plants that are several years old can be overwintered outdoors in many regions. However, it is advisable to cover the rootstock with brushwood, foliage, or some other protective material. Young plants in particular need this additional protection against the cold, because real hardiness is only achieved after a couple of years when the roots become fleshy. Try to plant this species in a sheltered location, perhaps beneath a projecting section of roof, and provide a permeable soil mix and good drainage. The aboveground part of this herbaceous species usually dies off in winter. However, when plants are overwintered at temperatures above the freezing point, the shoots will either not die or will die only partially.

In central Europe and in the native range of *Passiflora incarnata*, new growth appears in May to late June, depending on weather conditions. The first flowers can be expected toward the end of July. Normally there are numerous suckers growing out of the soil around the mother plant. These can be cut off and planted in pots. Unfortunately, nearly all clones of *P. incarnata* cultivated in Europe exhibit viral symptoms. The leaves have pale spots, with occasional deformations on the leaf margin.

In sunny locations *Passiflora incarnata* produces greenish yellow fruit about 6 cm long. These vary greatly in taste, depending on the plant. Sometimes they are processed into juice locally on a small economic scale, perhaps for use as local wines. The species has a much larger economic importance as a medicinal plant (see chapter 7).

Because of its many positive characteristics, including cold-hardiness, *Passiflora incarnata* has often been used in crossbreeding experiments, and therefore many beautiful hybrids are available (see chapters 17 and 18).

Vine: herbaceous, glabrous or finely pilosulous; stem: terete, younger parts angular; stipules: setaceous, 0.2–0.4 cm long, soon deciduous, entire; petioles: up to 8 cm long, with 2 glands mostly at apex; leaves: 3-lobed or rarely 5-lobed, 6–15 × 7–17 cm, membranous, serrate; peduncles: up to 10 cm, solitary, stout; bracts: oblong with 2 glands at base, 0.4–0.8 × 0.2–0.4 cm, slightly serrate, free; flowers: variable, white, pinkish, pale lavender, mauve, or violet, 6–8(–9) cm diam.; sepals: variable inside, green outside, oblong-lanceolate, ca. 3 × 1 cm, slightly keeled outside, keel terminating in an awn 0.2–0.4 cm long; petals: slightly shorter than sepals; corona: several series, white, pinkish, lavender, or violet, outer 2 series 1.5–2.5 cm long, wavy in upper 1/2, inner series 0.2–0.4 cm long; ovary: ovoid, densely tomentose; fruit: ovoid or subglobose, 5–6 × 3.5–5 cm, greenish yellow, edible; seeds: obovate or nearly obcordate, 0.4–0.5 × 0.3–0.4 cm, truncate at apex, reticulate.

Passiflora kermesina Link & Otto

Verhandl. Ver. Gartenb. Preuss. 2: 403 (1826)
kermesina: kermes scarlet
Supersection *Stipulata*, Section *Kermesinae*

For many years *Passiflora kermesina* (Plate 198) was ranked top on the wish lists of botanists and enthusiasts. This was due in part to ancient color drawings of this species, which promised a fascinating plant, and to the fact that it was in cultivation many years ago, as can be concluded from crossing formulas of hybrids that probably no longer exist.

During the last decades, various attempts were made to find *Passiflora kermesina* in its natural habitat in order to bring it back into cultivation. For many years this search was in vain. Either the indications of the old herbarium specimens were too vague or the plants had disappeared because of housing developments, industrial sites, or road construction. There were only 1 or 2 photos in which *P. kermesina* could be identified. Finally, some years ago Mauro Peixoto brought a specimen into cultivation in Brazil. Two clones with paler

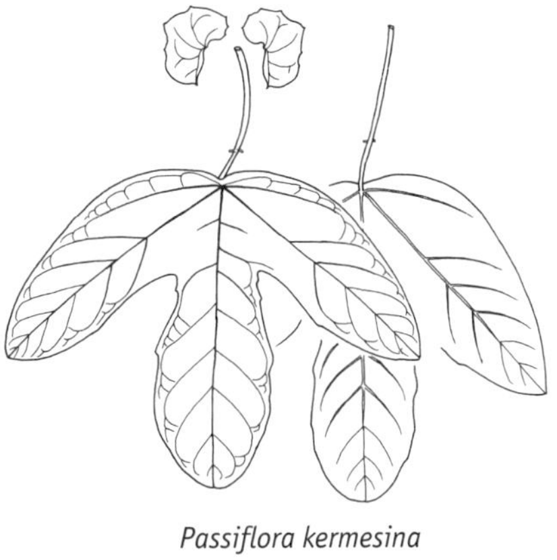

Passiflora kermesina

flowers were passed on to Roland Fischer. These plants have flowered abundantly since that time, and Fischer has been able to make various cross-breeding experiments. As a result of his studies, he has also determined that *P. kermesina* is indeed a parent plant of *P.* 'Amethyst', as has been speculated for several years.

Axel Frank and Reto Gamma have also brought cuttings to central Europe from a trip to Brazil. It seems to be only a matter of time until this wonderful species finds its way into enthusiasts' collections.

The flower color of *Passiflora kermesina* varies but does not measure up to the purple-red shown in ancient color drawings and suggested by the specific epithet. However, the diameter of a blossom, reaching about 11 cm immediately upon opening, is significantly wider than stated by Killip (1938). Because of the larger proportions of the blossom and the corona in 5 series, some doubts remain about the identification of the plants in cultivation.

As is usual with hummingbird-flowered species, the sepals and petals reflex widely so that the flower looks longer and not as wide. The purple-violet undersides of the leaves are also remarkable.

In its outward appearance, *Passiflora kermesina* is similar to *P. loefgrenii*. The only decisive difference becomes evident when the bracts are compared. In *P. kermesina* they are minute, located 2–2.5 cm below the base of the flower, and are soon deciduous, while in *P. loefgrenii* they are elliptic and 0.5–0.7 cm wide.

So far *Passiflora kermesina* has turned out to be quite free-flowering in cultivation, although one or another flower bud aborts. It prefers a location that is protected and sunny but not too hot, preferably under glass. During overwintering the temperature should not fall below 10°C. The main flowering period occurs before and after the height of summer, when daily temperatures are around 25°C.

Vine: glabrous; stem: terete, slender; stipules: subreniform, 1–2.5 × 0.5–1.2 cm; petioles: 2.8–4 cm long, with 2–4 elongated glands above middle, glands 0.1–0.3 cm long; leaves: 3-lobed, 4–8 × 5–11.3 cm, membranous, purplish violet beneath, entire; peduncles: 8–15 cm, solitary; bracts: minute, 2–2.5 cm below base of flower, soon deciduous; floral tube: cylindrical-campanulate, 0.9–1 cm long, whitish green with pink outside; flowers: deep pinkish purple to pinkish violet, (6–)8–9.5 (–11) cm diam.; sepals: pale pinkish purple outside, deep pinkish purple to pinkish violet inside, oblong-lanceolate, 3.5–5.6 × 0.5–1 cm; petals: deep pinkish purple to pinkish violet, subequal to sepals; corona: 5 series, coronal filaments thin, outer 2 series 1.4 cm long, violet with a small paler violet to white band beneath middle, 3rd series 0.8 cm long, 4th series 0.6 cm long, innermost series 0.9–1 cm long, pale violet at base, deep violet in upper 1/2; ovary: oblong-ovoid, green, glabrous; fruit: elliptical to oblong-ovoid, olive-green when ripe, 7.5–9.5 × 2.2–3 cm.

Passiflora laurifolia L.

Sp. Pl. 956 (1753)
laurifolius: laurel-leafed
Synonyms: *P. oblongifolia* Pulle, *P. tinifolia* Juss.
Supersection *Laurifolia*, Series *Laurifoliae*

In many parts of Europe and the United States, *Passiflora laurifolia* (Plate 199) can be cultivated satisfactorily only in a greenhouse, but in sunny places like Florida it can spend most of its time outdoors. It is among the most fastidious and warmth-loving passionflowers. During the cooler, darker months of the year, *P. laurifolia* will react sensitively to the worsened conditions. Some plants will have difficulties getting through winter or will not survive it at all. Additional lighting is needed. It will tolerate a minimum temperature of 10°C.

The fragrant flowers of about 7 cm diam. are impressive and hang down from the foliage. In addition to warmth and a high atmospheric humidity,

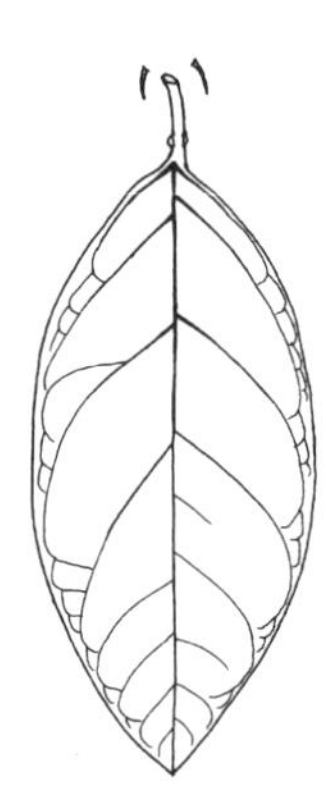

Passiflora laurifolia

Passiflora laurifolia also needs to be planted directly into an earth bed in a greenhouse in order to flower. Only on large plants with long stems can flowers be expected.

This species' original native range is restricted to the Guianas, the West Indies, Trinidad, Venezuela, Peru, and Brazil. Because of its edible fruit, it is valued as an economic plant and is grown on plantations outside of South America.

Passiflora laurifolia is the best-known species from the series that bears its name, *Laurifoliae*. Representatives from this series are characterized by unlobed, glabrous leaves, linear stipules, and pendent flowers. Two important distinguishing features within this section are the length of the outermost coronal series and the presence or absence of pubescence on the ovary. See *P. acuminata*, *P. ambigua*, and *P. popenovii* for descriptions of how *P. laurifolia* differs from these 3 closely related species.

Vine: glabrous except ovary and bracts; stem: terete; stipules: linear, 0.3–0.4 cm long, coriaceous; petioles: 0.5–1.5 cm, with 2 glands paired near apex; leaves: unlobed, ovate-oblong or oblong, 6–12 × 3.5–8 cm, thick-coriaceous, entire; peduncles: 2–3(–8) cm, solitary; bracts: ovate-oblong, 2.5–4 × 2–2.5 cm, glandular-serrate, free; flowers: purplish, violet, and white, 5–7 cm diam.; sepals: white, mottled with purple spots inside, whitish green outside, oblong, 2–2.5 × 1 cm; petals: white, densely mottled with purple spots, oblong, 1.8–2.4 × 0.8–0.9 cm; corona: 6 series, banded purple and white in lower 1/3, banded violet and white above, outermost series ca. 2 cm long, 2nd series 3–4 × 0.2 cm, ligulate, inner series ca. 0.1 cm long; ovary: ovoid, tomentose; fruit: ovoid, 5–8 × 4–5 cm, lemon-yellow to pale orange, edible; seeds: obcordate, 0.5–0.7 × 0.3–0.5 cm, finely reticulate.

Passiflora ligularis Juss.

Ann. Mus. Natl. Hist. Nat. 6: 113, t. 40 (1805)
ligularis: tongue-shaped (in reference to the petiole glands)
Synonym: *P. serratistipula* DC.
Supersection *Laurifolia*, Series *Tiliifolia*

Passiflora ligularis (Plate 200) is cultivated as a commercial plant in the subtropics almost as much as *P. edulis*. The yellow to orange fruits are among the tastiest of all passionflowers and are also available in Europe. The flavor of the yellow-white aril is similar to the taste of gooseberries. The fruit resembles a small, smooth orange but has a droplike extension toward the stalk. Contrary to general belief, the fruit, despite its hard shell, cannot be stored for long and should be eaten as fresh as possible. New plants can easily be grown from the numerous seeds contained in the fruit. They will germinate as soon as 2–4 weeks after sowing. However, it usually takes these plants 3–4 years before they produce flowers.

The habitats of this species extend from western Bolivia over the mountain ranges of the Andes to the northwest of South America and proceed through the mountains of Central America to Mexico, at altitudes of 1000–3200 m. Experiments to cultivate this species in lower regions with a warmer climate have failed. *Passiflora ligularis* is characterized by its conspicuous, filiform nectar glands, 3–13 mm long, on the petioles, to which it owes its name. The stem and the heart-shaped leaves are completely glabrous; the waxy coating gives the plant a greasy luster. Though in recent years there have been more and more reports about flowering plants under cultivation, *P. ligularis* is still regarded as being reluctant to flower unless very large. The light violet, pendent flowers, about 8 cm diam., appear only rarely. Nevertheless, *P. ligularis* is cultivated very often. This may be due in part to this plant's very impressive, rampant, heavy foliage, and it may have something to do with the challenge of getting the plant to flower. In view of the conditions prevailing in its natural surround-

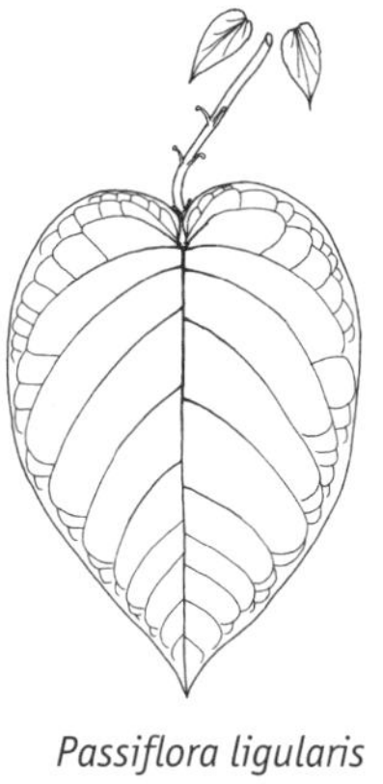

Passiflora ligularis

ings, the plant should be kept neither too warm nor too cold. Mild day temperatures around 22°C and nightly drops to 12°C will have a positive effect on flower bud formation. During the overwintering period, temperatures around 10°C will suit the plant very well, although even 5°C will be tolerated for a short time.

Vine: glabrous; stem: terete, striate; stipules: oblong-lanceolate, 1.5–3.3 × 0.7–1.2 cm; petioles: 2–10 cm, stout, with 3–6 scattered, liguliform to filiform glands, 0.3–1.3 cm long; leaves: unlobed or rarely 3-lobed (forma *lobata*), broadly ovate, cordate, 8–20 × 7–17 cm, entire; peduncles: 2–5 cm, solitary or in pairs; bracts: ovate, 3–6 × 1.5–2.5 cm, minutely glandular-serrate, connate at base; flowers: white and purple or violet, (4–)6–7(–9) cm diam.; sepals: white inside, green outside, oblong, 2–3 × 0.7–1 cm, aristate, awn 0.2–0.4 cm long; petals: white or mottled with purple spots, oblong, 2–3 × 0.7–1 cm; corona: 5 or 6 series, banded white and purple or violet, outer series ca. 3 cm long, 2nd series 2.5–2.7 cm long, inner series ca. 0.1 cm long; ovary: ovoid, glabrous; fruit: ovoid to globose, 6–10 × 4–7 cm, green then purple when immature, orange with pale spots when ripe, hard brittle shell, edible; seeds: narrowly obcordate, 0.5–0.6 × 0.3–0.4 cm, reticulate, black.

Passiflora loefgrenii Vitta

Novon 7: 210, fig. 2 (1997)
named after Albert Loefgren, collector of the paratype
Supersection *Stipulata*, Section *Granadillastrum*

Passiflora loefgrenii (Plate 201) is an unquestionably exceptional species. In a manner similar to *P. gritensis*, the flowers hang on peduncles up to 20 cm long but bend upward at their ends so that it is possible to look from above into the open flower. The floral ecology of *P. loefgrenii* deserves special mention. Its flowers are pollinated by both hummingbirds and bees. Consequently it unites bird-appropriate characteristics—for example, a brilliant purple flower, reflexed sepals and petals, and a long androgynophore—with bee-appropriate characteristics, such as a marked corona in several series. In the original description by Fabio Augusto Vitta (1997), it was pointed out that during the first hours after the opening of the flower, the stigmas and anthers are radially disposed. During this stage, *P. loefgrenii* is pollinated by hummingbirds (*Phaethornis eurynome* and *Phaethornis squalidus*). Later the styles and anthers bend downward, facing the corona. During this 2nd stage, carpenter bees (*Xylocopa*) are capable of pollinating the flowers. The flowers, reaching about 12 cm diam., are among the most beautiful of this subgenus. This is all the more enjoyable as this species, endemic in the Ribeira region of southeastern Brazil, has been under cultivation in Europe since the end of 2000. Because of its moderate demands and its willingness to flower, *P. loefgrenii* is a valuable addition to the collections of many enthusiasts. When it is grown in a pyramid-like fashion in a 20-cm pot, the first flowers can be produced at a height of 90 cm. This species is another short-day plant, most of its flowers set when light intensity decreases in late fall and winter. It does well as a container plant outdoors in summer. In winter it is satisfied with temperatures above 10°C.

Vine: glabrous; stem: terete, striate; stipules: reniform, (0.8–)2–3.5 × 0.5–1.5 cm, glandular-serrate near base; petioles: (2–)3–7 cm, with 6 stipitate, paired glands; leaves: 3-lobed, 5–9.5 × 5.5–14 cm, entire; peduncles: 11–20 cm, solitary; bracts: elliptic, 1–1.5 × 0.5–0.9 cm, soon deciduous, glandular-serrate, free; flowers: purple, bluish violet and white, 9–12 cm diam.; sepals: bright purple inside, purple outside, oblong-lanceolate, 4–5.5 × 0.9–1.5 cm, aristate, awn (0.4–)0.7–1 cm long; petals: bright purple, oblong, subequal to sepals; corona: 6 or 7 series, white at base, bluish violet

Passiflora loefgrenii

above, outer 2 series 1.7–2 cm long, inner series 1–1.2 cm long, erect; ovary: ellipsoidal, glabrous; fruit: ellipsoidal, ca. 6 × 4 cm, greenish yellow.

Passiflora luetzelburgii Harms

Repert. Sp. Nov. 19: 32 (1923)
named after Philipp von Luetzelburg, collector of the type specimen
Supersection *Passiflora*, Series *Setaceae*

Passiflora luetzelburgii (Plate 202) is exceptional in several respects, its most conspicuous feature being its brilliant red flowers, which reach a diameter of about 6 cm The combination of a short, upright, white corona with a widely protruding androgynophore and the absence of any scent suggest that its flowers are pollinated by hummingbirds. Also remarkable are the leaves, which are 3-lobed, densely villose on the underside, and relatively small, at a length of 4–5 cm.

This combination of characteristics caused difficulties when this species was incorporated into the taxonomic system. Harms (1925) regarded *Passiflora luetzelburgii* as belonging to the *Tacsonioides* group and thought it to be a close relative of *P. reflexiflora*, *P. mendoncae*, and *P. umbilicata*. In contrast to these, however, *P. luetzelburgii* is villose, has linear, soon deciduous stipules, and lacerate bracts. Although the flowers of all these species are similar because they are adapted to pollination by hummingbirds, *P. luetzelburgii* does not fit well into subgenus *Tacsonioides*. A look at Plate 202 suggests a similarity to the representatives of supersection *Coccinea*, but Feuillet and MacDougal place it without comment in series *Setacea*.

This quite rare species, at home in eastern Brazil, has yet to come into cultivation.

Vine: more or less villosulous except ovary; stem: subterete; stipules: linear, soon deciduous; petioles: 0.3–1 cm, with 2 glands at or below middle; leaves: 3-lobed, lateral lobes occasionally reduced, 4–5 × 2–4 cm, subcoriaceous, densely villosulous beneath, short-pilose above, slightly serrate; peduncles: 2–7.5 cm, solitary; bracts: lanceolate-oblong, 1–1.5 × 0.5–0.7 cm, lacerate-pectinate, reddish, free; floral tube: broadly cylindrical, 1.5–2 cm long, greenish without; flowers: red, 5–6 cm diam.; sepals: red, oblong-lanceolate, 2–3 × 0.7–0.9 cm, aristate, awn ca. 0.3 cm long; petals: red, subequal to sepals; corona: 3 series, outer series 0.4–0.7 cm long, erect, white, inner series ca. 0.2 cm long, white; ovary: narrowly ovoid, glabrous; fruit: not seen.

Passiflora magnifica Escobar

Phytologia 69: 365 (1990)
magnificus: noble, splendid
Supersection *Laurifolia*, Series *Tiliifolia*

Passiflora magnifica (Plate 203) has only been found in the Cordillera Occidental in the department of Antioquia, Colombia, where it grows on the rainy western slopes at elevations of 1250–1750 m. It is closely related to *P. triloba*, from Bolivia and Peru, and to the common and much-cultivated *P. ligularis*. It can easily be distinguished from *P. ligularis* by its flower color, 3-lobed leaves, and considerably larger bracts. Although *P. triloba* has a similar flower color, the outermost coronal series is an important distinguishing criterion. While in *P. mag-*

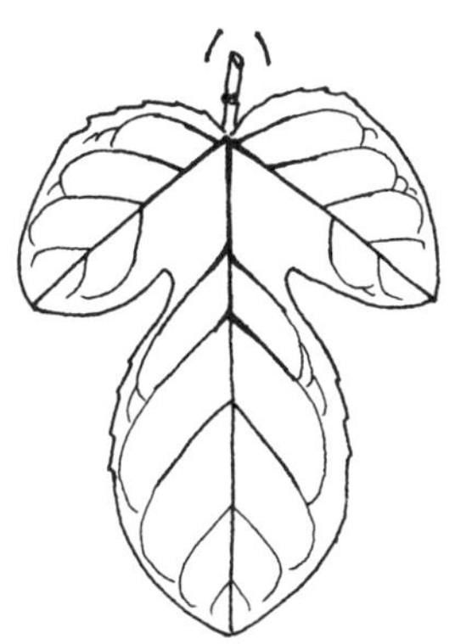

Passiflora luetzelburgii

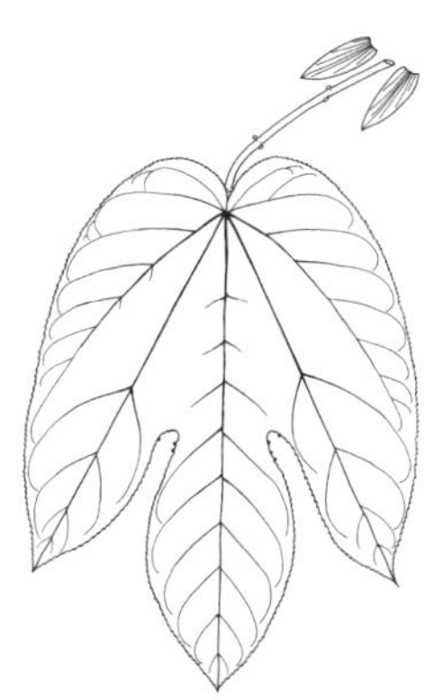

Passiflora magnifica

nifica the 2 outermost coronal series are equally long, the outermost series in *P. triloba* is significantly shorter than the following one.

Passiflora magnifica produces pendent red-purple flowers that reach a diameter of about 5 cm. The very large, connate bracts that enclose the bud are quite imposing. Although several seedlings were introduced by John MacDougal and James Zarucchi around 1990, this species has been cultivated only rarely and may now be lost from cultivation. It needs high humidity and well-balanced temperatures of around 25°C during the day and at least 12°C at night. Moreover, the plant must reach a certain height before it can set flowers. This species should be planted directly in the ground in a suitable location or grown in a large container.

Vine: glabrous; stem: terete to subangular, striate; stipules: ligulate-oblong, 2.7–3 × ca. 1 cm; petioles: 7–11 cm, with 2–6 mostly paired stipitate glands; leaves: 3-lobed, 13–23 × 15–27.5 cm, cordate at base, membranous to subcoriaceous, entire to very shallowly glandular-serrate; peduncles: 9.5–12 cm, solitary; bracts: ovate, 4–8.7 × 2.5–5 cm, connate 1–2 cm at base; flowers: red-purple and white, 4.5–5 cm diam.; sepals: white, densely mottled with purple spots inside, greenish outside, lanceolate, 2.5–3 × 0.8 cm, aristate, awn 0.6–0.8 cm long; petals: purple, lanceolate, 2.3–2.9 × 0.7 cm; corona: 4 series, purple banded with white, outer 2 series filamentous, ca. 2 cm long, inner series 0.3–0.4 cm long; ovary: ellipsoid, glabrous; fruit: ovoid to obovoid, 5.5–6.5 × 3.5–5 cm, light yellow to orange-yellow, with many pale spots; seeds: obovate, ca. 0.5 × 0.3–0.4 cm, reticulate, dark brown.

Passiflora maliformis L.

Sp. Pl. 956 (1753)
maliformis: apple-shaped (in reference to the fruit)
Synonyms: *P. caudata* Gentry, *P. ornata* Kunth
Supersection *Laurifolia*, Series *Laurifoliae*

Since *Passiflora maliformis* (Plate 204) will generally set flowers only when cultivated under glass, this warmth-loving species is recommended for a greenhouse or conservatory. In warmer regions it can also be kept outdoors during summer months. Since *P. maliformis* needs a certain height to develop its fragrant flowers, it is advisable to cultivate it in a spacious container or to plant it directly into an earth bed in a greenhouse.

The charming flowers, hanging down from a canopy of foliage, reflex the sepals and petals immediately upon opening. The diameter of a flower is therefore comparatively small, at 5–6 cm, although it actually looks much larger. The purple and white-zoned corona, slightly bent inward at the apex, surrounds the ovary in a campanulate fashion. At a length of 4–6 cm, the bracts are remarkably large. They enclose both the bud and the ripening fruit.

The name of the species refers to the edible fruit, which is about 5 cm diam. and resembles a small apple. When ripe, it is greenish yellow to light orange. Its shell is so hard that it can hardly be opened with bare hands. *Passiflora maliformis* is at home in Colombia, Ecuador, Venezuela, and the West Indies, where it grows near sea level as well as at altitudes up to 1700 m. It should be overwintered at temperatures above 12°C, while planted-out specimens might well tolerate lower temperatures.

Vine: glabrous or occasionally finely pubescent; stem: terete, striate; stipules: setaceous, 0.5–1.5 × 0.1–0.2 cm; petioles: (1.5–)3–6 cm, with 2 or occasionally 3–4 ovate glands; leaves: unlobed, ovate, 6–15(–25) × 4–10(–15) cm, slightly serrate or subentire; peduncles: 5–7 cm, solitary; bracts: ovate, 4–5 × 3–5 cm, connate at base, entire; flowers: greenish white, purple, and violet, 6–7.5(–9) cm diam.; sepals: greenish white, densely mottled with purple spots inside, green outside, oblong, 3.5–4.5 × 1.5–2 cm, keeled, keel terminating in an awn

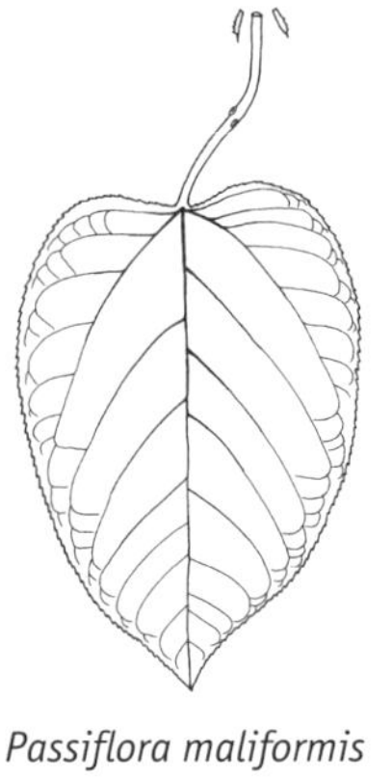

Passiflora maliformis

0.2–0.7 cm long; petals: greenish white, mottled with purple spots, oblong, 3–3.5 × 0.5–0.7 cm; corona: 4 series, purple banded with white near base, violet banded with white above, outer series 0.8–1.2 cm long, 2nd series 1.5–3(–4) cm, inner series ca. 0.1 cm long; ovary: ovoid, glabrous or puberulous; fruit: ovoid to globose, 4–5 × 3.5–5 cm, greenish yellow to light orange, pericarp very hard, edible; seeds: oblong-obcordate, 0.5–0.6 × 0.4 cm, finely reticulate.

Passiflora mayarum MacDougal

Ann. Missouri Bot. Gard. 76: 615, figs. 1a, 1c, 2 (1989)
named for the native distribution, centered in the heart of the Mayan Old Empire
Supersection *Passiflora*, Section *Passiflora*, Series *Passiflora*

The native range of *Passiflora mayarum* (Plate 205) covers northern Guatemala, many regions of Belize, and the bordering areas of Mexico. Collections have often been made along roadsides through forested areas on limestone at altitudes of 20–550 m.

Passiflora mayarum produces purple and white flowers that reach a diameter of about 6 cm. The remarkable innermost coronal series, banded purple and white, is unusually long and inclined toward the androgynophore. *Passiflora mayarum* is closely related to *P. prolata*, which occurs in Chiapas, Mexico, and western Guatemala at altitudes of 700–2000 m. Its different habitat is one of the most reliable distinguishing features. John MacDougal points out that *P. mayarum* is also rather closely related to *P. edulis*.

Passiflora mayarum has been under cultivation for several years. Taking a hint from its native habitat, it should be planted in a warm, protected location. Hence, it can be cultivated only in a greenhouse year-round. So far it has turned out to be quite reluctant to flower. It remains to be seen how this will change as the plants age. During the overwintering period, temperatures above 15°C are recommended. At prolonged temperatures around 10°C, the leaves turn yellow and fall off.

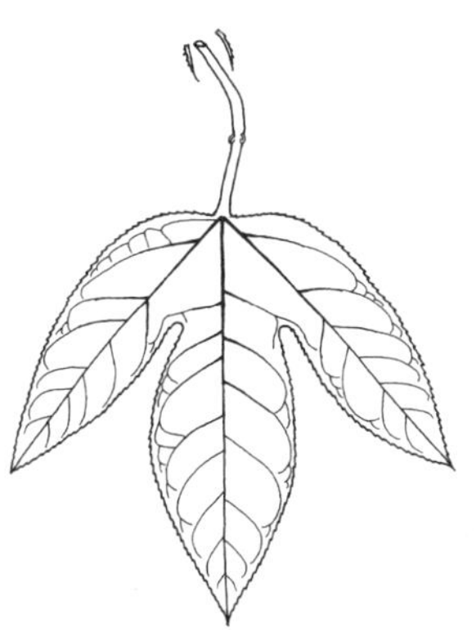

Passiflora mayarum

Vine: puberulent except upper leaf surface and flowers; stem: subterete to subangular, striate; stipules: linear-lanceolate, (1–)1.3–1.8(–2.2) × 0.1 cm, soon deciduous, glandular-denticulate; petioles: ca. 2–4 cm, with 2(–4) glands; leaves: 3-lobed, 10–13 × 9–14 cm, glandular-serrate; peduncles: 2–5(–6) cm, solitary; bracts: ovate-elliptic, 4.8–6.6 × 2–3.4 cm, entire with 2–6 glands at margin, free; flowers: purple and white, 5.5–7.5 cm diam.; sepals: white inside, oblong-lanceolate, 2.4–3.7 × 0.7–1.2 cm, aristate, awn 0.8–1.9 cm long; petals: white, oblong-lanceolate, 2.4–3.7 × 0.6–0.9 cm; corona: 6–8 series, dark purple banded with white near base, purple above, whitish near apex, outer 2 series 2.6–3.8 cm long, inner series 0.1–0.5(–0.7) cm, innermost series 0.6–1 cm long; ovary: ovoid-ellipsoid, glabrous; fruit: ellipsoid, 7.5–9 × 4.5–5 cm, greenish yellow to yellow; seeds: obovate, 0.7–0.8 × 0.4–0.5 cm, flattened, reticulate, brown.

Passiflora mendoncae Harms

Repert. Sp. Nov. 18: 297 (1922)
named after Francisco Ribiero de Mendonça, collector of the type specimen
Synonym: *P. rhodoptera* Dusén ex Harms
Supersection *Stipulata*, Section *Tacsonioides*

Passiflora mendoncae (Plate 206) is another exceptional species within subgenus *Passiflora*. Its habitats are at moderate altitudes of 1000–1300 m in southern Brazil. The markedly reflexed sepals and petals of the purple-red flowers are about twice as long as the comparatively short floral tube. Because of its particular floral ecology, including the large purple-red bracts and its overall appearance, the hummingbird-pollinated *P. mendoncae* can hardly be confused with another species.

The filaments of the outer coronal series are only 0.5 cm long and are slightly bent toward the androgynophore. This barrier hinders insects and

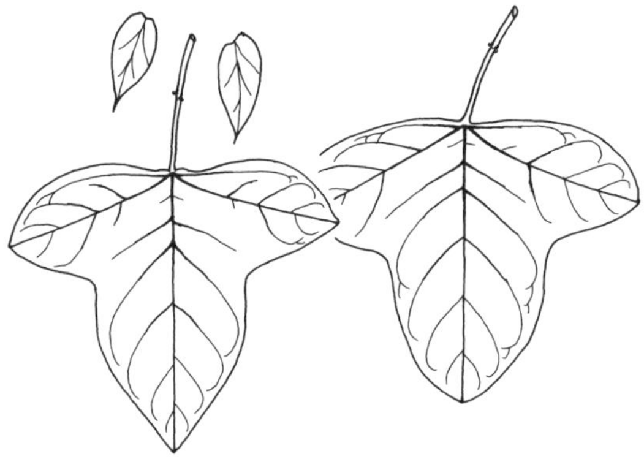

Passiflora mendoncae

other nectar robbers from getting to the nectar at the bottom of the floral tube. In 2002, Reto Gamma and Maurizio Vecchia succeeded in bringing this species into cultivation. *Passiflora mendoncae* is now grown in several collections.

At first, *Passiflora mendoncae* was assigned to subgenus *Passiflora* (formerly *Granadilla*) by Harms. Because of its flower, Killip (1938) placed it in the controversial subgenus *Tacsonioides*. Finally, Feuillet and MacDougal recognized the real kinship relations, dissolved the status of an independent subgenus, and included its representatives as section *Tacsonioides* of supersection *Stipulata* in subgenus *Passiflora*.

Vine: herbaceous, glabrous, membranous in most parts; stem: terete, striate, slender; stipules: ovate to oblong-ovate, 1–2.2 × 0.5–1.3 cm, mucronulate, entire; petioles: slender, 1.2–3.8 cm, with 2 or occasionally 4 short stipitate glands in upper 1/2; leaves: 3-lobed, 2–5.5 × 3–7.5 cm, lustrous above, glaucous or violet beneath, entire; peduncles: slender, 5–13 cm long, solitary; bracts: ovate to ovate-oblong, 2.7–4.2 × 1.5–2.8 cm, reddish purple, free; flowers: purple to reddish purple, 5–7 cm long and diam.; floral tube: short-cylindrical, 1.3–2 cm long, 1–1.6 cm wide; sepals: reddish purple inside and out, lanceolate, 3–5 × 0.4–0.7 cm, keeled, keel terminating in an awn 0.4–0.9 cm long; petals: purple, lanceolate, subequal to sepals; corona: 2 series, violet at base, outer series filamentous, 0.4–0.7 cm long; ovary: narrowly ovoid, glabrous; fruit: ovate-oblong, 3.5 × 2 cm.

Passiflora menispermifolia Kunth

Nov. Gen. Sp. 2: 137 (1817)
menispermifolia: having leaves like species of genus *Menispermum*
Synonym: *P. cuellensis* Goudot ex Triana & Planch.
Supersection *Stipulata*, Section *Granadillastrum*

Passiflora menispermifolia (Plate 207) is a charming and very warmth-loving species. Its cultivation can be problematic, particularly during winter. Short days and temperatures below 12°C can cause a partial loss of the leaves and a growth stoppage. The dropped and dried-up leaves turn blackish, a characteristic of this and some related species. *Passiflora menispermifolia* is recommended for cultivation under glass or outdoors during summer months in warmer regions, but it is at its best when planted directly into an earth bed in a greenhouse. It develops very quickly into a large plant with shoots that are several meters long and on which up to a hundred flowers can be open at the same time. Depending on the size of the plant, such a batch can last 1–3 days. Afterward the plant gathers new buds for several weeks until the next big bang. The fragrant violet flowers reach a diameter of 6–7.5 cm.

Passiflora menispermifolia occurs mainly in tropical regions from Nicaragua to Colombia but is also at home in Ecuador, Peru, Bolivia, and Brazil. Although its native range is very large, it is fairly uncommon across its range. Recent collections from various habitats show that *P. menispermifolia* is quite variable with regard to leaf form and vestiture density. Lawrence E. Gilbert discovered a single hairless plant of this species in Costa Rica. The fruits are apparently eaten by bats.

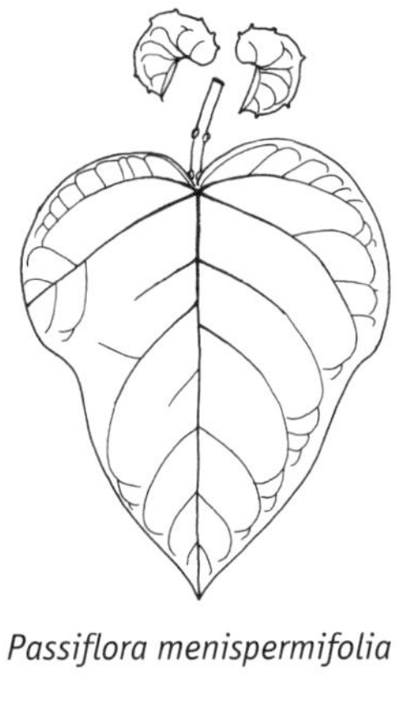

Passiflora menispermifolia

Vine: hispid-hirsute, except flower and ovary; stem: terete, striate; stipules: subreniform, 1–3.5 × 0.5–1 cm, glandular-serrate; petioles: 1.5–4 cm, with (2–)4–6(–11) shortly stipitate glands; leaves: 3-lobed or obscurely so, 5–20 × 9–19 cm, coriaceous, serrate to dentate; peduncles: (2.5–)4–6(–8) cm, solitary; bracts: oblong-lanceolate, 0.7–1.5(–2) × 0.3–0.5 cm, glandular-serrate, free; flowers: violet with reddish purple center, 6–7.5 cm diam.; sepals: violet inside, green outside, oblong-ovate, (2–)2.5–3.5 × 1–2 cm, aristate, awn 0.1–0.5 cm long; petals: violet, oblong, 2.5–3.5 × 0.8–1.5 cm; corona: several series, purple or reddish purple at base, banded violet and white above, outer 2 series 2–4 cm long, inner series 0.1–0.7 cm long; ovary: ovoid, glabrous; fruit: obovoid, 6–7 × (2–)3–4 cm; seeds: oblong-ovoid, ca. 0.6 × 0.3 cm, flattened, finely reticulate, dark brown.

Passiflora miersii Mast.

Fl. Bras. 13 (1): 599 (1872)
named after John Miers, collector of the type specimen
Supersection *Stipulata*, Section *Kermesinae*

Passiflora miersii (Plate 208), at home in eastern Brazil, belongs to the small section *Kermesinae*. Representatives of this section are characterized by foliaceous stipules, very narrow, mostly scattered bracts along the peduncle, and an operculum that is filamentous at its tip, resembling an additional series of coronal filaments. With regard to other characteristics, such as leaf form, species of *Kermesinae* can vary considerably. The section is conceivably not a "natural" one in that several lineages may be represented. Thus, the floral morphology and floral ecology of *P. miersii* has little in common with that of the red-flowering *P. kermesina*, to which this section owes its name. Instead, the flowers of *P. miersii*, 4–5 cm diam., have a markedly purple-and-white-banded corona that is typical for pollination by bees and other insects. In its natural surroundings the main flowering period is September to January (Cervi 1997).

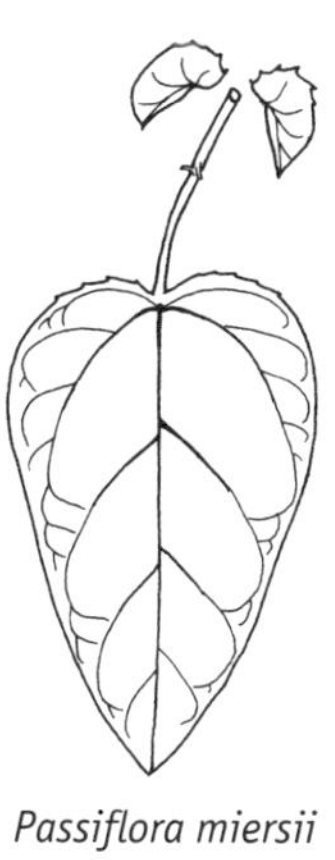

Passiflora miersii

Passiflora miersii can hardly be confused with the other 2 species from this section with unlobed leaves: *Passiflora eggersii* from Ecuador and *P. guentheri* from Bolivia. Besides the different geographical ranges, *P. eggersii* is distinguished by its red flowers and *P. guentheri* by its wider petiole glands and longer leaves.

For several years *Passiflora miersii* has been cultivated mainly in the United States. It is relatively easy to grow and will flower reliably in a 30-cm pot. In winter, temperatures around 10°C are ideal.

Vine: glabrous; stem: terete, wiry; stipules: semi-ovate, 1–2.5 × 0.5–1.2 cm, crenulate or subentire; petioles: 1.5–2.5 cm long, slender, with 2 stipitate glands in upper 1/3; leaves: unlobed, lanceolate or ovate-lanceolate, 3–6(–8) × 2–3 cm, subcoriaceous, entire; peduncles: slender, 3–5 cm long, solitary; bracts: variable, narrowly ovate to linear-setaceous, 0.5–0.8 × 0.2 cm, born 1 cm below base of flower, soon deciduous, free; flowers: white and purple, 4–5 cm diam.; sepals: white inside, green outside, oblong, 1.5–2.5 × 0.5–0.7 cm, aristate, awn ca. 0.2 cm long; petals: white, oblong, subequal to sepals; corona: 4 series, dark purple at base, banded purple and white above, white at apex, outer 2 series 1–1.5 cm long, 3rd series 0.1–0.2 cm long, 4th series 0.5–0.7 cm long; ovary: ovoid, glabrous; fruit: obovoid or ellipsoid, 3–4 × 1.5–2 cm; seeds: obovoid, reticulate.

Passiflora mooreana Hook.

Bot. Mag. 66: pl. 3773 (1840)
named after Thomas Moore
Supersection *Stipulata*, Section *Granadillastrum*

Passiflora mooreana (Plate 209) occurs naturally in southern Bolivia, Paraguay, and northern Argentina, where it grows in dry areas at low elevations. This species can easily be identified by its deeply 3-

Passiflora mooreana

lobed leaves, which are greenish gray like the whole plant. The leaves have a distinctly cuneate base, followed by a very short petiole, which gives the impression of sessile leaves. The stipules are conspicuous because of their large size and because of the presence of a nectary gland at their very tip. Because of their form, the flowers are a little reminiscent of *P. caerulea*, though they are somewhat smaller, at a diameter of 4–7 cm, and are not as intense in color. In addition, the flowers of *P. mooreana* are characterized by an amazingly strong fragrance similar to gardenia or jasmine. The corona is dark purple-violet at its base and otherwise pale violet with white bands. It is hard to confuse this with any other species.

Passiflora mooreana has been under cultivation for several years and is well recommended. It flowers reliably even in a small pot and can be cultivated on a windowsill as long as it does not become too large. It will also flourish in a container outdoors during summer months. Although it is said to withstand a slight frost, it should be overwintered at temperatures beyond 7°C.

Vine: glabrous; stem: subangular, glaucous; stipules: ovate-lanceolate, 2–4.5 × 0.6–1.6 cm, coriaceous, entire or slightly serrate; petioles: 0.5–1.5 cm long, with 2 sessile glands above middle; leaves: deeply 3-lobed, 5–12 × 2–10 cm, coriaceous, entire; peduncles: 1.3–2.5 cm long, solitary; bracts: ovate-lanceolate, 1.2–3 × 0.7–2 cm, serrate, free; flowers: white and purple to pale violet, 4–7 cm diam.; sepals: white inside, greenish gray outside, oblong, 2.2–3.5 × 0.6–0.8 cm, slightly keeled, keel terminating in an awn 0.2–0.3 cm long; petals: white, oblong, 2–3.5 × 0.6–1 cm; corona: 3 series, dark purplish violet at base, banded pale violet and white above, whitish at apex, outer 2 series (0.6–)1.2–2.5 cm long, inner series less than 0.2 cm long; ovary: globose to ellipsoid, glabrous; fruit: obovoid, 3–5.5 × 2.5–3.5 cm, yellow to orange-yellow; seeds: ovate, 4.5–6 × 0.3 cm, reticulate.

Passiflora mucronata Lam.

Encycl. 3: 33 (1789)
mucronatus: provided with a tip
Synonyms: *P. aetheoantha* Barb. Rodr., *P. albida* Ker, *P. pallida* Vell., non L.
Supersection *Stipulata*, Section *Granadillastrum*

Passiflora mucronata (Plate 210), endemic in eastern Brazil, is most interesting with regard to its floral ecology. The whitish flowers of this species open in the middle of the night and remain open until late morning. This unusual flowering can only be observed in a few other species that belong mostly to other systematic groups, such as *P. galbana* (supersection *Stipulata*), *P. ovata* (subgenus *Deidamioides*), and *P. trisecta* (supersection *Tacsonia*).

The zygomorphic flowers of *Passiflora mucronata* are characterized by a semicircular alignment of the anthers. They reach a diameter of 8–9 cm and come out of the foliage on long, robust peduncles. Immediately upon the opening of the flowers, the 2 coronal series effuse an intense, sweetish scent that soon fades. The wide calyx tube is not covered by the short, white corona. These adaptive features indicate that pollination is affected by night-active bats (Glossophaginae).

Alas, *Passiflora mucronata* has turned out to be rather reluctant to flower under cultivation. Although the plant grows vigorously under glass, as well as outdoors during summer months, it does not set flowers. Our own specimen flowered only once in 6 years. After it was taken to its winter quarters, 5 flowers opened in late fall. It seems that in order to set flowers the plant must reach a certain size, with shoots several meters long. *Passiflora mucronata* prefers temperatures beyond 10°C, although temperatures around 5°C can be tolerated for a short time.

Vine: glabrous; stem: terete, stout, slightly flexuous; stipules: ovate-lanceolate, 1.5–2.5 × 0.5–1.5

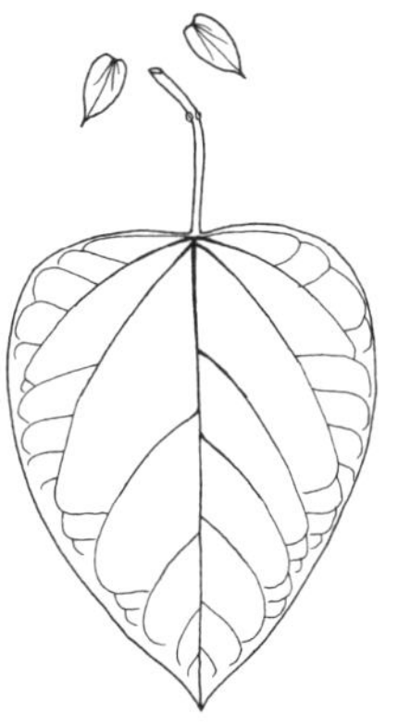

Passiflora mucronata

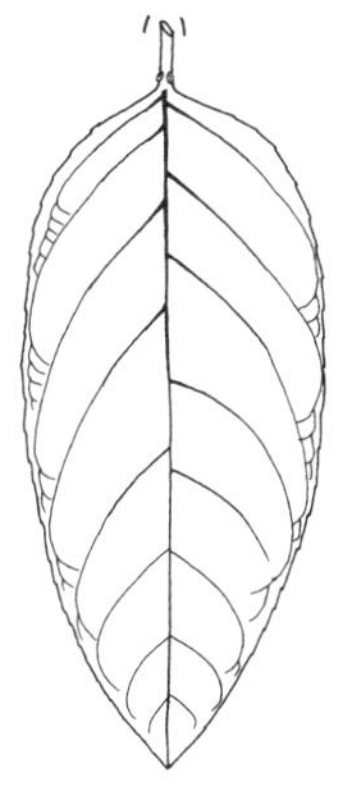

Passiflora nitida

cm, coriaceous, often glaucous, entire; petioles: 1–2 cm, with 2–4 glands near middle; leaves: unlobed, ovate-cordate, 4–12 × 2.5–6 cm, thick-coriaceous, entire; peduncles: up to 8 cm long, solitary, very stout; bracts: oblong-lanceolate, 2–2.5 × 1, membranous, entire, free; flowers: white, 8–9 cm diam.; sepals: white inside, green outside, linear, 2.7–3.5 × 0.5–0.8 cm, broadly keeled, keel terminating in an awn 0.2–0.3 cm long; petals: white, oblong, (1.9–) 2.3–3 × 0.5–0.7 cm; corona: 2 series, white, outer series very slender, ca. 1 cm long, inner series 0.2–0.4 cm long; ovary: narrowly ovoid, glabrous; fruit: ovoid, 4–5 × 2–2.5 cm; seeds: oblong-obcordate, ca. 0.5 × 0.4 cm, strongly flattened, reticulate.

Passiflora nitida Kunth

Nov. Gen. Sp. 2: 130 (1817)
nitidus: shiny, glossy (in reference to the upper side of the leaf)
Synonym: *P. nympheoides* Karst.
Supersection *Laurifolia*, Series *Laurifoliae*

Of all the species of *Laurifoliae*, *Passiflora nitida* (Plate 211) has one of the widest geographic ranges. It grows in the tropical lowlands from Costa Rica in the north and French Guiana in the northeast, through wide parts of Brazil to Bolivia and Peru. In these countries it is sometimes grown on a small economic scale because of its edible fruit (see Plate 16), and plants that are growing in the wild are exploited as a natural resource. The main flowering period in Brazil is December to March. The pendent flowers of this warmth-loving species effuse a pleasant scent and reach a considerable diameter, 8–12 cm. The corona, banded purple and white, surrounds the ovary in a campanulate fashion. A characteristic of *P. nitida* is the innermost coronal series, about 1 cm long, that is inclined toward the androgynophore. A similar coronal series can also be found in *P. acuminata*. The presently cultivated clones of *P. nitida*, for the most part imported from French Guiana, have turned out to be reluctant to flower. The plants must first overcome the critical adolescent stage and reach a certain size before they can set flower buds. This development takes time because the long winters in central Europe and many parts of the United States always lead to a setback. *Passiflora nitida* should be cultivated in a warm greenhouse year-round. As with many of the large passionflowers, it is advisable to plant it directly into an earth bed. During winter, temperatures of at least 15°C are required. Unfortunately, the same holds true for *P. nitida* as for many other tropical species: the more sumptuous the flower, the smaller the number of enthusiasts who have the facilities to cultivate such a plant successfully.

Vine: glabrous except ovary; stem: terete, younger portions subangular; stipules: linear-subulate, 0.3–1.1 cm long, deciduous; petioles: 3–4 cm, with 2 sessile glands at middle or near apex; leaves: unlobed, ovate-oblong, 9–23 × 6–11.5 cm, lustrous on upper surface, coriaceous or subcoriaceous, subentire or minutely glandular-denticulate or -serrulate; peduncles: 2–6 cm, solitary, stout; bracts: ovate-oblong, 3.5–5.5 × 2–4.5, subentire or glandular-crenate or -serrate, or with 2 larger pairs of glands near middle; flowers: white or pale pink, with purple and violet, 8–12 cm diam.; sepals:

white or pale pink inside, green outside, oblong-lanceolate, 4–5.2 × 1–1.8 cm, aristate, awn 0.15–0.7 cm long; petals: white, oblong, 4–5 × 0.8–1.5 cm; corona: 6–7 series, banded purple and white at base, violet and white above, outer 2 series 2.5–4.8 cm long, wavy at apex, following series 0.2–0.3 cm long, innermost series ca. 0.1 cm long; ovary: obovoid, glabrous or minutely puberulent; fruit: subglobose to ellipsoid to ovoid, 5–11 × (3–)4–8 cm, orange, red, or dark purplish red, with paler spots, edible; seeds: obcordate, ca. 0.5–0.9 × 0.3–0.7 cm, tridentate at apex, reticulate.

Passiflora oerstedii Mast.

Fl. Bras. 13 (1): 562 (1872)
named after Anders Sandoe Oersted, collector of the type plant
Synonyms: *P. populifolia* Triana & Planch., *P. praeacuta* Mast., *P. purpusii* Killip, *P. rojasii* Hassl. ex Harms
Supersection *Stipulata*, Section *Granadillastrum*

Passiflora oerstedii grows from Veracruz, Mexico, to northern Paraguay and is quite variable with regard to flower color, leaf shape, pubescence, and other features. It was originally named from Costa Rica. When different clones are cultivated and compared directly, the question arises as to whether the differences exceed the normal variability allowed for one species (see Plates 212–214). In additional to the typical species, with mostly oblong-ovate leaves, there are 3-lobed forms. One of these, from near the Gulf of Honduras, was treated as *P. oerstedii* var. *choconiana* by Killip (1938) but is now thought to be a valid species under the original name *Passiflora choconiana*. Another form with 3-lobed leaves was distributed in Europe in the 1990s under the erroneous name of *P. porphyretica*. This plant, imported from Mexico by Alfred Lau, is best placed under the name *P. oerstedii* but differs in some parts of its floral morphology. Perhaps it is a 3-lobed variant of the race represented by Killip's *P. purpusii*. It is very free-flowering and an excellent partner in crossbreeding.

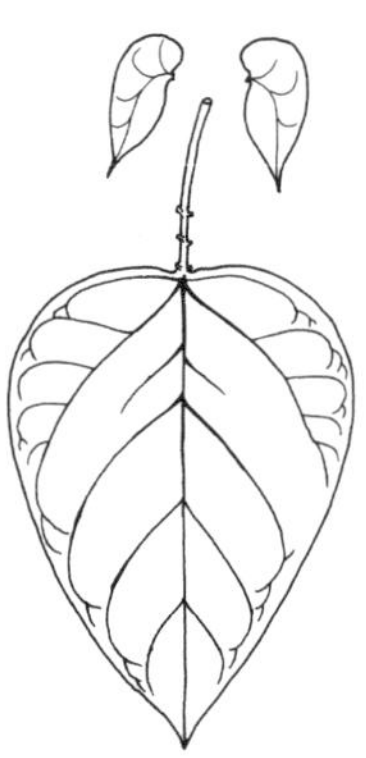

Passiflora oerstedii

A white-flowered variant from Costa Rica has a characteristic mix of lobed, asymmetrically 2-lobed, and 3-lobed leaves on the same branch, and was distributed from the 1970s to 2000 under the name "pseudo-oerstedii." This is now known to be a completely separate species, probably the same as the "white-flowered Venezuelan variety" illustrated in *Passion Flowers* (Vanderplank 1996a), and appears to have already been named by Killip as *Passiflora dispar*.

The flowers of *Passiflora oerstedii* are typically an intense purple but can also be purplish pink or whitish with pink. At a diameter of 4–5.5 cm, they are among the smallest of section *Granadillastrum*. It would be worthwhile to do field studies on this complex around *P. oerstedii*, to which such species as *P. trisulca* and *P. lehmannii* can be assigned. In its wide native range it grows in the lowlands as well as at altitudes up to 1800 m.

Cultivation in a greenhouse or outdoors during summer is unproblematic. However, since some clones have turned out to be rather reluctant to flower, *Passiflora oerstedii* is not a favorite among passionflower enthusiasts. It is overwintered at temperatures beyond 10°C.

Vine: glabrous or sparsely pilose, upper leaf surface glabrous; stem: terete or subangular, striate, glabrous or lightly pilose; stipules: subreniform, 0.5–3.5 × 0.5–1.8 cm, entire or serrate; petioles: 1.5–4.3 cm, with 4–8 stipitate glands; leaves: unlobed or occasionally 3-lobed, ovate-oblong to oblong-lanceolate, 6–15(–19) × 3–9(–11.5) cm, entire; peduncles: 2–6 cm, solitary, stout; bracts: ovate-oblong, (0.5–)1–1.5 × (0.2–)0.5–0.8, glandular-serrate, free; flowers: pinkish, pale purple, and white, (3–) 4–5.5 cm diam.; sepals: whitish inside, green outside, oblong, 2–2.5(–3) × 0.8–1 cm, aristate, awn ca. 0.2–1 cm long; petals: whitish to pale pink, oblong, 2–2.5(–3) × 0.8–1 cm; corona: several series, pink to purple, banded white at base, outer 2 series 1.5–3 cm long, following series 0.2–0.4 cm long; ovary: obovoid, glabrous; fruit: ellipsoid to obovoid or ovoid, 4–9 × 2–4.5 cm; seeds: narrowly obovate, 0.4–0.8 × 0.2–0.3 cm, shallowly tridentate, reticulate.

Passiflora pectinata Griseb.

Fl. Brit. W. I. 294 (1859)
pectinatus: comblike (in reference to the bracts)
Supersection *Stipulata*, Section *Dysosmia*

Passiflora pectinata (Plate 215) is another relatively little-known species from section *Dysosmia*. Representatives of this section, which was regarded as a subgenus of its own until a few years ago, are characterized by deeply cleft stipules, 1- to 3-pinnatifid bracts, and missing petiole glands that are often replaced by gland-tipped hairs. The section is sometimes called the *P. foetida* group.

Passiflora pectinata is at home in Bermuda, the Bahamas, Haiti, and the Dominican Republic. The coriaceous, shiny, nearly unlobed leaves with a deep cordate base are characteristic of this species, as are the pectinate or 1-pinnatifid bracts. At a diameter of up to 8 cm, its pure white flowers are larger than those of most other representatives from section *Dysosmia*. The form and color of the flower are most similar to *P. palmeri*, from which *P. pectinata* can easily be distinguished by its lack of pubescence. *Passiflora pectinata* was grown once by John MacDougal in an outdoor garden but is not presently represented in any collections. MacDougal's plant was easy to grow and flourished in the warm sun of summer. The self-incompatible white flowers were strictly day-blooming, remained open most of the day, and had no odor. As with a number of species of *Dysosmia*, the petals are often slightly longer than the sepals.

Vine: glabrous; stem: subangular, striate; stipules: deeply cleft into linear or subulate segments; petioles: 1–2 cm, often with a few stiff, gland-tipped hairs; leaves: unlobed to slightly 3-lobed, cordate-deltoid, 4–7 × 3–6 cm, coriaceous, repand crenulate; peduncles: 2–3 cm, solitary; bracts: pectinate or 1-pinnatifid, 2–3 × 0.8–1.2 cm, free; flowers: white, 5–8 cm diam.; sepals: white inside, greenish outside, linear to lanceolate, 2.5–3.5 × 0.5–0.8 cm, aristate, awn up to 0.7 cm long; petals: white, linear, 2.5–3.8 × 0.4–0.6 cm; corona: several series, white, filaments of outer series narrowly ligulate, 1–15 cm long, 2nd series 0.6–0.8 cm long, filaments of inner series 0.2–0.3 cm long; ovary: ovoid, glabrous; fruit: subglobose, 2–3 cm diam., deep pink; seeds: oblong-ovoid, ca. 0.5 × 0.4 cm, tridentate at apex, reticulate.

Passiflora pergrandis Holm-Nielsen & Lawesson

Ann. Missouri Bot. Gard. 74: 501, fig. 4 (1987)
pergrandis: very large (in reference to the flowers)
Supersection *Laurifolia*, Series *Laurifoliae*

Since its first description in 1987, this fascinating species (Plate 216) has been much sought after by enthusiasts. Only Alfred Lau was able to collect seeds in Ecuador, where this plant is endemic, and grow it in his garden in Mexico through the 1980s and 1990s, until it was finally brought into general cultivation by Michael Schwerdtfeger. Unfortunately, it turns out to be difficult to grow. Like many closely related species, it flowers only when the stems have reached a length of several meters. In

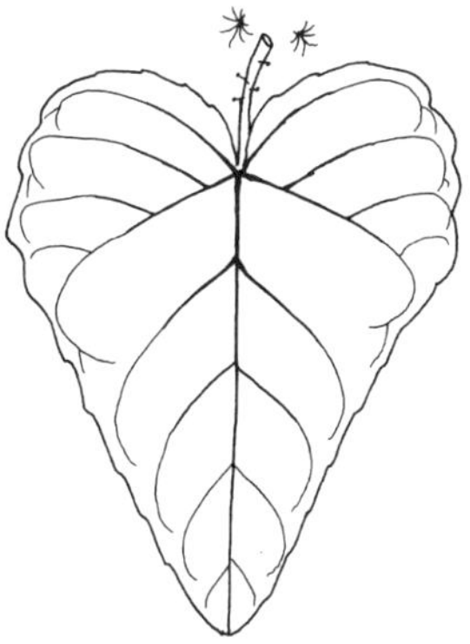

Passiflora pectinata

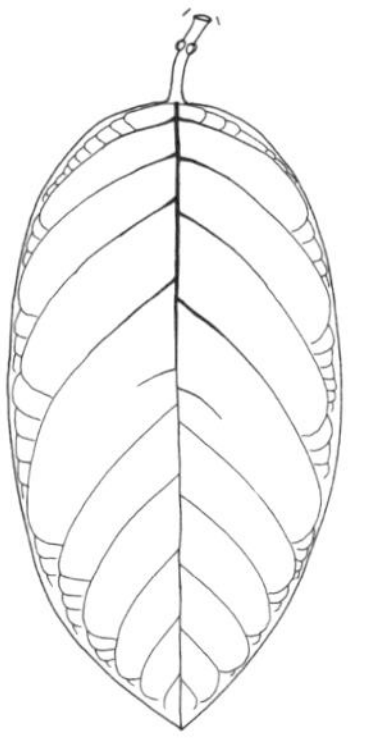

Passiflora pergrandis

addition, it needs a great deal of warmth and should be planted directly in the soil in a warm greenhouse. *Passiflora pergrandis* can develop into flower-bearing size only in the showrooms of botanical gardens or similar institutions. The Botanical Gardens of Munich and Göttingen have succeeded in bringing their specimens to flower every year, with flowering time occurring from November to January. Since few enthusiasts can offer adequate conditions, cultivation often ends in failure. Due to a lack of trace elements the leaves often become chlorotic, so a sufficient supply of nutrients and trace elements is essential. At the same time, the content of peat in the soil mix should be reduced.

By reduction of the leaves, *Passiflora pergrandis* can produce racemes that are up to 30 cm long and often bear 6 or more flowers. The large flowers do not open completely, so the theoretically possible diameter of 15–16 cm is not reached. Sometimes the plant also produces solitary flowers. With the exception of the outer coronal series, *P. pergrandis* is in all parts larger than the closely related *P. ambigua*. The 2 species can be distinguished by the different coloration of their flowers.

Vine: glabrous, except inflorescence and ovary; stem: terete, striate; stipules: soon deciduous; petioles: 2–3 cm, with 2 large glands below middle; leaves: unlobed, narrowly ovate, 11–20 × 5–10 cm, coriaceous, entire; peduncles: 3–9 cm, solitary; bracts: ovate, 4–5 × 2.2–4 cm, entire with several stout glands, purplish, free; flowers: white and dark violet, 9–11 cm diam., several-flowered inflorescence, up to 30 cm long; sepals: white inside, oblong-ovate, 6–7.5 × 3.5–4 cm, aristate, awn 0.2–0.5 cm long; petals: white, oblong, 5.5–6 × 1.3–2 cm; corona: 3 series, banded dark violet and white, outer series ca. 0.2 cm long, 2nd series 4.8–5 cm long, inner series 0.1–0.2 cm near base of floral tube; ovary: ovate, densely tomentose; fruit: not seen.

Passiflora phoenicea Lindley

Bot. Reg. 19 (1833)
phoeniceus: purple, scarlet-red
Synonym: *P. alata* 'Ruby Glow'
Supersection *Laurifolia*, Series *Quadrangulares*

Passiflora phoenicea (Plate 217), a close relative of *P. alata*, grows wild in eastern Peru. Until a few years ago this independent species was known under the name *P. alata* 'Ruby Glow'. Though it has had a reputation as being very free-flowering, it does not often rise to such expectations. Its delightfully fragrant flowers appear in late summer and fall. The sepals and petals are deep red, while the long coronal filaments are banded violet, white, and purple. The diameter of a flower can reach 13 cm. The number of flowers set depends on the plant's location, which should be protected and warm, and on the size of the plant. It is advisable to grow *P. phoenicea* in a spacious container or to plant it directly into an earth bed in a greenhouse. Provided with a good root run it will develop into an imposing plant with shoots that are several meters long. When the roots have permeated the potting mix completely after 1 or 2 years and the plant has reached its optimum size, the number of flowers increases considerably.

Passiflora phoenicea is characterized by the 2, rarely 4, bright yellow nectar glands (butterfly egg mimics) in the upper part of its petiole, a feature not seen in *P. alata*, *P.* ×*decaisneana*, or *P. quadrangularis*.

This rather easy-to-grow species is suited for cultivation outdoors during summer or inside a greenhouse or conservatory. *Passiflora phoenicea* should be overwintered at temperatures above 10°C, although older specimens can tolerate temperatures down to the freezing point by losing the tips of their shoots.

Vine: glabrous; stem: 4-angular, the angles winged, stout; stipules: ovate, 1–1.2 × 0.3–0.4 cm, entire or slightly serrate; petioles: 3–4 cm, with

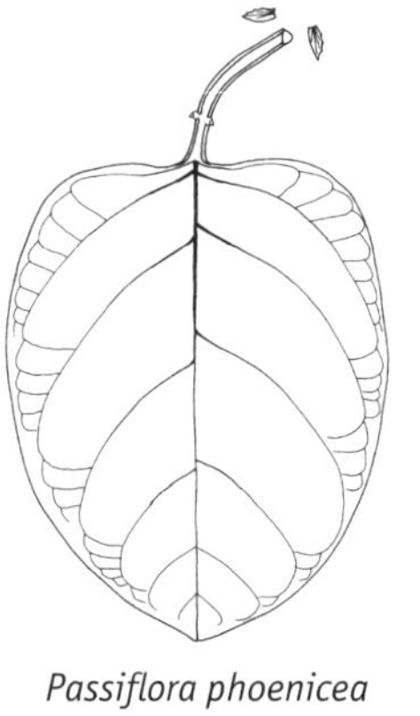

Passiflora phoenicea

2(–4) large, bright yellow glands near apex; leaves: unlobed, ovate, 10–17.5 × 7.5–15 cm, entire; peduncles: 2.5–3.5 cm, solitary, stout; bracts: ovate, 2–3.5 × 1.2–2 cm, glandular-serrate, free; flowers: purple-red, violet, and white, 11–13 cm diam.; sepals: purple-red inside, greenish outside, oblong-ovate, 4.5–5.5 × 2–3 cm, slightly awned below apex; petals: purple-red inside, oblong, 4.8–6 × 2–2.5 cm; corona: 5 or 6 series, banded purple and white at base, banded violet and white above, outer 2 series 4.5–5.5 cm long, inner series 0.1–0.3 cm long; ovary: ovoid, glabrous; fruit: ovoid, ca. 12.5 × 7.5 cm, yellow, edible.

Passiflora platyloba Killip

J. Wash. Acad. Sci. 12: 260 (1922)
platylobus: having a broad lobe
Supersection *Laurifolia*, Series *Tiliifolia*

The flowers of this species (Plate 218), which grows in the lowlands from Chiapas, Mexico, to Panama, hang down from the canopy of foliage. The sepals and petals of its fragrant flowers, up to 5 cm diam., are greenish with purple spots and reflex immediately upon opening. The large bracts, which surround the buds (and later the fruits) like a balloon, give the flower a particular charm. The corona is banded white and purple at its base, while the upper parts are banded violet and white.

Sometimes *Passiflora platyloba* var. *pubescens* (syn. *Passiflora williamsii*) is recognized. Very little is known of this variant from Panama and northern Colombia, however, and it is not recognized here.

Passiflora platyloba is closely related to *P. maliformis* and should also be cultivated in a greenhouse. These species will produce flowers only when they have reached a certain size. Cultivate *P. platyloba* in a large pot or plant it directly in the ground in a greenhouse. Under favorable conditions this species will flower reliably, unlike other species of the series *Tiliifolia* and *Laurifoliae*. At temperatures above 12°C it will get through winter without problems. Its epithet refers to the 3-lobed leaves that make it distinguishable from *P. maliformis*. The edible fruit has a hard shell. In order to set fruit, cross-pollination with another clone or another suitable species from the series *Tiliifolia* is required.

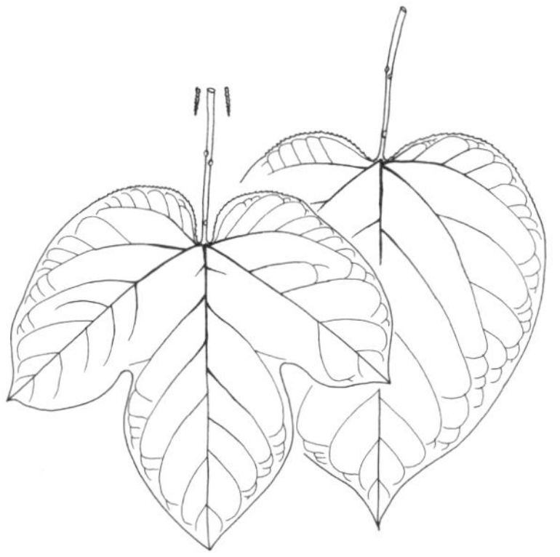

Passiflora platyloba

Vine: glabrous, except bracts; stem: terete, striate, stout; stipules: linear, 0.6–1.5 cm long, orange-yellow in upper 1/2, finely serrate, deciduous; petioles: 5–7 cm, with 2 sessile glands in lower 1/3; leaves: 3-lobed or rarely unlobed, 10–17 × 10–23 cm, membranous, entire; peduncles: 4–9.5 cm, solitary, stout; bracts: ovate, 4.5–8.5 × 3–6 cm, entire, connate, 1–2.5 cm at base; flowers: greenish purple, violet, and white, (4–)5–6 cm diam.; sepals: greenish, mottled with purple spots inside, green outside, oblong-lanceolate, 2–3.5 × 0.8–1 cm, keeled, keel terminating in an awn 0.5–1.2 cm long; petals: green, mottled with purple spots, linear-lanceolate, 1.5–2.8 × 0.5–0.7 cm; corona: several series, banded purple and white at base, banded violet and white above, outer series 0.3–1.2 cm long, 2nd series 2–3 cm, long, wavy in upper 1/2, inner series 0.1 cm long; ovary: ellipsoid, glabrous; fruit: 3.5–5.5 cm diam., pale greenish yellow, edible; seeds: 0.4–0.5 × 0.3–0.4 cm.

Passiflora popenovii Killip

J. Wash. Acad. Sci. 12: 332 (1922)
named after Wilson Popenoe, collector of the type plant and expert in tropical fruits
Supersection *Laurifolia*, Series *Laurifoliae*

This interesting species (Plate 219) belongs to the series *Laurifoliae*, whose representatives are characterized by entire, unlobed leaves, linear-lanceolate stipules, free bracts, and pendent flowers. As a rule they also produce tasty fruits. The yellow fruits of *Passiflora popenovii* are offered in local markets in the Andes of central and northern Ecuador, where this species grows at altitudes of 1500–2000 m. Its fragrant flowers are white and have a pur-

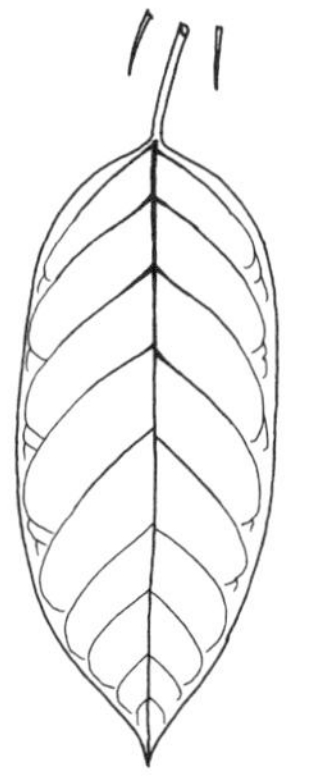

Passiflora popenovii

plish violet, white-banded corona. Flowers are 7–8 cm diam. The outermost coronal series is significantly shorter than the following one, indicating a close relationship to *P. laurifolia* and *P. nigradenia*. *Passiflora popenovii* can be distinguished from these species by its usually missing petiole glands and by its thinner, narrower leaves.

This species is under cultivation in Ecuador and Colombia but evidently has yet to be introduced into other collections. Judging from other cultivated species of series *Laurifoliae*, *Passiflora popenovii* can be expected to flower only as a large plant. However, it does not seem to need tropical conditions: it grows at moderate altitudes of the Andes and not in the lowlands, like the other species from this series.

Vine: glabrous, except ovary; stem: terete, striate; stipules: setaceous, ca. 1 cm long, soon deciduous; petioles: 1–2 cm, slender, without glands; leaves: unlobed, narrowly oval, 5.5–13 × 1.5–4 cm, papyraceous, entire; peduncles: 2–3 cm, solitary or in pairs; bracts: ovate, 2–2.5 × 1.2–1.5 cm, entire, free; flowers: white, purple, and violet, 7–8 cm diam.; sepals: white (or rose) inside, green outside, ovate-oblong, 3.5–4 × 1.5–1.8 cm, keeled, keel terminating in an awn ca. 0.4 cm long; petals: white, oblong, 2–3(–3.5) × 0.5–1 cm; corona: 4 series, banded purple and white at base, banded violet and white above, outer series ca. 1.5 cm long, 2nd series ca. 3 cm long, inner series 0.1 cm long; ovary: globose, white puberulent; fruit: ovoid, yellow, edible; seeds: obovoid, ca. 0.7 × 0.6 cm, tridentate, coarsely reticulate.

Passiflora quadrangularis L.

Syst. Nat. (ed. 10) 2: 1248 (1759)

quadrangulatus: having 4 angles (in reference to the stem)

Synonyms: *P. grandiflora* Salisb., *P. sulcata* Jacq., *P. tetragona* Roemer

Supersection *Laurifolia*, Series *Quadrangulares*

Little by little the genuine *Passiflora quadrangularis* (Plate 220) reenters the collections from which it had disappeared for many years. For a very long time *P.* ×*decaisneana* was erroneously regarded as *P. quadrangularis*. Although the error has been pointed out by John Vanderplank (1996a), it still true that quite a number of plants cultivated under the name *P. quadrangularis* are actually specimens of *P.* ×*decaisneana*, a hybrid between *P. alata* and *P. quadrangularis*. It is relatively easy to check whether a plant is the botanical species or the hybrid. In *P. quadrangularis* the inside of the sepals is white or at times violet or light pink, while the pale petals are a pink shade (Killip 1938). When the sepals and petals are an intense red this suggests that the plant is *P.* ×*decaisneana* or *P. alata*. Moreover, *P. quadrangularis* has significantly larger stipules, at a length of 2–4.5 cm. Further differences result from the coloration and structure of the corona. There are different varieties of *P. quadrangularis*, differing by geographic range, size of petioles, and coloration and structure of corona, and some of these varieties can be regarded as species of their own (Feuillet and MacDougal 1996). Among these is what we take to be *P. macrocarpa*, a variety from northern South America with light blue-violet and white flowers, often put in synonymy under *P.*

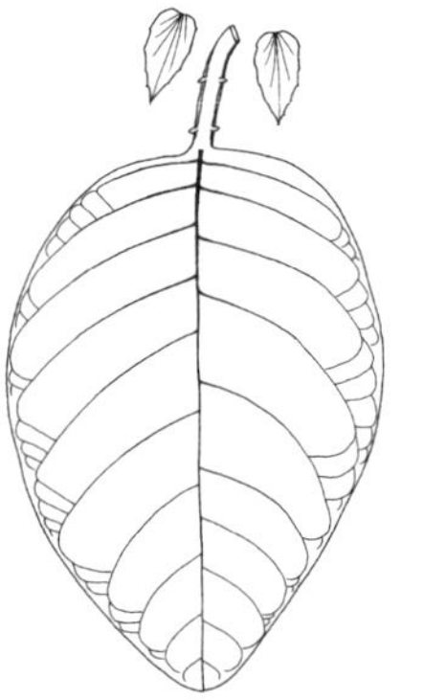

Passiflora quadrangularis

quadrangularis. The flower lacks any pink or purple, and the corona is more wide-spreading. This plant is grown for fruit in native gardens in Ecuador and other neighboring countries, as is *P. quadrangularis*. The plants traded in the 1990s as *P. quadrangularis* 'Warmlands' appear to be this species, and *P.* 'Blue Moon' is either this species or a hybrid with it. Despite its specific epithet, which means "large-fruited," the fruits of *P. macrocarpa*, though large, are no bigger than typical *P. quadrangularis* and are actually smaller than some forms.

Passiflora quadrangularis is among the most beautiful passionflowers. Because of its sumptuous flowers and tasty fruits, which are the largest of the whole genus at a weight of 1.5–4 kg, it is cultivated in many regions of the tropics. Its original native range is not exactly known but is surmised to have been in northwestern South America. A wild form with relatively small stipules is known to be native from Nicaragua to Colombia.

In comparison with *Passiflora* ×*decaisneana*, *P. quadrangularis* is more exacting and so should be cultivated in a warm greenhouse. When planted in a large container or directly in the ground it rapidly develops into a huge plant with shoots that are several meters long. At first the plant uses all its energy for vegetative growth. After a year or two it develops numerous flowers when temperatures are 25°–30°C and drop only slightly during the night. Winter temperatures should not fall below 12°C for a long period.

Vine: glabrous; stem: 4-angular, the angles winged, stout; stipules: ovate or ovate-lanceolate, 2–4 × 1–2 cm, entire or slightly serrate; petioles: 2–7 cm, with (2–)6(–8) large, mostly paired glands; leaves: unlobed, ovate, 8–20 × 7–15 cm, entire; peduncles: 1.5–4(–5.3) cm, solitary, stout; bracts: cordate-ovate, (2–)3–5.5 × (1–)2–4 cm, glandular, entire or serrate near base, free; flowers: pinkish, purple, violet, and white, (8–)10–12 cm diam.; sepals: whitish or pinkish white inside, green outside, oblong to ovate, 3–4.7 × 1.5–2.5 cm; petals: pale purple to reddish, oblong, 3–4.7 × 1–2 cm; corona: 5–6 series, banded purple and white at base, banded violet and white above, outer 2 series (4–)6–7 cm long, very wavy, inner series 0.1–0.4 cm long, the innermost series connate or irregularly connate into an operculum-like flap to 1 cm long; ovary: oblong or ovoid, glabrous; fruit: oblong-ovoid, often longitudinally grooved, 20–30 × 10–15 cm, yellow or greenish yellow, with a thick rind, edible; seeds: obcordate or suborbicular, 0.7–1 × 0.5–0.8 cm, strongly flattened, reticulate at center, radiately striate at margin.

Passiflora quadriglandulosa Rodschied

Med. Bem. Kol. Rio Essequebo 77 (1796 [1794])
quadriglandulosus: having 4 glands
Synonyms: *P. vitifolia* var. *minor* Mast., *P. yacumensis* Rusby
Supersection *Coccinea*

Passiflora quadriglandulosa (Plate 221), a beautiful and interesting species, is somewhat controversial in that it is sometimes difficult to distinguish from its closest relatives. It produces reddish flowers that are typical of supersection *Coccinea* and reach a diameter of 9–15 cm. As in other upright, hummingbird-pollinated flowers, the sepals and petals recline immediately upon opening so that the diameter is considerably reduced. *Passiflora quadriglandulosa* occurs frequently near the Amazon River and its tributaries. It is also found in some places in Venezuela, the Guianas, and Trinidad-Tobago that are not connected to the Amazon. Plate 221 was taken by Günter Gerlach near the Auyantepui River in Venezuela.

As the species belonging to this supersection are very similar, especially with regard to their flowers, distinctions are sometimes difficult. Characteristic of *Passiflora quadriglandulosa* are the mostly unlobed leaves that are slightly glabrous, although sometimes 3-lobed or asymmetrically 2-lobed

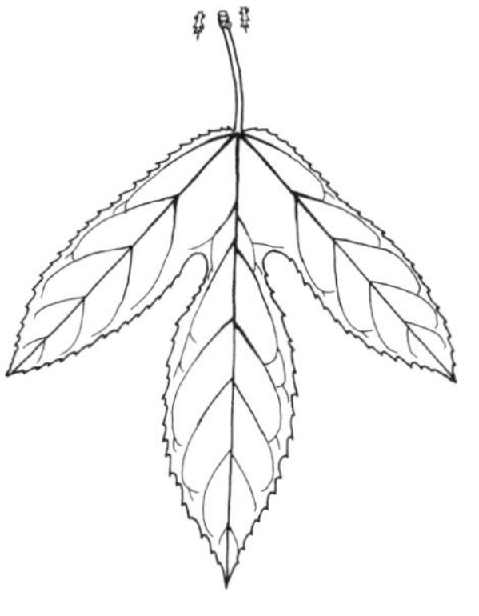

Passiflora quadriglandulosa

leaves can occur. Furthermore, the unusually small bracts and the trace of a red and white banding of the outermost coronal series are useful distinguishing criteria. Thus, a distinction from the standard clones of *P. vitifolia* that are under cultivation does not seem very difficult. However, in consideration of the variability of *P. vitifolia* in its assorted habitats, one discovers more things in common than one likes. Nevertheless *P. quadriglandulosa* can usually be distinguished from *P. coccinea* and *P. vitifolia*, 2 closely related species whose native ranges overlap with that of *P. quadriglandulosa*, by its combination of unlobed leaves and small bracts.

So far *Passiflora quadriglandulosa* has remained a rarity and little is known about its cultivation. It might be assumed that this tropical species can be cultivated only in a warm greenhouse. It is hoped that it becomes more widespread in the future.

Vine: nearly glabrous to slightly pubescent; stem: terete, occasionally angulate, striate; stipules: linear, 0.3–0.5 cm long, with 2 glands at base, soon deciduous, serrate; petioles: 1–2.5 cm, with 2 glands at base; leaves: unlobed, occasionally 3-lobed or asymmetrically 2-lobed, 6–10 × 3–5 cm, serrate; peduncles: 2.4–6 cm, solitary; bracts: linear or oblong-lanceolate, 1.1–1.7 × 0.3–0.6 cm, green to reddish green, with 2–6 glands at base, serrate, free; flowers: bright red, 9–15 cm diam.; sepals: red inside, red or greenish red outside, oblong-lanceolate, 4.2–6(–8) × 0.8–1.4 cm, aristate, awn 0.2–0.5 cm; petals: red, oblong-lanceolate, 3.4–5.4(–7.6) × 0.9–1.5 cm; corona: 3 series, outer series 1.2–1.8 cm long, slightly banded white and red, 2nd series 0.5–1 cm long, reddish, inner series 0.8–1.1 cm long, connate for 1/2–2/3 of their length; ovary: ovoid to ellipsoid, tomentose; fruit: ellipsoid, 4–7.2 × 2.5–5 cm, yellowish green with white spots; seeds: oblong-ovoid, ca. 0.8 × 0.4 cm, reticulate, black.

Passiflora racemosa Brot.

Trans. Linn. Soc. London 12: 71, t. 6 (1871)
racemosus: having clusters (in reference to the inflorescence)
Synonym: *P. princeps* Lodd.
Supersection *Passiflora*, Section *Calopathanthus*

This charming species (Plate 222), indigenous to the state of Rio de Janeiro, southeastern Brazil, is often cultivated in the showrooms of botanical gardens but also enjoys great popularity among enthusiasts. It is among the few species that produce very beautiful racemose inflorescences by reduction of the leaves. These racemes can bear 5–30 brilliant red flowers, very rarely a few more, of which only 3 at the utmost are open at one time. Flowers are 8–10 cm diam. In contrast to almost all passionflowers under cultivation that produce flower buds on the tips of the shoots, *Passiflora racemosa* produces its racemes as short lateral shoots on old wood. Pruning, therefore, should be undertaken with care; otherwise the plant will flower poorly in the next year, if at all. Characteristic of this species are the keel-like ridge that runs along the back of the sepals, the floral tube, which is about 1.5 cm long, and the white pollen. Because of its many peculiarities this passionflower occupies a poorly understood position with regard to systematics and is the only species within section *Calopathanthus*.

Its leaves are slightly asymmetric, leathery, and dark green. Propagation from cuttings is often difficult. Cuttings from the middle of a shoot are preferred, as cuttings that are either too soft or too hard root poorly. Sometimes weeks may pass before new growth appears.

The different clones under cultivation are distinguished mainly by the coloration of the outermost coronal series. The filaments are either pure white or white with a broad reddish purple band. A greenhouse or conservatory is an ideal location for *Passiflora racemosa*. It is also possible to grow it in a pot outdoors or on a windowsill, but the likelihood that it will produce flowers in such a location will

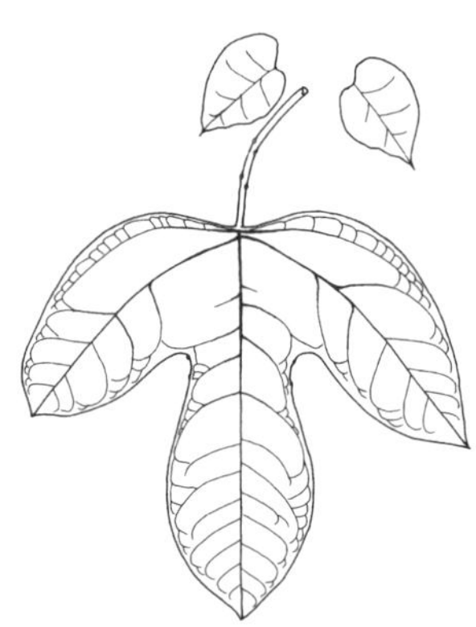

Passiflora racemosa

depend on the skill of the enthusiast. Although there have been a growing number of reports about plants that flowered in a pot on a windowsill or in a container outdoors, the warmth-loving *P. racemosa* will display all its splendor only when grown under glass. This plant will tolerate some nights with temperatures down to 7°C, but the overwintering temperature should stay above 12°C.

Vine: glabrous; stem: terete to subquadrangular, striate; stipules: broadly ovate, 1–1.5 × 0.7–1 cm, deciduous, entire; petioles: 2–4.2 cm, with 2–4 sessile glands; leaves: 3-lobed, occasionally unlobed or asymmetrically lobed, 6–12.5 × 6–14 cm, coriaceous, entire; pedicels: ca. 1 cm long, solitary or in pairs, born on pseudoracemose inflorescence, 25–75 cm long; bracts: ovate, 1–1.5 × 0.8–1 cm, reddish, membranous, entire, free; flowers: red with partly or pure white corona, 8–10 cm diam.; sepals: red inside and outside, oblong, 3.5–4.5 × 1–1.3 cm, keeled; petals: red, oblong, 3.2–4.3 × 1 cm; corona: 3 series, white and/or white with reddish purple band, outer series (0.3–)0.5–1.2 cm long, inner series 0.2–0.3 cm long, greenish; ovary: ovoid, glabrous; fruit: oblong, slightly angular, 5–7 × 1.5–3 cm, green; seeds: obovate, 0.5–0.6 × 0.3–0.4 cm, reticulate.

Passiflora recurva Mast.

Fl. Bras. 13 (1): 608 (1872)
recurvus: recurved (in reference to the peduncle)
Supersection *Passiflora*, Series *Passiflora*

The very rare *Passiflora recurva* (Plate 223) is another bat-flowered species. The native range of bat-flowered passionflowers of subgenus *Passiflora* shows that, apart from a few exceptions, they occur in 2 regions located between the 8th and 24th parallel of latitude south. Four presumably bat-flowered species from section *Manicata* occur in the Andes between central Peru and northern Bolivia. The region that hosts the most species, however, is located in the Atlantic coastal rain forest of eastern and southern Brazil. This is also the home of *P. recurva*. The type was collected in the state of Pernambuco, while further specimens have been found in the state of Bahia. Together with the Peruvian *P. peduncularis*, this species marks the northernmost occurrence of bat-flowered species from subgenus *Passiflora*.

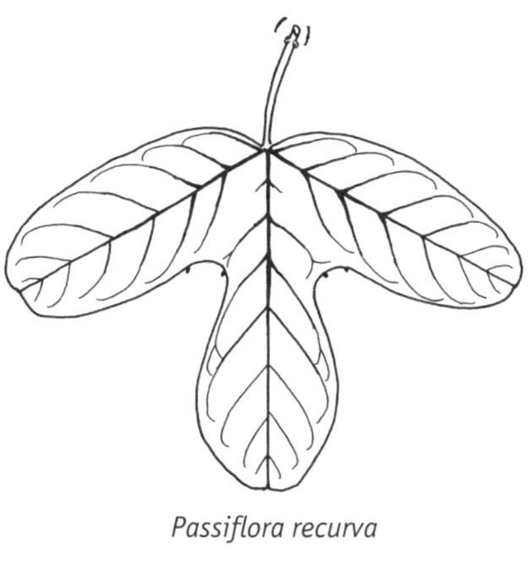

Passiflora recurva

Passiflora recurva produces white flowers that, at a diameter of 4–5 cm, are the smallest bat-pollinated flowers in the subgenus. The 2 coronal series are also white and do not cover the wide throat. The androgynophore is curved and pressed against the edge of the floral tube. The anthers, bearing a large amount of pollen, are directed toward the opposite side. *Passiflora recurva* can be distinguished from other bat-flowered species by its 3-lobed leaves with an entire margin and also by its tiny stipules. The peduncle is also remarkable, curved at the tip when the flowers are in bud. The open flowers are borne on straight peduncles. This species is not under cultivation.

Vine: glabrous; stem: terete; stipules: linear-lanceolate, about 0.3 cm long, soon deciduous; petioles: 1–2.5 cm long, with 2 sessile glands near base; leaves: 3-lobed, 4.5–8 × 5.5–10 cm, subcoriaceous, entire; peduncles: 4–5 cm, solitary, recurved; bracts: oblong-lanceolate, ca. 1 × 0.5 cm, subentire to slightly serrate, free; flowers: white, 4–5 cm diam.; sepals: greenish white outside, white inside, oblong, 1.5–2 × 0.5–0.7 cm, keeled, winged keel terminating in an awn 0.4–0.5 cm long; petals: white, linear-oblong, 1.3–1.8 × 0.4–0.5 cm; corona: 2 series, outer series ca. 1 cm long, white, inner series 0.2–0.3 cm long, white, erect; ovary: ovoid, glabrous; fruit: unknown.

Passiflora reflexiflora Cav.

Icon. 5: 15, t. 425 (1799)
reflexiflorus: having reflexed flowers
Synonyms: *P. lorifera* Mast., *P. reflexa* J. E. Sm.
Supersection *Stipulata*, Section *Tacsonioides*

The few known habitats of *Passiflora reflexiflora*

(Plate 224), endemic in Ecuador, are restricted to the surroundings of Port Guayaquil in the drainage of the Río Guaya. This region is also the rising commercial center of the country. *Passiflora reflexiflora* is extremely endangered, because its habitats are converted into industrial sites and agricultural acreage. Consequently, the loss of this species in its natural surroundings is imminent, and the species is an immediate conservation concern. Michael Schwerdtfeger is to be thanked for his success in bringing it into cultivation. This fantastic species has turned out to be an absolutely wonderful and easy-to-grow vine. It flowers even in a small pot and can be grown on a windowsill or outdoors in summer. Some problems can arise during the overwintering season in the temperate zone. The leaves become deformed and discolored and some fall off. If the plant is kept rather dry during this period it will recover the next spring. The minimum temperature for *P. reflexiflora* is around 10°C.

Since this species has come into cultivation, growers have tried many times to use it in breeding hybrids. So far, these experiments have been unsuccessful. Although *Passiflora reflexiflora* sets fruit easily after cross-pollination, the seeds are usually hollow.

The magnificent flowers, 5–7 cm diam., are borne by conspicuous peduncles up to 11 cm long. They are purplish red and upright to horizontal. The androgynophore juts far out, giving the flower a unique appearance. The specific epithet refers to the reflexed sepals and petals that veil most of the floral tube, which is 2–3 cm long. The corona in 2–4 series is violet to lilac, with white tips. The filaments of the outer rank reach a length of 0.3 cm. The pollen is whitish, which is true for only a few other species.

Vine: herbaceous, glabrous, membranous in most parts; stem: terete, striate, slender; stipules: subreniform, 1–2.2 × 0.5–1.3 cm, mucronulate, entire to slightly serrulate; petioles: slender, 1.5–5.3 cm, with 2–4 stipitate glands in upper 1/2; leaves: 3-lobed up to 2/3 of their length, 2.8–7.5 × 3–10 cm, entire or remotely serrulate; peduncles: 4–11 cm long, solitary or in pairs; bracts: ovate to ovate-oblong, 0.9–2.2 × 0.4–1.5 cm, entire, free; flowers: purplish red, 5–7 cm diam., about 6 cm long; floral tube: cylindrical, 1.9–3.1 cm long, about 0.7 cm wide, green; sepals: lanceolate, 2.7–4.3 × 0.5–0.8 cm, aristate, awn up to 0.5 cm long; petals: lanceolate, subequal to sepals; corona: 2–4 series, violet to lilac, green at base, outer series filamentous, 0.1–0.3 cm long; ovary: ovoid, green, glabrous; fruit: globose to ovoid, 4.6–5.6 × 4–4.5 cm, green; seeds: oblong-ovoid, 0.6–0.7 × 0.3–0.4 cm, tridentate at apex, coarsely reticulate, brown.

Passiflora resticulata Mast. & André

J. Linn. Soc. Bot. 20: 42 (1883)
resticulatus: having thin threads
Supersection *Stipulata*, Section *Granadillastrum*

This species (Plate 225), at home in Ecuador and Colombia, mainly occurs at altitudes of 1000–2500 m, although *Flora of Ecuador* (Holm-Nielsen et al. 1988) mentions findings from lower regions. Because of the partially upright, bushy corona, the flowers look slightly disheveled. They are about 5 cm diam. and white, while the thin coronal fila-

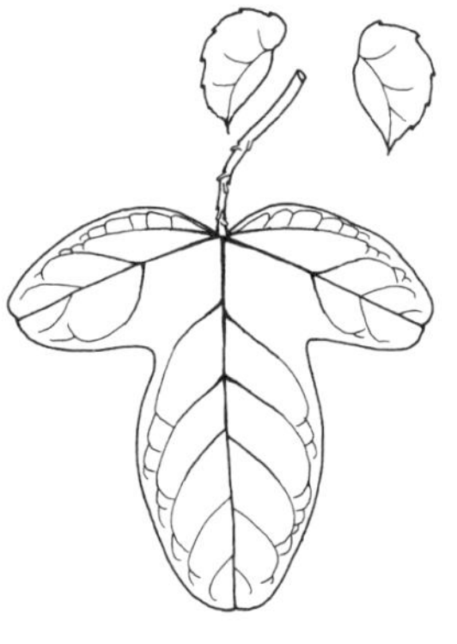

Passiflora reflexiflora

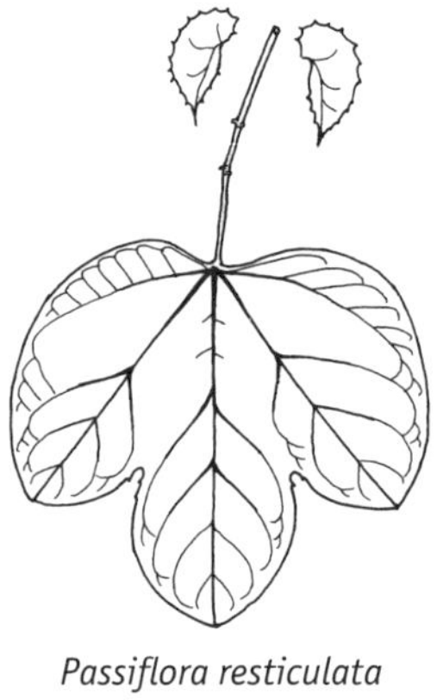

Passiflora resticulata

ments have a light violet banding. Extending from the keel on the back of the sepals are conspicuous awns that are 2–2.5 cm long. In cool weather the flowers stay open for 2 or even 3 days, most unusual in this section.

Passiflora resticulata is closely related to *P. sprucei* and *P. montana*. However, it can easily be distinguished from *P. sprucei*, in which the operculum stands out like an additional coronal series in the middle of the flower and surrounds the androgynophore at a distance of about 1 cm. The operculum is erect in *P. resticulata* and recurved in *P. montana*.

Passiflora resticulata is an excellent foodplant for *Heliconius* butterflies, and so far this easy-to-grow species has mainly been cultivated by butterfly enthusiasts. In a greenhouse it will begin flowering in July; outdoors, in August. It should be kept at temperatures above 10°C in winter, although short periods down to 5°C are tolerated.

Vine: glabrous; stem: terete, striate; stipules: reniform, 1–2.5 × 0.7–1 cm, glandular-serrate; petioles: slender, 2–5 cm, with 2–5 shortly stipitate or sessile glands; leaves: 3-lobed, 4–7.5 × 4.5–8 cm, entire or occasionally glandular-serrate; peduncles: 2.5–6 cm long, solitary, slender; bracts: ovate-oblong, 1.5–2 × 0.8–1 cm, glandular-serrate, free; flowers: white, 4–6 cm diam.; sepals: white inside, green outside, ovate-oblong, 1.5–2.5 × 0.5–1 cm, keeled, keel terminating in an awn 2–2.5 cm long; petals: white, oblong, 1–2.2 × 0.5–1 cm; corona: several series, very slender, white banded with violet, outer series (0.8–)1.5–2.2 cm long, inner series 0.2–0.5 cm long; ovary: ovoid, glabrous; fruit: globose to ovoid, 4–5 × 3 cm, green; seeds: ovoid, ca. 0.6–0.3 cm, flattened, reticulate.

Passiflora retipetala Mast.

Bull. Misc. Inform. Kew 1893: 12 (1893)
retipetalus: having reclined petals
Synonym: *P. lonchophora* Harms
Supersection *Stipulata*, Section *Granadillastrum*

Time and again it happens that plants are brought into circulation and cultivated under an erroneous name that passes from collector to collector, achieving a life of its own. This is the case with a clone collected by Lawrence E. Gilbert in the 1970s in Trinidad. It has been cultivated in recent years in the United States and Europe under the name of *Passiflora cyanea* though it is in fact *P. retipetala* (Plate 226). Although the 2 species look very similar, they differ with regard to the coloration of the corona (Herklots 1976, as *P. lonchophora*). While in *P. cyanea* the corona is purple or bluish at the base and white from the middle upward, it is exactly the other way around in *P. retipetala*. Moreover, in *P. retipetala*, the bracts are larger and the stipules have much longer tips. Adding to the story, Emil Kugler brought our attention to 2 articles by Noel Y. Sandwith. In *Icones Plantarum* (1956), Sandwith presented a detailed description of what was thought to be another species in this group, *P. lonchophora*, and worked out the differences with *P. cyanea*. In 1960, after comparative studies of the herbarium material, he declared *P. lonchophora* to be a synonym of the earlier name, *P. retipetala*. This seems to be rather confusing, but the result is that the correct name of the species cultivated in many enthusiasts' collections is not *P. cyanea* but *P. retipetala*.

The yellow-orange mimics of butterfly eggs on the filiform appendices of the stipules are a characteristic feature of *Passiflora retipetala*. Its natural habitats are in Trinidad and Tobago, the Guianas, and northern Brazil, where it occurs at altitudes of 500–1000 m. Similar to some other species from supersection *Passiflora*, it can flower in batches. The white sepals and petals of the very fragrant flowers, 6–8 cm diam., reflex completely. The filaments of the outer coronal series are blue-violet for about 1/2 of their length when the flower opens and turn purple-violet over the course of time. The

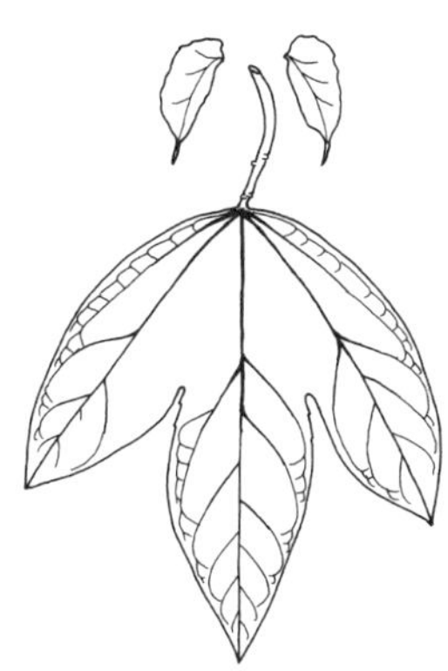

Passiflora retipetala

warmth-loving *P. retipetala* can be cultivated successfully in a protected, sunny location outdoors during summer. In winter, temperatures above 12°C are recommended.

Vine: glabrous; stem: subterete; stipules: ovate or ovate-oblong, (1–)1.5–3.5 × (0.6–)1–1.8 cm, with yellow tip 0.3–0.8 cm long at apex, entire; petioles: 3–4 cm, with 5–8 glands; leaves: 3- or unlobed, (6–)10–13 long, unlobed leaves 4–7 cm wide, lobed leaves 12–14 cm wide, entire; peduncles: 2.5–5 (–8.8) cm long, solitary; bracts: ovate-lanceolate, 2–2.7 × 0.9–1.5 cm, entire, free; flowers: white and bluish purple or violet, 6–8 cm diam.; sepals: white inside, green outside, oblong, 2–3 × 0.8–1 cm, slightly keeled outside, keel terminating in an awn 0.3–0.4 cm long; petals: white, oblong, 2.5–3.2 × 0.8–1.2 cm; corona: several series, white in lower 1/2, bluish purple or violet in upper 1/2, outer 2 series 1.6–2.5 cm long, inner series 0.2–0.6 cm long; ovary: ovoid to ellipsoid, glabrous; fruit: ovoid to ellipsoid, 6–7 × 4–5 cm.

Passiflora seemannii Griseb.

Bonplandia 6: 7 (1858)
named after botanist Berthold Carl Seemann
Synonym: *P. orbifolia* Planch. & Linden
Supersection *Laurifolia*, Series *Tiliifolia*

The native range of *Passiflora seemannii* (Plate 227) reaches from Chiapas, Mexico, through wide parts of Central America to Colombia, Venezuela, and Peru in northwestern South America. Because of its tasty fruit, it is occasionally cultivated in Mexico, Nicaragua, and Hawaii. *Passiflora seemannii* grows mainly in the warm lowlands but occasionally advances to altitudes of up to 1600 m in South America.

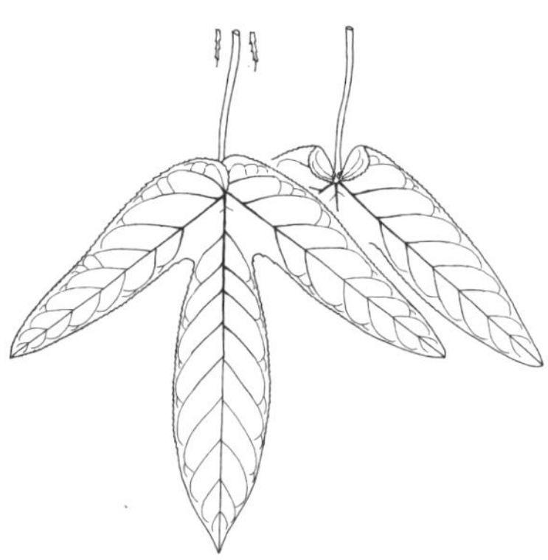

Passiflora seemannii

The fragrant flowers reach a diameter of 8–10 cm. They are more or less intensely purple, at times slightly bluish. The heart-shaped to egg-shaped leaves are mostly unlobed, though 3-lobed or asymmetrically 2-lobed leaves may also occur, especially on younger plants. *Passiflora seemannii* has been cultivated in the United States and Europe for some years. However, it is very difficult to get this species to flower under cultivation. Growth is often interrupted, and the plant can suffer, particularly in winter, so that it does not reach the necessary size to set flower buds. Jan Meerman reports that in Belize, *P. seemannii* is one of his most vigorous vines that flowers regularly. Conditions are certainly ideal in Belize, but it should be possible to imitate similar conditions in climatically less favorable zones. To cultivate *P. seemannii*, provide it with a heated greenhouse and additional lighting in winter. Instead of growing it in a container, plant it directly into an earth bed that can be heated by cables. If it does not flower under these conditions, assume the clone is reluctant to flower and replace it with a more free-flowering one.

Vine: glabrous; stem: terete, striate, glaucous; stipules: narrowly linear, 0.6–1.8 cm long, soon deciduous, serrate toward apex; petioles: 3–7 cm long, with 2 sessile glands at apex, occasionally 2 further glands near middle; leaves: unlobed, cordate-ovate, 6–15 × 5–15 cm, or occasionally 3- or unsymmetrically 2-lobed, membranous, glaucous, subentire; peduncles: 6–11 cm, pendent, solitary; bracts: cream, purple-tinged, ovate-lanceolate, 2.5–6 cm long, connate for about 1/2 of their length, entire; flowers: purple, purple-violet, and white, 8–10 cm diam.; sepals: white tinged with purple inside, ovate-lanceolate, 2.5–4 × 1–1.5 cm; petals: purple, oblong-lanceolate, 2.5–3.5 × 1.2 cm; corona: 2 series, banded purple and white at base, purple-violet and white above, outer series 1–1.7 cm long, 2nd series 2–3.6 cm long, succeeding rows of tubercles less than 0.1 cm long; ovary: ovoid, glabrous; fruit: ovoid, 4–8 × 2.5–5 cm, edible; seeds: orbicular-ovoid, 0.35–0.4 × 0.3 cm, reticulate.

Passiflora serratifolia L.

Sp. Pl. 955 (1753)
serratifolius: having serrate leaves
Synonym: *P. denticulata* Sessé & Mociño
Supersection *Passiflora*, Section *Passiflora*, Series *Passiflora*

This free-flowering species (Plate 228) is indigenous from Tamaulipas, Mexico, to central Costa Rica. Its colorful and very fragrant flowers reach a diameter of 5–7 cm. The intensely purple-violet corona, undulate and up to 3.5 cm long, covers the reclining sepals and petals almost completely. *Passiflora serratifolia* sets flower buds even in a small pot but can display all its splendor only when grown in a 30-cm container. In this situation the flower buds that have been building for several weeks can open at nearly the same time. Such gorgeous flowering batches, which last 2–3 days, occur only a few times a year. Between these big bangs the plant can flower quite normally, with flowers opening every day.

The ripe fruits of *Passiflora serratifolia* are sweet but quite rare under cultivation. This species does not fruit with its own pollen but needs cross-pollination. When a 2nd clone is not available, another species must be taken as a crossing partner. Oftentimes, even this method does not bring about the desired success. Apparently, most other species are not suited to pollinate *P. serratifolia*. This is also made clear by the fact that there are as yet no hybrids with *P. serratifolia* as a parent plant.

This species should be overwintered at temperatures above 10°C.

Vine: partly pubescent or hirtellous; stem: terete, striate; stipules: linear-subulate, 0.5–0.9 cm long, serrate toward apex; petioles: 5–8(–12.5) cm long with (4–)6–9 glands; leaves: unlobed, ovate to ovate-oblong, 5–15(–18.5) × 2–7.5(–9.5) cm, short-hirtellous beneath, serrate; peduncles: 3.5–11.5 cm, solitary; bracts: oblong-lanceolate, 2–3.8 × 0.6–1.5 cm, finely pubescent, serrate, free; flowers: purple, purple-violet, and white, 5–7 cm diam.; sepals: white, occasionally tinged with purple inside, green outside, lanceolate, 2.5–3.6 × 0.5–1 cm, keeled, keel terminating in an awn 0.2–1.2 cm long; petals: purple, oblong-lanceolate, 1.5–3.6 × 0.4–0.8 cm; corona: 4–8 series, deep purple with some whitish purple bands near base, purple or purplish violet above, white at apex, outer 2 series 2–4.5 cm long, wavy in upper 1/2, inner series tuberculate to 0.5 cm long, innermost series 0.5–1 cm long; ovary: ellipsoidal, glabrous or very rarely pubescent; fruit: ovoid or subglobose, 5–9 × 3.5–5.5 cm, lemon-yellow, edible; seeds: cuneate, 0.6–0.9 × 0.4–0.6 cm, flattened, tridentate at apex, reticulate.

Passiflora serratodigitata L.

Sp. Pl. 960 (1753)
serratus: serrate, *digitatus*: having fingers (in reference to the leaves)
Synonyms: *P. cearensis* Barb. Rodr., *P. digitata* L., *P. palmata* Lodd., *P. serrata* L.
Supersection *Laurifolia*, Series *Tiliifolia*

The native range of this vigorous species (Plate 229) reaches from easternmost Panama through the Chocó region of Colombia to eastern Ecuador,

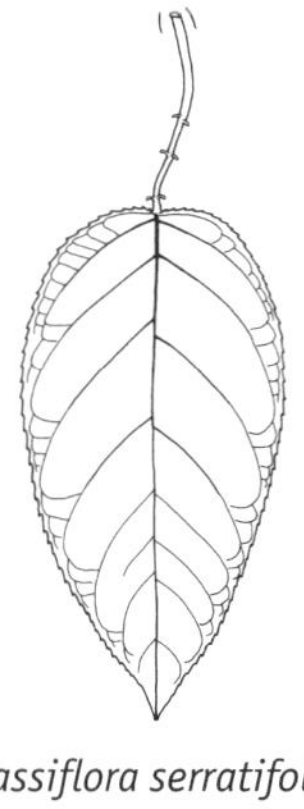

Passiflora serratifolia

Passiflora serratodigitata

Peru, Bolivia, and Brazil, and in the north of South America to Venezuela, the Guianas, and the West Indies, growing in the tropical lowlands of these countries. Occasionally it is grown for its edible fruit. The white aril, similar in taste to guavas, is covered by a yellow, brittle shell. By means of its large, palmately 5-lobed or 7-lobed leaves, it can easily be distinguished from all other species. It is most closely related to *Passiflora seemannii.*

Like most species from section *Laurifolia*—above all the well-known *Passiflora laurifolia, P. maliformis,* and *P. quadrangularis*—*P. serratodigitata* must reach a certain size before flower buds are set on shoots that are several meters long. Again, it is advisable to plant this species directly in the ground in a spacious heated greenhouse. With a good supply of nutrients, *P. serratodigitata* grows rapidly; and when provided temperatures of 25°–30°C in summer and a high atmospheric humidity, it produces numerous pendent flowers. These are bluish violet and reach a diameter of 7–9 cm. This species can display all its splendor only in the large showrooms of botanical gardens or similar institutions. Only a few enthusiasts have adequate facilities, so it is cultivated rarely. Overwintering temperatures should be at least 12°C, although 15°C is more suitable.

Vine: glabrous; stem: terete, striate; stipules: linear-subulate, 0.7–2 cm long, minutely glandular-serrate at apex; petioles: 4–11 cm long, with 3–7 funnel-shaped or ligulate glands; leaves: deeply 5- or 7-lobed, 8–15 × 13–18 cm, membranous, serrate; peduncles: 2–8.5 cm, solitary; bracts: ovate-lanceolate, 3–5 × 1.5–2 cm, white inside, subentire or serrate, connate up to 1/2 of their length; flowers: pale violet, bluish violet, and white, (6–)7–9 (–10) cm diam.; sepals: white tinged with violet inside, greenish white outside, oblong, (2.5–)3–4 × 1–1.5 cm, with a short awn 0.2–0.3 mm; petals: pale violet, oblong, 2–3 × 1–1.3 cm; corona: 3 series, deep bluish violet banded with white, white at apex, outer series 1.2–2 cm long, 2nd series 2–3.5 cm long, many rows of obscure tubercles are followed by an innermost series 0.1–0.2 cm long; ovary: ovoid to subglobose, glabrous; fruit: globose, 4–9 × 4–6 cm, yellow, edible; seeds: obovoid, 0.5–0.7 × 0.3–0.5 cm, reticulate, black.

Passiflora setacea DC.

Prodr. 3: 329 (1828)
setaceus: bristle-shaped
Synonym: *P. sururuca* Vell.
Supersection *Passiflora*, Series *Setacea*

The native range of *Passiflora setacea* (Plate 230) reaches from Bahia along the east coast of Brazil to Rio de Janeiro. A collection is also known from the more centrally located Mato Grosso. This species, quite frequent in the Atlantic rain forest, produces white flowers that reach a diameter of about 10 cm. It is characterized by the glands on the outside of the sepals and by the corona in only one series. The typically white filaments reach a length of about 1 cm. The brownish purple variegation on the outside of the otherwise greenish sepals is also remarkable. However, such a conspicuous variegation may also be absent, as can be seen in Plate 230, taken in the plant's habitat.

The floral ecology suggests that *Passiflora setacea* is pollinated by bats, a common phenomenon among passionflowers endemic in the Atlantic rain forest. The sturdy peduncles, 8–10 cm long, raise the flowers well out of the mass of foliage, enabling bats to pollinate the flowers during their flight.

Harms (1925) assigned *Passiflora setacea* to the monotypic subsection *Setacea* of subgenus *Passiflora* (formerly *Granadilla*), from where it was taken over by Killip as series *Setacea.* When Armando Carlos Cervi described *P. hatschbachii* in 1994, another species was added to this series. That rare species, only known from the state of Minas Gerais, is distinguished from *P. setacea* by 2 coronal series and a pubescent ovary.

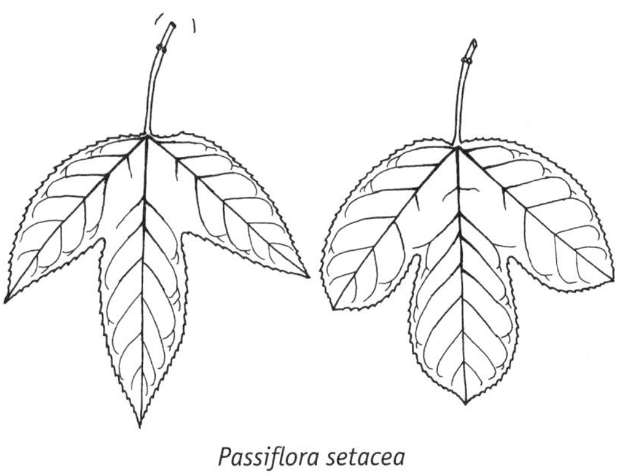

Passiflora setacea

Passiflora setacea does not seem to be under cultivation in Europe or the United States. Plants were seen recently by John MacDougal in an experimental fruit garden at an agricultural university in Costa Rica. It may be that it has also come into cultivation in Brazil. It is hoped that one day enthusiasts in other countries will also be able to enjoy this species.

Vine: brownish tomentose in parts; stem: terete; stipules: linear-lanceolate to setaceous, about 0.5 cm long, soon deciduous; petioles: up to 3 cm long, with 2 sessile glands near base, occasionally with an additional 3rd gland near middle; leaves: 3-lobed, 5–8 × 6–10 cm, nearly glabrous above, densely brown-tomentose beneath, serrate; peduncles: 8–10 cm, solitary, stout, often curved near apex; bracts: oblong-lanceolate, 1.5–2 × 0.5–0.8 cm, serrulate or laciniate-serrulate, free; floral tube: cylindrical-campanulate, about 1.5 cm long, green outside, conspicuously purple-lined without; flowers: white, 9–10 cm diam.; sepals: green, conspicuously purple-lined outside, white inside, oblong-lanceolate, 3.5–4 × 0.5–0.7 cm, keeled, keel terminating in an awn 1–1.5 cm long, bearing outside numerous curved, sessile glands; petals: white, oblong-lanceolate, 2–2.5 × 0.5–0.6 cm; corona: 1 series, white, about 1 cm long; ovary: ellipsoid, glabrous; fruit: ovoid, ca. 4 × 3 cm, coriaceous pericarp; seeds: obovate, ca. 0.5 × 0.3 cm, finely reticulate.

Passiflora speciosa Gardn.

Fielding and Gardn. Sert. Pl.: pl. 17 (1844)
speciosus: beautiful, showy
Supersection *Coccinea*

Passiflora speciosa (Plates 231–234) is a relatively little-known Brazilian species from supersection *Coccinea* with large flowers and slender sepals. While in this hummingbird-pollinated species the odor is nearly absent, other visual cues are important. An erect flower position and a variable intensity of coloration enable the main pollinator, *Phaethornis idaliae* (see Plate 36), to become aware of it at a great distance (Galarda Varassim et al. 2001). When the flower has completely unfolded, it reaches a diameter of 10–12 cm, although one report gives an astounding maximum of 22 cm (Frank and Gamma 2002). However, the sepals and petals quickly recline, so that the full diameter can only be seen immediately upon the opening of the flower at 5:30 a.m. and once again before it closes in the early evening.

Alwyn Howard Gentry (1981) wrote an interesting article about the distribution patterns of the *Passiflora vitifolia* complex. He noted that with the exception of *P. speciosa*, collections of all other species are known from the Amazonian lowlands even though the centers of their respective native ranges may be in other regions. This holds true for *P. vitifolia*, for example, which is very common in southern Central America. *Passiflora speciosa*, however, occurs mainly in coastal eastern Brazil, although a few collections are known from central and western Brazil near the border to Bolivia. As a result, the easiest criterion to distinguish *P. speciosa* from the closely related *P. vitifolia* is by the area of origin, or native provenance. Sometimes the fruit of *P. speciosa* differs by being greenish gray with a purple-brown mottling that is most intense at the base and along the 5 longitudinal stripes; fruits are poorly known, however. In comparison with *P. vitifolia*, the leaf lobes in *P. speciosa* are narrower, and in parts of Brazil the leaves are predominately 5-lobed. In view of the variability of *P. vitifolia*, other distinguishing features are difficult to define. *Passiflora speciosa* is not known to be under cultivation.

Vine: villosulous-tomentose nearly throughout; stem: terete, striate; stipules: setaceous, 0.4–0.5 cm long, soon deciduous; petioles: 1.5–2 cm, with 4 glands; leaves: 3-lobed, 4–12 × 4–10 cm, membranaceous, lustrous above, serrate; peduncles: 5–12 cm, solitary, stout; bracts: oblong, 2–4 × 0.8–1

Passiflora speciosa

cm, 2–4 glands at base, red, slightly serrate, glandular, free; flowers: bright red, scarlet, or crimson, 10–12(–22) cm diam.; sepals: reddish, narrowly oblong-lanceolate, 4–6(–9) × 0.8–1.4 cm, slightly keeled, keel terminating in an awn 0.3–0.5 cm long; petals: reddish, narrowly oblong-lanceolate, 4–6(–9) × 0.7–1 cm; corona: 2 series, outer series ca. 1.5 cm long, white or white with purplish band, filaments narrowly awl-shaped, inner series ca. 0.8 cm long, connate at base; ovary: ovoid, densely villosulous; fruit: ellipsoid-ovoid, ca. 6 × 2.5 cm, greenish gray, sometimes mottled with purplish brown mostly at base and at 5 longitudinal stripes; seeds: strongly compressed, ca. 0.5 × 0.3 cm, minutely reticulate.

Passiflora sprucei Mast.

Fl. Bras. 13 (1): 568 (1872)
named after Richard Spruce, collector of the type voucher
Supersection *Stipulata*, Section *Granadillastrum*

Since *Passiflora sprucei* (Plate 235) is at home in western Ecuador, where it mainly grows in the lowlands, it is astonishing that it has also been found at altitudes of up to 2300 m in the province of Loja, southeastern Ecuador. It may also occur in some parts of southern Colombia and northern Peru.

A clone of *Passiflora sprucei* with an unusually great number of petiole glands has been under cultivation for several years. This species will flourish in a greenhouse and can also be grown outdoors in summer. Flower bud formation will be earlier and more abundant when it is grown under glass. A plant that has reached a height of about 1 m can produce its first flowers. The more lateral shoots a plant has, the more flowers it can set, so that a 1.5- or 2-m specimen in a 30- or 40-cm pot can have a great number of them. The diameter of the rose-purple and white flowers is around 6 cm. Flower color intensity is quite variable. This species is characterized by its erect operculum, whose tip juts from the center of the flower like an additional coronal series and adjoins at a distance of up to 1 cm to the androgynophore. It differs thus from closely related species *P. resticulata* and *P. montana*.

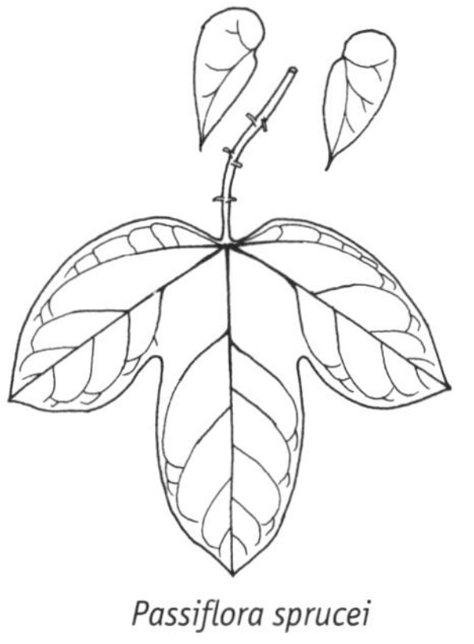

Passiflora sprucei

Vine: glabrous; stem: terete, striate; stipules: subreniform, 1–2.5(–3.5) × 0.5–1.5 cm, entire or glandular-serrate; petioles: 1.5–4.5(–7) cm, slender, with 3–6 short stipitate glands; leaves: 3-lobed or occasionally 4-lobed, 2–8 × 5–12 cm, entire; peduncles: 3–9 cm, solitary; bracts: ovate to oblong, 1.5 × 0.5–1 cm, glandular-dentate, white, free; flowers: pale rose-purple to purplish violet, 6–7 cm diam.; sepals: rose-purple inside, greenish white outside, oblong to lanceolate, 3–3.5 × 0.5–0.7 cm, aristate, awn ca. 0.5 cm; petals: rose-purple to purplish violet, linear-lanceolate, 2.7–3.3 × 0.4–0.5 cm; corona: 4 series, banded purple and white, outer 2 series 1–1.5 cm long, inner series 0.1–0.3 cm long; ovary: ovoid, glabrous; fruit: ovoid to subglobose, 3–4.5 × 2–3 cm, greenish yellow; seeds: ovoid, ca. 0.5 × 0.3 cm, flattened, reticulate.

Passiflora sublanceolata (Killip) MacDougal

Novon 14 (2004)
sublanceolata: nearly lance-shaped (in reference to the leaves)
Synonym: *P. palmeri* var. *sublanceolata* Killip
Supersection *Stipulata*, Section *Dysosmia*

This brilliantly flowered species (Plate 236) was considered a variety of *Passiflora palmeri* until very recently. That species is endemic to the deserts of Baja California and has a white flower with a short violet corona. The pollinator of the real *P. palmeri* is undocumented, but the plant now called *P. sublanceolata* is known to be pollinated by hummingbirds (Janzen 1968). This unusual species, distributed from Tabasco, Mexico, through the Yucatán Peninsula to northern Belize, has enriched the collections of many passionflower enthusiasts. The striking characteristic of this species is the intense pink or purple-red flower. In its natural habitats, this rather slender vine grows in swamp thickets on fine white clay. These soil conditions provide much moisture during the rainy season, while the soil dries out in the dry season. Nevertheless, it can be cultivated in a customary, standard potting mix that contains peat. However, in order to set flower buds, *P. sublanceolata* should be grown in a warm and sunny location, preferably under glass. It is among the few species that withstand hot spells without much damage. Under favorable conditions even young plants in a 12-cm pot can produce a few flowers, though plants will flower more reliably and persistently in a 20- or 25-cm pot. At a diameter of about 8 cm, the flowers are 2–4 times as large as those from other species from section *Dysosmia*, and they also have a very different shape and appearance. The corona is short. The blooms open very early in the morning, reaching their maximum expansion very soon before the sepals and petals reflex, and close in the early afternoon hours. Like the closely related *P. foetida*, the body of *P. sublanceolata* effuses a characteristic odor. When one touches the plant, one's hands become sticky from contact with the numerous gland-tipped hairs. In winter, temperatures around 15°C are recommended, although the plant will tolerate a few cooler nights.

Vine: slender, densely pilose or tomentose, except flowers; stem: terete; stipules: semiannular with central lobe ca. 1 cm long, deeply cleft into filiform, gland-tipped segments; petioles: 1–1.5 cm without glands, but with gland-tipped hairs; leaves: nearly unlobed to shallowly 3-lobed, 3–8.5 × 2–6 cm, coriaceous or subcoriaceous, serrate; peduncles: 4–8 cm, solitary; bracts: oblong, 1.5–3 × 0.7–2.5 cm, deeply 2-pinnatisect into gland-tipped segments, free; flowers: bright pink to purplish red, 7–8 cm diam.; sepals: pink or purplish red inside, greenish outside, linear, 3–4 × 0.5–0.8 cm, aristate, awn 0.3–0.8 cm; petals: pink or purplish red, linear, 3–4 × 0.4–0.6 cm; corona: ca. 5 series, white or white with a pink or purple band, outer series erect, 0.3–1 cm, inner series 0.1–0.5 cm; ovary: subglobose, densely tomentose; fruit: ovoid to widely ellipsoid, 3–4 × 2–3 cm, pilose, red; seeds: oblong or narrowly cuneate, ca. 0.5–0.6 × 0.2–0.3 cm, 3-toothed at apex, reticulate.

Passiflora subpeltata Ortega

Nov. Pl. Descr. Dec. 6: 78 (1798)
subpeltatus: slightly scutiform, having the shape of a shield (in reference to the leaf base)
Synonyms: *P. adenophylla* Mast., *P. alba* Link & Otto, *P. atomaria* Planch.
Supersection *Stipulata*, Section *Granadillastrum*

Passiflora subpeltata (Plate 237) is undoubtedly among the most rewarding passionflowers. Unlike most species of subgenus *Passiflora*, it can be cultivated successfully in a small pot on a windowsill. Many more of its brilliant white flowers can be expected when it is placed outdoors during summer. In many cases it will have already set flower buds when it is taken out in spring, and by June the first of its flowers will appear, about 5 cm diam. With a good supply of nutrients, *P. subpeltata* can produce a fruit from every flower by self-pollination. However, too many fruits sap the plant of the energy needed to set new flower buds, so it is advisable to

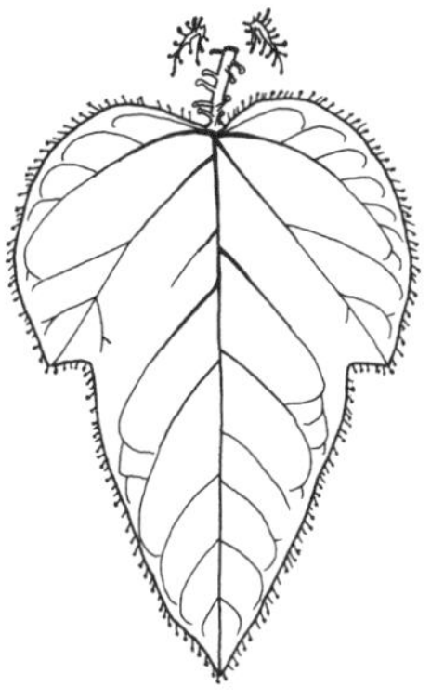

Passiflora sublanceolata

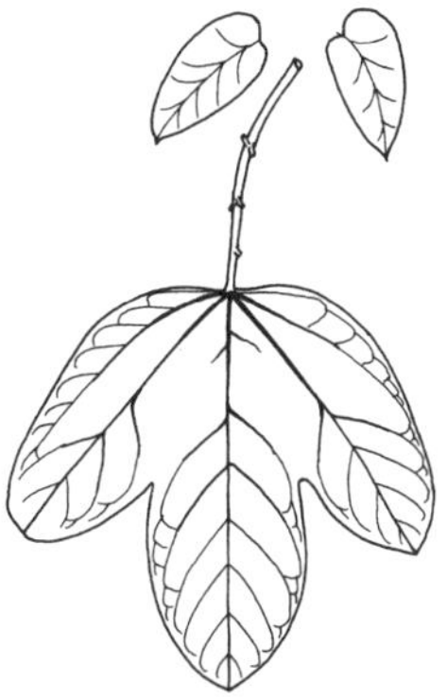

Passiflora subpeltata

remove some of them. When fruits are ripe, they turn yellowish green. In winter *P. subpeltata* can be kept rather cool as long as temperatures do not fall below 7°C.

Passiflora subpeltata is at home from central Mexico to Guatemala, where it grows in mountainous, seasonally dry habitats at 1100–3100 m. The flowers are almost always pure white, with the outermost coronal row noticeably the longest. A closely related species that parades under the name *P. subpeltata* (but is not that species) occurs from Panama to Colombia and Venezuela from sea level to 1000 m elevation. The lowland species has a purple-banded corona, 2 outer coronal series that are subequal, more glandular stipules, larger bracts, and seeds that are tridentate at the end. It is unclear if this species has been named yet—is it *P. atomaria*?—but it is not *P. subpeltata*.

Passiflora subpeltata is often confused with the closely related *P. eichleriana*, from which it is difficult to distinguish. However, *P. eichleriana* is at home in the Southern Hemisphere, in southeastern Brazil and Paraguay.

Vine: glabrous; stem: terete, striate; stipules: semioblong, 1.5–4 × 1–2.5 cm long, entire, glandular-crenulate at base; petioles: 4–6 cm long, slender, with 2–4(–6) minute, ligulate glands; leaves: 3-lobed, 4–10(–12) × 5–12(–15) cm, membranous, entire; peduncles: 3–6 cm, solitary, rarely in pairs; bracts: ovate-oblong, 1–1.5 × 0.7–1 cm, minutely serrate at base, free; flowers: white, 4–5.5 cm diam.; sepals: white inside, green outside, oblong, 2–3 × 0.5–1 cm, keeled, keel terminating in an awn 0.6–1.3 long; petals: white, linear-oblong, 2–3 × 0.4–0.7 cm; corona: 5 series, white, outer 2 series 1.5–2 cm long, the outermost definitely longest, inner series 0.2–0.4 cm long; ovary: ovoid, glabrous; fruit: ellipsoid or subglobose, 4–6 × 2.5–4 cm, greenish yellow; seeds: obovate, ca. 0.4–0.5 × 0.3 cm, flattened, finely reticulate.

Passiflora tarapotina Harms

Verh. Bot. Vereins Prov. Brandenb. 48: 185 (1906)
named after Tarapoto, Peru, location of the type species
Supersection *Stipulata*, Section *Tacsonioides*

The small native range of this charming little species (Plate 238) is around the town of Tarapoto in northern Peru. There it grows in the gradual transitions between the westernmost foothills of the Andes at altitudes of 300–500 m and in the adjoining Amazonian lowlands.

Passiflora tarapotina produces very long peduncles, up to 17 cm. Despite the considerable length of these relatively thin peduncles, the flowers stand upright. The long awns on the back of the sepals are remarkable, as are the serrate bracts. In addition to the normally solitary flowers near the tips of the long stems, there are conspicuous short racemes of flowers that sometimes appear on lateral shoots of the main stem; these are produced by reduction of the leaves on the short shoot.

Passiflora tarapotina belongs to the small group of species within supersection *Passiflora* that are pollinated by hummingbirds. In most cases the flowers of these species are reddish; they have a long, cylindrical floral tube, short coronal series, a relatively long androgynophore that juts from the flower, and more or less reflexed sepals and petals. Unfortunately, this little-known species is not yet under cultivation.

Vine: herbaceous, glabrous, membranous in most parts; stem: terete, striate, slender; stipules: subreniform to ovate-oblong, 1.4–2.5 × 0.5–1 cm, mucronulate, entire to slightly serrate; petioles: slender, 1.5–4.3 cm, with 2 stipitate glands near middle; leaves: 3-lobed up to 2/3 of their length, 4.7–8 × 5.5–10.7 cm, entire; peduncles: 8.5–17 cm long, slender, solitary, erect; bracts: ovate-oblong to oblong-lanceolate, 1–1.5 × 0.3–0.6 cm, serrate, free; flowers: red, 4–5 cm diam., about 5 cm long; floral

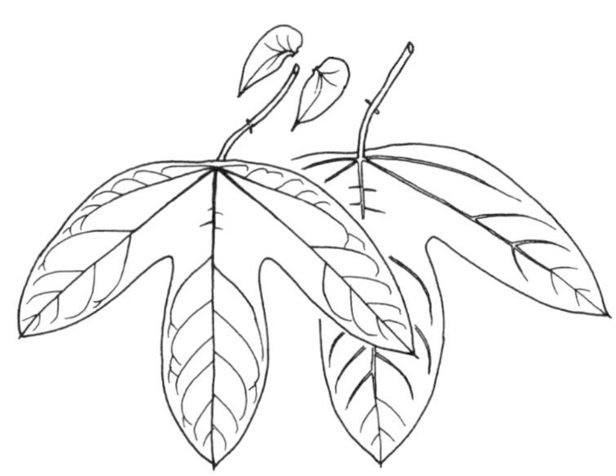

Passiflora tarapotina

tube: cylindrical, 2–2.5 cm long, about 0.6 cm wide; sepals: lanceolate, 2.5 × 0.5–0.6 cm, keeled, keel terminating in an awn 0.7–1 cm long; petals: lanceolate, subequal to sepals; corona: 3 or 4 series, violet, outer series filamentous, 0.1–0.2 cm long; ovary: ovoid, glabrous; fruit: not seen.

Passiflora tenuifila Killip

J. Wash. Acad. Sci. 17: 430 (1927)
tenuifilus: having thin coronal filaments
Supersection *Stipulata*, Section *Granadillastrum*

The native range of *Passiflora tenuifila* (Plate 239) reaches from southern Brazil and Bolivia to Paraguay and northern Argentina. Several years ago this species was brought into cultivation in the Botanical Garden of Bochum in Germany and later in many parts of Europe. It is an unpretentious and very easy-to-grow species. It flowers even as a small plant and can be kept on a windowsill. Grown in a 30-cm container outdoors during summer, it can develop optimally and will produce countless flowers. In central Europe the flowering season starts when it is taken outdoors in May and continues until October. Its rather inconspicuous flowers are borne on horizontal peduncles that are up to 8 cm long. The white flowers with a light violet, seamed corona can reach a diameter of up to 5 cm and characterize it as a small-flowered representative of subgenus *Passiflora*. It owes its specific epithet to the fine filaments on the outermost and innermost coronal series. *Passiflora tenuifila* is self-fertile and produces numerous fruits that hang on the plant alongside additional flowers. However, the plant should be prevented from setting too many fruits, as this costs a great deal of energy, resulting in fewer flower buds.

Overwintering is unproblematic. Although this species stands temperatures down to the freezing point for a short period of time, temperatures should not fall below 7°C.

Vine: glabrous; stem: terete or subangular; stipules: semioblong or subreniform, (1–)2.5–4 × (0.5–)1.2–2 cm, subentire; petioles: 1.5–5(–11) cm, with 2–6(–8) glands; leaves: 3-lobed, 3.5–9 × 5–14 cm, membranous, entire; peduncles: 2–11 cm long, slender, solitary; bracts: cordate-ovate, 0.8–1.8 × 0.5–1.3 cm, glandular-serrate at base, free; flowers: white and pale violet, 4–5 cm diam.; sepals: white inside, green outside, linear-oblong, (1–)1.5–2 × 0.4–0.8 cm, keeled, keel terminating in an awn 0.4–0.8 cm long; petals: white, linear-oblong, (0.7–)1–1.8 × 0.3–0.5 cm; corona: 4–5 series, white and pale violet at apex, outer 2 series 0.4–0.8 cm long, inner series 0.2–0.5 cm long; ovary: ovoid, subglobose, or ellipsoid, glaucous; fruit: subglobose, 3.5–6 × 3.5–5 cm, greenish yellow to yellow; seeds: oblong-obovoid, 0.5–0.6 × 0.3–0.4 cm, closely reticulate.

Passiflora trialata Feuillet & MacDougal

Novon 6: 351, figs. 1, 2 (1996)
trialatus: 3-edged (in reference to the stem)
Supersection *Laurifolia*, Series *Quadrangulares*

Passiflora trialata (Plate 240) was discovered by Christian Feuillet during his field studies in French Guiana in 1988. At an altitude of 450 m on the northern summit of Montagnes Torture, he found 2 sterile plants. Some years later he took cuttings

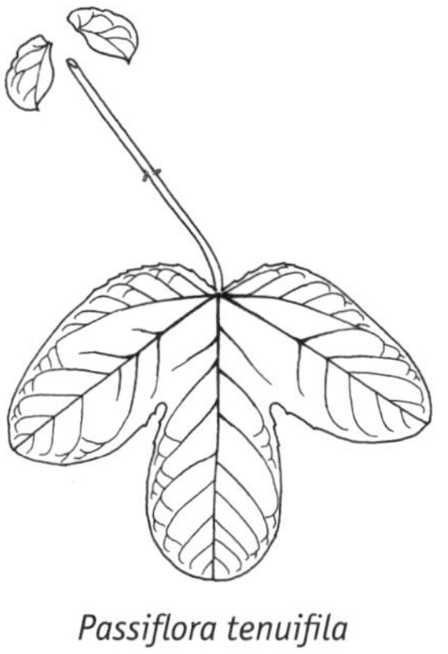

Passiflora tenuifila

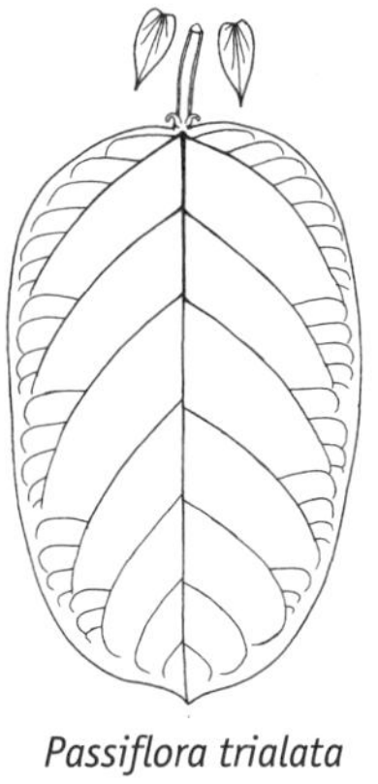

Passiflora trialata

from another sterile plant and succeeded in bringing them into cultivation in the United States. The first flower appeared in 1994 on a cutting given to and grown by John MacDougal at the Missouri Botanical Garden. Detailed information of this flower led to the first complete description. *Passiflora trialata* has one of the largest, most beautiful flowers of the whole genus. The bracts are gigantic. The sepals and petals are whitish with purple-red mottles and contrast with the violet, white-banded corona. The pendent flowers reach a diameter of up to 15 cm, although they are often somewhat smaller. Feuillet and MacDougal describe the floral odor as spicy, like anise with terpene overtones.

Together with *Passiflora alata*, *P. phoenicea*, *P. quadrangularis*, *P. macrocarpa*, and a couple of other species, *P. trialata* belongs to the series *Quadrangulares*. It is easily distinguished from the other species by its 3-angled stem, gigantic bracts, and flower color.

Passiflora trialata has been cultivated with varying success by enthusiasts in Europe and the United States. Its cultivation is an art that requires much in the way of instinct. Like most passionflowers from the same region in French Guiana, especially *P. amoena*, *P. cirrhiflora*, and *P. glandulosa*, *P. trialata* needs much warmth, a high atmospheric humidity, and at least 2 or 3 years to reach the necessary size to flower. A prerequisite for successful cultivation is a heated greenhouse, where it should be planted in an earth bed. In winter, temperatures should not fall below 15°C. Additional lighting is recommended.

Vine: glabrous; stem: sharply 3-angular, becoming 4-angular on very old plants, the angles winged; stipules: ovate to oblong-ovate, 2.3–3.2 × 0.8–1.3 cm, entire; petioles: 3-angular, 2.8–5 cm, with 2 large pale yellow or whitish glands near apex; leaves: unlobed, ovate, 15–26.5 × 8.5–16 cm, entire; peduncles: 4–6(–7.5) cm, solitary, 3-angular; bracts: ovate-triangular, (6–)7–8.5(–9.4) × (3.5–)4.5–6.5 cm, yellowish green, entire, free; flowers: purple-red, violet, and white, 13–15 cm diam.; sepals: whitish, densely mottled with purple-red spots inside, yellowish green outside, ovate-triangular, 4.4–6 × 1.8–2.5 cm; petals: more or less intense purplish red, oblong-lanceolate, 4.5–6.1 × 1.3–1.7 cm; corona: 10–12 series, banded purple and white at base, violet and white above, white at apex, outer 2 series 5.5–8 cm long, wavy at apex, inner series up to 0.3 cm long; ovary: ellipsoid, glabrous, cream to whitish; fruit: unknown.

Passiflora triloba Ruíz & Pavón ex DC.

Prodr. 3: 330 (1828)
trilobus: 3-lobed (in reference to the leaves]
Synonym: *P. colubrina* Poepp. & Endl.
Supersection *Laurifolia*, Series *Tiliifolia*

Passiflora triloba (Plate 241), at home in Peru and Bolivia, may have the most beautiful flowers of the series *Tiliifolia*. Its habitats extend from the moist eastern foothills of the Andes up to 1000 m down to the tropical lowlands. Its bright purple-red to red sepals and petals contrast with the corona, which is banded violet and white. The flower, reaching a length of about 8 cm, is a little reminiscent of *P. alata*. The number of sepals and petals is unique within the whole genus. Instead of the usual 5, *P. triloba* has 6 of each. All clones that are in cultivation have this characteristic. Whether it applies to the whole species is not known.

Passiflora triloba has been cultivated in enthusiasts' collections for a few years. It is a vigorous vine with thick, imposing leaves that are either unlobed or 3-lobed. Like all species from the series *Tiliifolia*, it needs a good root run to develop its vigorousness. A 50-cm container is a good choice, but it is even better to plant it directly into an earth bed in a greenhouse. Although experience is limited regarding its cultivation, this plant should not be too difficult to induce to flower. It thrives optimally with day temperatures around 25°C and nightly drops to 15°C. Winter temperatures above 12°C are

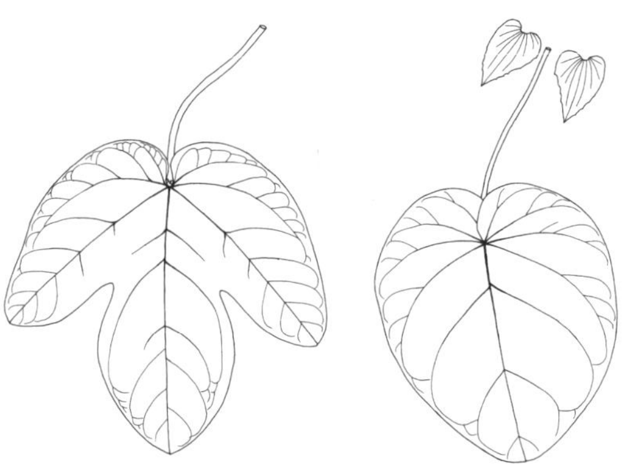

Passiflora triloba

necessary, although temperatures around 5°C are tolerated for a short period.

Vine: glabrous; stem: terete, stout; stipules: ovate-lanceolate, 2–3 × 1.5–2.5 cm, minutely serrate; petioles: up to 10 cm long, with 2 stipitate glands at apex; leaves: unlobed or 3-lobed, 8–17 × 9–20 cm, thick-membranous, glaucous beneath, minutely denticulate; peduncles: up to 5 cm, solitary; bracts: ovate, 4.5–7 × 3–5.5 cm, glandular-serrate, connate at very base; flowers: purple-red to red, 7–9 cm diam.; sepals: densely mottled with purple-red spots inside, green outside, oblong-lanceolate, 3–4 × 1–1.8 cm; petals: purplish red, oblong-lanceolate, 2.5–3.5 × 1–1.4 cm; corona: 3 series, banded purple-violet and white at base, violet and white above, outer series 1–1.5 cm long, 2nd series 2.5–3 cm long; ovary: ovoid, glabrous; fruit: unknown.

Passiflora trisulca Mast.

Bot. Jahrb. Syst. 8: 218 (1887)
trisulcus: having 3 grooves
Supersection *Stipulata*, Section *Granadillastrum*

Passiflora trisulca (Plate 242) occurs at moderate altitudes around 1500 m on the Western and Central Cordillera in Colombia. Particular characteristics of this species are the curved petioles and the operculum whose tip juts from the center of the flower like an additional coronal series.

In 1989, John MacDougal succeeded in bringing this species, with a general relationship to the *Passiflora oerstedii* complex, into cultivation in the United States. At first the distributed clones were erroneously thought to be specimens of *P. lehmannii*, which grows on the Eastern Cordillera of Colombia. A distinction from *P. trisulca* is made by its differently shaped, less deeply lobed leaves and shorter peduncles. The mistake was noticed only after a few years, the plants having already been passed on under the erroneous name. Presumably, all plants currently cultivated under the name *P. lehmannii* are actually *P. trisulca*, including those that are increasingly grown in Europe.

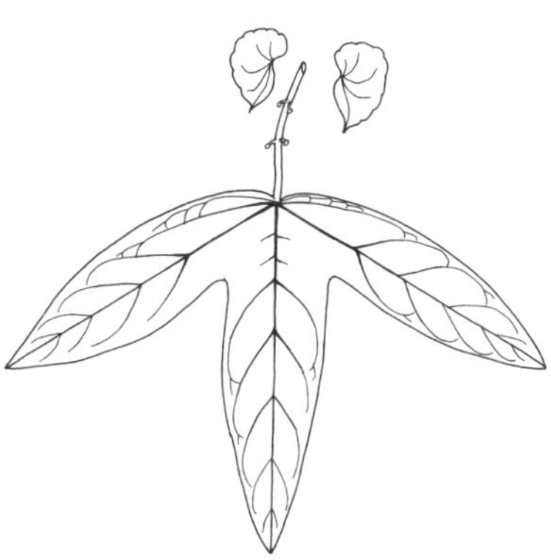

Passiflora trisulca

At a diameter of 4–6 cm, the white and purple flowers of this species are among the less spectacular of the genus. Such a typical form and flower color can be found in dozens of species from subgenus *Passiflora*.

Its demands are comparable with those of *Passiflora gritensis*. Outdoors during summer, *P. trisulca* prefers a location with some shade, finding excessive sunshine and temperatures beyond 30°C disagreeable. Grown in a 20- or 30-cm pot, it will reach the necessary size at which it can flower. The overwintering temperature should be kept beyond 10°C, although temperatures down to the freezing point will be tolerated.

Vine: glabrous; stem: terete, pale reddish brown; stipules: semiovate, 1.5–3 × 0.5–1.5 cm, entire; petioles: 3.5–5 cm, with 4 liguliform, often curved glands, 0.1–0.2 cm long; leaves: 3-lobed, 8–13 × 12–16 cm, membranous or subcoriaceous, entire; peduncles: 6–7 cm, solitary; bracts: ca. 0.8 × 0.2–0.3 cm, deciduous; flowers: white and purple or reddish purple, 4–6 cm diam.; sepals: white inside, green outside, oblong, 2–2.8 × 0.7–1 cm; petals: white or occasionally white with pale purple near margin, narrowly oblong, 2–2.5 × 0.5–0.7 cm; corona: several series, banded purple and white, purplish violet in upper 1/3, paler at very apex, filaments of 2 outer series 1–1.8 cm long, filaments of inner series 0.2–0.5 cm long, purple with white apex; ovary: ellipsoidal, glaucous; fruit: unknown.

Passiflora tucumanensis Hook.

Bot. Mag. 65: pl. 3636 (1839)
named after Tucuman, Argentina, the type location
Synonym: *P. naviculata* Griseb.
Supersection *Stipulata*, Section *Granadillastrum*

Recent studies have shown that the much-cultivated *Passiflora naviculata*, well known among enthusiasts, is conspecific with *P. tucumanensis* (Plate 243) (Deginani 2001). In accordance with the regulations of nomenclature, *P. naviculata*, first described in 1874, was declared a synonym of *P. tucumanensis*, which had been described some decades earlier.

The native range of this species reaches from northwestern Argentina to southern Bolivia and Paraguay, where it proceeds to altitudes of up to 2800 m. Although it is said to have no petiole glands (Killip 1938), a reddish brown pair can occasionally be found at the apex of the petiole.

Under cultivation *Passiflora tucumanensis* has turned out to be a rewarding and recommendable species. Even when it is grown in a small pot on a windowsill, it will produce pendent flowers 3–5.5 cm diam. The sepals and petals are white, and the campanulate corona is banded violet and white. The plant effuses a characteristic odor when its leaves are touched.

Passiflora tucumanensis can also be cultivated successfully as a robust container plant outdoors. A 1.5- or 2-m plant in a 30-cm container can produce more than a hundred flowers over the course of summer. Moreover, it is remarkably resistant to cold. It will survive mild winters with slight frosts when the rootstock has some protection. At any rate, temperatures beyond 5°C are more agreeable. Because of its many positive qualities, it has often been used in crossbreeding experiments, and a number of hybrids have become established during the last several years.

Vine: glabrous; stem: terete or subangular; stipules: semiovate, 1.2–3 × 0.5–1.6 cm, glandular-serrate; petioles: up to 3 cm long without glands or rarely 2 glands paired at apex; leaves: 3-lobed or rarely 5-lobed, 2.5–6(–9) × 2–8(–13) cm, membranous, entire, finely serrate at base; peduncles: 1.5–3 cm, slender, pendent, solitary or occasionally in pairs; bracts: cordate, 0.5–2 × 0.5–1.5 cm, serrate, free; flowers: white and violet, 3–5.5 cm diam.; sepals: white inside, green outside, oblong-lanceolate, 1–2.8 × 0.5–0.8 cm, aristate, awn 0.5–1.4 cm long; petals: white, oblong, 1–2.6 × 0.4–0.7 cm; corona: 4–5 series, banded violet and white, filaments of outer 2 series ca. 0.6–2 cm long, filaments of inner series 0.1–0.3 cm long; ovary: subglobose, glabrous; fruit: ovoid to globose, 2–5 × 1–4 cm, yellowish; seeds: obovate, 0.3–0.5 × 0.2–0.4 cm, reticulate.

Passiflora umbilicata (Griseb.) Harms

Nat. Pflanzenfam. 3 (6a): 91 (1893)
umbilicatus: resembling a navel
Synonym: *P. ianthina* Mast.
Supersection *Stipulata*, Section *Tacsonioides*

This magnificent species (Plate 244) hails from the high altitudes beyond 2500 m in Argentina, Bolivia, and Paraguay. Its long floral tube, in combination with a short, upright corona and widely reflexed sepals and petals, indicate that it is pollinated by hummingbirds. With its violet to dark violet flowers that contrast with its reddish purple bracts, *Passiflora umbilicata* is exceptional within the group of hummingbird-pollinated species characterized by pink and red flowers. When the perianth is reflexed, the diameter of the flower is about 4 cm. At times the upper end of the floral tube is torn apart when the sepals and petals reflex and build up too much tension.

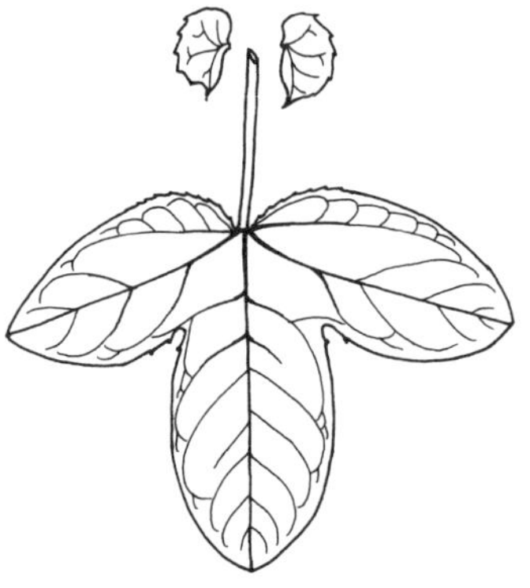

Passiflora tucumanensis

Passiflora umbilicata

The occurrence of *Passiflora umbilicata* in the highlands of the Andes, its long floral tube, and its short corona all suggest an affinity to the species of supersection *Tacsonia*. However, the entire leaves, the form of the stipules, and the corona in several series suggest a closer affiliation to supersection *Stipulata*. Apart from the fact that it is pollinated by hummingbirds, it does not have much in common with the quite heterogeneous species from section *Tacsonioides*, which grow in lower regions.

Under cultivation, *Passiflora umbilicata* has turned out to be free-flowering and relatively easy. It should be cultivated outdoors during summer, as temperatures under glass are too high for such a high-altitude species. Nevertheless, it stands short periods of more than 30°C without problems. A 25-cm pot is sufficient to let the plant reach a size at which it can flower. When provided a good supply of nutrients, *P. umbilicata* is able to produce up to a hundred flowers in one season. Because it is self-fertile, it can also set some fruit. Unfortunately, since propagation by cuttings is difficult, it can be difficult to find a specimen of *P. umbilicata*.

Vine: glabrous; stem: subterete, striate; stipules: semiovate, 1–2 × 0.5–1.5 cm, glandular-serrate; petioles: up to 4 cm long, slender, wiry, with no or occasionally 2 glands near middle; leaves: 3-lobed, 2–7.5 × 3.5–7.5 cm, subcoriaceous, entire or slightly undulate; peduncles: 6–10 cm, stout, solitary or occasionally in pairs; bracts: cordate-ovate, 1.5–3.5 × 1.3–3 cm, purple or reddish purple, crenate-serrate, free; flowers: violet, 3.5–5 cm diam.; floral tube: 2.2–5.1 cm long; sepals: violet, linear-oblong, 2–4.5 × 0.4–0.8 cm, keeled, keel terminating in an awn ca. 0.5 cm long; petals: violet, subequal to sepals; corona: 3–5 series, outer 2 series 0.2–0.5 cm long, violet with white apex, succeeding 1–3 series ca. 0.1–0.3 cm long; ovary: ovoid to elliptic, glabrous; fruit: ovoid to subglobose, 3.5–7 × 3–6 cm, yellowish; seeds: obovate, ca. 0.5–0.7 × 0.4 cm, shallowly tridentate, coarsely reticulate.

Passiflora urbaniana Killip

J. Wash. Acad. Sci. 17: 426 (1927)
named after German botanist Ignatz Urban
Supersection *Passiflora*, Section *Dysosmia*

Passiflora urbaniana (Plate 245), another representative from section *Dysosmia*, is endemic to the low pine ridges of Belize. It has been cultivated for quite some time. In contrast to the related and well-known *P. foetida*, it does not exude an unpleasant odor. Overall, it can be described as warmth-loving but easy to grow. If in winter it is not provided a well-lighted location and temperatures higher than 12°C, it will suffer, although it should recover in spring. *Passiflora urbaniana* can spend the summer outdoors in a protected, sunny location. A 15-cm pot is large enough to get it flowering. Flower bud formation seems to be induced as days become shorter and temperatures fall a little. The main flowering period in Europe is September to November. Its lavender to purplish pink or light violet flowers are about 5 cm diam. They open early in the morning and close by noon. This species was named in honor of Ignatz Urban, one of the most important authors to write about the flora of the Antilles.

Vine: tomentose; stem: terete; stipules: inconspicuous, 0.1–0.5 cm long, deeply cleft, occasionally with 1–2 glandular or nonglandular cilia; petioles: 0.5–1 cm long without glands; leaves: unlobed or obscurely 3-lobed, narrowly ovate to oblong, 3–9 × 2–4(–7) cm, coriaceous or subcoriaceous, subentire or undulate-crenulate; peduncles: 2.7–8 cm, solitary or occasionally in pairs; bracts: deeply 2-pinnatisect, 2–3.5 × 1–2.5 cm, free; flowers: lavender to purplish pink or pale violet, 5–6 cm diam.; sepals: pale lavender or purplish violet inside, green outside, oblong, 1.5–2.7 × 0.6–0.8 cm, slightly keeled, keel terminating in an awn ca. 0.2 cm long; petals: lavender or purplish violet, linear,

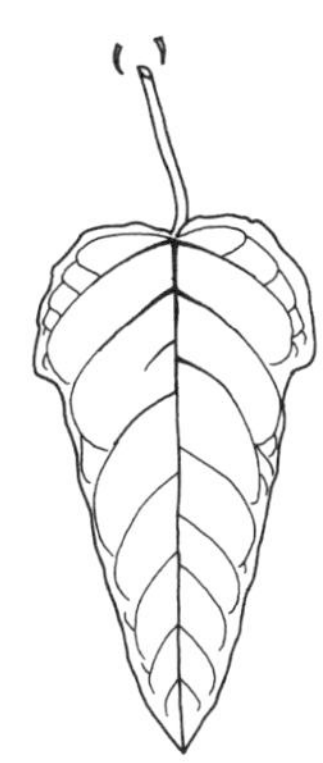

Passiflora urbaniana

1.5–2.5 × 0.5–0.6 cm; corona: 5 series, purplish violet at base, white near middle and violet at apex, outer 2 series 1.2–1.5 cm long, inner series 0.1–0.2 cm long; ovary: subglobose, tomentose; fruit: globose to ovoid, 2.5–4 × 2.5–3.5 cm, red; seeds: oblong-cuneate, 0.4–0.5 × 0.3–0.35 cm, shallowly tridentate, finely reticulate.

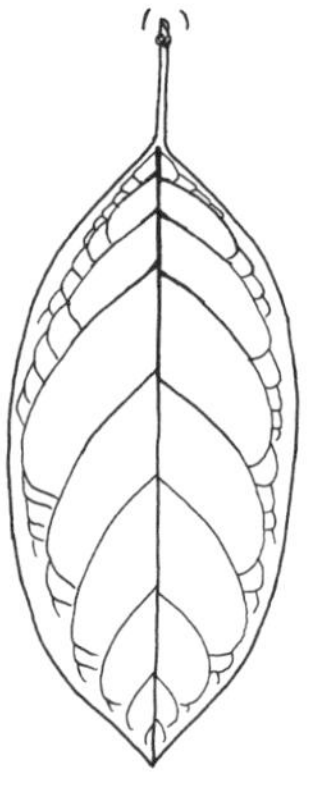

Passiflora variolata

Passiflora variolata Poepp. & Endl.

Nov. Gen. Sp. Pl. 2: 58, t. 179 (1835)
variolatus: having marks resembling those of smallpox (in reference to the glands on the sepals)
Supersection *Distephana*

The hitherto little-known *Passiflora variolata* (Plate 246) is at home in the humid tropical regions of Amazonian Venezuela, Colombia, Brazil, Peru, and the Guianas at altitudes of 50–1200 m. Together with *P. glandulosa*, *P. bomareifolia*, *P. ernestii*, and *P. amicorum* it belongs to supersection *Distephana*. It is distinguished from *P. glandulosa* by its outermost coronal series, which is connate at its base and inclined toward the androgynophore. The white flowers of *P. amicorum* make confusion impossible. *Passiflora ernestii*, at home in the Amazonian lowlands, has a similar vegetative habit but produces smaller, red flowers. The differences with regard to *P. bomareifolia* are outlined in the description of that species.

While Ellsworth Killip (1938) notes that the flowers of *Passiflora variolata* are red, Linda Escobar (1988) points out that the specimens collected in southeastern Colombia are rose-pink. Flowers are 8–10 cm diam. and distinctive for the tiny bracts, the short hypanthium, and above all the sepals that are glandular on the outside near the margin.

Until quite recently it was speculated whether specimens of *Passiflora variolata* from French Guiana might be assigned to a completely new species named *P. wurdackii*. A number of plants were brought into circulation under this name. However, recent comparisons of specimens from varying origins did not show decisive differences, so the subject was dropped.

During the last few years seeds from this species have been collected in French Guiana. At present, *Passiflora variolata* is cultivated by only a few enthusiasts. Like *P. glandulosa*, this is a warmth-loving, fastidious species. During overwintering, additional lighting and a heating mat are recommended. Temperatures should not drop below 15°C.

Vine: woody, glabrous; stem: terete; stipules: linear-falcate, 0.2–0.4 cm long, soon deciduous; petioles: 0.6–2 cm long, with 2 sessile glands near base; leaves: unlobed, oblong or oblong-ovate, 5–14.5 × 1.5–6 cm, membranous or coriaceous, entire; peduncles: 2.4–8.5 cm, solitary; bracts: linear-lanceolate, 0.3–0.8 × 0.1 cm, 2–4 glands near base, free; flowers: red, 8–10 cm diam.; sepals: red, linear-oblong, 3.1–4.9 × 0.7–1.4 cm, glandular at margin, slightly keeled, keel terminating in an awn 0.2–0.7 cm long; petals: red, subequal to sepals; corona: 2 series, outer series tubular, filamentous at apex, 0.8–1 cm long, white, inner series 0.2–0.3 cm long; ovary: ovoid, glabrous; fruit: ovoid, 4–7 × 2–4 cm, green, mottled with white, with 6 green longitudinal lines.

Passiflora vitifolia Kunth

Nov. Gen. Sp. 2: 138 (1817)
vitifolius: grapevine-leafed
Synonyms: *P. punicea* (DC.) Ruíz & Pavón, *P. sanguinea* J. E. Sm., *P. servitensis* Karst.
Supersection *Coccinea*

Passiflora vitifolia (Plate 247) is valued by enthusiasts all over the world for its conspicuous brilliant red flowers. It is among the most beautiful and widespread passionflowers. At home from Nicaragua to Venezuela and Peru, it grows at altitudes of 200–1100 m. In the West Indies it is also cultivated for its fruit on a small economic scale. The whole

Passiflora vitifolia

plant is densely pubescent with rusty-brown hair and grows vigorously. Its very beautiful flowers reach an extraordinary diameter of 12–17 cm. The outer rank of the upright, short corona is red, while the 2 inner series are usually white, though the corona is variable and can be red, yellow, or white. *Passiflora vitifolia* is pollinated by hermit hummingbirds of the genus *Phaethornis*, and thus is odorless and elongate. It does not set fruit under cultivation; the job must be done by hand if one wants to enjoy the beautifully mottled fruit in fall. The 3-lobed, dark green leaves, which are brownish when young, are reminiscent of grapevine-leaves.

Passiflora vitifolia belongs to supersection *Coccinea*, which was recently assigned to subgenus *Passiflora*. This supersection contains almost exclusively red-flowering species, such as *P. coccinea*, *P. quadrifaria*, and *P. quadriglandulosa*. Most of these species are quite fastidious and suited only for a greenhouse. *Passiflora vitifolia*, however, is very robust and can be cultivated in a protected, sunny location outdoors. Plants grown in central Europe are real eye-catchers from July to September, when they are covered with numerous flowers. They will not flower so freely, however, during a cool, rainy summer. As a greenhouse offers optimum conditions, *P. vitifolia* will grow more vigorously there and will set flower buds as early as late spring.

Garden centers occasionally offer this species tied around a circular wire, giving the impression that *Passiflora vitifolia* is suitable for cultivation indoors. This is not so. This species grows too vigorously indoors and suffers from the extremely dry air. Such a plant is a pleasure for only some weeks What applies to most passionflowers holds true for *P. vitifolia*: overwintering is the most critical period of its cultivation. Although the plant may have developed wonderfully during summer, problems will occur in winter if it is not flooded with daylight and provided temperatures beyond 15°C. Unfortunately, many enthusiasts lack such facilities, and so a compromise must be found. This species should be taken to a location that is well lit and guaranteed atmospheric humidity, which can be achieved by spraying the plant regularly. The location can be an unused room, corridor, staircase, or something similar. In winter the leaves often dry up from the margin inward; they can even fall off, leaving the plant looking very desolate. In such a case, watering should be reduced to a minimum to prevent the rootstock becoming too moist. The plant will recover in early spring, and the leaves will gradually look healthier.

Vine: puberulent to tomentose in many parts; stem: terete to subangulate, striate; stipules: linear, 0.4–0.9 cm long, with 2–3 or occasionally no glands at base; petioles: 0.8–6.3 cm, with 2 glands at base, rarely with 2–3 additional glands near middle; leaves: 3-lobed, 5–18 × 4–16 cm, puberulent above, tomentose beneath, serrate; peduncles: 3–9 cm, solitary or rarely in pairs; bracts: linear to oblong-lanceolate 1.5–4.5 × 0.2–1.2 cm, with 2–6 glands at base, glandular-serrate above, free; flowers: bright red, (9–)12–16(–17) cm diam.; sepals: red, oblong-lanceolate, 5–8.5 × 0.8–1.7 cm, slightly keeled, keel terminating in an awn 0.3–1.5 cm long; petals: red, oblong-lanceolate, 4.5–7.8 × 0.7–1.7 cm; corona: 3 series, outer series 1.4–3 cm long, red (or white), filaments of inner series 0.8–2.2 cm long, white (or red); ovary: ellipsoid, tomentose; fruit: subglobose to ovoid, 5–7 × 4–6 cm, dark green to brownish green with white spots in longitudinal direction, edible; seeds: obovoid, 0.5–0.7 × 0.3–0.5 cm, reticulate, dark brown to black.

16

Subgenus *Passiflora*, Supersection *Tacsonia*

The eye-catching species of supersection *Tacsonia* are among the most magnificent and large-flowered representatives of the genus *Passiflora*, and they deserve special attention. The typically pendent flowers of the more than 60 colorful species of this supersection differ from those of most other passifloras by their long floral tube. The corona, so conspicuous in other groups, is generally reduced to a colored ring with tiny elements only a few millimeters long. The reduced corona and long tube, as well as the predominantly rose-pink sepals and petals, are floral ecological adaptations to pollination by certain hummingbirds with a long, straight bill (see, for example, Plate 248). Moreover, there is a small group of species that produce white flowers and belong to section *Manicata*. These species are pollinated by nocturnal bats.

The native range of species from supersection *Tacsonia* reaches from Venezuela through Colombia, Ecuador, and Peru, to Bolivia and Chile. Tacsonias grow on the mountain ranges of the Andes at altitudes of 1400–4400 m, where fog and drizzle are quite natural, as is continuous rain; night temperatures often drop to near the freezing point or a little below. In southern Peru some species also grow in dry scrub and grassland vegetation. Quite a number of tacsonias produce edible fruit. However, only 4 or 5 species are of major economic importance, principally *Passiflora tarminiana* and *P. tripartita* var. *mollissima*, which are grown on large plantations. These fruits are regularly exported to Europe, while fruits from *P. pinnatistipula*, *P. cumbalensis*, and *P. antioquiensis* are also produced on plantations but are rarely available in other countries. The popularity of other species with edible fruit is confined to the local area where they are grown in gardens or harvested from the wild. Except for *P. tarminiana* and *P. tripartita* var. *mollissima*, tacsonias are infertile with their own pollen. However, Giselher Behnken succeeded in transferring self-compatibility to the offspring when he crossed *P. tarminiana* with *P. tripartita* var. *azuayensis*. Otherwise, all species of this supersection need cross-pollination with pollen from another clone. Tacsonias also cross freely among themselves. Accordingly, hybrids frequently occur in the natural surroundings and particularly in the transition zones between the individual species' native ranges.

Systematics—Subdivision in Sections

In 1781 the first 2 species of supersection *Tacsonia* were described by Linné fil., son of Linnaeus. From José Celestino Mutis he had received drawings of plants frequently found near Bogotá. Mutis had gone to Colombia as a young man, and there he remained to study the flora of the Andes until his death in 1808. By naming them, Linné fil. recognized these species as different from any previously described passionflowers: *Passiflora adulterina*, the fake or false passionflower, and *P. mixta*, which he thought to be a combination of *P. adulterina* and the other species known at that time.

In his *Genera Plantarum* (1789), Antoine Laurent de Jussieu went one step further when he created a genus, *Tacsonia*, for the long-tubed species that he found to be so different and deviating from other passionflowers. He adopted the name *Tacsonia* from Quichua, a language still spoken by the indigenous people of the Andes. It is a phonetic spelling of the word that the Indians use for this plant. *Tacsonia* was recognized as a genus for several decades, and more species were named. In 1873, Triana and Planchon (in their *Prodromus florae novo-granatensis*) moved tacsonias to their own subgenus in the genus *Passiflora*. Here they stayed for a century, and most older books cite *Tacsonia* as a subgenus. As a consequence of the most recent revision of the systematics by Feuillet and MacDougal, *Tacsonia* loses its status as a subgenus and is classified into subgenus *Passiflora* together with most of the other large-fruited, large-flowered passionflowers. They are still called tacsonias, but the rank is changed to supersection instead of subgenus, only a technical point. Supersection *Tacsonia* now also includes the old subgenus *Rathea* and representatives of the former subgenus *Granadillastrum*.

The first comprehensive presentation of the then-known 37 species of subgenus *Tacsonia* was published in Killip's *The American Species of Passifloraceae* (1938). The 2nd and most recent comprehensive treatment of all species from subgenus *Tacsonia* was set forth by Linda Escobar in her outstanding doctoral thesis (1980). In it, she declared some species to be synonyms or reduced them to varieties. On the whole, she presented 40 species with 8 varieties, classifying them into 10 sections.

In 1988, 2 works were published that deal exclusively with the passionflowers of Ecuador and Colombia, among them many tacsonias. Both are entitled "Passifloraceae," the first written for *Flora of Ecuador* by Lauritz Holm-Nielsen and his coworkers, the second written for *Flora de Colombia* by Linda Escobar. At the end of her article, Escobar gives a tabular list classifying the species of *Tacsonia* in a system that is still basically followed, though the names have different ranks. Feuillet and MacDougal followed Escobar's system in their new systematics but changed the sections to series.

This chapter provides a short survey of the sections of *Tacsonia* with regard to their characteristic features. Section *Colombiana* is the most rich in species, comprising those species that have filiform or lanceolate stipules (except *Passiflora quindiensis*), typically free but rarely connate bracts with a serrate or fimbriate margin, and deeply 3-lobed (except *P. coactilis*) or unlobed leaves (see Plate 249). With a few exceptions their native range is limited to Colombia, and several subdivisions, or series, are recognized. The long peduncles and cernuous shoot apices of some species in the series *Leptomischae* are worth mentioning.

The representatives of the 2nd-largest section, *Elkea* (formerly *Bracteogama*), are plants with mostly large, entire, distinctly connate bracts, subreniform or ovate, serrate to lacerate stipules, and 3-lobed leaves. Further studies will show whether *Passiflora zamorana* and particularly *P. jamesonii*, which have free, serrate, or lacerate bracts, can remain within section *Elkea*. The main native range of species belonging to this section extends over Ecuador and Peru.

Species from section *Insignes* (formerly *Poggendorfia*) are characterized by pinnatisect stipules and a filiform corona with filaments up to 2 cm long. Except for *Passiflora pinnatistipula*, which can be found in many parts of the Andes, the native range of the other representatives of section *Insignes* is restricted to Bolivia and Peru.

Section *Tacsonia* comprises all species with horizontal or upright flowers. Apart from this characteristic, the representatives of this section do not differ from those of section *Elkea*. A fusion of both sections into one might be more genealogically appropriate. In addition, the species of the formerly independent subgenera *Rathea* and *Manicata* are now placed in sections in *Tacsonia*. The 3 species from section *Rathea* are distinguished by their linear-lanceolate stipules, free bracts, and particularly by their uncommon floral morphology. The sepals, connate at their base, form a tubular to slightly funnel-shaped flower. The section *Manicata* comprises 4 species with white flowers and one species with red flowers. They all have upright flowers with a shorter floral tube as well as a corona in several series and are thus distinguished from the original long-tubed tacsonias. An account of the smaller, often monotypic sections can be given in a shortened form.

Section *Boliviana*: very small flowers and leaves, free bracts, no or 2 petiole glands (Peru and Bolivia); section *Fimbriatistipula* (see Plate 250): ovate, fimbriate stipules, large, free or almost free bracts, orange-pink to salmon-pink flowers (Colombia); section *Parritana*: lanceolate stipules, free or almost free bracts, prominent keel ridge on the back of the orange to salmon-pink sepals (Colombia); section *Tacsoniopsis*: oblong-ovate stipules, large and free bracts, tubular to funnel-shaped flowers (Colombia and Venezuela); *Trifoliata*: oblong-ovate stipules, tripartite leaves, large, glandular-fimbriate, and free bracts (Peru).

Cultivation

To cultivate tacsonias successfully, one has to bear in mind the conditions in their natural surroundings, the high mountain ranges of the Andes: annual rainfall of 1500–2000 mm, nightly drops in temperature down to near the freezing point or even below that at higher altitudes, and day temperatures of 12°–25°C. In principle, the resistance to frost increases in accordance with the altitude of the habitat. Tacsonias growing near the equator are accustomed to a daytime rhythm and not to a seasonal rhythm. Hence, longer spells of cold weather or even frost should be avoided. Because they are adapted to such extreme climatic conditions, tacsonias are not suited for indoor cultivation or for cultivation in a heated greenhouse. Anyone lucky enough to live in an area with climatic conditions like those of this plants' habitats, or anyone lucky enough to own a cool greenhouse in which tacsonias can be planted directly in the ground, will be rewarded with a wealth of flowers. In the United States, only those on the west coast can successfully cultivate tacsonias. An almost paradisiacal spot is at the Strybing Arboretum in San Francisco, one of the finest botanical gardens in the country. The plants can be grown outdoors at the foot of the Golden Gate Bridge and develop into magnificent specimens with an abundance of flowers. Few enthusiasts, however, have such an opportunity, so it is usually necessary to find a compromise.

Tacsonias should be cultivated outdoors as long as possible and only be taken to a greenhouse, conservatory, or other light and protected location during times when night frosts are expected. The plants must be transportable. Cultivation in pots or containers is impaired by the gigantism of many tacsonias and by the fact that most species will only flower when the shoots are several meters long. However, quite a number of species, above all *Passiflora luzmarina*, *P. mathewsii*, *P. mixta*, *P. tarminiana*, and their hybrids, can be induced to flower as container plants when their shoots have reached a length of about 2 m. Despite adverse conditions in northern Germany, Giselher Behnken achieves good results by planting his tacsonias close together directly in the ground in the garden. After a generous pruning in fall, a simple, heatable greenhouse construction is mounted around the plants. The structure can be removed easily and quickly in spring. This way the plants reach the necessary size to flower abundantly every year. Behnken suggests inducing flower bud formation by removing the lower lateral shoots and leaving only one main branch, which should be at least 1.5–2 m long. Avoid letting temperatures get above 30°C, and make sure plants do not dry out.

Cuttings from various tacsonias sometimes have difficulty rooting. At soil temperatures of 15°–20°C, propagation is quite feasible, but there is a higher failure rate in comparison with cuttings from the species of the subgenera *Decaloba* and *Passiflora*. Propagation from seeds is more reliable. When the seeds are fresh they may germinate at a temperature of 20°C within a week.

Passiflora amazonica Escobar

Ann. Missouri Bot. Gard. 76: 877 (1989)
named after the department of Amazonas, Peru, site of the type collection
Section *Elkea*

The native range of the rare *Passiflora amazonica*

(Plate 251) is limited to a small region in northern Peru. Its habitats are the last foothills of the Eastern Cordillera at altitudes of 1800–2300 m, in sight of the adjoining Amazonian lowlands. This means that it is at the lower end of the altitudinal range for tacsonias. Its pink flowers with a diameter of 6–8 cm indicate a close relationship with *P. mixta* and *P. tripartita* var. *mollissima*. *Passiflora amazonica* can be distinguished from both by means of its coriaceous, dentate leaves, which are wrinkled on the upper side and have a deeply heart-shaped base, and by its longitudinally 5-grooved fruit. Another difference is that *P. mixta* produces horizontal to upright flowers.

Like most species indigenous to Ecuador and northern Peru, *Passiflora amazonica* belongs to section *Elkea*, whose representatives are characterized by typically pendent flowers, connate bracts, reniform to oblong-lanceolate stipules, and 3-lobed leaves. The general impression of the representatives from this section corresponds to the typical notion most enthusiasts have of tacsonias.

Maximilian Weigend recently succeeded in bringing *Passiflora amazonica* into cultivation. The plants are developing very well but are still too young to be judged for their willingness to flower. In comparison with many other tacsonias, this species gets along quite well with temperatures above 30°C, although it prefers a cooler climate. Growth is most vigorous when day temperatures are 20°–25°C and drop to about 10°C at night.

Vine: pubescent; stem: subangular, striate, tip erect; stipules: subreniform, 0.8–1.5 × 0.5–0.7 cm, with long arista at apex, lacerate; petioles: 2.1–2.6 cm with up to 10 elongated glands; leaves: 3-lobed, 7–11.7 × 9–15.2 cm, rugose, coriaceous, dentate; peduncles: 5–6.3 cm, solitary; bracts: oblong-lanceolate, 3–4 × 1.2–3.5 cm, connate for about 2/3 of their length, entire; floral tube: cylindrical, 6.6–7 cm long, 1–1.2 cm wide, green; flowers: pink, 6–8 cm diam., 9.6–13 cm long; sepals: pink, oblong, 4–4.5 × 1.2–1.8 cm, aristate, awn ca. 0.1 cm; petals: pink, oblong, 3.9–4.1 × 1.2–1.6 cm; corona: 1 series, reduced to an undulate ring, dark purple; ovary: ellipsoid, pubescent; fruit: ellipsoid, ca. 7.5–8.5 × 2.5–3.5 cm, greenish yellow, longitudinally 5-grooved; seeds: oblong-ovate, 0.6–0.7 × 0.4 cm, reticulate, blackish brown.

Passiflora ampullacea (Mast.) Harms

Nat. Pflanzenfam. 3 (6a): 91 (1893)
ampullaceus: bottlelike
Synonym: *P. hieronymi* Harms
Section *Colombiana*, Series *Leptomischae*

Passiflora ampullacea (Plates 252 and 253), indigenous to central Ecuador at altitudes of 2000–3250 m, occupies a special position within supersection *Tacsonia*. It is distinguished from other tacsonias by its combination of unusual yellowish green flowers and green-orange to light red bracts. It is further characterized by the horizontal position of its zygomorphic flowers, which are raised out of the foliage on long, stout peduncles. On the basis of its unique flower morphology and flower color, it was once speculated that this species might be pollinated by bats. This conjecture had to be refuted, however, because it could be established that *P. ampullacea* is pollinated by hummingbirds, as is

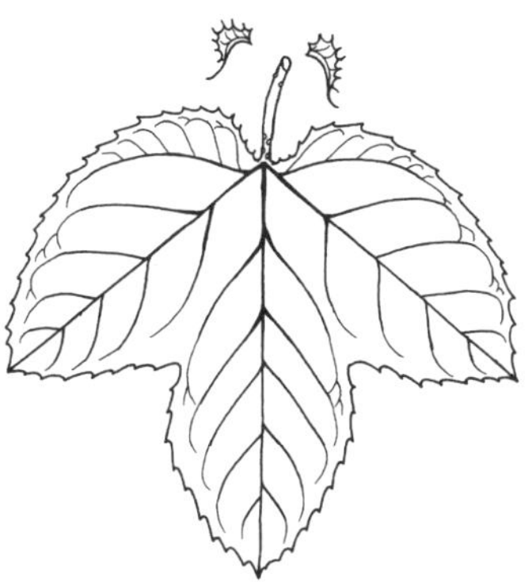

Passiflora amazonica

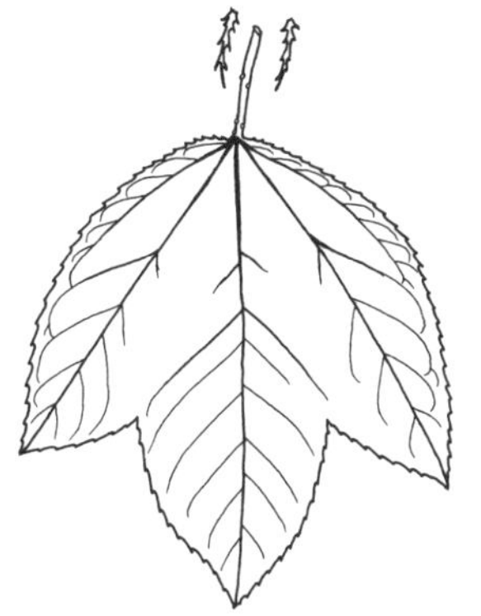

Passiflora ampullacea

true of many green-flowered representatives from other plant families. The diameter of the flower is relatively small at 4.5–6 cm. The length of the funnel-shaped floral tube is quite variable. Variations with a long floral tube are distinguished by a smaller throat, variations with a short floral tube by a wide throat.

For a long time *Passiflora ampullacea* occupied a special position within the systematics because of its unusual morphology. It was finally assigned to the unpublished, monotypic section *Ampullacea* by Linda Escobar. However, in accordance with recent considerations, it was incorporated into section *Colombiana*, series *Leptomischae* because of its stipules, bracts, and nodding shoot tips.

The oblong-ovoid fruits are edible and are offered in local markets. The remarkable, pulpous, thickened pericarp is found in few other species of *Tacsonia*, *Passiflora tenerifensis* being one example.

Some clones of this species have been cultivated over the last few years, particularly in Europe. They turned out to be rather unproblematic and grew quite vigorously. However, they are quite unwilling to flower. Only a few large specimens have produced any flowers.

Even without flowers *Passiflora ampullacea* can be distinguished easily from other tacsonias by its nodding shoot tips and deeply serrate or slightly pinnatisect stipules. This species should be planted out directly in the ground of a cold greenhouse. It can also spend summers outdoors, where it will tolerate temperatures exceeding 30°C for some days. In winter it will tolerate short spells of slight frost.

Vine: densely pubescent; stem: terete, slightly striate, tip flexed; stipules: linear-lanceolate, slightly pinnatisect, 1–1.8 × 0.2 cm; petioles: 2–4.5 cm, with 5–9 subsessile glands, occasionally covered by the indument; leaves: 3-lobed, 6–12 × 7.5–14 cm, serrate; peduncles: 7–13 cm, stout, solitary, horizontal; bracts: ovate to oblong-ovate, 3–5.5 × 1.2–3.5 cm, yellow to orange-red, entire, free; floral tube: funnel-shaped, 5–7.5(–10) cm long, 0.8–2 cm wide at throat, green; flowers: yellowish green, 4.5–6(–8) cm diam., 9.5–11(–12) cm long; sepals: yellowish green inside, green outside, oblong, 2.4–3.6 × 1.2–2 cm, aristate, awn ca. 0.2 cm; petals: yellowish green, oblong, 2–3 × 1–1.4 cm; corona: 1 series, purple or whitish, filaments 0.1 cm, rarely up to 0.5 cm long; ovary: ellipsoid, densely pubescent; fruit: ovoid, 5.5–10 × 3–5 cm, greenish yellow, edible; seeds: ovoid, 0.8 × 0.5–0.6 cm, reticulate, dark brown.

Passiflora anastomosans (Lambert) Killip

J. Wash. Acad. Sci. 17: 428 (1927)
anastomosans: flowing together
Section *Tacsonia*

This species (Plate 254), once regarded as rare, has been encountered in central and northern Peru quite frequently during the last few years. It is found at altitudes of 2300–3100 m on the Cordillera Occidental y Azul. It only grows in scrub over sandstone and stays away from any other soil (Maximilian Weigend, personal communication). In contrast with other species of *Tacsonia*, *Passiflora anastomosans* is a rather small plant with only a few lateral shoots. A characteristic of this species is the horizontal flower with a zygomorphic androecium and short sepals and petals that form a small pink ball at the end of the floral tube. When the flower has completely opened, its diameter is only 2–4 cm. *Passiflora anastomosans*, remarkable in several respects, can easily be distinguished from other species of section *Tacsonia* that are at home in Peru, such as *P. mixta* and *P. mathewsii*, as well as from *P. schlimiana*, endemic in northern Colombia. It normally has conspicuously large stipules and bracts, which are entire as well, and leaves that are thick-coriaceous with a red vein and a lustrous upper side. This is an exceptional species to bring into cul-

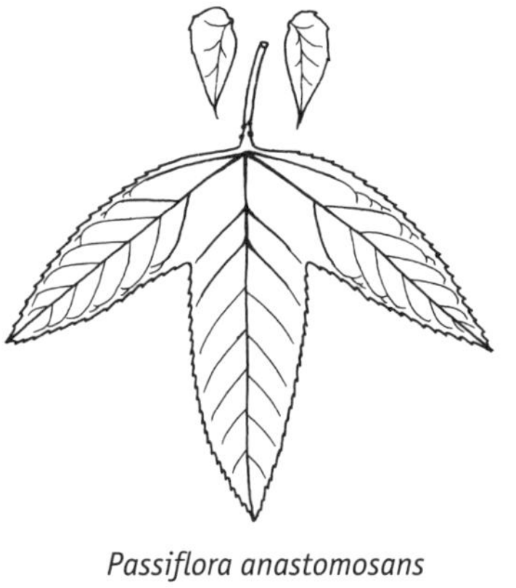

Passiflora anastomosans

tivation. In this case special attention should be paid to the potting mix. Because of its smaller growth form, *P. anastomosans* can be kept under better control than many other tacsonias.

Vine: glabrous; stem: terete, younger parts subangular, tip erect; stipules: oblong-ovate, 1.9–3.6 × (0.6–)1–1.7 cm, coriaceous, nearly entire; petioles: 1.5–2 cm, with 2–5 subsessile glands near apex; leaves: 3-lobed, 4–9 × 5–13 cm, thick-coriaceous, very smooth and lustrous above, serrate; peduncles: 2.5–8.3 cm, stout, solitary; bracts: (2.8–)3.7–7 × (1–)1.8–3 cm, connate 2/3 of their length, tomentose within, free parts entire; floral tube: cylindrical, 5.8–10 cm long, 0.6–1 cm wide, greenish turning pinkish green toward apex; flowers: pink, occasionally reddish orange, 2–4 cm diam., 8–12 cm long; sepals: pink, oblong-lanceolate, 1.5–3 × 1–2 cm, aristate, awn ca. 0.1 cm; petals: pink, oblong-lanceolate, 1.3–2.8 × 0.7–0.9 cm; corona: 1 series, purple ring with white filaments, ca. 0.1 cm long; ovary: ellipsoid, glabrous; fruit: ovoid-ellipsoid.

Passiflora andina Killip

Publ. Field. Mus. Nat. Hist., Bot. Ser. 19: 256 (1938)
andinus: belonging to the Andes
Section *Rathea*

Passiflora andina (Plate 255) is endemic in the province of Pichincha, northern Ecuador, at altitudes of 2500–3600 m. Its flowers do not open in the usual way, but its sepals and petals form, together with the floral tube, a greenish white funnel that becomes increasingly orange toward its base. The diameter of an open flower is only 3–5 cm. While the sepals are connate at their base and free for some centimeters only in the upper part, the innermost part of the flower is formed by the well-developed petals, which are almost twice as large. The corona, reduced to distant, tiny teeth, is hidden inside the flower, about 1.5 cm above the base of the flower. The inner walls below the corona are purplish.

Passiflora andina

Because of its particular flower morphology, *Passiflora andina* has previously been assigned to the independent subgenus *Rathea*, together with the closely related *P. harlingii* and *P. colombiana*. Since their other characteristics correspond with those of tacsonias, they are now assigned to supersection *Tacsonia*. *Passiflora andina* is distinguished from *P. harlingii* by its significantly longer sepals and shorter, funnel-shaped flowers. *Passiflora colombiana*, on the other hand, has considerably smaller leaves with narrow lobes and orange-pink flowers.

Passiflora andina was brought into cultivation in Germany in the 1990s. Unfortunately the plants, grown from seeds, turned out to be very susceptible, so it is possible this species is no longer under cultivation.

Vine: glabrous, except ovary and upper leaf veins; stem: subquinquangular, tip flexed; stipules: linear-lanceolate, 0.7–1 × 0.1–0.2 cm, glandular-serrate; petioles: 1.5–2 cm, with 2–6 minute glands; leaves: deeply 3-lobed, 4–7 × 4–8 cm, margin serrulate and revolute; peduncles: 10–12 cm, solitary; bracts: oblong-ovate, 1.5–3.5 × 1.5–2 cm, serrate, free; floral tube: ca. 2 cm long, tapering in a funnel-shaped calyx; flowers: yellowish green, orange at base, 3–5 cm diam., 11–16 cm long; sepals: yellowish green, oblong, free for ca. 5 cm, aristate, awn ca. 0.2 cm; petals: yellowish green, oblong to obovate-oblong, 6–9 × 1.2–1.7 cm; corona: 1 series, very slender filaments, inserted 1–1.5 cm from base of flower; ovary: ellipsoid, densely villous-tomentose; fruit: ovoid, 5–7 × 3–4 cm, green.

Passiflora antioquiensis Karst.

Linnaea 30: 162 (1859)
named after Antioquia, Colombia, location of the type plant
Synonym: *P. van-volxemii* (Funk) Triana & Planch.
Section *Colombiana*, Series *Leptomischae*

Passiflora antioquiensis (Plate 256) is a most magnificent species. Its brilliant purplish red flowers,

8–12 cm diam., are pendent on peduncles that are 20–40 cm long, and it is distinguished by its short floral tube, only 2.3–4 cm long.

Passiflora antioquiensis is at home mainly on the Central Cordillera in Colombia at altitudes of 1800–2700 m. That said, there are some wild occurrences on the Western Cordillera, and this species is often cultivated in gardens on the Eastern Cordillera in the region around Bogotá. It is cultivated for its edible fruit not only in Colombia and its bordering states but also in New Zealand, where it is grown commercially on minor plantations.

Together with *Passiflora coactilis, P. cremastantha, P. flexipes, P. leptomischa,* and *P. tenerifensis, P. antioquiensis* belongs to the series *Leptomischae*, which is of particular interest with regard to systematics. All these species have long peduncles, free bracts, and linear-lanceolate stipules. *Passiflora antioquiensis* can easily be distinguished from these species by its short floral tube and reddish flowers. The only exception is *P. cremastantha*. Although its floral tube is longer, at a length of 4–6 cm, it is nevertheless rather short for tacsonias of this size. There are a few collections of *P. cremastantha*, but there is not enough material at hand to make general assertions about the species, let alone assertions about its distinctiveness. Hence the differentiation from the closely related *P. antioquiensis* is a little difficult, though it seems to be possible by means of the bracts, which have an entire margin.

During the last few years *Passiflora antioquiensis* has reentered the collections of many enthusiasts from which it had disappeared for many years. For a long time *P.* ×*exoniensis* had erroneously been regarded as *P. antioquiensis*. However, *P.* ×*exoniensis* is a cross between *P. antioquiensis* and *P. tripartita* var. *mollissima*. These can be distinguished at first sight. At a length of up to 12 cm, *P.* ×*exoniensis* has significantly shorter peduncles. But even without flowers, a correct identification is possible. *Passiflora antioquiensis* has a nodding shoot tip, while the hybrid's shoot tip grows upright. So far experience has shown that *P. antioquiensis* is not as unpretentious and willing to flower as *P.* ×*exoniensis*.

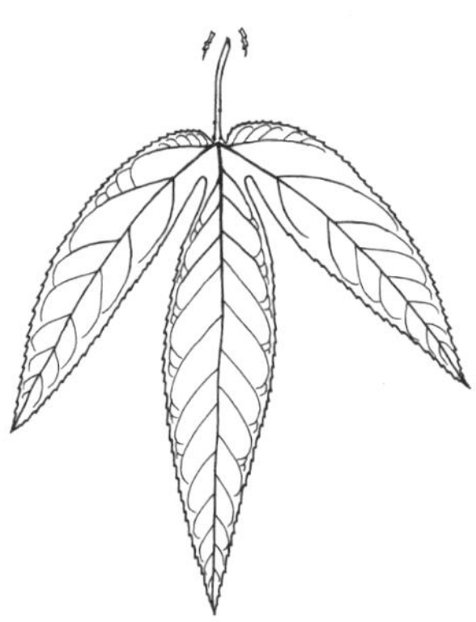

Passiflora antioquiensis

Vine: pubescent; stem: terete, younger parts subangular, striate, tip flexed; stipules: linear-lanceolate, 0.5–0.8 cm long, entire; petioles: 1.5–4.5 cm, with 7–12 subsessile glands; leaves: polymorphic, deeply 3-lobed or unlobed, 3-lobed leaves 6–17 × 4.5–12.5 cm, unlobed leaves 7–17.5 × 3.5–8 cm, serrate; peduncles: 20–40 cm, slender, solitary; bracts: lanceolate, 1.3–2 × 0.4–0.8 cm, free, serrate; floral tube: cylindrical, 2.3–4 cm, 0.5–1 cm wide, green or occasionally brownish purple; flowers: red to purplish red, 8–12 cm diam., 7–9 cm long; sepals: red to purplish red, oblong, 4–6 × 1.5–2.5 cm, aristate, awn ca. 0.3 cm; petals: red to purplish red, oblong, 4–6 × 1–1.5 cm; corona: 2(–3) series, filaments of outer series 0.1(–0.5) cm, violet, inner series inside the floral tube; ovary: ellipsoid, glabrous, rarely pubescent; fruit: ellipsoid, 9–11.8 × 3–3.5 cm, yellow, edible; seeds: oblong, ca. 0.8 × 0.3–0.4 cm.

Passiflora bracteosa Planch. & Linden

Ann. Sci. Nat. Bot. 5 (17): 137 (1873)
bracteosus: having bracts
Section *Tacsoniopsis*

The small native range of this exceptional species reaches along the Central Cordillera from northwestern Colombia into Venezuela. The collection sites are at altitudes of 2300–3000 m. Unfortunately, *Passiflora bracteosa* is not under cultivation. At first sight it becomes evident that this is quite an unusual species. As Plate 257 illustrates, the flowers do not open in the way tacsonias usually do; rather, their tubular form is reminiscent of the 3 representatives from section *Rathea*. The formation of such an apricot to salmon-pink tube, 10–12 cm long, is due to the unusual arrangement of petals and sepals on the floral tube. The sepals are connate for about 1/2 of their length and separate only about 1 cm above the corona. At this spot the small, free petals join. In this way the flowers reach a diameter of 1.2–2 cm only when they have opened completely.

Passiflora bracteosa also has peduncles that are often paired, large, free bracts, and unique fruit. No other species from supersection *Tacsonia* produces fruit with such an uneven and rough surface. *Passiflora bracteosa* is closely related to *P. purdiei*, the only other species from section *Tacsoniopsis*. Unfortunately, only a single herbarium specimen is available, without any indication of the type location. It might, however, originate from the Cordillera Oriental in northern Bogotá. In *P. purdiei*, the sepals join directly above the corona and are much longer than in *P. bracteosa*. Moreover, *P. purdiei* has narrower leaf lobes and a shorter floral tube.

Vine: glabrous; stem: subangular to terete, striate; stipules: ovate, 1–1.8 × 0.7–1.3 cm, glandular-serrate; petioles: 1.3–4.2 cm with up to 18 small glands; leaves: 3-lobed, 5.3–11.5 × 6.5–15.6 cm, membranous, serrate; peduncles: 9.5–14.5 cm, stout, solitary or often in pairs; bracts: ovate to oblong-ovate, 3.5–5.4 × 1.4–3 cm, free, serrate; floral tube: cylindrical to funnel-shaped, 7–9 cm long, orange to salmon; flowers: orange to salmon, 1.5–2 cm diam., 10–12 cm long; sepals: orange to salmon outside, crème inside, oblong-ovate, ca. 2.6–2.8 × 1 cm, connate for 1/2 of their length, aristate, awn 0.3–0.6 cm long; petals: crème, ovate, ca. 1.3 × 0.9 cm; corona: 1 series, reduced to a violet band; ovary: ovate, glabrous; fruit: ovate, ca. 7 × 3.5 cm, greenish yellow, covered by small tubercles; seeds: obovoid, 0.7–0.8 × 0.5 cm.

Passiflora coactilis (Mast.) Killip

J. Wash. Acad. Sci. 14: 212 (1924)
coactilis: folded
Synonym: *P. mariae* (Sodiro) Harms
Section *Colombiana*, Series *Leptomischae*

For its gigantic growth and for its flowers, *Passiflora coactilis* (Plate 258) is among the mightiest tacsonias. It is at home in central to northern Ecuador, where it can be found in the wet montane forests of the western slopes of the Cordillera Occidental at altitudes of 2200–3000 m. It can easily be distinguished from other *Tacsonia* species by its linear-lanceolate stipules, connate bracts, and cinnamon-colored vestiture. With regard to systematics, *P. coactilis* is the species that proceeds furthest to the south in comparison with other representative from section *Colombiana*, series *Leptomischae*, which are noted for long peduncles, linear to lanceolate stipules, and nodding shoot tip; it is possible that this species is the connecting link to series *Colombianae*, whose representatives are characterized by a dense, brown tomentum, connate bracts, and rugose leaves.

This beautiful species is especially conspicuous for its flower, 11–16 cm diam., which is pendent on threadlike peduncles about 25 cm long. This feature makes *Passiflora coactilis* particularly attractive to enthusiasts, as there are only a few passionflowers with peduncles of this length. A few years ago Giselher Behnken succeeded in bringing *P. coactilis* into cultivation in Germany, and now almost all *Tacsonia* species from Ecuador can be admired in cultivation. When the summer is cool and wet and temperatures are around 20°C, *P. coactilis* is

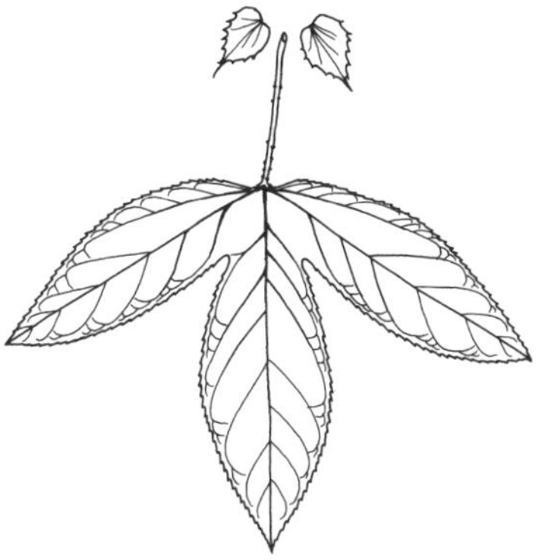

Passiflora bracteosa

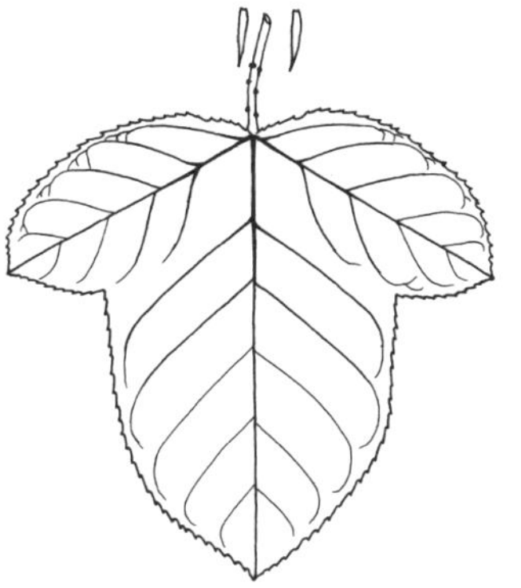

Passiflora coactilis

happy. It is more sensitive to higher temperatures than other tacsonias, but when it feels comfortable in its location it will develop into an imposing plant, displaying all its beauty only when planted out with a good root run. Consequently this species is not very suitable for cultivation in a container but should be planted directly in the ground in a large, cold greenhouse.

Vine: villous, except upper leaf surface; stem: terete, tip flexed; stipules: linear, 0.5–1.2 cm long, entire; petioles: (1–)2–3.2 cm, with 4–8 subsessile glands covered by the brownish indument; leaves: 3-lobed or rarely 2-lobed or unlobed, (5–)6.5–13.5 × (4.7–)6–15.5 cm, rugose, serrate; peduncles: (9–)15–25(–35) cm, slender, solitary; bracts: oblong-ovate, 2.9–5.5 × 0.9–2.7 cm, connate for 0.5–1 cm at base, entire or slightly serrate; floral tube: cylindrical, 7.5–10.9 cm long, 0.7–1.1 cm wide, green turning pinkish toward apex; flowers: pink to light red, 11–16 cm diam., 14.5–17 cm long; sepals: pink to light red, obovate, 5.3–7.5 × (1–)1.8–2.7 cm, aristate, awn 0.3–0.5 cm long; petals: pink to light red, oblong, 5–6 × (1–)2–3 cm; corona: 1 series, tubercles or filaments 0.1 cm or rarely up to 0.6 cm, bluish violet, pale to whitish toward apex; ovary: ellipsoid, densely villous; fruit: ellipsoid, 8–10 × 4–4.5 cm, yellow; seeds: obovoid, 0.9–1 × 0.6–0.7 cm, reticulate.

Passiflora crispolanata Uribe

Mutisia 21: 5 (1954)
crispus: irregularly wavy, *lanatus*: woolly
Synonym: *P. boyacana* Killip
Section *Colombiana*, Series *Colombianae*

The small native range of *Passiflora crispolanata* (Plate 259) is located northeast of Bogotá at altitudes of 2600–3100 m. Its pink flowers, pendent from short peduncles, reach a diameter of 5–7 cm only because the sepals and petals remain in a bell-shaped position. This rare species is a very close relative of *P. lanata* and *P. adulterina*, endemic in the Cordillera Oriental in Colombia. A definite distinction between these species cannot always be made solely by means of herbarium material, because they are too similar. In addition, there is a limited variability with regard to the stipules and degree of pilosity. This becomes clear when comparing the typical occurrences of *P. lanata* and *P. adulterina* from the department of Cundinamarca with those from Boyaca. In an ideal case, *P. crispolanata* can be distinguished from these species by its polymorphic leaves: in the parts that bear flowers, there are both unlobed and asymmetrically 2- or 3-lobed leaves. All other criteria, above all pilosity, will be helpful only at times. Further type collections and investigations on the spot are necessary in order to settle the questions regarding this group of species. It is hoped that such field studies can soon be pursued in Colombia. In view of its background, it is not surprising that this species has not yet come into cultivation.

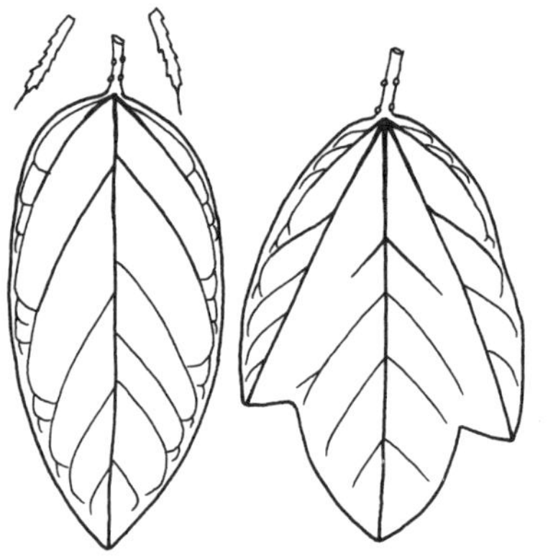

Passiflora crispolanata

Vine: densely pubescent, except upper leaf surface and flowers; stem: subangular to terete, striate; stipules: linear-lanceolate, 0.7–1.6 × 0.1–0.3 cm, minutely serrate; petioles: 0.7–1.7 cm with up to 10 subsessile glands; leaves: polymorphic, unlobed or variously lobed, unlobed leaves 5–9 × 1.8–4.1 cm, 3-lobed leaves 4.5–8.3 × 2.3–8.5 cm, coriaceous, slightly serrate; peduncles: 2.3–5 cm, solitary; bracts: lanceolate to oblong-ovate, 2–3.6 × 0.6–1.4 cm, free; floral tube: cylindrical, 6.2–9 cm long, 0.7–1.5 cm wide, green turning light brownish purple toward apex; flowers: pink, 5.5–7.5 cm diam., (9–)11–12.5 cm long; sepals: pink, narrowly oblong-ovate, 2.3–3 × 0.7–1.5 cm, aristate, awn ca. 0.2 cm long; petals: pink, subequal to sepals; corona: 1 series, reduced to a violet band with white teeth, ca. 0.1 cm; ovary: ellipsoid, glabrous; fruit: obovate, 5–7.5 × 3–4 cm, yellow with white speckles; seeds: obovoid, ca. 0.7 × 0.4 cm.

Passiflora cumbalensis (Karst.) Harms

Bot. Jahrb. Syst. 18 Beibl. 46: 13 (1894)
named after Volcán Cumbal, Colombia, the type location
Synonyms: *P. ecuadorica* Killip, *P. goudotiana* Triana & Planch., *P. macrochlamys* Killip, *P. mesadenia* Killip, *P. sparrei* Holm-Nielsen, *P. tungurahuae* Sodiro
Section *Elkea*

Passiflora cumbalensis has one of the most colorful flowers within supersection *Tacsonia* (see Plates 260–262). Together with the closely related *P. glaberrima, P. loxensis, P. luzmarina, P. parvifolia*, and *P. roseorum*, this species represents a complex of species characterized by a different touch of blue in the flowers. Depending on the intensity of this touch, the flowers can be described as pinkish, lavender-pink, or purplish blue.

Passiflora cumbalensis has a relatively large native range, which is quite unusual for tacsonias. Its habitats reach from northern Colombia to central Peru and are located at altitudes of 1800–4100 m. Consequently, the individual occurrences of this species vary considerably. With regard to specific features, clones from Colombia may differ considerably from clones that are growing in Peru. At times the differences are so conspicuous that one might regard them as being separate species. Upon a closer look, however, one notices that there are numerous transitional forms between both extremes. These transitional forms do not follow any geographical orientation; there does not seem to be a line of development from the north to the south.

As a consequence, this species was split up into a number of varieties. In her 1987 revision of *Passiflora cumbalensis*, Linda Escobar distinguished 7 varieties. Another 3 were presented in *Flora of Ecuador* (Holm-Nielsen et al. 1988) one year later.

Unequivocal distinctions between individual varieties were easier to make in the past due to the limited known ranges of the varieties. But additional collections are expanding the ranges and blurring the varieties. Without knowing the origin of a plant, it becomes very difficult to define it by means of leaf form, pilosity, and color of the floral tube. A revision of this species complex seems necessary, and it is possible some varieties will need to be moved. In the end, it is not easy to differentiate *Passiflora cumbalensis* from closely related species. In some cases the variability within this species is greater than the difference between *P. cumbalensis* and some other species. It can be distinguished from the Ecuadorian *P. loxensis* and *P. roseorum* by its subreniform stipules. The species of northern Peru can be distinguished by the smaller flowers of *P. glaberrima*, by the smaller leaves of *P. parvifolia*, and by the largerer stipules of *P. runa*.

The mostly red but occasionally yellowish fruits of *Passiflora cumbalensis* are edible, so this species is occasionally cultivated or harvested where it grows in the wild. Several clones from Ecuador have come into cultivation during the last few years. Despite recent reports about flowering plants, *P. cumbalensis* is not considered free-flowering. It is advisable to plant it out directly in the ground of a cold greenhouse. The lateral shoots should be removed up to a height of about 1.5 m. This promotes the development of a vigorous main shoot on which flowers are more likely to appear.

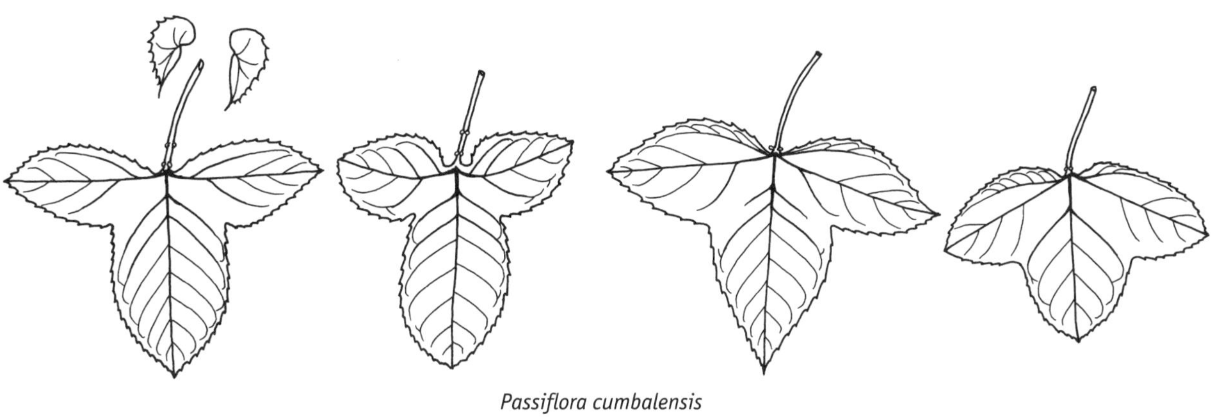

Passiflora cumbalensis

Vine: glabrous or occasionally partly pubescent; stem: subangulate to terete, tip erect; stipules: subreniform, 0.7–2.4 × 0.3–1.3 cm, glandular-serrate; petioles: 1.1–4.5 cm, with 2–4 glands, generally near apex; leaves: quite variable, 3-lobed, 7.6–14 × 3.5–16 cm, upper leaf surface glabrous, glabrous or pubescent beneath, serrate; peduncles: 3–10.5 cm, solitary; bracts: oblong-ovate, 2–6 × 0.9–2.7 cm, connate for 2/3–4/5 of their length, entire; floral tube: cylindrical, 5–13 cm long, 0.6–1.3 cm wide, purplish blue, violet, pink, or greenish; flowers: lavender-pink, purplish blue, or pinkish, (6–)8–10 (–12) cm diam., (8–)10.5–15 cm long; sepals: lavender-pink, purplish blue, or pinkish, oblong, 2.5–5.6 × 1.2–2.4 cm, aristate, awn ca. 0.2 cm long; petals: lavender-pink, purplish blue, or pinkish, oblong, 2.5–5.6 × 1–1.5 cm; corona: 1 series, purple ring with white filaments, ca. 0.1 cm; ovary: ellipsoid, glabrous; fruit: oblong-ovate, 5.5–9 × 1.5–4.5 cm, red or rarely yellow, edible; seeds: obovoid, 0.6–0.7 × 0.4 cm, reticulate, dark brown.

Passiflora gracilens (A. Gray) Harms

Nat. Pflanzenfam. 3 (6a): 91 (1893)
gracilens: very slender, dainty
Synonym: *P. boliviana* Harms
Section *Boliviana* group

At a length of 3.5–5 cm and a diameter of just 2–3.5 cm, the flowers of *Passiflora gracilens* (Plate 263) are the smallest of any tacsonia. The small fruits are also unique: greenish, flushed with red at maturity, and with a soft pericarp. There is another peculiarity in *P. gracilens*: in addition to the stipules, there are small stipulelike leaves at the bifurcations of the shoot. The native range of this species reaches over wide parts of Peru and extends into northern Bolivia. It grows in dry scrub vegetation at altitudes of 2400–3500 m. Unfortunately, it does not seem to be under cultivation.

Passiflora gracilens

It is quite astonishing that there is such a dainty plant among tacsonias, which are characterized by their vigorous growth. This species may belong in a different part of the genus. Its appearance suggests a certain proximity to the species of supersection *Decaloba*. Although its combination of characteristics is unique in comparison with *Decaloba* species, there are representatives of that supersection that show one or several of the characteristics typical of *Passiflora gracilens*: flower form, petiole glands, bracts, 3-lobed leaves, and fruit. Some day a genetic examination will bring about a clear picture of the real systematic relationships.

Passiflora gracilens is another exceptional passionflower that illustrates the enormous biodiversity within this genus. *Passiflora raimondii*, described in 1960 (Killip and Cuatrecasas), is distinguished from *P. gracilens* by its larger proportions.

Vine: glabrous; stem: 4-angulate, tip erect; stipules: lanceolate to slightly subreniform at base, 0.2–0.7 × 0.1–0.2 cm; petioles: 0.5–1(–3) cm, slender, with 2(–4) glands near apex, occasionally missing; leaves: variable, 3-lobed, 0.7–5.5 × 2–8.5 cm, subentire or slightly serrate; peduncles: 2–6 cm, very slender, solitary; bracts: ovate to lanceolate, 0.5–1.2 × 0.2–0.7 cm, free, subentire to slightly serrate; floral tube: cylindrical, 1.5–3.5 cm long, 0.3–0.8 cm wide, green turning pinkish toward apex; flowers: pink, 2–3.5 cm diam., 3.5–5 cm long; sepals: pink, oblong, 0.8–2 × 0.4–0.5(–0.9) cm, keeled, keel terminating in an awn 0.1–0.3 cm long; petals: pink, subequal to sepals; corona: 1 series, reduced to a purple ring with short tubercles; ovary: ellipsoid, glabrous; fruit: ovoid to ellipsoid, (2.3–)3.5–5 × 1–1.8 cm, greenish red; seeds: obovoid, ca. 0.4 × 0.2–0.3 cm, shallowly reticulate, brown.

Passiflora harlingii Holm-Nielsen

Bot. Not. 127: 339 (1974)
named after Gunnar Harling, botanist and editor
Section *Rathea*

Passiflora harlingii (Plates 30 and 264), *P. andina*, and *P. colombiana*, all previously assigned to subgenus *Rathea*, were moved into supersection *Tacsonia* by Feuillet and MacDougal. Despite all the similarities regarding vegetative appearance and native range, the representatives of section *Rathea* are conspicuously different from all other tacsonias because of their floral structure. For example, the flowers of *P. harlingii*, endemic in north-central Ecuador at altitudes of 2400–3500 m, do not open in the way that is usual for tacsonias. Instead, the sepals and petals are an elongation of the floral tube and form a slightly widened funnel attachment. The diameter of a completely opened flower is merely 1–3 cm. More impressive is its length: up to 18 cm. Thus its tubular flowers, orange for about 2/3 of their length and greenish in the distal 1/3, are longer than those of the closely related *P. andina*, which grows about 100 km farther to the north.

The occurrence of this unique flower morphology is restricted to southern Colombia (*Passiflora colombiana*) and northern to north-central Ecuador (*P. andina* and *P. harlingii*). Although *P. bracteosa* and *P. purdiei*, 2 species from section *Tacsoniopsis* that are at home in northern Colombia, seem to have a comparable flower at first sight, they are not close relatives of the species from section *Rathea*. Some years ago Giselher Behnken succeeded in raising plants from seeds of *P. harlingii* that he had collected during a trip to South America. With their nodding shoot tips and their deeply 3-lobed leaves, these plants look very similar to young plants of *P. ampullacea*. However, they have turned out to be quite fastidious and difficult to grow.

Vine: glabrous, except ovary and veins of upper leaf surface; stem: angular, tip flexed; stipules: linear-lanceolate, 0.8–1 × 0.2 cm, glandular-serrulate; petioles: 1.5–2 cm, with 2–6 minute glands; leaves: deeply 3-lobed, 5–7 × 5–8 cm, subcoriaceous, serrulate, revolute; peduncles: 10–12 cm, solitary; bracts: 2 or 3, ovate, 1.5–2 × 1.5–2 cm, free, serrate; floral tube: 2 cm, tapering into a long spongy calyx, orange turning green toward apex; flowers: orange and green to greenish yellow, 1–3 cm diam., 16–18 cm long; sepals: oblanceolate, ca. 1/4–1/3 of tube length, 1.5 cm wide, aristate, awn 0.1–0.2 cm long; petals: yellowish green, oblong to obovate-oblong, ca. 1/3 of flower length; corona: 2 series, filaments of outer series 0.1–0.3 cm long, inserted 1 cm above base of floral tube, filaments of inner series minute; ovary: ellipsoid, densely villous-tomentose; fruit: oblong-ovoid, 6.3–9.5 × 2.6–5.2 cm, green with grayish structure; seeds: oblong, 0.7 × 0.3 cm, shallowly reticulate at middle, dark brown.

Passiflora insignis (Mast.) Hook.

Bot. Mag. 99: pl. 6069 (1893)
insignis: excellent
Synonym: *P. callimorpha* Harms
Section *Insignes*

The flowers of *Passiflora insignis* (Plate 265), 13–17 cm diam., are among the largest of the whole genus. They usually appear somewhat smaller, however, because the corolla is spread out completely only upon opening and is folded back afterward. This species is limited to northern and north-central Bolivia, where it can be found on the eastern slopes of the Andes near La Paz at altitudes of

Passiflora harlingii

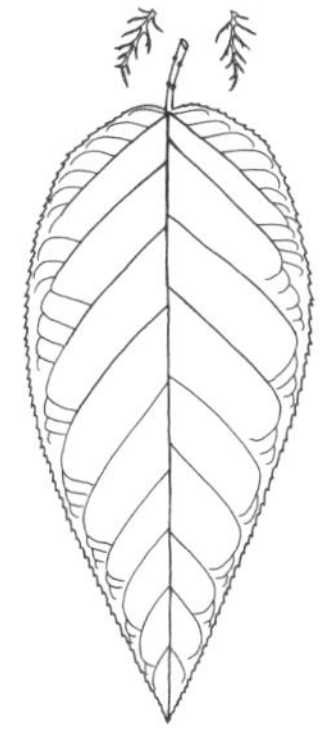

Passiflora insignis

1300–3000 m. This imposing plant, which belongs to section *Insignes* together with 4 other species, is characterized by its pinnatisect stipules, thick brownish pilosity, and short floral tube. The outer, violet coronal series is also remarkable at a length of up to 1 cm. *Passiflora insignis* can be distinguished from the closely related *P. pilosicorona*, also endemic in Bolivia, by its unlobed leaves.

The exact status of this species was only made clear a few years ago. The first description, published in 1873, was based on a plant that had been grown from seeds at Kew Gardens in London. In his monograph, Killip (1938) noted that the vegetative characteristics of *Passiflora insignis* and *P. callimorpha* are very similar and doubted the independence of this species. It seemed to him far more probable that *P. insignis* was a cultivar, possibly a cross between *P. callimorpha* and an unidentified species. About 50 years later, Linda Escobar finally ascertained that *P. insignis* is conspecific with *P. callimorpha*, endemic in Bolivia.

It is known from these descriptions that the magnificent *Passiflora insignis* flowered under cultivation at Kew Gardens about 130 years ago, but it has not been represented in collections for many decades. This is all the more astonishing because *P. insignis* is not at all rare in its habitats, and these regions are easily accessible. Nevertheless, Thomas Schneider was able to bring this species into cultivation in 2003. The first flowers were reported within a year of sowing the seeds.

Passiflora insignis tolerates temperatures around 30°C better than other tacsonias.

Vine: villous, except upper leaf surface; stem: terete; stipules: deeply pinnatisect; 1.8–2.5 × 1 cm; petioles: 1–2 cm, with 4 slightly elongated glands; leaves: unlobed, oblong-ovate, 6.5–25 × 4–12 cm, coriaceous, rugose, serrate; peduncles: 7.2–20 cm, stout, solitary; bracts: oblong-ovate, (2.5–)3.5–4.7 × 1.4–2.5 cm, free, short-fimbriate; floral tube: cylindrical, (1.8–)2.4–4 cm long, 0.5–0.8 cm wide, brownish; flowers: deep rose, 13–17 cm diam., 8–14 cm long; sepals: deep rose, oblong, 5.4–8.5 × 1.8–2.4 cm, keeled, keel terminating in an awn (0.8–)1.5–2 cm long; petals: deep rose, oblong, 6–7.5 × 1.5–2.5 cm; corona: 1(–2) series, filaments of outer series 0.8–1 cm long, violet, white at apex, inner series (if present) borne ca. 0.7 cm below outer; ovary: ovoid, villous; fruit: unknown.

Passiflora jamesonii (Mast.) Bailey

Rhodora 18: 156 (1916)
named after Hampden Gurney Jameson, collector of the type
Section *Elkea*

This lovely species (Plate 266) is among the few tacsonias that proceed to altitudes of 3000–4000 m in central and northern Ecuador. Even near the equator the climate is rough at such altitudes. Night temperatures often drop below the freezing point, and maximum day temperatures are around 15°C. The soil temperature, measured 30 cm beneath the surface, does not exceed 3°–4°C year-round. It is difficult to believe a passionflower could grow in such a place. Nevertheless, *Passiflora jamesonii* survives in such adverse surroundings, producing its spectacular pink to magenta flowers with a diameter of 10–12 cm.

It can be distinguished from other species of section *Elkea* by its lack of hair and by its free bracts with a deeply lacerate margin. It is also characterized by its stipules, which are similar in form to the bracts but a little smaller. The 3-lobed leaves are also remarkable, comparatively small at a length of 2.5–6.5 cm. The flower and androecium are slightly zygomorphic, placing the pollen on the hummingbird's chin. Because of its distinctive morphology, this species does not fit well into section *Elkea*, whose representatives are all distinguished by their connate bracts (with the exception of *Passi-*

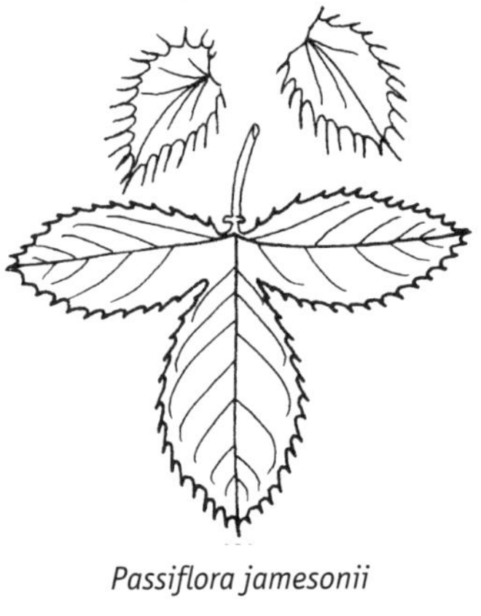

Passiflora jamesonii

flora zamorana). Like most species from such extreme altitudes, *P. jamesonii* is difficult to cultivate. In comparison with many other tacsonias, it causes far more problems. At temperatures beyond 22°C in summer, it feels uncomfortable and begins to ail. Temperatures are not a problem in winter, but light conditions can be insufficient. Hence, *P. jamesonii* is only recommended when it can be cultivated in a spacious, well-lighted, cold greenhouse. *Tacsonia* hybrids that are less fastidious and more willing to flower are occasionally offered under the name *P. jamesonii*. These are neither a cross with this species nor do they have anything in common with *P. jamesonii*. They can be distinguished from the real species at first sight by means of their stipules.

Vine: glabrous; stem: terete or subangular, tip flexed; stipules: ovate-lanceolate, 1.4–2.6 × 0.6–1.4 cm, lacerate; petioles: 1–2.5 cm, with 2 elongated glands near apex; leaves: deeply 3-lobed, 2.5–6.5 × 5.5–9 cm, serrate, revolute; peduncles: 4–8 cm, solitary; bracts: oblong-ovate, 2–3.5 × 0.8–1.7 cm, free, deeply lacerate; floral tube: cylindrical, 8.5–12 cm long, (0.8–)1.2–1.7 cm wide, purplish, greenish at base; flowers: pink to magenta, 9.5–12(–14) cm diam., 14–17.5(–19) cm long; sepals: pink to magenta, oblong, 3.9–6.5 × 1.4–2.7 cm, aristate, awn ca. 0.2 cm; petals: pink to magenta, oblong, 4.3–6.7 × 2.3–3.3 cm; corona: 1 series, filaments 0.1–0.2 (–1.4) cm long, dark purple; ovary: ellipsoid, glabrous; fruit: ellipsoid to oblong-ovate, ca. 7 × 3–4 cm, brownish violet; seeds: elliptic, 0.6 × 0.3–0.4 cm, weakly reticulate, black.

Passiflora jardinensis Escobar

Mutisia 71: 4 (1988)
named after Municipio de Jardín, Colombia, location of the type plant
Section *Parritana*

It is always a particular pleasure to find among the more or less pink tacsonias one of the few species with orange flowers, such as *Passiflora jardinensis* (Plate 267). This endangered species only grows in a tiny area on the top of one hill in the Municipio de Jardín in Colombia, on the Cordillera Occidental, at an altitude of 2750–2900 m. The occurrence of orange to salmon-pink flowers within supersection *Tacsonia* is limited almost exclusively to Colombia and is completed only by *P. sanctae-barbarae* from northern Ecuador.

The flower of *Passiflora jardinensis* is average in size, 8–10 cm diam. This species is characterized by its extremely deeply 3-lobed leaves with lanceolate or linear lobes and by the keel ridge on the back of its orange sepals. It also has lanceolate stipules, a typical feature of many species endemic in Colombia. With regard to these characteristics, Linda Escobar pointed out in the first description of the species its kinship with *P. parritae* from the eastern slopes of the Central Cordillera in central Colombia. *Passiflora jardinensis* can be distinguished from *P. parritae* not only by its different geographical origin but also by its shorter awns, lack of pubescence, and considerably narrower leaf lobes. This species has not yet come into cultivation.

Vine: glabrous; stem: subangular to terete, striate; stipules: linear-lanceolate, 0.8–1.2 cm, entire; petioles: 0.8–1.5 cm, with 2–4 subsessile glands; leaves: deeply 3-lobed, 3–9.7 × 4.5–9 cm, leaf lobes lanceolate to linear, 0.3–0.7 cm wide, revolute, subentire; peduncles: 6–14 cm, solitary; bracts: oblong-ovate, 1.9–2.6 × 0.7–0.9 cm, glandular-serrate, connate at base, rarely free; floral tube: cylindrical, 6.6–7.7 cm long, 1–1.2 cm wide, orange, greenish yellow at base; flowers: orange to salmon, 8–10 cm diam., 11.5–12.5 cm long; sepals: orange to salmon, oblong, 4.1–5.3 × 1.3–2.2 cm, keeled, keel terminating in an awn 0.3–0.6 cm long; petals: orange to salmon, oblong, 4–5 × 1.5–2.3 cm; coro-

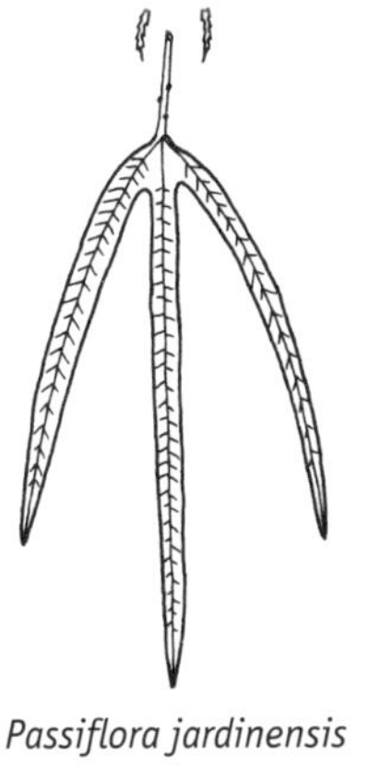

Passiflora jardinensis

na: 1 series, reduced to a ring; ovary: ellipsoid, glabrous or pubescent; fruit: oblong-ovate, 6.4–8.3 × 2.7–4.5 cm, yellow; seeds: ovoid, 0.8–0.9 × 0.4–0.5 cm, reticulate, dark brown.

Passiflora lanata (Juss.) Poir.

Encycl. 2: 843 (1811)
lanatus: woolly
Section *Colombiana*, Series *Colombianae*

Passiflora lanata (Plate 268) has the thickest pubescence of all *Tacsonia* species. The undersides of the leaves are rugose, the young shoots are covered with a brownish coat of hair, and the petiole glands are concealed by the thick hairiness. Michael Schwerdtfeger has given it a fitting pet name: teddybear.

The native range of *Passiflora lanata* is on the Cordillera Oriental in Colombia at altitudes of 2200–3500 m. It reaches from the department of Boyaca down to Bogotá and some kilometers farther to the south of the capital. Its bright pinkish rose flowers are pendent on short peduncles and reach a diameter of 5.5–7.5 cm.

Passiflora lanata is distinguished from the closely related *P. adulterina*, at home in the same region, by the acuminate apex of its leaves and by its thicker vestiture. However, there are often plants, presumably natural hybrids, that cannot be assigned unanimously to one of the 2 species. *Passiflora crispolanata*, another representative from this complicated complex of species, is characterized by its polymorphic leaves; this leads to further identification problems. The whole dilemma is complicated by the fact that clones of *P. lanata* from the department of Cundinamarca are typical of this species, while those from Boyaca are significantly different with regard to their vestiture, less acuminate leaf apex, and width of their stipules.

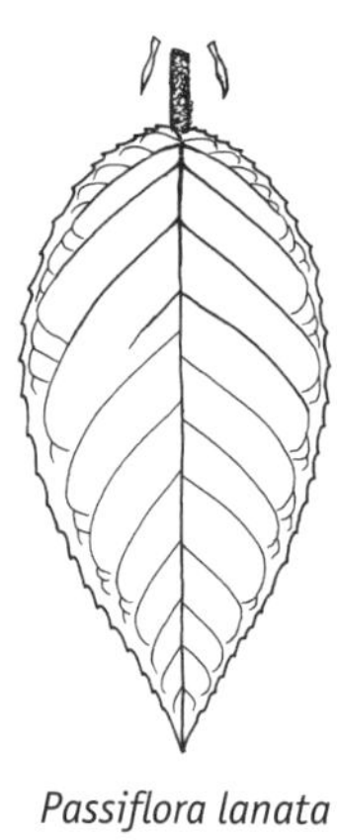
Passiflora lanata

Passiflora lanata is occasionally cultivated in Colombia but has not yet found its way into enthusiasts' collections in the United States and Europe.

Vine: lanate, except upper leaf surface and flower; stem: subangular to terete, striate; stipules: linear-lanceolate, 0.8–1.2 × 0.1–0.2 cm; petioles: 0.5–1.8 cm, with 2–4 minute glands; leaves: unlobed, rarely 3-lobed, oblong-ovate to oblong-lanceolate, 4.5–15 × 2.7–8 cm, coriaceous, rugose, serrate; peduncles: 2–7(–8) cm, solitary or occasionally in pairs; bracts: lanceolate, 1.7–3.5 × 0.6–1, subentire, free; floral tube: cylindrical, 6–9.5 cm long, (0.5–)0.7–0.9 cm wide, pink; flowers: pinkish rose, 5.5–7.5 cm diam., (8.5–)11.5–13 cm long; sepals: pinkish rose, oblong, 2.7–3.7 × 0.9–1.5 cm, aristate, awn, 0.2–0.4 cm long; petals: subequal to sepals; corona: 1 series, reduced to a purple ring with minute tubercles; ovary: ellipsoid, glabrous; fruit: oblong-ovate, (4.2–)7–10 × (2.5–)3.7–4.5 cm, brownish yellow with white speckles; seeds: ovoid, 0.7–0.9 × 0.5–0.6 cm, reticulate.

Passiflora leptomischa Harms

Repert. Sp. Nov. 18: 295 (1922)
leptomischus: having a thin stalk (in reference to the peduncle)
Section *Colombiana*, Series *Leptomischae*

Passiflora leptomischa (Plates 269 and 270) has imposing flowers that hang from peduncles up to 25 cm long, a feature shared by 5 other species belonging to the series *Leptomischae*. It bears a great resemblance to *P. flexipes*, *P. antioquiensis*, and *P. cremastantha*. It differs from the latter 2 by its longer floral tube and by the characteristic contrast of light and dark on the stipules, which is most obvious on the herbarium material. This bicolor distinctness is due to the fact that the tips of the stipules dry out on the living plant.

In contrast to *Passiflora flexipes*, whose corona filaments reach a length of 0.3–0.4 cm, the corona of *P. leptomischa* is reduced to tiny knots. Furthermore, the relatively isolated native range of *P. leptomischa* in southern and western Colombia is a criterion of demarcation from related species. Its habitats are in wet, montane forests of the western slopes of the Cordillera Occidental, especially in the region around Munchique at altitudes of 2000–2800 m. This raises some questions about an as-yet-unidentified *Tacsonia* species found in Ecuador by Giselher Behnken. The yellow fruit was hanging on long peduncles. The large seeds germinated so well that the plant was easily cultivated. Its rugose, 3-lobed leaves are similar to those of *P. leptomischa*, and the stipules indicate that it might be this species. The herbarium in Quito also has several collections of this plant, all without flowers, which were mostly identified as *P. coactilis* (Michael Schwerdtfeger, personal communication). The true identity of this plant remains to be seen. Apart from this unidentified clone, *P. leptomischa* does not seem to be under cultivation.

Vine: pubescent; stem: subangular, striate, tip flexed; stipules: linear-lanceolate, 0.8–1 cm long; petioles: 0.9–2.5 cm, with 8–10 glands; leaves: unlobed or deeply 3-lobed, unlobed leaves 5.5–17 × 2.4–8.3 cm, 3-lobed leaves 5.2–15 × 7–16 cm, membranous, rugose, serrate; peduncles: (15–)24–27 cm, slender, solitary; bracts: oblong-lanceolate, 2–3 × 0.8–1 cm, free, serrate to fimbriate; floral tube: cylindrical, 7–8.5 cm long, 1–1.2 cm wide, purple turning greenish toward base; flowers: pink to purplish pink, 10–14 cm diam., 11.5–13 cm long; sepals: pink, oblong, 4.3–6(–6.8) × 1.5–2.9 cm, aristate, awn ca. 0.3 cm; petals: pink, oblong, 3.7–5.2(–6.3) × 1.8–3.3 cm; corona: 1 series, reduced to a violet ring, often with short tubercles; ovary: ellipsoid, pubescent; fruit: ellipsoid, ca. 11 × 3 cm, yellow with white speckles; seeds: unknown.

Passiflora loxensis Killip & Cuatrecasas

Contr. U.S. Nat. Herb. 35: 12 (1960)
named after the province of Loja, Ecuador, the type location
Section *Elkea*

The purplish violet flower of the spectacular *Passiflora loxensis* (Plate 271) is one of the largest in the whole genus. The sepals and petals of the pendent flowers can reach a length of up to 8 cm. The diameter of the campanulate corolla is 10–14 cm when it has opened completely. The flowers also reach a considerable length of 14–18 cm.

Due to an increasing number of herbarium specimens, the picture of *Passiflora loxensis* has become clearer, although in some aspects it does not differ from the closely related *P. cumbalensis* and *P. roseorum*. Often it is very difficult to distinguish these species from one another, especially in view of some specimens of *P. cumbalensis* with very large flowers from some regions of Ecuador. As a rule, however, *P. loxensis* can be identified by its larger flowers, larger bracts, and wide floral tube. Upon closer inspection, one notices the oblong-reniform, deeply dentate stipules and thickened venation on the undersides of the leaves. This venation is also characteristic of *P. roseorum*.

Passiflora loxensis is endemic to the provinces of Loja and Zamora-Chinchipe, southern Ecuador. It is found at altitudes of 2750–3400 m. As it does not grow alongside roads but instead along more remote mule tracks, one must deliberately search for it. In 2001, Giselher Behnken found a plant with ripe fruit, and it is possible he may introduce this species in Europe. Its cultivation looks very prom-

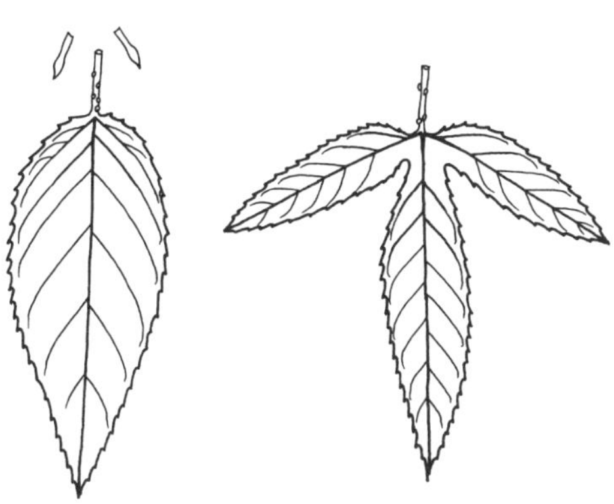

Passiflora leptomischa

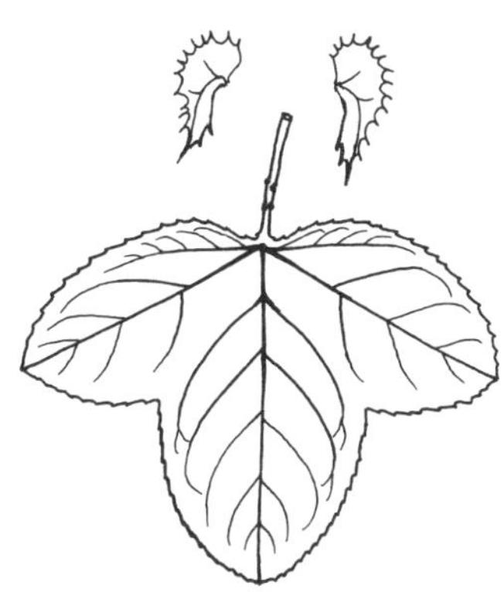

Passiflora loxensis

ising. It grows without problems, similar to *P. luzmarina* and other species from the *P. cumbalensis* complex that are at home in southern Ecuador. One year after sowing the seeds, the plants have not yet reached the size at which flowers can be expected. It remains to be seen how willing they will be to flower. This species suffers a little at temperatures above 25°C but does not seem to be as susceptible to warmer temperatures as other representatives from the high altitudes of the Andes. During overwintering, it tolerates temperatures down to the freezing point for a short while, but in general temperatures should not fall below 5°C.

Vine: nearly glabrous; stem: angular, striate; stipules: oblong-reniform, 1.5–2.5 × 0.7–0.8 cm, deeply dentate to slightly fimbriate at margin; petioles: 1.3–3.3 cm, with 3 or 4 subsessile glands at upper 1/2; leaves: 3-lobed, 3.5–7 × 5.5–9.7 cm, subcoriaceous to coriaceous underneath with thickened secondary veins, entire; peduncles: 1.5–6(–9) cm, solitary; bracts: oblong-ovate, 4.7–7 × 2.2–3 cm, connate for 2.3–3.5 cm, free parts entire; floral tube: cylindrical, 7.5–10 cm long, 1.9–2.5 cm wide, purplish violet outside; flowers: purplish violet, 10–14 cm diam., 14–18 cm long; sepals: purplish violet, oblong, 6–8 × 2–2.8 cm, keeled, keel terminating in an awn 0.2–0.5 cm; petals: purplish violet, oblong, 6–8 × 2.5–3.3 cm; corona: 1 series, reduced to a colored band with white teeth; ovary: oblong-ovoid to ellipsoid, glabrous; fruit: oblong-ovoid to ellipsoid, 6.5–7 × 2–3 cm, deep red; seeds: obovoid, 0.5–0.6 × 0.3–0.4 cm, reticulate.

Passiflora luzmarina Jørgensen

Novon 7: 385 (1997)
named after Luz Marina Unda, niece of the author
Section *Elkea*

Passiflora luzmarina (Plate 272), a small-flowered relative of *P. cumbalensis*, was discovered just a few years ago in southern Ecuador at altitudes of 2450–2700 m. Part of its known habitats near La Argentina in the province of Loja have since been destroyed by building projects and landslides, and the few plants left are endangered. Giselher Behnken and Michael Schwerdtfeger have been able to bring several clones into cultivation. Within 9 months after sowing the seeds, the first flowers appeared on shoots about 1 m long. Accordingly, the free-flowering *P. luzmarina* can be cultivated successfully in a 20-cm pot, which is normally not recommended for *Tacsonia* species. In contrast with its vigorously growing relatives, this species tends to branch out, yet this does not hinder it from setting flower buds. *Passiflora luzmarina* is therefore another highly recommended tacsonia for enthusiasts. It is the first free-flowering representative of the *P. cumbalensis* group to be grown under cultivation.

The flowers of *Passiflora luzmarina* reach a length of 4.5–7 cm and a diameter of 3.5–5 cm. Their color varies from light pink to light purple. It is remarkable that tacsonias with a shorter floral tube occur comparatively frequently in southern Ecuador and northern Peru. Besides *P. luzmarina*, 4 other species with a short floral tube occur in this region: *P. mathewsii*, *P. linda*, *P. glaberrima*, and the doubtful *P. brachyantha*, which seems to be a cross between *P. luzmarina* and *P. mathewsii*. This is especially interesting because some of these species come from different lineages but have nevertheless adopted the same strategy. Presumably this development is due to coevolution with short-billed hummingbirds. If so, we expect relatively larger populations of short-billed hummingbirds to occur in that region.

Vine: glabrous; stem: terete to angular, striate, tip erect; stipules: reniform, 0.7–1.7(–2.7 cm) × 0.3–1 cm, entire or serrate; petioles: 1.2–6 cm, with (2–)4(–5) stipitate glands, one pair near apex; leaves: deeply 3-lobed, 2.8–11.7 × 5.2–20 cm, subentire or serrate; peduncles: 2.5–6(–10) cm, slender, solitary; bracts: oblong-ovate to elliptic, 1.6–2.8

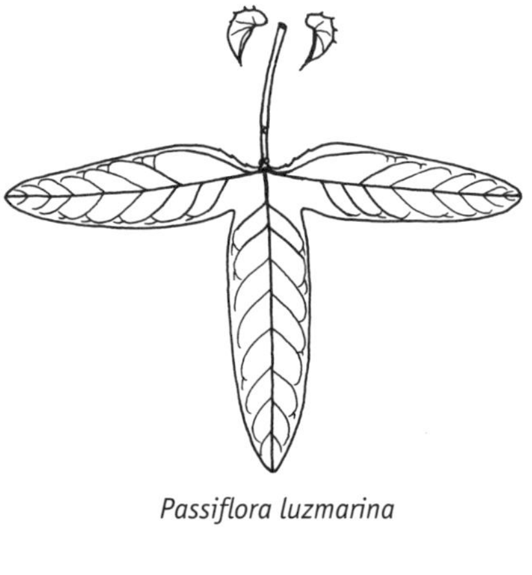

Passiflora luzmarina

× 1.3–1.4 cm, connate for 0.8–1.6 cm, entire; floral tube: cylindrical, 2–4.3 cm long, 0.4–0.7 cm wide, light pink to light purple; flowers: pink to light purple, 3.5–5 cm diam., 4.5–7 cm long; sepals: pink, oblong, 2.2–2.7 × 1–1.2 cm, aristate, awn ca. 0.1 cm; petals: pink, oblong, 1.8–2.6 × 0.4–0.9 cm; corona: 1 series, reduced to a purple ring with white teeth; ovary: ellipsoid, glabrous; fruit: oblong-ovate, 4–7.2 × 1.5–2.7 cm, orange-red; seeds: obovoid, 0.5–0.6 × 0.3–0.5 cm, reticulate, dark brown.

Passiflora macropoda Killip

Contr. U. S. Natl. Herb. 35: 14 (1960)
macropodus: having a long stalk
Section *Manicata*

Passiflora macropoda (Plate 273) belongs to the former subgenus *Manicata*, which is now incorporated in supersection *Tacsonia* as a section. At first sight this species differs from typical tacsonias by its upright white flowers, short floral tube, and outer coronal series, which is up to 2.5 cm long. *Passiflora macropoda* is at home in northern Bolivia at altitudes of 2900–3100 m, where day temperatures do not exceed 20°C.

Its flowers reach a maximum diameter of 12 cm, and it has remarkable large bracts with incised to lacerate margins. The floral morphology suggests that *Passiflora macropoda* is pollinated by nocturnal bats. This is indicated by the long stout peduncles, wide-open calyx, long androgynophore that is curved toward the upper end, zygomorphic androecium, and flower color.

A comparison of flowers reveals that *Passiflora macropoda* is closely related to *P. trisecta* and *P. weberbaueri*. Nevertheless, it is possible to differentiate these 3 species at first glance. While *P. trisecta* and *P. weberbaueri* have 3-foliate leaves, the leaves of *P. macropoda* are 3-lobed only to their middle.

At least 2 different clones of this species are infrequently cultivated. Thomas Schneider reports that his specimen grows 2.5 m within one year from a small plant, but without developing the typical adult leaf form of flowering plants.

Vine: partly pubescent; stem: terete, stout; stipules: cordate-ovate, 2–3 × 1.5–2 cm, laciniate-fimbriate; petioles: stout, 1.5–3 cm, with 2–3 short stipitate glands near apex and additional 1–2 glands below; leaves: 3-lobed, 10–12 × 16–18 cm, subcoriaceous, serrate; peduncles: 12–14 cm, stout, solitary; bracts: ovate-lanceolate, 4–5 × 2 cm, free, laciniate-fimbriate; floral tube: cylindrical-campanulate, 1.6–2.5 cm long, ca. 1.3 cm wide; flowers: white, 8–12 cm diam., 6–7.5 cm long; sepals: white, narrowly lanceolate, 4–5.5 × 1–1.5 cm, keeled, keel terminating in an awn ca. 0.8 cm long; petals: white, narrowly lanceolate, 3.5–5 × 1 cm; corona: 4 series, white, filaments of outer 2 series 1.5–2.5 cm long, inner series a short, thick membrane; ovary: ovoid, white-velutinous; fruit: unknown.

Passiflora mandonii (Mast.) Killip

J. Wash. Acad. Sci. 14: 213 (1924)
named after Gilbert Mandon, collector of the type
Synonym: *P. steinbachii* Harms
Section *Insignes*

Passiflora mandonii (Plate 274) is another species from section *Insignes*, whose representatives are

Passiflora macropoda

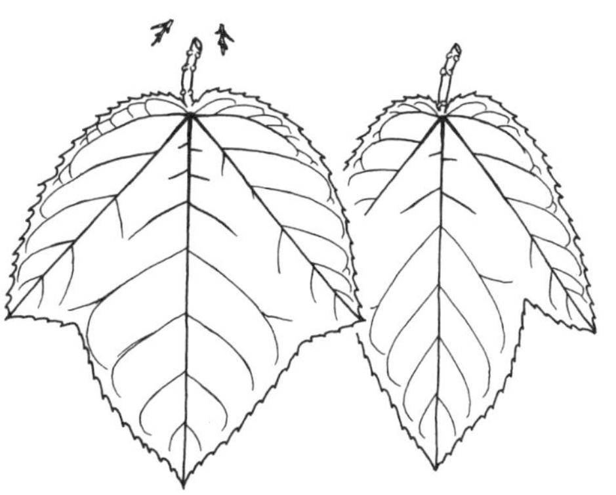

Passiflora mandonii

characterized by pinnatisect stipules, free, serrate to fimbriate bracts, and an exceptionally long outer coronal series. While the corona is usually reduced to a colored ring or little knots about 0.1 cm long, in *P. mandonii* it reaches a length of almost 2 cm. Another characteristic of this species is the form of its leaves. Its variable leaves, extremely rugose on the upper side, are often 3-lobed only toward the apex and only shallowly so.

This species, at home in central and northern Bolivia and occasionally in southern Peru at altitudes of 2500–4000 m, is sometimes confused with *Passiflora pinnatistipula* and *P. pilosicorona*. However, the quite rare *P. mandonii* can be distinguished from the more common and widespread *P. pinnatistipula* by its leaf form and degree of pilosity. *Passiflora pilosicorona* has similar leaves, but its corona filaments and floral tube are significantly shorter.

The light pinkish purple flowers of *Passiflora mandonii* reach a diameter of 7–10 cm. The oblong, dark green fruit is edible, as is the fruit of most tacsonias. However, the fruit of this species has been without any major importance and is unknown outside of Bolivia. Sometimes the fruit is collected from wild plants; sometimes it is grown in fruit gardens. At present this species does not seem to be under cultivation.

Vine: densely pubescent, except upper leaf surface; stem: subangular to terete; stipules: pinnatisect, 0.5–0.6 cm long; petioles: (1–)1.5–3.5 cm, with 5–11 subsessile glands, mostly in upper 1/3; leaves: variable, 3-lobed, 5–13.5 × 7.5–15 cm, coriaceous, rugose, serrate; peduncles: 4–15 cm, slender, solitary; bracts: oblong to oblong-ovate, 1.8–3 × 0.9–1.5 cm, free, glandular-serrate; floral tube: cylindrical, 5.5–9 cm long, 0.8–1.2 cm wide, greenish brown; flowers: pink to light purple, 7–10 cm diam., ca. 9–13 cm long; sepals: pink to light purple, oblong, 3–5.5 × 0.8–1.9 cm, aristate, awn 0.6–1 cm; petals: pink to light purple, oblong, 2.5–4.8 × 0.8–1.4 cm; corona: 1 series, violet, 0.8–2 cm; ovary: ellipsoid, densely tomentose; fruit: oblong, 5–7.5 × 4.5–5.5 cm, dark green to blackish; seeds: obovoid, 0.7–0.8 × 0.4–0.5 cm, reticulate, dark brown.

Passiflora manicata (Juss.) Pers.

Syn. Pl. 2: 221 (1807 [1806])
manicatus: cufflike (in reference to the stipules]
Synonyms: *P. meridensis* Karst., *P. rhodantha* Harms
Section *Manicata*

The brilliant red flowers of *Passiflora manicata* (Plate 275), up to 9 cm diam., are among the most beautiful of the entire genus. The species is clearly distinguished from the original tacsonias by its shorter floral tube and the corona in several series. It and 4 other species belong to section *Manicata*, which was named after *P. manicata*. The other species of this section range from Peru to Bolivia, including *P. trisecta* and *P. peduncularis*, and have white flowers that are adapted to pollination by nocturnal bats. Although this has not been explicitly proven for all 4 species, their floral morphology hardly admits another conclusion. So the hummingbird-pollinated *P. manicata*, with its red flowers, occupies a special position and can be easily identified.

The native range of *Passiflora manicata* is one of the largest of all *Tacsonia* species, reaching from Venezuela through Colombia to Ecuador and Peru. It very often occurs in high valleys at altitudes of 1400–3000 m.

Passiflora manicata is extremely well known not only because of its spectacular flowers but also because of its wide native range and long history in cultivation. Much to the enthusiast's regret, it is not particularly free-flowering. It prefers a location in a cold greenhouse, where it should be planted out directly in the ground. When it is allowed to reach

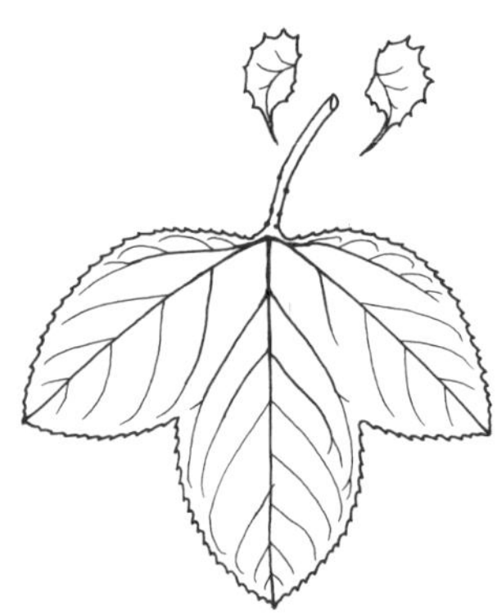

Passiflora manicata

an appropriate height, flowers can be expected on shoots that have reached a length of several meters. The predominant flowering period is late summer and fall. Cultivation in a transportable container during summer and in a greenhouse or conservatory in winter is only the 2nd best choice. Although the plant will grow well under these conditions, it will not reach the size necessary to induce flower formation. To this end, pruning should be effected with caution so that enough growth remains. On the whole, this species will tolerate higher temperatures than most tacsonias.

Unfortunately, *Passiflora manicata* is often confused with its hybrids, including *P.* 'Coral Glow', *P.* 'Coral Sea', and *P.* 'Mum's Gate'. The pale red of these hybrids' flowers does not compare to the unique brilliant red of the botanical species. However, the hybrids will flower more easily, even when cultivated in pots.

Vine: more or less pubescent; stem: terete, striate, tip erect; stipules: reniform, 0.5–2.5 × 0.5–2 cm, glandular-serrate; petioles: (0.5–)1–3(–5) cm, with 4–9 subsessile to slightly stipitate glands; leaves: 3-lobed, 4–11 × 4.5–14 cm, glabrous on upper surface, serrate; peduncles: 5–10 cm, stout, solitary; bracts: ovate, 2–4 × 1–2 cm, free or connate at base; floral tube: cylindrical-campanulate, 1–2 cm long, ca. 1 cm wide, green; flowers: red, 6–9 cm diam., 5–6 cm long; sepals: red inside, greenish outside, oblong, 3–4.3 × 0.5–1 cm, aristate, awn ca. 0.5 cm; petals: red, oblong, 2.5–4 × 0.5–1 cm; corona: 3–4(–5) series, filaments of outer 2 series 0.2–0.4 cm long, dark purple; ovary: ovoid, glabrous to puberulent; fruit: ovoid or ovoid-oblong, 3.5–5.5 × 2–3.7 cm, greenish; seeds: ovoid, 0.4–0.5 × 0.2–0.3 cm, reticulate, blackish brown.

Passiflora mathewsii (Mast.) Killip

J. Wash. Acad. Sci. 17: 428 (1927)
named after Andrew Mathews, collector of the type plant
Section *Tacsonia*

Although *Passiflora mathewsii* (Plates 276 and 277) belongs to the small-flowered species of supersection *Tacsonia*, it is no less attractive. Its pink flowers reach a diameter of about 5.5 cm. They are not pendent but usually horizontal or even slightly erect at times. Therefore, *P. mathewsii* might be called the little sister of *P. mixta*, from which it can be distinguished by its shorter floral tube, smaller flowers, and zygomorphic androecium. Its native range facilitates a distinction from the closely related and also small-flowered *P. schlimiana*. While the latter is endemic in a small, restricted region in northern Colombia, *P. mathewsii* can be found in southern Ecuador and northern Peru at altitudes of 2200–3500 m. The form of the stipules is remarkable, varying over the native range. The northernmost plants found in Ecuador, as well as many of those from Peru, differ from those in the region around Loja by their stipules with conspicuous, deeply lacerate-ciliate margins. Furthermore, in northern Peru, Maximilian Weigend came across an interesting population of *P. mathewsii* with a thick stem of up to 0.8 cm diam. and white, partly pink-tinted flowers. This illustrates the variability that sometimes occurs within a single species. Such examples extend our range of knowledge and allow us to expand our concept of a species that is often based only on a solitary cultivated clone.

During the last few years, *Passiflora mathewsii* has become an established species from supersection *Tacsonia* in many collections. Every now and then, a new variation from a different source of origin is brought into cultivation. Generally speaking, all clones develop quite well under cultivation, although there are differences regarding their willingness to flower. Judging from experience, a variation with white flowers does seem to be a very

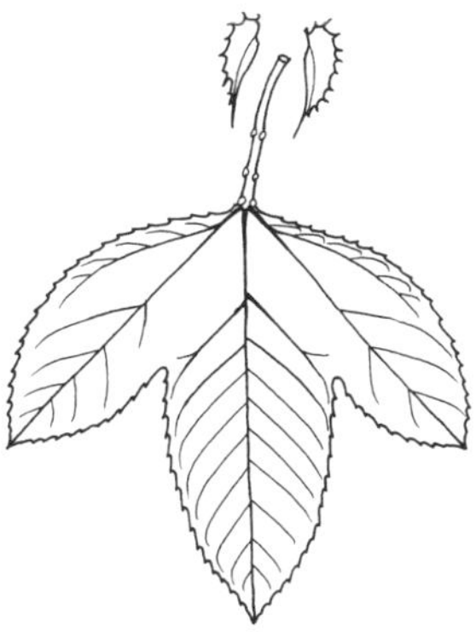

Passiflora mathewsii

good choice—grown in a 30-cm container, it flowers without problems every year. The first flower buds appear in late summer at a height of 1 m. Like *P. mixta*, *P. mathewsii* tolerates higher temperatures a little better than other tacsonias.

Vine: pubescent; stem: angulate, tip erect; stipules: reniform, 0.8–1.8 × 0.2–0.5 cm, glandular-serrate or -lacerate; petioles: 0.7–2.5(–3.5) cm, with 6–8 mostly paired, subsessile glands; leaves: 3-lobed, 3.5–8(–11.5) × 4–11(–13.5) cm, serrate; peduncles: 1.4–3.5(–4.5) cm, solitary; bracts: ovate, (2–)3–5.5 × 1.2–3.5 cm, connate for 2/3 of their length, entire; floral tube: cylindrical, 3–5.5 cm long, 1–1.2 cm wide, slightly narrowing toward base, green; flowers: pink to purplish pink, occasionally white, 4–5.5 cm diam., 7–8.5 cm long; sepals: pink to purplish pink, oblong, 2.2–4 × 0.8–1.6 cm, aristate, awn ca. 0.2 cm; petals: pink to purplish pink, oblong, 2.2–4 × 0.9–1.6 cm; corona: 1 series, reduced to a purple ring with white teeth, rarely up to 0.3 cm long; ovary: ellipsoid, densely white-pubescent; fruit: obovoid, 4–5.5 × 1.8–2.5 cm, green; seeds: obovoid, 0.4–0.5 × 0.3 cm, reticulate, dark brown.

Passiflora mixta L. fil.

Suppl. Pl. 408 (1781)
mixtus: mixed
Synonyms: *P. brachychlamys* Harms, *P. longiflora* Lam., *P. tacso* Cav., *P. tomentosa* Lam., *P. urceolata* (Mast.) Killip
Section *Tacsonia*

Passiflora mixta (Plates 248, 278, and 279) is a surprise among tacsonias for its horizontal to erect flowers that can sometimes reach a length of up to 18 cm. Because of the distinctive campanulate arrangement of the sepals and petals, the diameter is only 5–9 cm. Due to its wide native range, which reaches from western Venezuela to northern Bolivia and Chile, this species is quite variable. The clones of *P. mixta* differ from one another mainly with regard to the length of the floral tube, the varying pubescence, the size of the bracts, and the size of the leaves. These differences induced the authors of *Flora of Ecuador* (Holm-Nielsen et al. 1988) to subdivide the Ecuadorian species into 3 varieties. *Passiflora mixta* var. *eriantha* is distinguished by the dense, white pubescence on the undersides of its leaves and on the outside of its flowers. One characteristic of *P. mixta* var. *pilaloensis* is the floral tube, up to 15 cm long, that is surrounded at its base by a close-fitting, long tube formed by the bracts. All other types are assigned to *P. mixta* var. *mixta*.

Passiflora mixta is closely related to *P. mathewsii* and *P. schlimiana*, thus constituting section *Tacsonia* together with *P. anastomosans*. These 4 species all have horizontal to upright flowers. However, the flowers of *P. mathewsii* and *P. schlimiana* are much shorter than those of *P. mixta* (additional differences are described under *P. mathewsii*). Apart from the positioning of the flowers, the species of section *Tacsonia* correspond in every respect to the representatives of section *Elkea*. Hence, the distinction between *P. mixta* and *P. tripartita* var. *mollissima* is often a problem. Although the difference may be obvious in the field or when 2 cultivated plants are compared directly, they can hardly be distinguished when flowers are missing or when judging strictly from the herbarium material. Relying on the forms of the leaves, length of the bracts, or pubescence often leads to wrong results.

Where the natural range of *Passiflora mixta* overlaps with that of another *Tacsonia*, it is generally *P. mixta* that is found in the lower regions, from 1700 m upward, while other species, often *P. cumbalensis*, prefer the cooler, moister mountaintops. At times, hybrid forms occur in the transition zones (Michael Schwerdtfeger, personal communication). Very often, *P. mixta* proceeds to altitudes of up to 3300 m but rarely beyond that. The record holder is a collection from Apurimac, Peru, found at an altitude of 3800 m. *Passiflora mixta* is of no economic importance, but fruit is harvested from

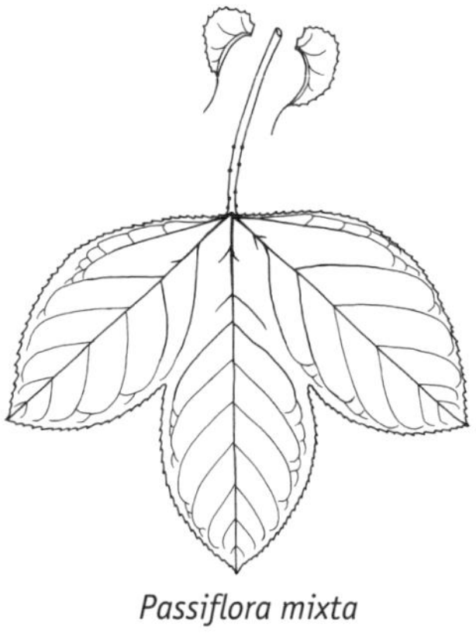

Passiflora mixta

wild plants and occasionally grown in fruit gardens.

Clones of different origin, especially from Ecuador, have been cultivated in recent years. Among these are several variations that rank among the most free-flowering tacsonias. There is another positive quality as well: the cultivated clones of *Passiflora mixta* will tolerate higher temperatures better than other tacsonias. In short, *P. mixta* is an absolutely recommendable species that will enrich every collection.

Vine: glabrous or pubescent; stem: in general angulate, tip erect; stipules: reniform, 0.6–3.3 × 0.2–2 cm, glandular-serrate; petioles: 0.9–8 cm, with 4–10 subsessile or stipitate glands; leaves: 3-lobed, 3.5–12 × 5–19 cm, glabrous on upper surface, serrate; peduncles: 1–9 cm, stout, solitary or rarely in pairs; bracts: oblong to oblong-ovate, 1.2–7.5 × 0.5–2.5 cm, connate for at least 3/4 of their length, entire; floral tube: cylindrical, 7–15 cm long, 0.5–1.5 cm wide, green; flowers: pink, occasionally pinkish orange, or rarely white, 5–9 cm diam., 9.3–19.5 cm long; sepals: pink, oblong to oblong-ovate, 2.3–5.5 × 0.9–2.1 cm, aristate, awn 0.2–0.5 cm; petals: pink, oblong to oblong-ovate, 2.5–5 × 1–2 cm; corona: 1 series, rarely 2 series, reduced to a purple ring with white teeth; ovary: ellipsoid, densely pubescent or glabrous; fruit: oblong-ovate, 4–7.2 × 2–3.5 cm, green; seeds: obcordate, 0.3–0.5 × 0.2–0.4 cm, reticulate, dark brown.

Passiflora parritae (Mast.) Bailey

Rhodora 18: 156 (1916)
named after Parra, collector of the type
Synonym: *P. salmonea* Harms
Section *Parritana*

One of the most beautiful tacsonias, *Passiflora parritae* (Plate 280) is also one of the rarest. Its few known habitats on the eastern slopes of the Central Cordillera in central Colombia, at altitudes of 1900–2800 m, are endangered. In view of this, gratitude is due to Linda Escobar and her students, who found a ripe fruit on a plant in the department of Caldas, Municipio of Pensilvana, at an altitude of about 2000 m. The seeds germinated, and Escobar passed on a seedling to John MacDougal, who took it to the United States. All other seedlings died, and the mother plant was cut down in its habitat. MacDougal gave the last remaining clone to the Strybing Arboretum in Golden Gate Park, San Francisco—an ideal refuge. Under the climatic conditions near the foot of the Golden Gate Bridge, tacsonias can be planted outdoors, and here the plant is flourishing magnificently, reaching the dimensions known from its natural surroundings and rewarding the public with countless flowers (see Plate 280, taken at the arboretum in August when the plant was in full bloom). *Passiflora parritae* is cultivated by a few enthusiasts in the United States, and some cuttings have found their way to Europe. So far, cultivation of this species has failed due to excessively high summer temperatures. In 2003, Giselher Behnken received seeds of this rare species that were collected in Colombia. The seedlings have grown well, and it is hoped that different clones will be established in European collections.

Passiflora parritae is distinguished by its large, brilliant orange flowers. A contrast is formed by a small pale spot at the base of the petals that gives the impression of a little yellow eye in the center of the flower. The diameter is 10–12 cm, rarely a little larger. Particularly striking are the enormous keel ridges on the back of the sepals, ending in very large awns. The closely related *P. jardinensis* has a comparable balloonlike flower bud and flower. Both species, constituting section *Parritana*, have characteristic, deeply 3-lobed leaves with narrow lobes. (For a description of their differences, see *P. jardinensis*). Other features, such as the linear stipules and free, serrate bracts, indicate a close kinship with the representatives of section *Colombiana*.

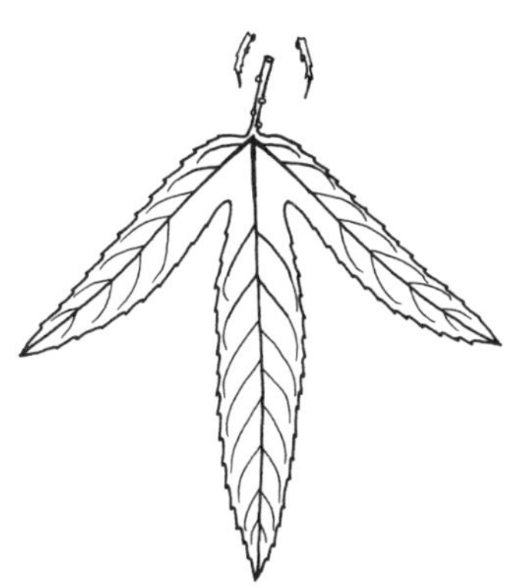

Passiflora parritae

Vine: pubescent; stem: terete, striate; stipules: linear, ca. 0.6 cm long; petioles: 1–2 cm, with 4–6 elongated glands; leaves: deeply 3-lobed, 4–8 cm × 4.5–13 cm, leaf lobes lanceolate, 0.8–1.6 cm wide, revolute, serrate; peduncles: 12–15 cm, solitary, rarely in pairs; bracts: lanceolate to oblong-ovate, 1.2–1.5 × 0.5 cm, greenish red, free or connate at very base, serrate; floral tube: cylindrical, 5.3–9 cm long, 0.7–1.1 cm wide, purplish orange outside, turning greenish toward base; flowers: orange to salmon, 10–12 cm diam., ca 11–17.2 cm long; sepals: orange to salmon, oblong-ovate, 5.2–6.2 × 1.6–3 cm, keeled, keel terminating in an awn 1–2 cm long; petals: orange to salmon, cream-yellow at base, oblong-ovate, ca. 4.5 × 2.2 cm; corona: 1 series, reduced to a purple ring with 0.1 cm teeth; ovary: obovoid, densely pubescent; fruit: ellipsoid, 6 × 4 cm, yellow, longitudinally grooved; seeds: obovoid, 0.7–0.8 × 0.4 cm.

Passiflora peduncularis Cav.

Icon. Pl. 5: 15 (1799)
peduncularis: conspicuous peduncle
Section *Manicata*

Passiflora peduncularis (Plate 281) is a typical representative of the species from the western Andes in Peru. Its native range extends from the department of La Libertad in north-central Peru to the department of Moquagua in the south, where it occurs at altitudes of 2500–3500 m. Besides the typical pubescent form, some individuals are nearly glabrous.

Even at a distance, the white flowers of *Passiflora peduncularis* are eye-catching. They are borne out of the foliage on stiff, upright peduncles that can be up to 20 cm long, and make this species unmistakable. The flowers reach a considerable diameter of 7–11 cm, but only for a short while, as the sepals and petals reflex widely. The arrangement of the short, white corona is remarkable and distinguishes this species from other representatives of section *Manicata*. The 2 outermost series are located at the tip of the floral tube. Beneath them are up to 6 indefinite series, with filaments up to 0.1 cm long, hidden in the floral tube. In addition, the unmistakable 3-lobed leaves of *P. peduncularis* distinguish it from the also white-flowering *P. trisecta* and *P. weberbaueri*, which are also endemic in Peru. Like many representatives from section *Manicata*, including those just mentioned, *P. peduncularis* is presumably pollinated by bats. Its flowers have the same bat-flowered characteristics as *P. trisecta*.

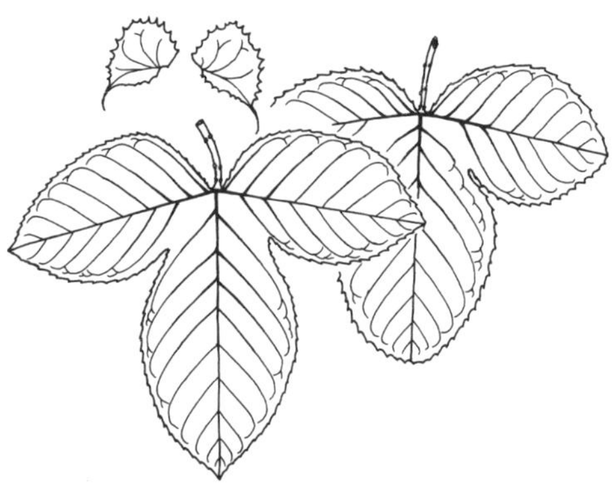

Passiflora peduncularis

Vine: densely pubescent or occasionally nearly glabrous; stem: subquadrangular, striate, stout; stipules: semiovate, 0.5–2.2 × 0.3–1.5 cm long, serrate, often purplish; petioles: 1–2.5 cm, with 3–4 glands; leaves: 3-lobed, 3.5–9 × 5–13.5 cm, glabrous above, densely pubescent beneath, subcoriaceous, serrate; peduncles: 10–20 cm, solitary; bracts: oblong-ovate, 2–4 × 0.8–1.5 cm, connate for 1/3–2/3 of their length, entire; floral tube: cylindrical, 1.5–2.7 cm long, green outside; flowers: white, 7–11 cm diam.; sepals: green outside, white inside, oblong-lanceolate, 3–5.5 × 0.8–1.3 cm, aristate, awn 0.3–0.5 cm long; petals: white, oblong-lanceolate, 2.7–5 × 0.7–1 cm; corona: ca. 8 series, outer series white, 0.2–0.3 cm long, erect, 2nd series an erect membrane, ca. 0.1 cm long, inner series indefinite rows inside the floral tube, up to 0.1 cm long; ovary: ovoid, tomentose; fruit: spherical, 3–4 cm diam.; seeds: unknown.

Passiflora pilosicorona Sacco

Bradea 1: 33 (1973)
pilosus: having soft hair
Section *Insignes*

Passiflora pilosicorona (Plate 282), at home in the province of Cochabamba, Bolivia, grows at altitudes of 2450–3100 m. Its purplish pink flowers, up to 16 cm long, are as spectacular as those of *P. insignis*, to which it is closely related. Both of these species from section *Insignes* are characterized by sepals and petals that are twice as long as the floral tube and by an outer coronal series that is inclined toward the androgynophore. A distinction is made exclusively on the basis of the leaves. While *P. insignis* produces unlobed leaves, those of *P. pilosicorona* are always 3-lobed.

Sometimes *Passiflora pilosicorona* is confused with *P. mandonii*, another representative of section *Insignes*. Although their 3-lobed, rugose leaves are somewhat similar, the floral structure is different. *Passiflora mandonii* has a significantly longer floral tube and usually has longer corona filaments.

The 5 representatives of section *Insignes* are endemic in an area that reaches from central Peru to central Bolivia, so that this region might be regarded as the cradle of these species with pinnatisect stipules and a filiform outer corona series. The frequently cultivated *Passiflora pinnatistipula* only proceeds farther north into Colombia. In his first description of *P. pilosicorona*, José da Costa Sacco recognized similarities to *P. callimorpha* and *P. buchtienii*, particularly with regard to the short floral tube and the corona in 3 series, and classified this species into subgenus *Distephana*. It was Linda Escobar (1980) who noticed that *P. callimorpha* was a synonym for *P. insignis*. After that, it was only a short step to assign *P. pilosicorona* to supersection *Tacsonia*. This species has not yet been brought into cultivation.

Vine: villous, except upper leaf surface; stem: terete, tip erect; stipules: deeply pinnatisect, 1–1.8 × 0.7–0.9 cm; petioles: 1.5–3.8 cm, with 2–4 slightly elongated glands; leaves: 3-lobed, 6–11.6 × 5.8–12.7 cm, rugose, serrate; peduncles: 6–9 cm, stout, solitary; bracts: ovate, 2.5–3.5 × 2–2.5 cm, free, glandular-serrate; floral tube: cylindrical, 2–3 cm long, 0.5–0.8 cm wide, greenish brown; flowers: rose to purplish pink, 11–16 cm diam., 7–12 cm long; sepals: rose to purplish pink, oblong, 4.8–8.5 × 1.2–2.1 cm, slightly keeled, keel terminating in an awn 0.4–0.7 cm long; petals: rose to purplish pink, oblong, 4–7 × 1–1.8 cm; corona: 3 series, filaments of outer series 0.5–0.8 cm long, bluish purple, inner series 0.1–0.3 cm; ovary: ovoid, villous; fruit: ovoid, ca. 7 × 5 cm; seeds: obovoid, 0.7 × 0.4 cm.

Passiflora pinnatistipula Cav.

Icon. Pl. 5: 16, t. 428 (1799)
pinnatistipulus: having pinnate stipules
Synonyms: *P. chilensis* Miers, *P. pennipes* J. E. Sm.
Section *Insignes*

Passiflora pinnatistipula (Plates 283 and 284) is the best-known and most widespread species from section *Insignes*. Though it probably originated in Bolivia, it is now cultivated from northern Colombia to southern Chile for its popular fruit. Offspring of many cultivated plants have made their way into the wild. This species is mostly found at altitudes of 2600–3300 m, but according to Linda Escobar (1980) there are populations at elevations up to 4000 m around Cochabamba in Bolivia. It also grows near sea level in Chile, demonstrating the extreme adaptability of the species. Although it has one of the largest native ranges within supersection *Tacsonia*, *P. pinnatistipula* does not show the variability typical of *P. cumbalensis*, *P. mixta*, or *P. tripartita*. As a rule, plants from northern Colombia do not differ from those of Bolivia or other regions of the Andes.

Passiflora pilosicorona

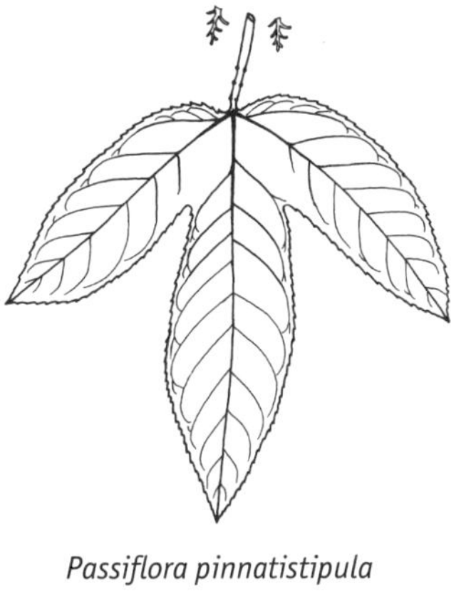

Passiflora pinnatistipula

The popularity of this species is due to its globular to oval, yellow fruit, up to 5 cm diam. A brittle pericarp encloses the grayish white pulp, which tastes sweetish to flavorless to slightly sour. It can be eaten fresh or used in drinks or desserts such as ice cream. The fruit is occasionally exported to Europe under the names "cholupa" or "gulupa."

Characteristic features of *Passiflora pinnatistipula* are its pinnatisect stipules, free bracts, and particularly its violet outer corona, which is up to 1.7 cm long. Its beautiful purplish violet flowers reach a diameter of 6–10 cm. The greenish brown to light pink floral tube is a little stouter than that of other tacsonias. The closest relative is *P. mandonii*, from which *P. pinnatistipula* can be distinguished by its more deeply 3-lobed leaves and different fruit. Clones of different origin were brought into cultivation only a few years ago. A clone with white flowers, found by Axel Frank and Emil Kugler in Bolivia, turned out to be very free-flowering. A 2-m-tall plant of this clone, cultivated in a 35-cm container, is able to produce about 30 flowers. Unfortunately, cuttings from this white-flowering variety are difficult to root.

Other pink clones, collected by Giselher Behnken in Ecuador, are still too young to set flowers. Nevertheless, it can be said that *Passiflora pinnatistipula* is among the less fastidious tacsonias and will tolerate higher temperatures better than most. Only extended spells of hot weather with temperatures beyond 30°C will cause problems. Flower bud formation starts in late summer and early fall when temperatures drop and there are marked differences between day and night temperatures.

Vine: densely pubescent, except upper leaf surfaces; stem: more or less 5-angular, tip erect; stipules: deeply pinnatisect, 0.6–1 × 0.4–0.6 cm; petioles: 1.5–3 cm, with (2–)5–10 minute glands; leaves: 3-lobed, (3–)4.5–12 × 5.5–14 cm, white-tomentose beneath, rugose, serrate; peduncles: (2.5–) 4–7(–9.5) cm, solitary; bracts: ovate to oblong-ovate, 1–2.4 × 0.7–1.5 cm, free, glandular-serrate; floral tube: cylindrical, 4–6.5 cm long, 0.8–1.8 cm wide, greenish to light brown; flowers: light purplish violet, occasionally white, 6–10 cm diam., 8–12.5 cm long; sepals: light purplish violet, 3–6 × 1.2–2.2 cm, aristate, awn 0.3–0.6 cm; petals: subequal to sepals; corona: 1 or 2 series, filaments of outer series 1.2–1.7 cm long, bluish violet, inner series (if present) inside of floral tube; ovary: ellipsoid, densely white-pubescent; fruit: ovoid, 5–6 × 3–5 cm, yellow, edible; seeds: obovate, 0.5–0.6 × 0.4 cm, dark brown.

Passiflora roseorum Killip

J. Wash. Acad. Sci. 17: 427 (1927)
named after Joseph Nelson Rose and Lou Beatrice Rose, collectors of the type plant
Section *Elkea*

Passiflora roseorum (Plate 285) is at home in a very small area between Loja and Saraguro in southern Ecuador at 2900–3100 m. It could be confused with *P. cumbalensis* at first sight. Upon closer inspection, however, it becomes evident that its oblong-ovate stipules and the thickened secondary veins on the undersides of its leaves are different. *Passiflora loxensis*, another representative from this complex of related species, also stands out because of its oblong-ovate stipules and thickened secondary veins. In contrast to *P. roseorum*, however, *P. loxensis* is usually completely glabrous, produces larger flowers with sepals 7–8 cm long, and is distinguished by its fruit, which turns red when it ripens.

The purplish lavender flowers of *Passiflora roseorum* reach a diameter of 7–9 cm. The dark purple-violet floral tube is striking. All in all, this flower is among the most intensely colored and has one of the largest admixtures of blue within the whole supersection. It is noticeable that species with a high portion of blue in the flower frequently occur in southern Ecuador and northern Peru. Six species

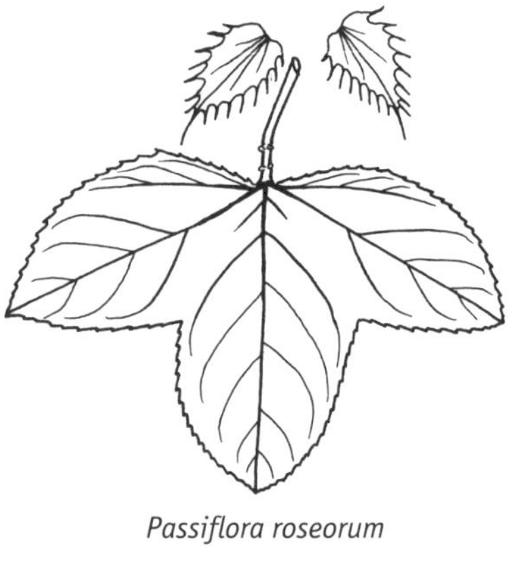

Passiflora roseorum

with similar flower colors can be found in this region: *P. cumbalensis*, *P. glaberrima*, *P. loxensis*, *P. luzmarina*, *P. parvifolia*, and *P. roseorum*. It is not always easy to describe these colors adequately. They range somewhere between purplish blue, pale violet, and pink-lavender. A few years ago, Michael Schwerdtfeger brought *P. roseorum* into cultivation. Unfortunately this clone has not become established, so it is hoped that one day mature seeds can be collected again in its habitat.

Vine: densely pubescent, except upper leaf surface; stem: terete or angular, tip erect; stipules: oblong-ovate to lanceolate, 0.8–2.3 × 0.5–1.2 cm, lacerate; petioles: 0.8–2.5 cm, with 2–4 subsessile glands in upper 1/3; leaves: 3-lobed, 2–5.5 × 2.5–9 cm, coriaceous, serrate; peduncles: 2.5–4.5(–6) cm, solitary; bracts: oblong-ovate, 2.5–4.5(–5.5) × 1.5–2.5 cm, connate for 3/4 of their length, green with purple veins; floral tube: cylindrical, 8–10 cm long, 0.9–1.3 cm wide, dark purple-violet; flowers: purplish violet, 7–9(–10) cm diam., 12–13.5 cm long; sepals: purplish violet, oblong, 3.3–5.5 × 1.6–2.5 cm, aristate, awn 0.3–0.5 cm; petals: purplish violet, oblong, 3.4–5 × 1.2–1.8 cm; corona: 1 series, reduced to a denticulate purple ring; ovary: ellipsoid, glabrous; fruit: ellipsoid, 4–7.5 × 2–3 cm, greenish with white speckles, longitudinally 5-grooved with dark green stripes; seeds: obovoid to cordate, 0.6–0.7 × 0.4–0.5 cm, reticulate, dark brown.

Passiflora sanctae-barbarae Holm-Nielsen & Jørgensen

Nord. J. Bot. 7: 132 (1987)
named after Santa Bárbara, Ecuador, the type location
Section *Elkea*

The very beautiful *Passiflora sanctae-barbarae* (Plates 33 and 286) is exceptional in several respects. Its sumptuous pink-orange flowers reach a diameter of up to 12 cm and are up to 15 cm long. Sometimes one flower, but generally 2 flowers, grow out of the leaf axils, an uncommon feature among tacsonias. It is also characterized by its ovate stipules, 3-foliate leaves, connate bracts, and zygomorphic androecium, a combination that is unique among tacsonias with pendent flowers. The sum of these characteristics is unmistakable.

Passiflora sanctae-barbarae is endemic in northern Ecuador at 1800–2500 m. Presumably there are also some occurrences in southern Colombia. Unfortunately, this rare species is extremely endangered. Upon the most recent visit to its habitat in the province of Napo, it turned out that the plants had disappeared due to road construction work. Among the orange-flowered tacsonias, this ranges farthest to the south. The other species with orange flowers are relatively unrelated and found only in Colombia. It is hoped that *P. sanctae-barbarae* can be brought into cultivation before its last habitats are destroyed. While many species of *Passiflora* from high elevations, above 3000 m, have turned out to be difficult to grow and reluctant to flower, there is a chance that *P. sanctae-barbarae* might flower in cultivation, coming as it does from lower mountain regions.

Vine: pubescent; stem: terete, striate; stipules: ovate, 1.7 × 1–1.3 cm, glandular-lacerate; petioles: 2–5 cm, with 6–8 subsessile glands scattered along petiole; leaves: 3-foliate, rarely 3-lobed, 7–13.5 × 8–16.5 cm, serrate; peduncles: 6–13 cm, in pairs or occasionally solitary; bracts: lanceolate to oblong, 3.5–6.5 × 2–2.5 cm, connate to 1 cm from base, minutely serrate; floral tube: cylindrical, 7–9 cm long, 1–1.8 cm wide, orange to salmon, turning greenish at base; flowers: orange to salmon, 10–12 cm diam., 11.7–15 cm long; sepals: orange to salmon, oblong, 4.7–6.5 × 2.5–3.2 cm, slightly keeled, keel terminating in an awn 0.2–0.4 cm long; petals: subequal to sepals; corona: 1 series, reduced to a denticulate, dark purple ring; ovary: ellipsoid, densely pubescent; fruit: unknown.

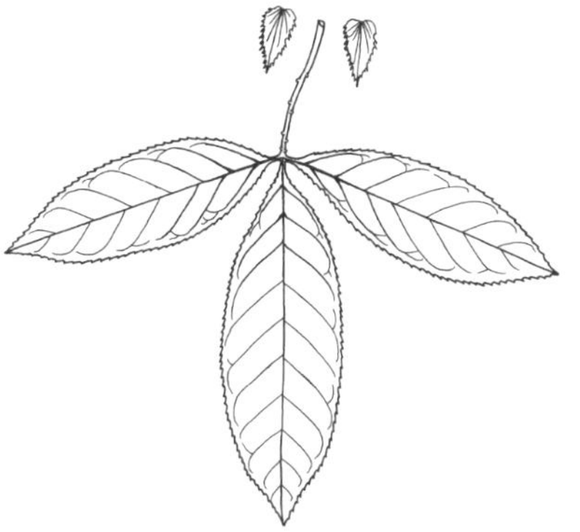

Passiflora sanctae-barbarae

Passiflora tarminiana Coppens & V. Barney

Novon 11: 9 (2001)
named after Tarmín Campos, Colombian agronomist and professor
Section *Elkea*

The recently described *Passiflora tarminiana* (Plate 287) is actually an old acquaintance. For many years plants of this species went by the name *P. mollissima* or *P. tripartita* var. *mollissima*, though it was noticed at times that 2 forms were subsumed under this name, one with reflexed petals and another with a campanulate corolla. Sometimes it was also speculated that the reflexed form might be a hybrid, possibly a cross with *P. cumbalensis*. Finally, Geo Coppens d'Eeckenbrugge and Victoria E. Barney, in the course of their "Project for Neotropical Fruits," investigated the 2 forms. As a result of recent morphological and genetic research, they showed that these are in fact 2 different species.

The large, reflexed corolla of *Passiflora tarminiana* is not the only thing distinguishing it from the very common, similar *P. tripartita* var. *mollissima* and *P. mixta*. It also has smaller, deciduous stipules, a light green floral tube, and a smaller length ratio between the floral tube and the sepals (the length of the tube is closer to the length of the sepals). *Passiflora mixta* is the only species among the 3 that typically has an angular stem and produces half-pendent, horizontal, or erect flowers. Due to the wide native ranges of these species, most other features are too variable to be used to distinguish them reliably.

Because of its tasty fruit, *Passiflora tarminiana* has a great economic importance, 2nd only to *P. tripartita* var. *mollissima* among the tacsonias. It is grown in almost all states of the Andes, though most often in Ecuador and Colombia. Plants mainly grow at altitudes of 2000–2700 m, both on plantations and in natural habitats. The ellipsoid, yellow fruit is offered in Europe and the United States as "curuba," "taxo," or "banana passionfruit." This species has been introduced into Mexico, the Canary Islands, Zimbabwe, Ethiopia, Réunion, New Guinea, and New Zealand. It has become an ecological nightmare in Hawaii, where it covers large forests and endangers local flora.

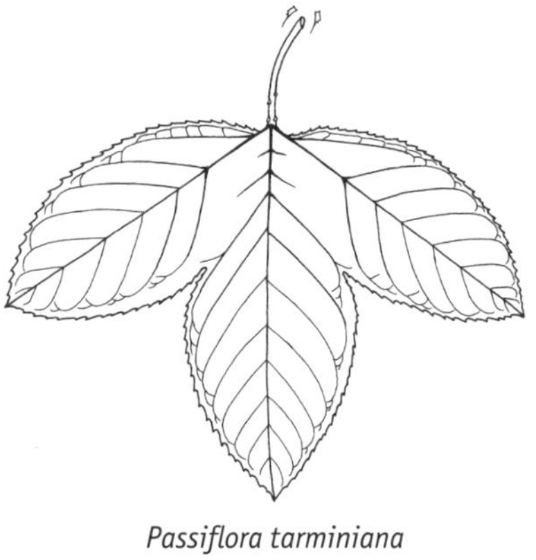

Passiflora tarminiana

Passiflora tarminiana has been represented in enthusiasts' collections for many years. It has become established as one of the most recommendable and free-flowering tacsonias. Nevertheless, because of its potential as a harmful weed, and because it is self-seeding, it should not be grown outdoors in tropical areas where it does not already occur. When it is grown in a 35-cm pot and has reached 2–3 m in height, it will flower reliably every year. At times it will also set flowers as a very young plant. *Passiflora tarminiana* is among the few species that will tolerate temperatures above 30°C for several days without suffering. The main flowering period is in late summer and fall.

Vine: pubescent; stem: terete, striate, tip erect; stipules: subreniform, 0.4–0.7 × 0.2–0.3 cm, with long arista at apex, serrate at base, soon deciduous; petioles: 1.5–4 cm, with 2–8 mostly paired glands; leaves: 3-lobed, 5.5–11(–16) × 7–16(–29) cm, upper surface nearly glabrous, serrate; peduncles: 3–10 cm, solitary; bracts: ovate, 3–5 × 2–3 cm, connate for about 1/2 of their length, entire; floral tube: cylindrical, 6–8 cm long, 0.7–1 cm wide, light green; flowers: light pink, 8–11 cm diam., 10–13 cm long; sepals: pink, oblong, 4.5–6 × 1.2–2.5 cm, aristate, awn 0.3–0.4 cm; petals: pink, oblong, 4–5.5 × 1.2–2 cm; corona: 1 series, reduced to a purple ring with white teeth; ovary: ellipsoid, pubescent; fruit: ellipsoid, 10–14 × 3.5–4.5 cm, yellow to orange-yellow, edible; seeds: asymmetrically ovoid, ca. 0.6 × 0.4 cm, reddish brown, reticulate.

Passiflora tenerifensis Escobar

Mutisia 71: 6 (1988)
named after Tenerife, Colombia, the type location
Section *Colombiana*, Series *Leptomischae*

Passiflora tenerifensis (Plate 288) is another species from the series *Leptomischae* and is distinguished by its exceptionally long, threadlike peduncles. Its spectacular pink flower, 12–14 cm diam., hangs resplendently on the tip of the peduncle, which reaches up to 30 cm long. The large greenish brown fruit is also remarkable; in contrast to that of most tacsonias, it has a thick, fleshy pericarp. The small native range of *P. tenerifensis* is in the Colombian department of Valle, where it grows at 2800–3000 m on the Central Cordillera. It is possible, however, that *P. tenerifensis* is a synonym for *P. flexipes*, whose habitats are only about 50 km farther to the north. The corona of 0.3 cm long, found in the herbarium specimens, supports this. In her first description, Linda Escobar mentions the stiffer, thicker leaves with a blunter tip as a primary feature distinguishing *P. tenerifensis* from *P. flexipes*.

Like almost all species endemic in Colombia, *Passiflora tenerifensis* has linear to linear-lanceolate stipules. It is worth noting that the development of this form of stipules is almost exclusively limited to Colombia. The southernmost occurrence of linear-lanceolate stipules can be found in *P. harlingii* and *P. coactilis* from central Ecuador. The form of the stipules is a most important criterion in identifying tacsonias. It is not known whether *P. tenerifensis* is under cultivation outside of Colombia.

Vine: pubescent, except inside of flower; stem: subangular, striate, tip flexed; stipules: linear-lanceolate, ca. 1 cm long; petioles: 1.9–2.4 cm, with 4–8 glands; leaves: unlobed, oblong-ovate, 7.1–10.5 × 3.8–6 cm, subcoriaceous, serrate; peduncles: 17.5–29 cm, slender, solitary; bracts: lanceolate, ca. 3.5 × 0.9 cm, lacerate, free; floral tube: cylindrical, 6.5–10 cm long, 0.8–1.2 cm wide, pinkish; flowers: pink, 12–14 cm diam., 12–15 cm long; sepals: pink, oblong, 5.3–6 × 1.8–2.3 cm, aristate, awn 0.5–0.7 cm; petals: subequal to sepals; corona: 1 series, up to 0.3 cm long, purple; ovary: ellipsoid, pubescent; fruit: oblong-ovate to ellipsoid, ca. 9 × 5 cm, greenish brown; seeds: oblong-ovoid, 0.7 × 0.3 cm, reticulate, dark brown.

Passiflora trifoliata Cav.

Icon. Pl. 5: 16, t. 427 (1799)
trifoliatus: having 3 leaves
Section *Trifoliata*

Passiflora trifoliata (Plates 289 and 290), at home in central and southern Peru, is unique in several respects. It is the only *Tacsonia* that reaches altitudes of up to 4300 m. At such altitudes it is confronted with frost every night. Occurrences below 3500 m are quite rare. *Passiflora trifoliata* mostly grows in dry scrub and grassland vegetation and rarely in the moist locations typical of other tacsonias.

Even at first sight this species looks different from other tacsonias because of its dense vestiture, 3-foliate leaves, glandular-fimbriate stipules and bracts, and ovate, longitudinally 3-grooved fruits. Consequently, Linda Escobar (1988a) intended to assign it to a section of its own, *Trifoliata*, but the

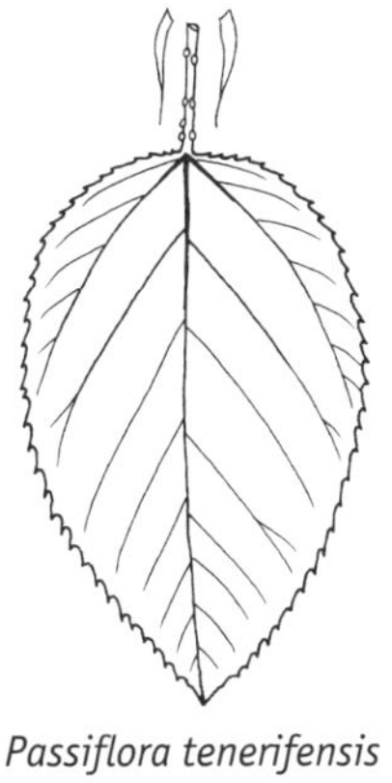

Passiflora tenerifensis

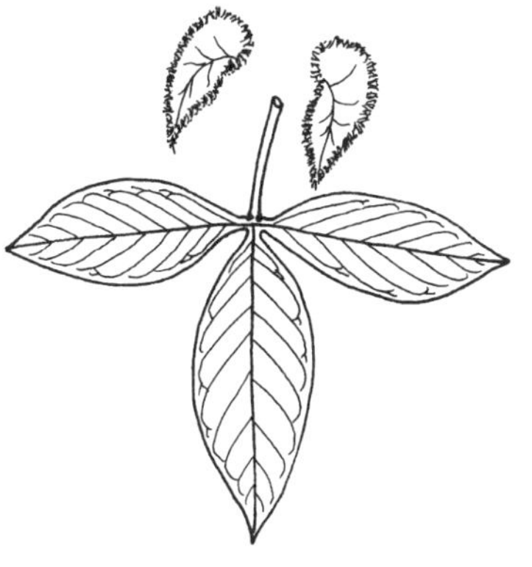

Passiflora trifoliata

name was published invalidly and has no official botanical standing. It remains to be seen how things will develop when the pending revision of supersection *Tacsonia* has been completed.

The pinkish flowers, often with an orange touch, are variable in color intensity. They are held in a horizontal to slightly nodding position by stout peduncles. The zygomorphic androecium is remarkable. The diameter of a completely opened flower is around 5 cm.

Two varieties of *Passiflora trifoliata* have been distinguished. In the departments of Junin and Pasco the plants are characterized by narrower leaflets and smaller stipules, so that Escobar described this variation under the name of variety *tarmensis*. The typical variety has correspondingly larger leaves. Although *P. trifoliata* is quite common in its habitats, and although there are many collections, it does not seem to be under cultivation.

Vine: short villous-hirsute, except inside of flower; stem: terete, striate, tip erect; stipules: oblong-ovate to lanceolate, 1–3 × 0.3–1.5 cm, glandular-fimbriate; petioles: (0.6–)1–3.5 cm, with 2–5 glands; leaves: 3-foliate, rarely deeply 3-lobed, 3–8.7 × 4.5–10 cm, rugose, subcoriaceous, slightly serrate; peduncles: 3.5–10 cm, stout, solitary; bracts: oblong-ovate to ovate, 1.5–4.2 × 1.2–3.5 cm, free, glandular-fimbriate; floral tube: cylindrical, 3–5.5 cm long, 0.7–1.1 cm wide, green with purple; flowers: pinkish, occasionally pinkish orange, 4–5.5 cm diam., 6.5–8 cm long; sepals: pinkish, oblong to oblong-ovate, 2–3.5 × 1–2 cm, slightly keeled, keel terminating in an awn 0.4–0.9 cm long; petals: pinkish, oblong, 1.9–3.3 × 0.8–1.6 cm; corona: 1 series, reduced to a tuberculate, violet ring; ovary: ovoid to ellipsoid, tomentose; fruit: ovate, 4–6 × 2.5–3.5 cm, 3-grooved, yellow; seeds: obovoid, 0.5–0.6 × 0.3–0.4 cm, black.

Passiflora tripartita (Juss.) Poir.

Encycl., Suppl. 2: 843 (1811)
tripartitus: divided into 3 parts (in reference to the leaves)
Synonyms: *P. mollissima* (Kunth) Bailey, *P. psilantha* (Sodiro) Killip, *P. tomentosa* var. *mollissima* (Kunth) Triana & Planch.
Section *Elkea*

In 1988 passionflower enthusiasts had to become accustomed to the fact that *Passiflora mollissima*, one of the best-known and most important cultivated species of the genus, had been reduced to a variety of *P. tripartita* (Holm-Nielsen et al.). Until then, *P. tripartita* had been of minor importance. This name change may be botanically correct and may better illustrate genetic relationships, but it continues to cause trouble. This species is often still called *P. mollissima* instead of *P. tripartita* var. *mollissima*, particularly in countries where it is grown commercially. Now, after many people who are interested in passionflowers have become accustomed to the new name, there is another major modification. Geo Coppens d'Eeckenbrugge and others (2001) found a new species hiding among plants grown as *P. tripartita* var. *mollissima*: on the one hand the true variety *mollissima*, with a campanulate corolla, and on the other hand a more frequently cultivated form with reflexed sepals and petals. These plants with reflexed sepals and petals are commonly represented in many enthusiasts' collections under the name of *P. tripartita* var. *mollissima*, and they actually belong to the new species, *P. tarminiana*.

The term *Passiflora tripartita* var. *mollissima* is properly used to speak of what was originally called *Passiflora mollissima*, which is characterized by a bell-shaped structure of the sepals and petals and by a longer floral tube. For a more detailed account, see *P. tarminiana*.

Depending on the leaf forms and the pubescence, 3 varieties are distinguished in *Flora of Ecuador* (Holm-Nielsen et al. 1988). Besides the well-known *Passiflora tripartita* var. *mollissima* (Plate

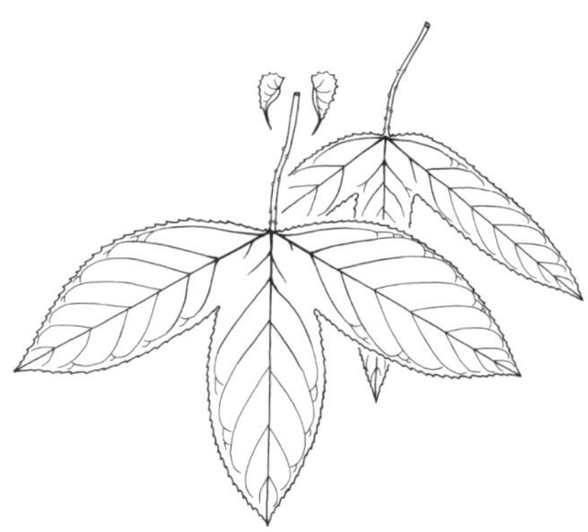

Passiflora tripartita

291), found in many regions of the Andes, there are the variety *azuayensis* (Plate 292) from Ecuador and the variety *tripartita* from Ecuador and Colombia, which has the narrowest leaf lobes of the 3. The flowers of these varieties can hardly be distinguished, if at all. They are pink to purplish pink and reach a diameter of only 5–9 cm due to the bell-shaped corolla. The edible fruit turns yellow as it ripens. These 3 varieties are cultivated mostly in fruit gardens. The few major plantations that exist are mainly reserved for *P. tarminiana*.

While *Passiflora tripartita* var. *tripartita* and *P. tripartita* var. *azuayensis* are rarely cultivated, *P. tripartita* var. *mollissima* is occasionally cultivated by enthusiasts. However, *P. tarminiana*, which is comparatively more free-flowering, is far more common in collections.

Vine: pubescent; stem: terete, tip erect; stipules: reniform, 0.9–3.3 × 0.4–1 cm, glandular-serrate; petioles: 1–5.5 cm, with 6–14 subsessile to elongated glands; leaves: 3-lobed, 4.5–17 × 6.5–25 cm, glabrous to pubescent above, pubescent beneath, serrate; peduncles: 1.5–11 cm, solitary; bracts: oblong-ovate to lanceolate, 2.5–6 × 0.5–2 cm, connate for 1/3–4/5 of their length, entire; floral tube: cylindrical, 7–12 cm long, 0.7–1.2 cm wide, green, often turning purplish toward apex; flowers: pink to purplish, 6–9 cm diam., 10–16 cm long; sepals: pink to purplish, oblong, 2.5–5.5 × 0.7–1 cm, aristate, awn ca. 0.2 cm; petals: subequal to sepals; corona: 1 series, reduced to a purple ring with white teeth; ovary: ellipsoid, densely pubescent; fruit: oblong-ovate to ellipsoid, 6–11 × 3–4.5 cm, yellow, edible; seeds: obovoid, 0.6 × 0.4–0.5 cm, reticulate, dark brown.

Passiflora trisecta Mast.

Fl. Bras. 13 (1): 564 (1872)
trisectus: cut into 3 parts
Synonym: *P. thaumasiantha* Harms
Section *Manicata*

Passiflora trisecta (Plate 293) occupies a special position with regard to its flower ecology. Its white flowers are adapted by their form to pollination by nocturnal bats and open at dusk. In warm weather the flowers close early in the morning, but in cold weather they stay open until early afternoon. Horizontal to erect, the flowers are raised from the foliage on long, stout peduncles. They have a wide floral tube that is not covered by the corona. This enables a bat to reach down to the nectar at the bottom of the blossom with its tongue, which is quite plump in comparison with an insect's rostrum or a hummingbird's bill. The androgynophore is inclined inward and adjoins to the edge of the floral tube. Furthermore, the flowers have a zygomorphic androecium. This plant is also remarkable for its velvet pubescence, which covers almost the whole plant, and for its 3-foliate leaves. Its growth structure is reminiscent of *P. trifoliata*. *Passiflora trisecta* is closely related to the rare *P. weberbaueri*, but the latter is completely glabrous and has longer outer corona filaments, 1.5–2 cm.

Passiflora trisecta is at home from south-central Peru to northern Bolivia at altitudes of 2400–2800 m. It can be regarded as a very free-flowering and recommendable species. It is relatively easy to cultivate and is not bothered by temperatures around 30°C or by short spells of even hotter weather. Nevertheless, it is best located in a cold greenhouse or outdoors during summer. This way it will flower in a 20- to 25-cm pot without problems.

Vine: villous; stem: terete, striate, tip erect; stipules: ovate-lanceolate, 1–2.5 × 0.5–1 cm, glandular-laciniate; petioles: 2–5 cm, with 2–3 filiform glands at apex and further 3–4 shorter glands near middle; leaves: 3-foliate, 7–10 × 10–13 cm, leaflets oblong-lanceolate, serrate; peduncles: 7.5–15(–18) cm, stout, solitary, erect to horizontal; bracts: ovate, 2.5–3.5 × 1.5–2.5 cm, free, fimbriate; floral tube: cylindrical-campanulate, 1–1.5 cm long, ca. 1.5 cm

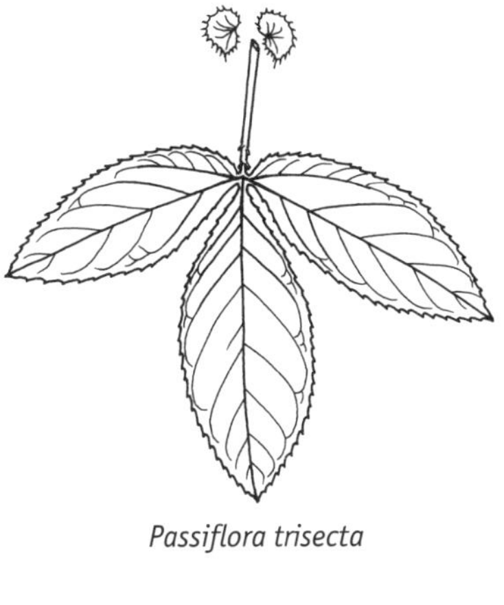

Passiflora trisecta

wide, green; flowers: white, 6–9 cm diam., 3.5–6 cm long; sepals: white inside, green outside, narrowly oblong, (2–)3–4.5 × 1–1.7 cm, keeled, keel terminating in an awn 0.8–1.1 cm long; petals: white, narrowly oblong, (2–)3–4 × 0.3–0.8 cm; corona: 3 series, outer series 0.3–0.4 cm, white, inner series minute; ovary: ovoid-oblong, densely villous; fruit: globose to ovoid, 4.5–6 × 3.5–5 cm, olive-green; seeds: reticulate.

Appreciating the Life and Contributions of Linda K. Escobar

Lawrence E. Gilbert

In early 1977 I presented an informal lecture on the natural history of *Passiflora* to a group of new graduate students in the Department of Botany at the University of Texas at Austin. Most of them were looking for Ph.D. projects. One such student, Linda K. Escobar, was soon a regular and enthusiastic visitor to our living collections of *Passiflora* atop the zoology building. Since none of my butterfly students had direct interest in these plants, I was excited to have a real botanist visit from time to time. Linda was initially oriented to economic and agricultural plants, which explains her eventual choice of tacsonias for study.

But why would a person from Ohio (b. 1940), educated at the University of New Hampshire (B.A. 1962) and Purdue University (M.S. 1971), and later, a graduate student working in Texas under a famous systematist of the Compositae, Billie L. Turner, want to work on a neotropical genus of plants? After Linda taught high school for a few years in the United States, she went to Colombia (in 1965) to teach secondary teachers in that country how to teach new biological science curricula being developed in the United States. While there, she met and married her late husband, Darío Escobar-Uribe, also a biologist.

After returning to the United States for her masters degree, Linda returned to Colombia to teach at the University of Antioquia in Medellín (1971–1976). When she arrived in Austin, she had a broad view of South American botany. Naturally, among the crops grown on her in-laws' family farm was *Passiflora edulis*. Linda was also very familiar with various cultivated members of the *Tacsonia* group of *Passiflora*. Initially, however, she was considering several other groups of economically valuable plants for her Ph.D. work.

Once Linda decided to focus on *Passiflora*, and specifically to revise the tacsonias, her progress was rapid. Her 647-page dissertation, "Interrelationships of the edible species of *Passiflora*, centering around *Passiflora mollissima* (H. B. K.) Bailey subgenus *Tacsonia*," was finished in just 3 years (1980), including extensive fieldwork in South America, museum surveys, and chemosystematic studies based on flavonoids. It was difficult to connect such a monumental effort to this modest, unassuming woman. Her earlier network of contacts in South America, her fluency in Spanish, and her intimacy with the culture were great advantages in the fieldwork phase of her study.

In the 1980s Linda taught once more at the University of Antioquia and conducted many collecting trips that expanded her considerable expertise in tropical botany. She was always willing to collect plant material for other researchers. These years of field study gave Linda important insights into *Passiflora* beyond *Tacsonia*. She had many research students in Medellín and directed some of them to do

Linda K. Escobar

early tissue culture on *Passiflora*. During this time she completed a draft of a revision of *Astrophea* and initiated molecular work in the laboratory of systematist Robert Jansen at the University of Texas at Austin.

After the untimely death of her husband in 1989, Linda was hired as assistant professor at the University of Puerto Rico. From there she continued collaborations back in Austin with myself, Jansen, and others. She had discovered, by the analysis of chloroplast DNA in some of our interspecific hybrids of *Passiflora*, that cytoplasmic inheritance may occur via the male gametophyte. She was also working on a more general molecular phylogeny of *Passiflora*.

Linda was an intensely private person. I knew little of her life, outside of our mutual interest in *Passiflora*, until after her tragic death from ovarian cancer in December 1993. She worked on and thought about *Passiflora* to the end. On her last trip to Austin in the early fall of 1993, she made a point to visit my greenhouse to see the flowers of *P. deidamioides*. Although she could hardly walk and was very near death, she showed the same great enthusiasm in seeing a *Passiflora* in flower that she had previously only read about.

Linda's field notes, specimens, and photographs are being cared for by appropriate individuals. I have care of her photographic slide collection of *Passiflora* and have provided some of these slides for this volume. Most of us will never lay eyes on some of the species she observed in the Andes except vicariously through her images.

I took Linda's existence for granted and wish I had asked her opinion on a few more things. In retrospect I realize how intensely satisfying it was to talk shop with her about our different perspectives on these plants. In her 16 short years working on the group, she had extensive exposure to many species of *Passiflora* in their native habitats and had many novel ideas about their relationships. I looked forward to getting seeds of new species from her fieldwork, and she enjoyed having a place where they could later be seen in flower and studied. When Linda Escobar departed the planet, everyone who is fascinated by the biodiversity of *Passiflora* had their worlds collectively diminished, even if most never experienced her wonderful spirit in person.

17
Hybrids and Hybridization

Roland Fischer

THE PASSIONFLOWERS offered in garden centers are more often hybrids than botanical species, and many plant fanciers are fascinated by them. Sometimes enthusiasts who are delighted by the splendor of the blossoms on a plant do not even know that it is a hybrid, let alone what it means to be a hybrid. Therefore, some terms need to be clarified, though to some it may seem too technical. In actual fact, every passionflower enthusiast can cultivate and even breed hybrids without any previous experience.

The information included in this chapter is intended to inspire rather than frighten. Gardeners can try breeding hybrids on their own, enriching the diversity of gardens everywhere with their own crosses. It can be quite fascinating and exciting to hand-pollinate passionflowers, see the fruits ripen, and experience the happy moment when the seeds germinate one morning. When the mature plants produce exceptional flowers, it could mark the beginning of a new favorite pursuit.

Terminology

Species have developed by natural selection over the course of evolution. Over thousands and millions of years, many small steps have led to genetic differences that distinguish one species from another. These differences are so great that the species normally cannot have a fertile offspring together. Since things are somewhat complicated in the plant kingdom, it may help clarify matters to take an example from the animal kingdom. A horse and a donkey belong to different species. Although they have the same ancestors, they have diverged so widely that they are now different species, and their common offspring is infertile. Although there are more extensive definitions of *hybrid*, the term is customarily used in one way with regard to passionflowers. In this context, hybrids are the result of crosses between different species. Referring back to the example from the animal kingdom, this means that a hinny (*Equus hinnus*) and a mule (*E. mulus*) are hybrids—the result of a cross between horse (*E. caballus*) and donkey (*E. asinus*).

Races are distinctive forms of a species. Although various races can look very different, their genetic differences are not great. A miniature pinscher and a mastiff are far more closely related than a horse and donkey. Hence, different races can easily have fertile offspring. Over the course of evolution, races will always develop when groups of a species are separated from one another for a long time and kept from blending. Races are a small step in evolution regarding the origin of species.

The term *clone* refers to a genetically identical living thing. Every individual seedling is genetically somewhat different from another, but its offspring raised from cuttings are genetically identical. All cuttings belong to the same clone. An example is the breeding of *Passiflora vitifolia* 'Scarlet Flame' by crossing a specimen of *P. vitifolia* from the high mountain regions with a plant from the lowlands. A particularly beautiful individual plant was selected from the seedlings and called 'Scarlet Flame' by the breeder. Since then, vegetatively produced offspring of this particular plant have been spread all over the world. The plants growing in London belong to the same clone as the plants growing in Los Angeles.

Although the term *hybrid* has been occasionally used in the natural sciences since the 17th century, it became popular only after Gregor Mendel's *Experiments in Plant Hybridization* (1865) gained influence early in the 20th century, after having been neglected for more than 30 years. The essentials of his experiments on the heredity of characteristics are summarized in the Mendelian laws of heredity. The parent plants of a cross are called the *parental generation*, now known in biological shorthand as "P," and its first offspring as the *first filial generation* (F_1). The result of a cross between individuals from the F_1 generation is called *second filial generation* (F_2), and so on. The cross of an F_1 plant with one of the 2 parent plants is called *backcrossing*.

Many species of *Passiflora* can be hybridized with one another without difficulties, and sometimes it is even easier to pollinate a plant with the pollen from another than with its own. The result of a cross between 2 pure species is often conspicuously different from the parent plants. Often the F_1 offspring of a cross shows an intermediate phenotype with regard to the characteristics of the parents.

In the following descriptions of hybridization procedures, the maternal plant is named first, the hybridization process is represented by "×," and the paternal partner (pollen donor) is named last. Thus, the hybridization scheme of the well-known hybrid *Passiflora* 'Incense' is *P. incarnata* × *P. cincinnata*. This means that the fruit has ripened on *P. incarnata* and that the fertilizing pollen came from *P. cincinnata*.

The History of Passionflower Hybridization

The first report of a man-made passionflower hybrid appeared in 1819. When Thomas Milne of Fulham, England, crossed *Passiflora caerulea* and *P. racemosa*, he obtained 7 plants called *P.* ×*violacea*, a hybrid that remains very popular. This cross has been repeated many times, so that now there are many different cultivars of this hybrid. In the first description by Joseph Sabine (1822), the hybridization formula *P.* ×*caeruleo-racemosa* was used, but now this name is considered inadmissible. The correct name, *P.* ×*violacea*, was established by Jean-Louis-Auguste Loiseleur-Deslongchamps in 1828. Interestingly enough, at that time the hybridization of plants was met with a distrust similar to that currently experienced by biotechnology. Several botanists wrote a public statement about Sabine's description, doubting whether the "mixing" of plants might result in any advantage for botany or horticulture. Their fears are now know to have been unjustified.

In the same year, further hybrids were bred that still exist. One of these is *Passiflora* ×*belotii*, the result of a cross between *P. alata* and *P. caerulea*. The then-selected cultivar 'Empress Eugenie' is still represented in many collections. It was raised by J. H. Masters of Canterbury, England. Another extant hybrid from 1824 is *P.* ×*colvillii*, bred by Colvill's Nursery in England. This cross between *P. incarnata* and *P. caerulea* was described by Robert Sweet in *The British Flower Garden* (1823–1829). *Passiflora* ×*allardii*, a 1907 cross between *P. quadrangularis* and *P. caerulea* 'Constance Eliott', also continues to enrich many collections.

Alas, many old hybrids are no longer under cultivation. In times of war, many private people have had to give up their collections, and even botanical gardens have had to cut back on costs incurred by overwintering plants in heated conservatories. A sad loss, for example, is the disappearance of *Passiflora* ×*kewensis* (*P. caerulea* × *P. kermesina*, 1888), along with its parent plant *P. kermesina*. As soon as *P. kermesina* is reestablished in cultivation—and concerted efforts are being made to find it—*P.* ×*kewensis* can again be bred. In many collections, other crosses are cultivated erroneously under the name *P.* ×*kewensis*. A hybrid of *P. racemosa* is widespread under this name.

After the boom in passionflower hybrids that occurred in the 19th century, the interest in hybridization died down for quite some time.

Quite recently, globalization and the progress of technology have had an increasing impact on the cultivation of passionflowers. In addition to certain negative effects, such as the destruction of primeval forests with their flora and fauna, there are positive effects. The boom in the tourist industry has brought about an increase in private botanical excursions. An increasing number of committed passionflower enthusiasts are introducing species into cultivation from their own expeditions. Since the number of established species increases the potential for hybridization, many new beautiful hybrids can be expected. As late as 1998, the gorgeous *Passiflora gritensis* was brought into cultivation, and it is presently represented in the collections of the most important European breeders. It will certainly meet with similar success in the United States. First hybrids with *P. gritensis*, such as *P.* 'Coocoochee', are already growing in European gardens. The influence of information technology is also increasing. The faster flow of information from one continent to another facilitates the exchange of plant material and instructions for care.

Biotechnology will also have a great influence on the cultivation of passionflowers. In renowned European nurseries, the recently biotechnologically created, larger, and more intensely colored *Passiflora* 'Temptation' and *P.* 'Inspiration' are beginning to supersede their little sister *P.* 'Incense'. The same holds true for *P.* 'Clear Sky', a large-flowered alternative to the hardy *P. caerulea*.

Natural Hybrids

Even without human interference, hybrids occur quite naturally. This happens when the pollinator, perhaps an insect, hummingbird, or bat, extends its flying range over the dispersal areas of different passionflower species. Sometimes there are also several compatible species in the same area. In many cases, the resulting hybrids are sterile and die away without any offspring; hence, a change in the natural flora does not occur. However, when the crossing partners are so closely related that the following F_1 is completely or partially fertile, this can have a lasting effect on the natural flora of the area.

One example of a natural hybrid is *Passiflora ×rosea* (Killip 1938), which resulted from a cross between *P. tripartita* var. *mollissima* and *P. pinnatistipula*. Other natural hybrids can result from a cross between *P. tripartita* var. *mollissima* and *P. cumbalensis* or *P. mixta*. Due to the so-called heterosis effect, these hybrids are often particularly robust and efficacious. Because of this, every responsible passionflower enthusiast should take care that his or her hybrids do not get a chance to escape into the wild and invade the local vegetation. This holds particularly true for climatically favored regions.

The Aim of Hybridization Experiments

The reasons for hybridizing are as manifold as the reasons why passionflowers are cultivated. Certainly, the most important economic target is the efficiency of fruit production. Much time, labor, and money are invested in pursuing this goal. Research institutions and professional breeders are trying to improve commercially profitable species, especially in countries where edible passionfruits are economically important, including the United States, Australia, the South American states, the Southeast Asian countries, India, Israel, and New Zealand. Some African states have also started special breeding programs. Although problems and aims may differ from one country to the other, the general goals are the same: to reduce costs and increase profits.

The Heterosis Effect

The term *heterosis*, introduced by George Shull in 1948, means "hybrid vigor." It can be observed in many plants and animals that the vigor of F_1 hybrids exceeds that of both parents. This phenomenon is used systematically in plant breeding.

Fruit Quality

Passiflora tarminiana, an important fruit plant, has a soft, sensitive skin that impairs export on a large scale. Breeders thus try to hybridize cultivars with a more resistant skin that will stand transportation without deterioration in quality. In many countries *P. edulis* f. *edulis* and *P. edulis* f. *flavicarpa* are of major economic importance. In Australia, where there are more than a hundred cultivars of *P. edulis* or its hybrids, fruit quality is improved by trying to breed plants that have larger fruit with a higher juice content, a high content in fructose in a harmonious relation to fruit acids, and an intense flavor.

Fruitfulness

The profitability of a plantation depends on the yield per unit area. In addition to the size of the fruit and the vigorousness of the plants, the percentage of fruits set plays a major part. Self-compatible cultivars are ideal, because a successful and more complete fertilization is guaranteed. Flower size and form must be adapted to the available pollinators. Fructification remains modest when the pollinators do not come into intimate contact with the pollen and stigmas. For this reason, it is common and economically profitable in Kenya, for example, for permanent agricultural workers to go through the plantations and hand-pollinate the flowers.

Resistance to Heat and Cold

Much time and money have been invested in improving the cold resistance of passionflower fruit plants. If this can be achieved, the areas of cultivation can reach farther into the temperate zones. Intense research is being undertaken on this problem in Florida and California in particular. Efforts are also being made to improve heat resistance, because in very hot weather flower formation and fructification is reduced. Research on this problem is given priority in Israel.

Resistance to Disease

In many areas of the world, soil-borne nematodes are a main cause for unproductive passionflower plantations. While some species react very sensitively to an infestation of nematodes, as do most cultivars of *Passiflora edulis* f. *edulis*, other species and cultivars are naturally resistant, such as *P. edulis* f. *flavicarpa*. Unfortunately the latter does not have the intense exotic flavor of *P. edulis* f. *edulis*. An obvious solution is to combine the positive characteristics of both in a hybrid. However, a real breakthrough has yet to come. In many regions, various pathogenic viruses cause severe problems. By hybridizing resistant species, these problems could be solved.

Medicinally Active Substances

The wishes of patients in recent years for natural medicines without serious side effects has induced scientists and the pharmaceutical industry to investigate the medicinal qualities of plants from the genus *Passiflora*. Although so far only *P. incarnata* has been used in the production of pharmaceutical products, scientists are trying to find hybrids with more potent ingredients or a higher concentration of ingredients.

Decorative Effects

While the aforementioned goals are mainly pursued by research institutions, many committed passionflower enthusiasts try to breed hybrids with a highly decorative effect. Many beautiful new hybrids have been bred in recent years to delight the plant fancier with their splendid colors and graceful flower forms.

Selection of Parent Plants

Depending on the goal of a cross, 2 parent plants are chosen whose offspring should have a combination of desired positive characteristics. For example, the goal might be to seek both a better resistance to extremely high or low temperatures and a high-quality fruit or colorful flower with a large corona. Many combinations of plants will inevitably lead to failures; they simply will not set fruit. Accordingly, what factors influence whether a cross succeeds or fails? Absolute factors in predicting a successful cross are the chromosome numbers and the genetic compatibility. Relative factors are the degree of kinship, time of pollination, and prevailing environmental conditions.

Number of Chromosomes

Most passionflowers are diploid. Diploid plants, or rather their cells, contain 2 homologous sets of chromosomes ($2n$), one set from each parent. When the gametes are produced, the chromosome number is reduced to a haploid chromosome set ($1n$). During this process, the distribution of chromosomes to the gametes, originating from the father and mother, is random. Later in the life cycle, at fertilization (after pollination), the 2 haploid nuclei fuse and the diploid number is restored. The number of chromosomes in subgenus *Passiflora* (including supersection *Tacsonia* and part of section *Dysosmia*) is $2n = 18$. The number in subgenus *Decaloba* is mostly $2n = 12$, and this number has been reported from subgenus *Astrophea* as well. Species that differ in the number of chromosomes can never be crossed naturally, although exceptions are known from other genera. An attempt at crossing a passionflower species that has 18 chromosomes with another that has 12 chromosomes will be a failure. Such a cross seems to be impossible. In 1993, Neil Snow and John MacDougal published a list that contains the passionflower species with known chromosome numbers; anyone interested can select a promising combination.

Incompatibility Factors

Even when the chromosome numbers are identical, some plants cannot be crossed with one another. This is often due to factors of genetic incompatibility. Pollination is also one of the most important factors affecting fruit set. It has been verified that fruit set, fruit size, seed number, and juice content are positively correlated with the number of pollen grains employed during pollination. Upon the opening of most passionflowers, the styles bend from their upright position until the stigmas have reached the level of the anthers where they can be touched by the pollinators. This is the best time to pollinate a flower. Physiological self-incompatibility also plays an important role. Self-incompatibility makes cross-pollination necessary, requiring 2 plants with different genotypes, or different clones. Cross-pollination can be achieved by either the activity of natural pollinators or by hand-pollination. Robert J. Knight and Harold F. Winters (1962) found that 30 minutes after pollination the pollen tube of incompatible genotypes seems to be blocked by the stigmatic tissue at a length similar to or slightly longer than the diameter of the pollen grain. In compatible genotypes, the pollen tube is 4–7 times wider than the diameter of a pollen grain and remains passable for about an hour. Three self-incompatible genotypes (S_1, S_2, and S_3) seem to be present in some species of *Passiflora*.

Bud Pollination

The incompatibility system in subgenus *Passiflora* is a gametophytic monofactorial system and can be overcome by bud pollination. The recognition proteins that elicit the incompatibility reaction are formed in the stigma late in the flower's development. Bud pollination is used to produce seeds of pure species when one possesses just a single clone. As a rule, flowers are pollinated 1–2 days prior to opening. With a razor blade, a cut is made through the perianth starting at 1/2 the height of the bud, cutting toward the tip and continuing to the other side of the bud. Then the top of the perianth is spread apart so that the reproductive organs are exposed. It turns out that bud pollination is of great help in hybridizing 2 species also. By making reciprocal crosses with 2 species one can identify the single direction that produces seeds. However, in complex hybrids that involve many species, the incompatibility systems are neutralized so that they seem to be receptive to the pollen of many species.

Bud pollination is not as difficult as it appears. Start by examining the plant for flowers that are expected to open the next day. Make sure to have on hand some fresh pollen from blooming flowers or flowers that have already bloomed. Instead of cutting, carefully pry the sepals and petals open near the end of the bud. Do not tear them apart more than necessary. They will help to protect the stigmas after pollination. Once the stigmas have been uncovered, transfer the pollen to them.

Degree of Kinship

In most cases, crosses are more successful the more closely related the parent plants are to one another. Success is even more probable if the parent plants belong to the same section. Nevertheless, it may be rewarding to attempt a cross between species that are not so closely related, as with the showy hybrids *Passiflora* 'Cary' (*P. cincinnata* × *P. umbilicata*) and *P.* 'Lady Margaret' (*P. coccinea* × *P. incarnata*).

Pollination Technique

As soon as the flowers on the crossing partners have opened, the pollen is collected from the anthers with a small soft brush and transferred as soon as possible to the stigmas of the flowers that are to be pollinated. The pollen should be gently massaged onto the surface of the stigmas to guarantee an intimate contact.

Securing Paternity

When pollen from different plants is transferred, a previously cleaned or fresh brush is always used to avoid any unintentional pollination with pollen that was previously applied. One should always make sure that paternity is unambiguously secured. One measure to guarantee this is to remove the anthers of the pollinated plant with a small pair of scissors, especially when it is a self-compatible clone.

Another measure is the covering of a pollinated flower. A small paper bag put over the flower hinders pollinators from transferring pollen originating from other plants. Plastic bags should not be used, because condensed water and heat could damage the pollinated flower.

Although securing paternity is an important issue, many breeders occasionally use a cocktail of pollen. This can make sense when trying to get seeds from plants that flower only very rarely or when trying to find efficient pollen-donor plants. An example of such an approach could be the use of 6 sorts of pollen in 2 different cocktails. When one cocktail turns out to be successful, each of the 3 sorts of pollen is individually transferred to the flowers of the same recipient species. It is not advisable to grow plants from seeds that stem from a pollination with a cocktail of pollen, as one cannot be sure which plant is the paternal plant.

The most interesting crosses are those between plants that have not been crossed before. The prospects of success increase if one tries to cross in both directions, administering the pollen of one species to the stigmas of another and vice versa. In most cases crosses within the genus *Passiflora* work only in one direction. It is easy, for example, to get fruit from *P. incarnata* with the help of pollen from *P. caerulea*. However, it does not work the other way around.

Pollination Outdoors

Pollination will be unsuccessful during wet weather unless the pollinated flowers are given protection from rain 6–12 hours after pollination.

The success or failure of pollination can be observed after just a few days. Upon carefully folding back the sepals and petals of a closed flower, a swelling of the ovary can be observed as early as 48 hours after fertilization. If pollination has not been successful, the flower will fall off within a few days. Even when the ovary has begun to swell, the flower or the young fruit can abort. This is because the embryos in the seeds abort.

Do not be discouraged by initial failures. Successful pollination is dependent on several environmental conditions: weather, day length, and so on. In spring of 2001 I pollinated 10 flowers of *Passiflora coccinea* with the pollen of *P. incarnata* 'Lorayne' over a period of several weeks without achieving a swelling of the ovaries. Only after the 11th attempt did a fruit develop; it contained more than a hundred mature seeds, of which about 50 germinated.

Marking

Every pollinated flower should be provided with a label indicating the relevant data regarding the cross. The labels should be placed near the stem close to the flower. Fixing them on the peduncle could damage the flower. Marking the pollinated flowers is a little tedious but very helpful in keeping track of things, as it is easy to quickly forget which flower was pollinated with which pollen.

Preserving Pollen

Unfortunately, open flowers of different species are often not available at the same time. Many species flower at different times of the year; *Passiflora actinia*, for example, flowers mainly in spring, while *P. phoenicea* flowers mainly in late summer. Many species flower for only a short period but in batches of up to 20 flowers at the same time, as with *P.*

retipetala. Even such banal reasons as the different ages of individual plants can bring about a situation in which the potential pollinator is in full bloom while the potential partner is just setting small flower buds. In such circumstances, pollen can be preserved for later use.

After the flower has opened, wait until the anthers have burst completely and released the mature pollen. Cautiously remove the pollen with a fine brush and place it on a clean white sheet of paper. The pollen from several flowers of the same plant can be put in one place. Take the paper holding the fresh pollen to a moderately warm place where the pollen can dry swiftly but gently. Many spots in a home will be suited for this: someplace warm but not too hot, with a very low atmospheric humidity. A location near a fluorescent lamp or near (not on) a furnace will be suitable. If there is no appropriate location to secure sufficient drying, store the paper holding the pollen in a little bag with a desiccant (available in computer shops) in a tight-fitting vessel like a jam jar. When this vessel is stored in a warm place, the pollen will dry very rapidly. Fold the paper holding the dry pollen and label it with the pertinent annotations: date, plant name, clone number, and so on. Put a fresh desiccant bag in a tight-fitting vessel and store it in a refrigerator at a temperature of 4°C. As soon as the flower that is to be treated is ready for pollination (when the styles begin to bend down), open the paper holding the pollen and spread the dry pollen onto the fresh stigma with a fine brush. Raise the atmospheric humidity, but avoid spraying the pollinated stigmas directly.

The Ripening Fruit

After pollination, the fruit takes 40 days to 6 months to ripen, depending on the species and the time of the year. Although some fruits appear large and normal at face value, they do not contain seeds at all, or only small, white, dead seeds. *Passiflora* ×*belotii* and *P.* 'Byron Beauty' regularly produce a great number of hollow fruits. In many hybrids, most seeds are dead, but the fruit sometimes contains a small number of mature and well-developed seeds that are recognizable by their dark color and by the enclosing pulpy aril. There are also crosses with normal fruit and many mature seeds. If possible, fruits should not be picked. Truly ripe fruits with viable seeds fall off the plant by themselves. Patience will be repaid. Seeds from fruits that are picked too early will often not germinate satisfactorily although they appear quite dark. If necessary, ripening fruits can be fixed to the stem with a net or thread to keep them from being eaten by animals when they fall off. The more patience one has with the fruit, the greater chance the seeds will have of germinating. This is why I let the fruit ripen in storage for a few days by putting them near some apples. Apples emit the plant hormone ethylene, which speeds up ripening.

Selection

Sow the seeds as described in chapter 5. Under favorable conditions seedlings will develop rapidly into large plants, often flowering even in the first year, though mostly in the 2nd or 3rd year. Bear in mind that every seedling differs genetically from its sibling. All vegetatively produced offspring are genetically identical.

The individual clones often differ in several ways. As soon as the plants have flowered, select only the best specimen for propagation. This procedure can be emotionally painful. Many breeders have high hopes for their crosses and do not evaluate the results realistically. Before new clones are named and put into circulation, every breeder should appraise the plants systematically. Serious breeding requires the extermination of inferior material. Newly bred plants should be rated with regard to certain criteria. Does the plant really differ significantly from other varieties? Does it have fundamental new positive characteristics? Is it free-flowering? Does it root easily? Is it resistant to disease? Can it be easily overwintered? In cases of doubt, the material should be judged by experienced breeders. The satisfaction of having put a hybrid into circulation is rather short-lived when the hybrid turns out to be of inferior quality.

Selected clones can only be propagated vegetatively, by cuttings. Seeds are often offered under the name of a cultivated variety. This is dubious because the plants grown from such seeds will differ in many respects from the selected variety. Seeds from diploid hybrid cultivars can by no means breed true.

The Fruitfulness of Hybrids

Hybrids between 2 species are very often pollen-sterile, meaning the pollen is not formed or is malformed and nonfunctional. The DNA sequence of the chromosomes shows a low homology when the parents of the hybrid are not closely related. Therefore, the chromosomes of paternal and maternal origin are unable to build pairs during meiosis. This pairing of chromosomes is very important in the exact division of the chromosomes (for example, 12 in *Astrophea*) when the diploid set of chromosomes in the adult cells is reduced to a haploid set in the gametes. When the chromosomes cannot pair, the gametes receive irregular numbers of chromosomes. Only a few gametes receive the correct number of 6 different chromosomes, purely by chance. The number of good cells rises in proportion with a greater homology of the parental chromosomes. Since the pollen-bearing anthers will not open sufficiently when there is only a small amount of pollen, the few good cells that may be there never get the chance to pollinate. The few good egg cells have a better chance. They can be found by the tubes from every compatible pollen cell. This is the reason for pollen sterility. A hybrid can be used as a maternal plant but not as a paternal plant. Many examples of ornamental hybrids can be found in passionflower collections. For example, there are a large number of crosses with *Passiflora* 'Amethyst' as a maternal plant but no hybrids with it as a paternal plant.

Hybridization with Laboratory Techniques

Modern laboratory techniques like tetraploidization and somatic hybridization have gained an increasing impact on the breeding of passionflowers. While diploid plants have 2 sets of chromosomes in each cell, one from the maternal plant and one from the paternal plant ($2n$), tetraploid plants have twice the usual set of chromosomes ($4n$). In the wild, tetraploids sometimes come into being by a defect during cell division, but this rarely happens. Tetraploids can be produced in the laboratory with the help of cell poisons like colchicine, derived from the autumn crocus plant. This dangerous chemical brings about profound chromosomal changes in the plant cell. When cells of meristematic tissue are treated with a combination of colchicine and DMSO, most of them die; some stay alive, however, and in a few specimens chromosome duplication has taken place. Colchicine is carcinogenic in animals and humans. Therefore the suppliers of the chemical provide serious warning labels and restrict its use. When the powder or solution is applied, tiny aerosols are released that can cause lung cancer and cancer of the liver. Colchicine is to be used only under the safety provisions of negative pressure in a laboratory of the category L1. The treated plants degrade the poison soon and are harmless.

The main reason for tetraploidizing plants is to get a set of homologous chromosomes. When there are 2 homologs of each kind, pairing during meiosis is easy and allows division into the gametes. The result is a sterile, diploid hybrid that is again fertile.

As soon as there is any imbalance in the genome, such as when a chromosome does not have a homologous partner or when a chromosome has 2 or 3 homologous partners, the correct pairing during meiosis is obstructed. This means that in most cases tetraploids with natural species as parent plants are sterile; backcrosses are also often sterile. The best way to use tetraploids is to breed fertile crosses between 2 species. Tetraploids often have larger cells, and so the parts of the plant are larger, which holds true for the flowers. This is true, for example, of *Passiflora* 'Clear Sky' and *P.* 'Byron Beauty'.

The chemical intervention in cell division also causes disorders in the plants. Rarely more than 10–20% survive the procedure. Only a few of these surviving specimens are completely tetraploid, more than 1/2 remain diploid, and the rest are chimeras, consisting of a mix of diploid and tetraploid tissue. The chromosome numbers of the surviving plants are determined by examining squash preparations of root tips. Because of the hazardous chemicals and the considerable costs of the procedure, the colchicine method is rarely used.

Somatic (from the Greek *soma*, meaning "body") hybridization is an efficient method used to overcome crossing barriers between sexually incompatible species and to increase the genetic variability of higher plants. In contrast to sexually propagated hybrids, the parents of a somatic hybrid contribute diploid sets of chromosomes from somatic cells. In this way an allotetraploid (from the Greek *allos*, meaning "other") is created. A somatic hybrid is very stable. It should be receptive to the introgression of genes from other plants when it is crossed with tetraploids that originate from sexual breeding or treatment with poison. Reports about somatic hybridization with *Passiflora edulis* f. *flavicarpa* are already available (Dornelas et al. 1995, Otoni et al. 1995, Vieira and Dornelas 1996). Marcelo Carnier Dornelas and others obtained somatic hybrids ($4n = 36$) via chemical protoplast fusion between *P. edulis* f. *flavicarpa* ($2n = 18$) and 4 wild diploid species: *P. amethystina*, *P. alata*, *P. gibertii*, and *P. cincinnata*.

Naming

Naming is regulated by international conventions. Only a few hybrid-related excerpts from this code are presented here. Plants should not be named twice. The priority rule states that the name that was given first is valid. A cultivar's name must be formed from the words of a living language. Latinized names or Latin-looking names have not been permitted since 1958. Abbreviations are not allowed, as in *Passiflora* 'RF 1'. Composite words consisting of the names of the parent plants are not permitted; for example, a hybrid of *P. vitifolia* and *P. incarnata* cannot be called *P.* 'Vitinca'. Single quotation marks before and after the cultivar's name are required, and each part of the name must be capitalized. The authors of this book preserve the historical names of very old hybrids—retaining *P.* ×*colvillii*, for example, rather than changing it to *P.* 'Colvillii'. Names should consist of no more than 30 letters. Duplicates of names in other languages are not allowed, and neither are names that are simply colors (*P.* 'Blue Star' would be allowed, but not *P.* 'Blue'). Names that are common in another genus or family are permitted so long as there is no chance of them being confused. *Passiflora* 'Roland Peach', *P.* 'Roland's Peaches', or *P.* 'Peach' would have to be rejected. Names that use words in a different sense are allowed, as with *P.* 'Peach Desires', but good taste should prevail. No names that are offensive in any language are allowed. In honoring someone by naming a hybrid after them, no abbreviations are allowed: *P.* 'T. Burnett' must be changed to *P.* 'Thomas Burnett'. The rules of botanical nomenclature are meant to keep us out of trouble by maintaining the stability of plant names.

Successful Passionflower Crosses

The following list of hybrids was drawn up by Roland Fischer in 2003. It does not claim completeness and does not guarantee or evaluate the correctness of the crossing procedure or the name. Hybrids of unknown origin are not included. Please note references to *Passiflora* ⟨×*kewensis*⟩, *P.* ⟨'Atropurpurea'⟩, and *P.* ⟨'St. Rule'⟩. Angle brackets indicate that although the plant is commonly known under this name, it is in fact a misidentification. Breeders are listed where known. Notes and information about new hybrids should be sent directly to the Passiflora Society International. A regularly updated list may be found at www.passiflora.org.

In the crossing formulas listed, *Passiflora caerulea* is found 121 times and *P.* 'Amethyst', a hybrid of *P. caerulea*, 45 times. This demonstrates how successful *P. caerulea* has been in passing on its genetic structure and how popular it is in gardens. *Passiflora incarnata* has 53 entries, *P. racemosa* 46, *P. cincinnata* 26, and *P. alata* 14. Among tacsonias, *P. tripartita* var. *mollissima* appears most often, with 17 entries, although the identity of this may be in most cases the close relative *P. tarminiana*.

A list of passionflower hybrids

NAME	MOTHER × FATHER	BREEDER
P 'Aafje'	P ⟨×*kewensis*⟩ × P *caerulea* 'Constance Eliott'	K. Oldegarm
P. 'Adularia'	P. *sanguinolenta* × P. *citrina*	J. Vanderplank
P. 'Adularia Variegata'	P. *sanguinolenta* × P. *citrina*	G. Van Herck
P. 'Agi'	= P. ×*colvillii*	E. Kugler
P. 'Aida'	P. ×*decaisneana* × P. aff. *oerstedii*	L. Ševčík
P. 'Albina'	(P. *racemosa* × P. *incarnata*) × P. aff. *oerstedii*	L. Ševčík
P. ×*albo-nigra*	P. *alata* × P. *kermesina*	F. Cardinaux
P. ×*allardii*	P. *quadrangularis* × P. *caerulea* 'Constance Eliott'	E. J. Allard
P. ×*allardii* 'Evatoria'	P. *quadrangularis* × P. *caerulea* 'Constance Eliott'	C. Laurens
P. ×*amabilis*	P. *racemosa* × P. *alata*	J. Schlachter
P. 'Amberg'	= P. ×*violacea*	
P. 'Amethyst'	P. *kermesina* × P. *caerulea*	
P. 'Amethyst Beauty'	= P. ×*violacea*	
P. 'Anastasia'	P. *gritensis* × P. *caerulea* 'Constance Eliott'	J. Brander
P. 'Andrea'	P. 'Katerina' × P. *caerulea* 'Constance Eliott'	L. Ševčík
P. 'Andy'	P. *caerulea* × ?	G. Van Herck
P. 'Anemona'	P. *actinia* × P. *caerulea* 'Constance Eliott'	C. Laurens
P. 'Angela'	P. 'Amethyst' × P. *caerulea* 'Constance Eliott'	L. Ševčík
P. 'Angelo Blue'	P. *amethystina* × P. *subpeltata*	M. Vecchia
P. 'Anna'	P. ×*decaisneana* × P. *tucumanensis*	L. Ševčík
P. 'Anna Christine'	P. ×*violacea* × P. *caerulea*	F. Wescott
P. 'Annette'	P. ⟨×*kewensis*⟩ × P. *racemosa*	A. Frank
P. 'Aragorn'	P. *kermesina* × P. aff. *oerstedii*	R. Fischer
P. 'Arcenciel'	= P. ×*violacea*	
P. 'Argenta'	P. 'Amethyst' × P. *subpeltata*	M. Vecchia
P. 'Asterix'	P. 'Barborea' × P. *caerulea* 'Constance Eliott'	M. Vecchia
P. ⟨'Atropurpurea'⟩	= P. ×*violacea*	
P. 'Atrorubens'	= P. ×*violacea*	
P. 'Aurora'	P. *foetida* × P. *sublanceolata*	M. Vecchia
P. 'Balam'	P. *coriacea* × P. *xiikzodz*	J. MacDougal
P. 'Barborea'	P. 'Amethyst' × ?	L. Ševčík
P. ×*belotii*	P. *alata* × P. *caerulea*	W. Masters
P. 'Bessy'	P. 'Amethyst' × P. *caerulea*	L. Ševčík
P. 'Betsie Greijmans'	P. *sanguinolenta* × P. *citrina*	H. Wouters
P. 'Blanka'	(P. 'Amethyst' × P. *caerulea*) × P. *caerulea* 'Constance Eliott'	L. Ševčík
P. 'Blaumilch'	P. *garckei* × P. *caerulea* 'Constance Eliott'	R. Fischer
P. 'Blue Bird'	P. *incarnata* × ?	J. Vanderplank
P. 'Blue Bouquet'	((P. *amethystina* × P. *caerulea*) × (P. *amethystina* × P. *caerulea*)) × P. *caerulea*	P. Worley
P. 'Blue Carnival'	P. *cincinnata* × P. *caerulea* 'Constance Eliott'	M. Gottschalk
P. 'Blue Crown'	P. *alata* × ?	J. Mráz
P. 'Blue Desire'	P. *tucumanensis* × P. *nephrodes*	B. Ulmer
P. 'Blue Moon'	P. *alata* 'Shannon' × P. *quadrangularis* 'Warmlands'	M. Kartuz
P. 'Blue Star'	P. 'Amethyst' × P. *subpeltata*	M. van der Steeg
P. 'Blue Velvet'	P. *incarnata* × P. *cincinnata*	R. Fischer
P. 'Brokwagter'	P. 'Amethyst' × P. *caerulea*	H. Wouters
P. ×*buonapartea*	P. *quadrangularis* × P. *alata*	
P. 'Byron Beauty'	(P. *edulis* f. *flavicarpa* × P. *edulis* f. *edulis*) × P. *incarnata* = 4*n*	R. Knight
P. 'Byte'	(P. *edulis* × P. *incarnata*) × (P. *incarnata* × P. *cincinnata*) = 4*n*	R. Fischer
P. 'Cacita'	P. 'Capsang' × P. *citrina*	S. Kamstra
P. 'Cannelle'	P. *coccinea* × P. *incarnata*	J. Segalen
P. ×*caponii*	P. *quadrangularis* × P. *racemosa* = 3*n*	W. Capon
P. 'Capsang'	P. *sanguinolenta* × P. *capsularis*	S. Kamstra
P. 'Carnival'	P. *racemosa* × P. *incarnata*	C. Daniels
P. 'Carolina'	P. *sanguinolenta* × P. *capsularis*	G. Brusa
P. 'Cary'	P. *cincinnata* × P. *umbilicata*	H. Gommer
P. 'Catherina Howard'	P. *cincinnata* × P. *caerulea*	R. McCain
P. 'Cecilie'	(P. *caerulea* × P. aff. *oerstedii*) × P. *oerstedii*	L. Ševčík
P. 'Celia Costen'	P. ×*violacea* × P. *caerulea* 'Constance Eliott'	D. Costen

NAME	MOTHER × FATHER	BREEDER
P. 'Cherries Jubilee'	(*P. mixta* × *P. manicata*) × *P. tripartita* var. *mollissima*	P. Worley & R. McCain
P. 'Chianti'	(*P.* 'Amethyst' × *P. subpeltata*) × *P. oerstedii*	R. Fischer
P. 'Clara'	= *P.* ×*colvillii*	M. Vecchia
P. 'Claudia'	*P.* ⟨×*kewensis*⟩ × *P. caerulea* 'Constance Eliott'	K. Oldegarm
P. 'Clear Sky'	((*P. amethystina* × *P. caerulea*) × *P. caerulea*) × *P. caerulea* = 4*n*	R. Fischer
P. 'Cleweriana'	= *P.* ×*violacea*	
P. 'Cold Blue'	*P.* 'Amethyst' × *P. caerulea*	G. Van Herck
P. 'Colet'	*P. racemosa* × *P. tucumanensis*	L. Ševčík
P. 'Columbus'	*P. amethystina* × *P. phoenicea*	M. Vecchia
P. ×*colvillii*	*P. incarnata* × *P. caerulea*	Colvill
P. 'Coocoochee'	(*P.* 'Amethyst' × *P. racemosa*) × *P. gritensis*	R. Fischer
P. 'Coral Glow'	*P. manicata*? × ((*P. mixta* × *P. tripartita* var. *mollissima*) × (*P. mixta* × *P. tripartita* var. *mollissima*))	P. Worley
P. 'Coral Sea'	*P. manicata* × ?	P. Worley
P. 'Cordelia'	*P. vitifolia* × *P. coccinea*	R. Knight
P. 'Corry Rooymans'	(*P.* 'Amethyst' × *P. subpeltata*) × *P. alata*	H. Wouters
P. 'Creamy'	*P.* 'Amethyst' × *P. caerulea*	G. Van Herck
P. 'Crimson Tears'	(((*P. kermesina* × *P. caerulea*) × *P. racemosa*) × *P. gritensis*) × *P. amethystina*	R. Fischer
P. 'Curiosa'	*P. suberosa* × *P. coriacea*	S. Kamstra
P. 'Dakota Ransom'	*P.* 'Amethyst' × *P. caerulea*	S. Ransom
P. 'Daniela'	(*P. racemosa* × *P. tucumanensis*) × *P. caerulea*	L. Ševčík
P. 'Darina'	*P. incarnata* × ?	L. Ševčík
P. 'Dark Star'	*P. edulis* f. *flavicarpa* × *P. caerulea*	F. Moser
P. 'Deanna'	*P. manicata* × ?	
P. 'Debby'	*P. caerulea* × *P.* aff. *oerstedii*	L. Ševčík
P. ×*decaisneana*	*P. alata* × *P. quadrangularis*	Gontier
P. 'Dedorina'	*P.* ×*violacea* × *P. caerulea*	S. Kamstra
P. 'Delicate Dancer'	*P. oerstedii* × *P. caerulea*	P. Worley & R. McCain
P. 'Denisa'	(*P. incarnata* × *P. racemosa*) × ?	L. Ševčík
P. 'Diana'	*P. gibertii* × *P. subpeltata*	F. Thuys
P. 'Doubravka'	*P. racemosa* × *P.* aff. *oerstedii*	L. Ševčík
P. 'Drahomira'	*P. cincinnata* × *P.* aff. *oerstedii*	L. Ševčík
P. 'Eclipse'	*P.* ×*violacea* × *P. caerulea*	F. Wescott
P. 'Eden'	*P.* 'Amethyst' × *P. caerulea*	M. Bradbury
P. 'Edita'	*P.* 'Sunburst' × *P. biflora*	L. Ševčík
P. 'EK1'	= invalid synonym for *P.* 'Dedorina'	S. Kamstra
P. 'Elizabeth'	*P. phoenicea* × *P. incarnata*	P. Worley
P. 'Ely'	*P.* 'Amethyst' × *P. caerulea*	H. Wouters
P. 'Emina'	(*P.* 'Incense' × *P. incarnata*) × *P. caerulea* 'Constance Eliott'	K. Delic
P. 'Enigma'	*P.* ×*decaisneana* × *P. alata*	J. Vanderplank
P. 'Eric'	*P.* ×*decaisneana* × *P. caerulea*	M. van der Steeg
P. 'Espero'	*P. incarnata* × ?	M. Vecchia
P. 'Etoile Blanche'	*P.* 'Amethyst' × *P. subpeltata*	M. van der Steeg
P. 'Evatoria'	*P.* ×*decaisneana* × *P. caerulea* 'Constance Eliott'	C. Laurens
P. 'Evelina'	*P. incarnata* × ?	L. Ševčík
P. 'Evelyne'	*P.* ×*decaisneana* × *P. caerulea*	M. van der Steeg
P. 'Excel'	*P. edulis* f. *edulis* × *P. caerulea*	L. King
P. ×*exoniensis*	*P. antioquiensis* × *P. tripartita* var. *mollissima*	
P. 'Eyleen'	*P. tucumanensis* × *P. racemosa*	L. Ševčík
P. 'Eynsford Gem'	= *P.* ×*violacea*	Geeson
P. 'Fertility'	*P. incarnata* × *P. cincinnata* = 4*n*	R. Fischer
P. 'Fixstern'	*P.* 'Amethyst' × *P. caerulea* 'Constance Eliott'	M. Gottschalk
P. 'Fledermouse'	*P. biflora* × *P. perfoliata*	P. Worley
P. 'Floral Fountain'	*P. phoenicea* × *P. actinia*	P. Worley & R. McCain
P. 'Flying V'	P. penduliflora × P. perfoliata	M. Cooper & M. Irvine
P. 'Franka'	*P.* 'Creamy' × *P.* aff. *oerstedii*	L. Ševčík
P. 'Freckle Face'	*P. racemosa* × *P. cincinnata*	P. Worley
P. edulis 'Frederick'	*P. edulis* cultivar × *P. edulis* cultivar (complex hybrid)	P. Worley
P. 'Frits'	= *P.* ×*violacea*	M. van der Steeg
P. 'Galaxy'	*P. racemosa* × *P. cincinnata*	P. Worley
P. 'Galina'	*P.* 'Incense' × ?	L. Ševčík

NAME	MOTHER × FATHER	BREEDER
P. 'Giftzwerg'	*P. citrina* × *P. conzattiana*	M. Wettges
P. 'Gizmo'	*P. foetida* × *P. palmeri*	J. Segalen
P. 'Golden Glow'	*P. manicata* × *P. tripartita* var. *mollissima*	P. Worley
P. 'Gratia'	*P. racemosa* × *P. cincinnata*	H. Reimann
P. 'Green Lace'	*P. edulis* × *P. cincinnata*	P. Worley & R. McCain
P. 'Guglielmo Betto'	*P. incarnata* × *P. tucumanensis*	M. Vecchia
P. 'Hana'	*P.* 'Shirley' × *P. caerulea*	L. Ševčík
P. 'Hardy Red'	*P. incarnata* × *P. racemosa*	G. Van Herck
P. 'Helena'	*P. tucumanensis* × *P.* aff. *oerstedii*	L. Ševčík
P. 'Hematitei'	*P. sanguinolenta* × *P. capsularis*	J. Vanderplank
P. 'Hermine'	*P.* ×*decaisneana* × *P. caerulea*	M. van der Steeg
P. 'Hot Shot'	*P. vitifolia* × *P. coccinea*	C. Laurens
P. 'Ida'	(*P. caerulea* × *P.* aff. *oerstedii*) × *P. oerstedii*	L. Ševčík
P. 'Incense'	*P. incarnata* × *P. cincinnata*	R. Knight
P. 'Indigo Dream'	*P.* 'Incense' × *P. caerulea*	P. Worley
P. ×*innesii*	*P. alata* × *P. quadrangularis*	W. Innes
P. 'Inspiration'	*P. incarnata* × *P. cincinnata* = 4*n*	R. Fischer
P. 'Intrigue'	(*P. ampullacea* × (*P. mixta* × *P. tripartita* var. *mollissima*)) × (*P. mixta* × *P. tripartita* var. *mollissima*)	P. Worley
P. 'Inverleith'	*P. vitifolia* × *P. incarnata*	A. Ensoll
P. 'Irma'	(*P.* 'Amethyst' × ?) × (*P. racemosa* × *P. incarnata*)	L. Ševčík
P. 'Iubilaeum'	= *P.* ×*colvillii*	M. Vecchia
P. 'Ivy Waves'	*P. coriacea* × *P. suberosa* = 4*n*	R. Knight
P. 'Jameson'	*P. mixta* × *P. tripartita* var. *mollissima*	
P. 'Jana'	= *P.* ×*colvillii*	L. Ševčík
P. 'Jara'	*P. caerulea* × *P.* 'Purple Haze' = 4*n*	R. Fischer
P. 'Jeannette'	((*P. amethystina* × *P. caerulea*) × (*P. amethystina* × *P. caerulea*)) × *P. caerulea*	P. Worley
P. 'Jeeny'	*P. cincinnata* × *P. caerulea*	L. Ševčík
P. 'Jelly Joker'	*P. cincinnata* × *P. umbilicata*	C. Laurens
P. 'John Innes'	*P. quadrangularis* × *P. racemosa*	W. Capon
P. 'Jolana'	(*P. racemosa* × *P. incarnata*) × *P. caerulea*	L. Ševčík
P. 'Jolie'	*P. racemosa* × *P. caerulea* 'Merryon'	H. Gommer
P. 'Josefine'	= *P.* ×*colvillii*	E. Kugler
P. 'Justina'	(*P.* 'Amethyst' × *P. caerulea*) × *P. caerulea*	L. Ševčík
P. 'Jutta'	*P. umbilicata* × *P. tucumanensis*	H. Wouters
P. 'Karen Jacobsen'	*P. incarnata* × *P. trisulca*	P. Worley
P. 'Karla'	*P. cincinnata* × ?	L. Ševčík
P. 'Kate Adie'	*P. mooreana* × *P. sprucei*	J. Vanderplank
P. 'Katerina'	*P.* ×*decaisneana* × *P.* aff. *oerstedii*	L. Ševčík
P. 'Kazi'	*P. incarnata* × *P. racemosa*	L. Ševčík
P. 'Ken Kera'	*P.* ⟨×*kewensis*⟩ × *P. racemosa*	G. Van Herck
P. ×*kewensis*	*P. caerulea* × *P. kermesina*	W. Watson
P. ⟨×*kewensis*⟩	*P. racemosa* × ?	
P. 'King Willem-Alexander'	*P. subpeltata* × *P. gritensis*	H. Wouters
P. 'Klara'	(*P. tucumanensis* × *P. racemosa*) × *P. caerulea*	L. Ševčík
P. 'Kunhuta'	*P.* ×*decaisneana* × *P. caerulea*	L. Ševčík
P. 'Kunigunde'	*P.* ×*decaisneana* × *P. caerulea* 'Constance Eliott'	L. Ševčík
P. 'Labgirl'	*P. umbilicata* × *P. oerstedii*	R. Fischer
P. 'Lada'	*P. cincinnata* × *P. caerulea*	L. Ševčík
P. 'Lady Margaret'	*P. coccinea* × *P. incarnata*	T. Skimina
P. 'Laura'	*P.* ×*decaisneana* × *P. caerulea* 'Constance Eliott'	L. Ševčík
P. 'Lavender Lady'	= *P.* 'Amethyst'	
P. ×*lawsoniana*	*P. alata* × *P. racemosa*	Lawson
P. 'Libuse'	*P. incarnata* × *P. racemosa*	L. Ševčík
P. 'Lilac Lady'	= *P.* ×*violacea*	
P. 'Linda'	*P. cincinnata* × *P. caerulea* 'Constance Eliott'	L. Ševčík
P. 'Linda'	*P. serrulata* × *P. maliformis*	C. Howell
P. ×*lineasrosea*	*P. quadrangularis* × *P. caerulea*	
P. 'Little Orchid'	unknown	G. Van Herck
P. 'Livie'	*P. cinnabarina* × *P. herbertiana*	L. Ševčík
P. 'Liza'	*P. incarnata* × *P. racemosa*	L. Ševčík

NAME	MOTHER × FATHER	BREEDER
P. 'Lobo'	*P. apetala* × *P. ornithoura*	P. Worley & R. McCain
P. ×*loudonii*	*P. racemosa* × *P. kermesina*	R. Sweet
P. 'Lucia'	*P.* 'Shirley' × ?	L. Ševčík
P. 'Ludmilla'	*P. incarnata* × *P. racemosa*	L. Ševčík
P. 'Machou'	*P.* 'Amethyst' × ?	J. Jannel
P. 'Madame Bruckhaus'	*P.* ×*loudonii* × *P. racemosa*	Haage & Schmidt
P. 'Madonna'	*P.* ×*buonapartea* × *P. racemosa*	E. Otto
P. 'Maeva'	*P.* 'Amethyst' × *P. subpeltata*	J. Segalen
P. 'Maggie'	*P. caerulea* × *P.* 'Elizabeth' (?)	G. Van Herck
P. 'Manapany'	(*P. edulis* × *P. incarnata*) × (*P. incarnata* × *P. cincinnata*) = 4*n*	R. Fischer & J. Segalen
P. 'Manta'	*P. xiikzodz* × *P. coriacea*	M. Vecchia
P. 'Marcella'	(*P. incarnata* × *P. racemosa*) × *P. incarnata*	L. Ševčík
P. 'Margaret Wilson'	*P.* ×*buonapartea* × *P. caerulea*	J. Wilson
P. 'Maria'	*P.* 'Amethyst' × *P. caerulea* 'Constance Eliott'	G. Van Herck
P. 'Marianne Timmer'	*P. cincinnata* × *P.* 'Pura Vida' (?)	H. Gommer
P. 'Marie'	((*P. amethystina* × *P. caerulea*) × (*P. amethystina* × *P. caerulea*)) × *P. caerulea*	P. Worley
P. 'Mariel'	= *P.* ×*violacea*	K. Oldegarm
P. 'Marijke'	*P.* ×*decaisneana* × *P. caerulea*	M. van der Steeg
P. 'Mario'	*P.* aff. *oerstedii* × *P. menispermifolia*	I. Reiber
P. 'Marlie'	*P. sanguinolenta* × *P. citrina*	H. Wouters
P. 'Mary Jane'	*P. vitifolia* × *P. glandulosa*	R. Boender
P. 'Mauve Madness'	*P. oerstedii* × *P. caerulea*	P. Worley & R. McCain
P. 'Mauvis Mastics'	= *P.* ×*violacea*	
P. 'Maxima'	(*P. amethystina* [long peduncle] × *P. caerulea*) × *P. caerulea* 'Constance Eliott'	C. Laurens
P. 'Medallion'	*P. incarnata* × *P. actinia*	P. Worley & R. McCain
P. 'Michael'	*P. vitifolia* × *P. caerulea*	M. van der Steeg
P. 'Michaela'	*P. coccinea* × ?	L. Ševčík
P. 'Michelle Noble'	(*P. mixta* × *P. tripartita* var. *mollissima*) × (*P. mixta* × *P. tripartita* var. *mollissima*)	P. Worley & R. McCain
P. 'Minai'	*P.* aff. *oerstedii* × *P. menispermifolia*	I. Reiber
P. 'Mirabilis'	*P. alata* × *P. cincinnata*	H. Reimann
P. 'Miranda'	*P.* ×*decaisneana* × *P. caerulea*	M. van der Steeg
P. 'Monique Klemann'	*P.* ×*decaisneana* × *P. caerulea*	C. Laurens
P. 'Moonburst'	*P. trifasciata* × *P.* 'Sunburst'	P. Worley & R. McCain
P. 'Mum's Gate'	*P. manicata* × *P. mixta*	P. Worley
P. 'Mystery'	*P. tripartita* var. *mollissima* × *P. ampullacea*	R. McCain
P. 'Natalia'	*P. incarnata* × *P. nephrodes*	M. Vecchia
P. 'Nebulae'	*P. incarnata* × *P. serratodigitata*	P. Worley & R. McCain
P. 'Nelly'	*P. racemosa* × *P. gibertii*	M. van der Steeg
P. 'New Amethyst'	*P. kermesina* × *P. caerulea* = 4*n*	R. Fischer
P. 'New Incense'	*P.* 'Incense' × *P. cincinnata*	J. Vanderplank
P. 'Nightfever'	= *P.* 'Amethyst'	
P. 'Nightmare'	*P. subpeltata* × *P. gritensis*	R. Fischer
P. 'Nina'	*P. biflora* × *P. cuneata*	L. Ševčík
P. 'Nocturne'	*P.* 'Amethyst' × *P. amethystina*	P. Worley & R. McCain
P. 'Nora'	= *P.* ×*violacea*	L. Ševčík
P. 'Nova Star'	*P.* 'Star of Kingston' × *P. caerulea*	M. van der Steeg
P. 'Olga'	*P. actinia* × *P. alata*	J. Lindsay
P. 'Olivie'	((*P.* 'Amethyst' × *P. caerulea*) × *P. caerulea* 'Constance Eliott') × *P. caerulea*	L. Ševčík
P. 'Onda Blu'	*P. incarnata* × *P.* 'Barborea'	M. Vecchia
P. 'Oriental Sunset'	*P. herbertiana* × *P. cinnabarina*	H. Wouters
P. 'Paganini'	*P. incarnata* × *P. gibertii*	M. Vecchia
P. 'Paola Gastaldo'	*P. incarnata* × *P. sprucei*	M. Vecchia
P. 'Patricie'	*P.* 'Colet' × *P. caerulea*	L. Ševčík
P. 'Patty Ann'	*P. incarnata* × *P. amethystina*	C. Daniels
P. 'Pavlina'	*P.* ⟨×*kewensis*⟩ × *P.* aff. *oerstedii*	L. Ševčík
P. 'Perfume'	? × *P. caerulea*	C. Laurens
P. 'Peter'	*P. coccinea* × *P. racemosa*	
P. 'Peter Lawrence'	*P. vitifolia* × *P. glandulosa*	R. Boender
P. 'Petra'	*P. incarnata* × (*P.* ×*violacea* × *P. caerulea*)	L. Ševčík
P. 'Philippine'	*P. racemosa* × *P. kermesina*	W. Neubert

NAME	MOTHER × FATHER	BREEDER
P. 'Pink Jewel'	*P. foetida* × *P. palmeri*	J. Vanderplank
P. 'Pink Nightmare'	*P. subpeltata* × *P. gritensis*	R. Fischer
P. 'Pink Pride'	*P.* 'Rosa Stern' × *P. gritensis*	H. Wouters
P. 'Pink Promenade'	*P. trisecta* × *P. mixta*	P. Worley & R. McCain
P. 'Pink Seduction'	(*P.* 'Amethyst' × *P. subpeltata*) × *P. gritensis*	H. Wouters
P. 'Pink Star'	= *P.* ×*violacea*	C. Laurens
P. ×*piresea*	*P. quadrifaria* × *P. vitifolia*	J. Vanderplank
P. ×*piresii*	= invalid name for *P.* ×*piresea*	
P. 'Pretty Ballerina'	*P. phoenicea* × *P. actinia*	P. Worley & R. McCain
P. 'Pura Vida'	*P.* 'Amethyst' × *P. racemosa*	S. Kamstra
P. 'Purple Haze'	*P. caerulea* × *P. amethystina* [long peduncle]	C. Laurens
P. 'Purple Rain'	= *P.* 'Jara'	R. Fischer
P. 'Purple Tiger'	*P. phoenicea* × *P. quadrangularis*	P. Worley & R. McCain
P. 'Queen Beatrix'	*P. subpeltata* × *P. gritensis*	H. Wouters
P. 'Radka'	*P. maliformis* × ?	L. Ševčík
P. 'Red Bird'	*P.* 'Coral Glow' × *P. manicata*	P. Worley
P. 'Red Cloud'	(*P.* 'Amethyst' × *P. racemosa*) × *P. gritensis*	R. Fischer
P. 'Red Inca'	*P. coccinea* × *P. incarnata*	C. Laurens
P. 'Red Infinity'	*P. herbertiana* × *P. cinnabarina*	H. Wouters
P. 'Regina'	(*P.* ×*violacea* × *P. caerulea* 'Constance Eliott') × *P.* aff. *oerstedii*	L. Ševčík
P. 'Rita'	*P.* ⟨×*kewensis*⟩ × *P. racemosa*	G. Van Herck
P. 'Rochelle'	*P. platyloba* × *P. incarnata*	P. Worley & R. McCain
P. 'Roedie'	*P. cincinnata* × *P.* 'Pura Vida' (?)	H. Gommer
P. 'Rosa Stern'	*P.* 'Amethyst' × *P. subpeltata*	M. van der Steeg
P. 'Rosa Traum'	*P. coccinea* × *P. caerulea* 'Constance Eliott'	M. Gottschalk
P. 'Rosea'	= *P.* ×*violacea*	K. Baum
P. ×*rosea*	*P. pinnatistipula* × *P. tripartita* var. *mollissima*	
P. 'Rossana'	(*P. amethystina* × *P. subpeltata*) × *P. phoenicea*	M. Vecchia
P. 'Saint Julien'	= *P.* ×*colvillii*	
P. 'Sancap'	*P. sanguinolenta* × *P. capsularis*	L. Ševčík
P. 'Sandra'	*P. incarnata* × *P.* aff. *oerstedii*	L. Ševčík
P. 'Sapphire'	*P. edulis* f. *flavicarpa* × *P. caerulea*	J. Vanderplank
P. 'Sarah Aimee'	*P. foetida* × *P. urbaniana*	J. Vanderplank
P. 'Scharka'	*P.* ×*decaisneana* × *P.* aff. *oerstedii*	L. Ševčík
P. 'Scribble'	*P. cincinnata* × *P. gibertii*	M. Vecchia
P. 'Senta'	*P.* ×*violacea* × *P. eichleriana*	L. Ševčík
P. 'Serena'	(*P. amethystina* × *P. subpeltata*) × *P. phoenicea*	M. Vecchia
P. 'Shelly'	*P.* 'Amethyst' × *P. caerulea*	H. Wouters
P. 'Shirley'	*P.* ×*violacea* × *P. eichleriana*	L. Ševčík
P. 'Showy Shadow'	*P.* 'Violetta' × *P. caerulea*	B. Ulmer
P. 'Silvie'	*P.* 'Purple Haze' × *P. caerulea*	H. Wouters
P. 'Simpin'	*P. caerulea* × *P. amethystina*	S. Kamstra
P. 'Simply Red'	*P. caerulea* × *P. amethystina* [long peduncle]	C. Laurens
P. 'SiSimona'	*P.* 'Purple Haze' × *P.* 'Simply Red'	C. Laurens
P. ×*smytheana*	*P. tripartita* var. *mollissima* × *P. manicata*	W. Smythe
P. 'Snow White'	(*P.* 'Amethyst' × *P. caerulea*) × *P.* 'Sapphire'	J. Thompson
P. 'Softleaf'	*P. vitifolia* × *P. incarnata*	D. Costen
P. 'Spider'	*P. incarnata* × *P.* ×*violacea*	M. Vecchia
P. 'Star of Bristol'	*P.* 'Amethyst' × *P. caerulea*	J. Vanderplank
P. 'Star of Clevedon'	*P.* 'Amethyst' × *P. caerulea*	J. Vanderplank
P. 'Star of Kingston'	*P.* 'Amethyst' × *P. caerulea*	J. Vanderplank
P. 'Star of Mikan'	= *P.* 'Amethyst'	
P. 'Star of Surbiton'	*P.* 'Amethyst' × *P. caerulea*	M. Irvine
P. 'Stella di Cremona'	*P.* ×*violacea* × *P. caerulea* 'Constance Eliott'	M. Vecchia
P. 'Stradivarius	*P. murucuja* × *P. cupraea*	M. Vecchia
P. 'Strasshof'	= *P.* ×*colvillii*	E. Kugler
P. 'Striker'	*P. phoenicea* × *P. racemosa*	P. Worley & R. McCain
P. 'St. Rule'	*P.* ×*buonapartea* × *P. subpeltata*	J. Wilson
P. ⟨'St. Rule'⟩	= *P. gibertii*	
P. 'Strybing Red'	*P. antioquiensis* × *P. tripartita* var. *mollissima*	G. Cunningham

NAME	MOTHER × FATHER	BREEDER
P. 'Sunburst'	*P. gilbertiana* × *P. jorullensis*	P. Worley
P. 'Sundance'	(*P.* 'Amethyst' × *P. caerulea*) × *P.* 'Sapphire'	J. Thompson
P. 'Surprise'	*P. actinia* × *P. caerulea* 'Constance Eliott'	B. Ulmer
P. 'Susan Brigham'	*P. tripartita* var. *mollissima* × *P. mixta*	P. Worley
P. 'Susanne'	*P. caerulea* × *P.* 'Star of Clevedon'	Voss-Grosch
P. 'Sweet Allure'	(*P. mixta* × *P. manicata*) × *P. tripartita* var. *mollissima*	P. Worley & R. McCain
P. 'Tangerine Cream'	*P. apetala* × *P. jorullensis*	P. Worley & R. McCain
P. 'Taza'	*P.* 'Amethyst' × ?	J. Jannel
P. 'Temptation'	*P. incarnata* × *P. cincinnata* = 4*n*	R. Fischer
P. 'Tereza'	unknown	L. Ševčík
P. 'Teta'	*P. incarnata* × *P. racemosa*	L. Ševčík
P. 'Tiffani'	*P. incarnata* × *P. kermesina*	R. Fischer
P. 'Tinalandia'	*P. edulis* f. *flavicarpa* × *P. incarnata*	C. Howell
P. 'Tresederi'	= *P.* ×*violacea*	
P. 'Twinkel Star'	(*P.* 'Amethyst' × *P. caerulea*) × *P. caerulea*	M. van der Steeg
P. 'Velvet Beauty'	*P.* ×*violacea* × *P. caerulea*	A. Schwarzwald
P. 'Venus'	*P.* 'Amethyst' × *P. caerulea* 'Constance Eliott'	M. Vecchia
P. 'Vera'	(*P. caerulea* × *P.* aff. *oerstedii*) × *P. caerulea*	L. Ševčík
P. 'Vibrant Pink'	*P. antioquiensis* × *P. tripartita* var. *mollissima*	G. Cunningham
P. 'Vicky'	*P.* (×*kewensis*) × *P. racemosa*	G. Van Herck
P. 'Victoria'	= *P.* ×*violacea*	
P. 'Viden'	= *P.* ×*violacea*	L. Ševčík
P. 'Vilma'	(*P. racemosa* × *P. tucumanensis*) × *P. incarnata*	L. Ševčík
P. 'Viola'	= invalid name for *P.* 'Violetta'	
P. ×*violacea*	*P. racemosa* × *P. caerulea*	T. Milne
P. 'Violet Selection'	*P.* 'Amethyst' × ?	B. Ulmer
P. 'Violet Star'	*P.* 'Amethyst' × *P. caerulea* 'Constance Eliott'	M. Gottschalk
P. 'Violetta'	*P. caerulea* × *P.* aff. *oerstedii*	L. Ševčík
P. 'Vladena'	*P. cincinnata* × *P. caerulea*	L. Ševčík
P. 'Vlasta'	*P. incarnata* × *P. caerulea*	L. Ševčík
P. 'White Star'	*P. caerulea* × ?	G. Van Herck
P. 'Wil'	*P. coccinea* × *P. racemosa*	
P. 'Wilgen Heintje'	*P. racemosa* × *P. alata*	P. Moermann
P. 'Wilgen Marieke'	*P. incarnata* × *P. amethystina*	P. Moermann
P. 'Yvonne Baron'	*P.* 'Amethyst' × *P. caerulea*	J. Jannel
P. 'Zita'	(*P. racemosa* × *P. tucumanensis*) × ?	L. Ševčík
P. 'Zlata'	*P.* 'Sancap' × *P. citrina*	L. Ševčík
P. 'Zofie'	(*P.* 'Amethyst' × *P. caerulea*) × *P.* aff. *oerstedii*	L. Ševčík

18

Descriptions of Selected Hybrids

FOR THE LAST 15 years hybrid passionflowers have gained increasing importance, and their numbers have increased in many enthusiasts' collections. Even flower shops and garden centers have discovered these easy-to-grow, free-flowering plants. This is not surprising, as these hybrids result from crossing species with spectacular flowers with other species (or, increasingly, with hybrids) that are also very robust. Such plants are easy to market and sell, especially since passionflowers are enjoying such a renaissance.

For the time being, the center of crossbreeding is central Europe. There are only a few breeders in the United States, while in Europe increasing numbers of amateurs and professional gardeners occupy themselves with crossbreeding. The result is an abundance of new hybrids, which for the most part are not much different from established ones. The long-yearned-for first flower of one's own cross often seems more extraordinary than it really is. Attributed with an attractive name, such crosses appear in collections only to be soon forgotten.

Only a few breeders are guided by a definite aim. One is to obtain larger flowers, which can be achieved by tetraploidization or by other slower methods (see, for example, Plate 294). Roland Fischer is a leading expert in this field. Among other renowned breeders are Hendrik Jan Gommer, Cor Laurens, and Henk Wouters (Netherlands); Leopold Ševčík (Czech Republic); John Vanderplank (Great Britain); Maurizio Vecchia (Italy); and Patrick J. Worley (United States), to name just a few. These and many others have created a great number of excellent hybrids that one cannot imagine missing from the collection of a true enthusiast.

The enthusiast's wish for exceptional colors and specific characteristics is the mainspring for hybridization. A free-flowering passionflower with large red blossoms that is cold-hardy and produces edible fruit would be the long-desired hot seller, the "killer application" of the passionflower world. It remains to be seen whether some day such a plant can be bred.

The following descriptions of hybrids are a subjective selection. We have tried to take equally into account the classical representatives that have been popular for decades and the novelties that have only recently become established. The plants included here were selected based on their availability and uniqueness and on there being a thorough knowledge of both origin and crossing formula. The female parent is given first in the formulas.

Passiflora 'Adularia'

named in reference to adular, a moonstone variety
P. sanguinolenta × *P. citrina*
Breeder: John Vanderplank (1993)

It was only a matter of time before someone crossed

the 2 popular species *Passiflora sanguinolenta* and *P. citrina*. The first successful attempt was made by John Vanderplank in Britain in the early 1990s.

Since then this cross has been repeated several times and showed similar results. Worth mentioning are *Passiflora* 'Adularia Variegata', raised by Guido Van Herck, which is characterized by leaves that are variegated along the 2 outer main nerves, and *P.* 'Betsie Greijmans', raised by Henk Wouters, which has somewhat larger flowers. Because of the Latinized epithet, *P.* 'Adularia Variegata' is not a valid name and should be changed to *P.* 'Adularia Variegated' (see chapter 17 for more information on naming).

Passiflora 'Adularia' (Plate 295) has entered many collections. The reasons for its popularity are evident. It is not picky about its location and is suited for cultivation on a windowsill as well as for cultivation in a greenhouse or in a container outdoors. It can be kept as a trailing plant, a ground cover, or a climber on a trellis. It will flower even in a pot small enough to fit any windowsill.

Outdoors this rewarding hybrid flowers June to September. In a heated greenhouse or conservatory it produces numerous apricot to light pink flowers throughout the year. Its flowers, 4–5.5 cm diam., are not among the largest of the genus. Even the corona is relatively short, and it does not effuse a scent since both parent plants are hummingbird-pollinated. Nevertheless, this plant is enjoyed for its unique flower color, wealth of blossoms, modest requirements, and ease of care. Demanding a minimum temperature of only 5°C, it is also very robust.

Vine: pubescent; stem: 3-angular, striate, green to occasionally reddish green; stipules: setaceous, 0.4–0.8 cm long; petioles: 1.7–5.5 cm long without glands; leaves: 2-lobed, or with a 3rd lobe in the middle, (2.5–)4.5–8(–11) × (–3)4.2–9(–14) cm, submembranous, entire; peduncles: 3–9 cm, solitary or in pairs; bracts: absent; floral tube: 1–1.5 cm long, apricot outside, purple inside; flowers: apricot to bright salmon, 4–5.5 cm diam.; sepals: apricot to bright salmon, oblong-lanceolate, 2–3.3 × 0.3–0.4 cm; petals: apricot to bright salmon, oblong-lanceolate, 2–2.7 × 0.2–0.3 cm; corona: one series, 0.7–1 cm long, purplish at base, whitish at middle, and apricot to cream at apex; ovary: ellipsoid, pubescent; fruit: ellipsoid, 6-carinate, ca. 3–4 × 1–2 cm, reddish.

Passiflora* ×*allardii Lynch

Gard. Chron. 42: 274 (1907)
named after the breeder
P. quadrangularis × *P. caerulea* 'Constance Eliott'
Breeder: E. J. Allard

Passiflora ×*allardii* (Plate 296) is another recommendable hybrid that is exceptionally suited for cultivation in a container. In the course of one summer, large specimens can produce countless flowers that are 8.5–11 cm diam. Their intense, fresh scent are reminiscent of citrus flowers. Even young plants in a 12-cm pot can set flower buds.

The sepals are whitish and occasionally pale pink toward the margin, while the petals are rich pink. Unfortunately, the dried-up flowers can remain on the plant for several weeks and blur the general impression.

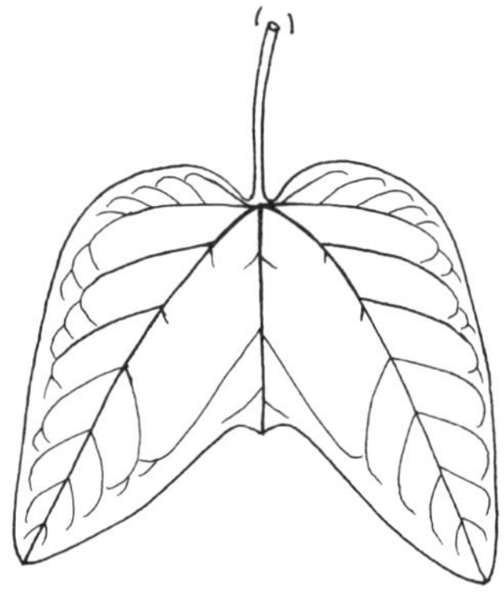

Passiflora 'Adularia'

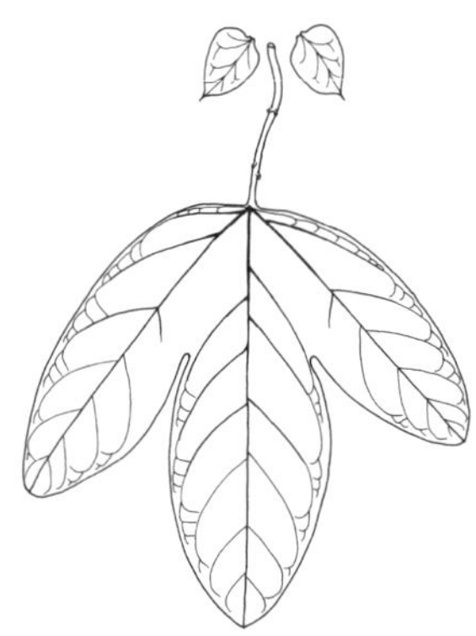

Passiflora ×*allardii*

It happens quite often that *Passiflora* ×*allardii* is confused with *P.* ×*belotii* or other comparable hybrids. With the help of the styles, stigmas, and leaf shape, these hybrids can easily be distinguished from one another. In *P.* ×*allardii*, the greenish white styles and stigmas are mottled intensely purplish brown. In contrast, these are green and only partially mottled with a few brown speckles in *P.* ×*belotii*. Furthermore, the individual leaf lobes of *P.* ×*allardii* are significantly narrower than those of *P.* ×*belotii*. *Passiflora* ×*allardii*, like some other hybrids, occasionally exhibits symptoms of a viral infection, especially during cooler weather. Speckled or even deformed leaves often occur when the plant suffers from insufficient conditions in its winter quarters. Make sure to thoroughly clean cutting tools after working on suspect plants. During overwintering, *P.* ×*allardii* should not be kept at temperatures below 7°C for long periods.

Vine: glabrous; stem: 4-angular, striate, green to occasionally reddish green; stipules: oblong-ovate, 1.5–2 × 1.1–1.5 cm long, entire; petioles: 2.5–6.5 cm long, with 4(–5) glands; leaves: 3-lobed, 7.5–17 × 8.5–18 cm, entire; peduncles: 3.5–6 cm, solitary; bracts: broadly ovate, 2.4–3.5 × 2–2.9 cm, whitish green, entire, free; flowers: white, pink, and violet, 8.5–11 cm diam.; sepals: whitish green outside, white inside, occasionally pinkish toward both sides, oblong, 3.4–4.5 × 1.5–1.9 cm, aristate, awn 0.3 cm; petals: pink, oblong, 3.9–4.8 × 1.3–1.7 cm; corona: 6 or 7 series, outer 2 series 2.8–3.5 cm long, purple in lower 1/3, banded violet and white below middle and bluish violet in upper 1/2, inner series 0.3–0.6 cm long, purple; ovary: ovoid, glabrous; fruit: ovoid, orange, up to 6 cm long.

Passiflora 'Amethyst'

named by John Vanderplank after the purple or bluish violet jeweler's stone
P. kermesina × *P. caerulea*
Synonym: *P.* 'Lavender Lady' (?)
Breeder: Unknown

Passiflora 'Amethyst' (Plate 297) is among the most rewarding and most popular passionflowers. It and *P. caerulea* are the top-ranking species offered by flower shops and garden centers. This is quite understandable in view of this plant's enormous willingness to flower and modest requirements. It owes its name to its purple-blue to violet flowers, which match the color of the well-known precious stone. It flowers throughout the summer and even in its winter quarters. Flowers up to 10 cm diam. may occasionally appear. Toward autumn the intensity in color diminishes, an effect seen in many species. *Passiflora* 'Amethyst' is particularly recommended for cultivation in containers. It is usually taken to its winter quarters in October before the first frost. A specimen that has been hardened off will even survive slight frost without damage. This plant can be overwintered at rather low temperatures, just above 2°C, although temperatures around 10°C are better.

Passiflora 'Amethyst' can also be cultivated as a houseplant, because it flowers reliably in a small pot of about 12 cm. At times it may be necessary to limit its vigorousness by a formative pruning.

The parentage of *Passiflora* 'Amethyst' was unclear for many years. Its origin and breeder are unknown, although there has been much speculation. Crossbreeding experiments with *P. caerulea* and *P. kermesina* conducted by Roland Fischer have shown that both species are parents. As a result, experts believe the plants known as *P.* 'Amethyst' represent the true *P.* ×*kewensis*, an old hybrid of the same parentage that was thought to be lost.

Passiflora 'Amethyst' was grown in Europe for many years under the erroneous names *P. amethystina* and *P. violacea*. After studying newly imported clones of the botanical species, John Vanderplank (1991) realized that the long-cultivated material was a hybrid and suggested the name

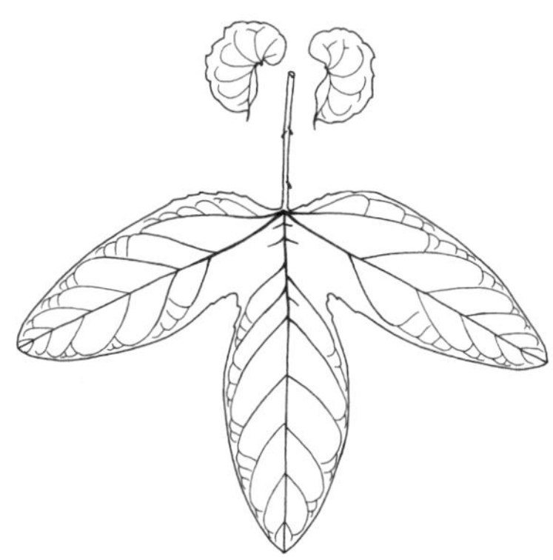

Passiflora 'Amethyst'

'Amethyst'. He also suggested that *P.* 'Lavender Lady' was a synonym. That hybrid, raised by Patrick J. Worley in the 1980s, is a multiple cross between *P. amethystina* and *P. caerulea*. However, there are photos of *P.* 'Amethyst' that were taken earlier, and although these 2 are extremely similar, they are apparently different.

Vine: glabrous; stem: terete; stipules: oblong-ovate to subreniform, (1.4–)1.7–2.8(–3.7) × 1.2–2 cm, entire or slightly serrate toward base; petioles: 2–6.5 cm long, with 2 elongated glands above middle and 1–2 additional glands above; leaves: 3-lobed, 6–10(–12.5) × 8.5–15(–17.5) cm, more or less reddish green beneath, entire; peduncles: 6–13 cm, solitary; bracts: ovate, 1.2–2 × 1–1.7 cm, entire, free; flowers: purplish to violet, 9–11 cm diam.; sepals: purplish green outside, purplish violet inside, oblong, 3.8–5.1 × 0.9–1.5 cm, aristate, awn 0.3–0.4 cm; petals: purplish violet, oblong, 4–4.7 × 1–1.4 cm; corona: generally 4 series, outer 2 series 1.4–2 cm long, violet with one paler band, inner series 0.3–0.6 cm long, dark purple; ovary: oblong-ovoid, glabrous; fruit: oblong-ovoid, bright orange, 4.5–6 × 1.5–2.3 cm.

Passiflora 'Anastasia'

P. gritensis × *P. caerulea* 'Constance Eliott'
Breeder: Joos Brander (2000)

One of the most fascinating passionflowers of all, *Passiflora gritensis* quickly became a favorite crossing partner after it was brought into cultivation toward the end of the 1990s. The demands of this species are as unusual as its flowers are spectacular. In summer it does not tolerate direct sunshine, and its main flowering period in central Europe occurs from September to May. Growers have tried to improve on this by crossing it with suitable species. The first hybrids were available only a few years after the species' introduction. *Passiflora* 'Pink Pride' and *P.* 'Pink Nightmare' are both worth cultivating, and *P.* 'Anastasia' (Plate 298) is among the most beautiful *P. gritensis* hybrids.

Passiflora 'Anastasia'

What makes *Passiflora* 'Anastasia' stand out is its combination of sumptuous rose-pink flowers and a reddish white corona. Flowers are 9–11 cm diam., but since the sepals and petals reflex quite soon, blossoms appear less wide and longer. Although this hybrid may produce some flowers during summer, the main flowering time in central Europe is late summer and early fall. It has turned out to be a free-flowering hybrid but does not compare with *P.* 'Amethyst', *P.* ×*belotii*, or *P.* ×*violacea* in this respect.

Passiflora 'Anastasia' grows in semishade or sun and should be cultivated in the open during summer. Even in a 20-cm pot, the plant reaches a size at which it can set flowers. In comparison with other rose-pink passionflowers, its resistance to cold is relatively great. Even short spells of slight frost are tolerated, although the tips of the shoots may be frostbitten. *Passiflora* 'Anastasia' should be overwintered at temperatures above 5°C.

Vine: glabrous; stem: terete to subangular, slightly striate; stipules: oblong-reniform, 1.5–2.5 × 0.8–1.2 cm, serrate; petioles: 2.2–5 cm, with 2–4 mostly paired, elongated glands generally in upper 1/2, glands 0.3–0.4 cm long; leaves: deeply 3-lobed, 5–15 × 7.8–20 cm, entire; peduncles: 7–12.5 cm, solitary; bracts: ovate, ca. 2 × 1.3 cm, serrate, free; flowers: pinkish red, 9–11 cm diam.; sepals: green outside, pinkish red inside, oblong, 4–5 × 1.7–2.2 cm, keeled, keel terminating in an awn 0.7–1 cm long; petals: pinkish red, subequal to sepals; corona: 4 or 5 series, outer 2 series ca. 2.5 cm long, reddish in lower 1/2 or 1/3, white to pinkish white above, inner series 0.6–1.3 cm long, reddish with white apex; ovary: oblong-ovoid, glabrous; fruit: not seen.

Passiflora 'Aurora'

P. foetida × *P. sublanceolata*
Breeder: Maurizio Vecchia (2001)

This remarkable hybrid (Plate 299) resulted from the cross of 2 species belonging to the *Dysosmia* complex. The first blossom of such a plant opened in the summer of 2001. The origin of the hybrid is immediately revealed by its flowers, which are about 8 cm diam. The paternal plant, *Passiflora sublanceolata* (formerly known as *P. palmeri* var. *sublanceolata*), provided the characteristic bright pink sepals and petals that reflex completely as soon as the flower opens.

The marked corona indicates the influence of the maternal plant. The combination of hummingbird-flowered elements with a corona typical of species pollinated by insects (in this case mainly bees) constitutes the particular charm of *Passiflora* 'Aurora'. Its pictorial name alludes to daybreak, when the sky is the same color as the sepals and petals of this plant.

Another characteristic of this hybrid is that, like most representatives from the *Dysosmia* group, its flowers open early in the morning and close around noon. However, in cool weather they may remain open until the afternoon.

Passiflora 'Aurora' is a rather warmth-loving plant. It may flower reliably in a sunny, protected location outdoors, but it will display its entire splendor only when cultivated under glass. Like its parent plants, it even flowers in a 15-cm pot and does not take up much space. It should be overwintered in a greenhouse or conservatory at temperatures above 10°C.

Vine: partly pubescent to subglabrous; stem: terete, slender, slightly pubescent; stipules: semi-annular, 0.6–0.8 × 0.4–0.5 cm, deeply cleft into filiform, gland-tipped divisions; petioles: 1–2 cm long with gland-tipped hairs; leaves: 3-lobed, 5.5–8 × 7–9 cm, nearly glabrous above, serrate; peduncles: 4–6.5 cm long, solitary; bracts: oblong, 2- or 3-pinnatisect, 4–5.2 × 2.9–3.5 cm, gland-tipped, free; flowers: pink to purplish, white, and violet, 7.5–8 cm diam.; sepals: greenish pink outside, pink to purplish inside, oblong-lanceolate, 3.4–4 × 1–1.5 cm, keeled outside, keel terminating in a short awn, 0.4–0.5 cm long; petals: pink to purplish, subequal to sepals; corona: 2 or 3 series, outer 2 series ca. 2 cm long, white with purplish violet base and violet apex; ovary: ovoid, glabrous; fruit: ovoid, 3.2–3.7 × 2–2.3 cm, red.

Passiflora ×*belotii* Pépin

Rev. Hort. 3 (3): 248 (1849)
named after Belot
P. alata × *P. caerulea*
Synonyms: *P. alato-caerulea* Lindley, *P. hybrida* Neubert, *P.* 'Impératrice Eugénie' Lemaire, *P.* ×*munroi* hort. Lawson ex Mast., *P.* ×*pfordtii* hort. ex J. Veitch
Breeder: William Masters

In 1989, John MacDougal came to the conclusion that one of the most popular hybrids of *Passiflora* had been cultivated under an incorrect name. The often-used *P.* ×*alato-caerulea* is the formula for a crossbreed and technically not a botanically correct name. Such are the wages of having official rules. Other names given to the same plant over the course of time, such as *P.* 'Impératrice Eugénie', must be regarded as synonyms. In accordance with the directions of nomenclature, the oldest botanically correct name is decisive. Hence, the correct name is *P.* ×*belotii* (see Plate 300).

The white sepals form a contrast to the pink petals. The showy corona is purple-blue with some lighter bands. Even as a young plant it produces

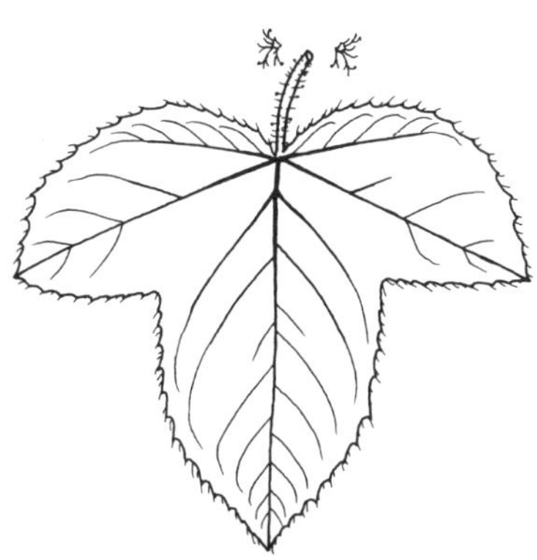

Passiflora 'Aurora'

Passiflora ×*belotii*

numerous very beautiful flowers up to 12 cm diam. Under optimum conditions this hybrid can flower throughout the year. Like *Passiflora* 'Amethyst', *P.* 'Incense', or *P. ×violacea*, its requirements are not too great, and thus it is included among the most rewarding passionflowers for containers. *Passiflora ×belotii* will also produce flowers when grown indoors, but over the course of time it will become too large for a windowsill.

Two characteristics of *Passiflora ×belotii* should be noted. One problem with this plant is that occasionally, particularly during overwintering, the leaves show yellowish speckles, indicating a viral infection. This infection is passed on with the cuttings and cannot be treated. However, the plant is not impaired by this disorder, and the discoloration disappears as conditions improve. In addition, during summer months some flower buds occasionally turn yellow and abort, a problem often caused by watering mistakes and a lack of trace elements. Nevertheless, this plant has many flower buds at once, so this problem does not seem much of a loss.

Numerous hybrids are similar to *Passiflora ×belotii* and can be difficult to distinguish, including *P. ×allardii*. Principal examples can be found among the crosses between *P. ×decaisneana* and *P. caerulea*: *P.* 'Eric', *P.* 'Evelyne', *P.* 'Hermine', *P.* 'Miranda', and *P.* 'Monique Klemann'. These hybrids are comparable with *P. ×belotii* with respects to their demands and their willingness to flower.

Vine: glabrous; stem: 4- or 5-angular, reddish green; stipules: oblong-ovate to subreniform, 2–3 × 1.3–1.8 cm, entire; petioles: 2.5–6.5 cm long, with 2 slightly elongated glands around middle, occasionally 1–2 additional glands; leaves: 3-lobed, (7.5–)10–16.5 × 10–22 cm, entire; peduncles: (2.5–) 3.5–6 cm, solitary; bracts: ovate, 2.3–3.1 × 1.9–2.7 cm, entire, free; flowers: white, pink, and purple, 9.5–12 cm diam.; sepals: green outside, white inside, oblong to oblong-ovate, 3.5–5.5 × 1.7–2.3 cm, aristate, awn 0.4–0.5 cm; petals: whitish outside, pink to occasionally pale purple inside, oblong to oblong-ovate, 4–5.7 × 1.8–2.5 cm; corona: generally in 4 series, outer 2 series 2.2–2.6 cm long, purple at base, few whitish bands near middle and purplish violet above becoming increasingly violet near apex, inner series 0.4–0.6 cm long, dark purple; ovary: ovoid, glabrous; fruit: unknown.

Passiflora 'Byron Beauty'

HortScience 30: 1112 (1995)
named after Byron, Florida
(*P. edulis* f. *flavicarpa* × *P. edulis* f. *edulis*) × *P. incarnata*
Breeder: Robert J. Knight

Passiflora 'Byron Beauty' (Plate 301) is a tetraploid hybrid resulting from the crossbreeding of *P. incarnata* with forms of *P. edulis*. It was selected in 1989. This hybrid, stemming from a breeding program by the U.S. Department of Agriculture, is self-incompatible and produces juicy and tasty fruit only with the pollen of other tetraploid passionflowers, such as *P.* 'Temptation'. Without such a suitable pollinating partner, the occasionally produced fruits are hollow.

Its very beautiful flowers, about 10 cm diam., are fascinating for the undulate corona, which is banded deep violet and white and covers the whitish sepals and petals almost completely. The scent of the flowers can be described, for lack of a suitable comparison, as being heavy. The dark green, shiny leaves give the plant a vigorous, healthy appearance.

Although specimens of *Passiflora* 'Byron Beauty' can become very large, even young plants may set numerous flower buds. In comparison with other passionflowers, this plant is more difficult to restrain. When the shoots are twisted they are easily damaged and the tips of the shoots can break off. Outdoors, *P.* 'Byron Beauty' will flower without

Passiflora 'Byron Beauty'

problems. Since it tolerates low temperatures, it can be taken outdoors early in spring and back to its winter quarters late in fall when frosts threaten. Overwintering is unproblematic at temperatures beyond 5°C.

All in all, *Passiflora* 'Byron Beauty' is among the most recommendable hybrids and should be represented in every collection. Quite recently, several new hybrids were bred that are similar to *P.* 'Byron Beauty': *P.* 'Byte', *P.* 'Marianne Timmer', and above all the somewhat darker *P.* 'Roedie'.

Vine: tetraploid, glabrous; stem: terete, striate; stipules: lanceolate, serrate to slightly pinnatifid, 0.7–1.1 × 0.1–0.2 cm; petioles: 1.5–4.2 cm long, with 2 paired glands near apex; leaves: 3-lobed, 7–16 × 9–20 cm, serrate; peduncles: 3–5(–6) cm, solitary; bracts: broadly ovate, 0.8–1.7 × 1–2 cm, serrate-lacerate, glandular near base, free; flowers: violet, (8–)10–12 cm diam.; sepals: green outside, whitish mottled with purplish violet speckles inside, oblong, 3.3–5 × 1.4–1.8 cm, aristate, awn 0.4–0.5 cm; petals: whitish mottled with purplish violet speckles, oblong, 3.7–5.2 × 1–1.4 cm; corona: several series, outer 2 series 3.7–4.5 cm long, purplish violet at base followed by a small, paler band, dark violet becoming paler toward apex, wavy in upper 1/2 or 1/3, 3rd series up to 0.7 cm long, banded violet and white, inner series 0.2–0.5 cm long, dark purple; ovary: ovoid, glabrous; fruit: subglobose to ovoid, 4.5 × 5 cm, green to greenish yellow, generally without seeds; seeds: unknown.

Passiflora 'Cary'

named after the breeder's sister
P. cincinnata × *P. umbilicata*
Breeder: Hendrik Jan Gommer (2000)

What makes this new hybrid (Plate 302) stand out is the deep violet of its flowers, which are 8–10 cm diam. However, like its parent plant *Passiflora umbilicata*, the sepals and petals reflex immediately upon opening. Because of the contribution of *P. cincinnata*, the flowers show a pretty corona in which the filaments of the 2 outermost series can reach a length of about 2 cm. Another popular hybrid, resulting from a cross with the same species, is *P.* 'Jelly Joker', raised by Cor Laurens. Although the 2 plants look very similar at first sight, *P.* 'Jelly Joker' can be distinguished from *P.* 'Cary' by 2 characteristics. The reproductive organs of *P.* 'Jelly Joker' often show slight deformations, and the upper margin of the floral tube tears as the sepals and petals reflex. Consequently the outward appearance of *P.* 'Jelly Joker' is somewhat impaired. In contrast to this, the flowers of *P.* 'Cary' are flawless.

Passiflora 'Cary' will flower reliably in a 15-cm pot. Even young plants with shoots no longer than about 0.5 m may set flower buds, so that *P.* 'Cary' can be cultivated successfully on a windowsill as well as in a container outdoors or in a greenhouse or conservatory. During overwintering it is satisfied with temperatures around 10°C, although it tolerates slightly lower night temperatures.

Vine: glabrous; stem: terete, striate; stipules: oblong-lanceolate, 0.6–0.8 × 0.2 cm; petioles: 1.8–3.7 cm long, with 2 paired glands above middle; leaves: 3-lobed, 4.5–10.5 × 6–11 cm, subentire to slightly serrate; peduncles: 4.5–6.5 cm, solitary or occasionally in pairs; bracts: ovate, 1.7–2.2 × 1.4–1.6 cm, serrate, free; floral tube: 0.9–1.2 cm long, green; flowers: deep violet, 8–10 cm diam.; sepals: greenish outside, violet inside, oblong, 3.4–3.8 × 0.7–0.9 cm, keeled, keel terminating in an awn 0.3–0.5 cm long; petals: violet, oblong to oblong-lanceolate, 3.5–3.9 × 0.5–0.7 cm; corona: several series, outer 2 series 1.6–2.1 cm long, whitish in lower 1/3, banded violet and white at middle, deep violet above middle, inner series 0.5–0.7 cm long, paler bluish violet; ovary: ovoid, glabrous; fruit: not seen.

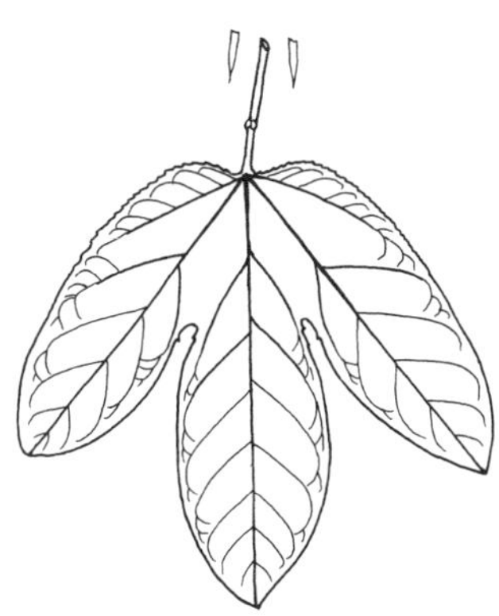

Passiflora 'Cary'

Passiflora ×*colvillii* Sweet

News Lit. Fashion 3: 94 (1825)
named after Colvill, nursery owner and breeder
P. incarnata × *P. caerulea*
Breeder: Colvill

Passiflora ×*colvillii* (Plate 303) is another popular hybrid of ancient origin. It resulted from a cross of 2 hardy species: *P. incarnata* and *P. caerulea*. In central Europe *P.* ×*colvillii* is not as hardy as *P. caerulea*, but it can nevertheless be overwintered outdoors in mild winters. The parentage of *P. incarnata* becomes evident as the aboveground parts of the plant die off in longer spells of frost. *Passiflora* ×*colvillii* is particularly recommended for cultivation in a container. Because of its resistance to cold, it can be taken outdoors early in spring and remain there until late fall. It flowers even as a young plant in a 12-cm pot and can be kept on a windowsill, but it will develop its full splendor only when cultivated in a 25-cm pot. It can be overwintered in a protected location adjacent to the wall of a building, possibly under the projecting section of a roof, or ideally in closed winter quarters at temperatures above the freezing point.

The flowers of *Passiflora* ×*colvillii* reach a diameter of 7–9 cm. The corona is nearly as long as the whitish petals. It is dark purple at the base, then white, and finally, in the upper 1/3, blue to purple-blue, occasionally with a little white. There are cultivars of *P.* ×*colvillii* that differ with respect to the intensity of coloration, the diameter of the blossom, and the undulation and banding of the corona. *Passiflora* ×*colvillii* 'Josefine', for example, raised by Emil Kugler, stands out for its relatively large flowers and densely undulated corona. Quite recently, crosses such as *P.* 'Darina' have been introduced whose flowers suggest a close relationship with *P.* ×*colvillii*.

Vine: glabrous; stem: terete, striate; stipules: linear-oblong, 0.4–0.6 cm long, in some forms up to 0.2 cm wide at base; petioles: 2.2–5 cm long, with 2 paired glands near apex or in upper 1/3 in some forms; leaves: deeply 5-lobed or occasionally 3- or 7-lobed, 6–12.5 × 7.5–15 cm, slightly serrate; peduncles: 4.7–7.5 cm, solitary; bracts: ovate to broadly ovate, 1–1.7 × 1–1.5 cm, serrate, free; flowers: purple, violet, and white, 7–9(–10) cm diam.; sepals: green outside, white inside, oblong, 3–3.8 × 1.3–1.6 cm, aristate, awn 0.4–0.7 cm long; petals: white, occasionally mottled with some violet speckles inside, oblong-lanceolate, 3.5–4.5 × 0.9–1.2 cm; corona: several series, outer 2 series 2.4–3.2 cm long, purplish in lower 1/3, white near middle, violet in upper 1/3, inner series 0.4–0.7 cm long, purple; ovary: ovoid to oblong-ovoid, glabrous; fruit: oblong-ovate, ca. 4–5 × 2 cm, yellow-orange.

Passiflora 'Coral Glow'

P. manicata? × ((*P. mixta* × *P. tripartita* var. *mollissima*) × (*P. mixta* × *P. tripartita* var. *mollissima*))
Breeder: Patrick J. Worley

Passiflora 'Coral Glow' (Plate 304) is the result of aligning several crosses. This experiment brought about a complex cross involving different species from supersection *Tacsonia*. One of the parents was originally identified as *P. jamesonii*, though it was

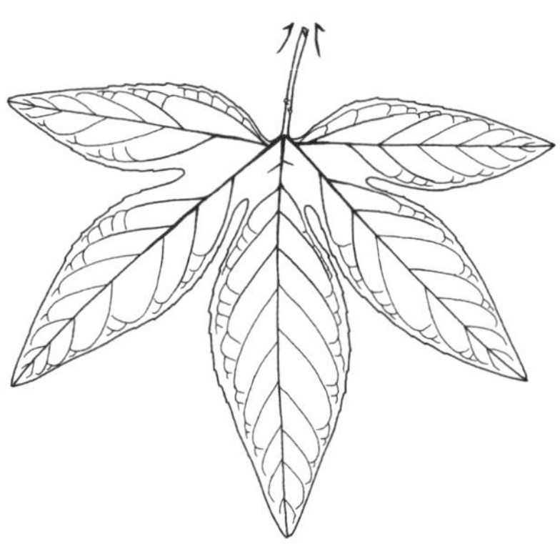

Passiflora ×*colvillii*

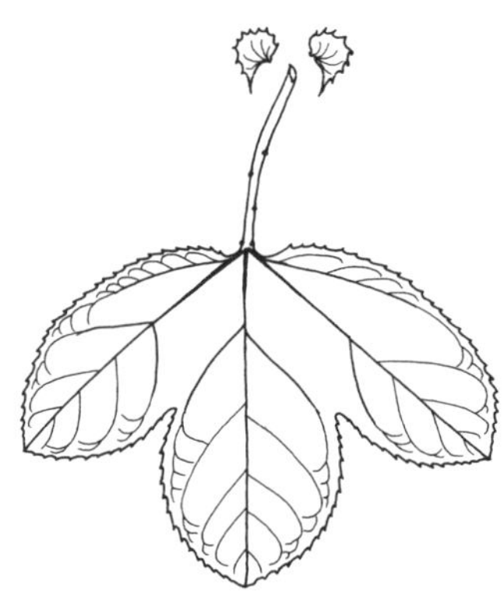

Passiflora 'Coral Glow'

really a *P. manicata* hybrid that went by an erroneous name. This contribution can be seen in the short flower tube of *P.* 'Coral Glow'. Apart from its dubious ancestry, this charming hybrid is worthy of presentation.

Although *Passiflora* 'Coral Glow' tolerates temperatures above 30°C better than its parent plants, which stem from the high altitudes of the Andes, the main flowering time starts in September when the hot midsummer days are over. A first batch of flowers may appear in May or June, but with rising temperatures flower bud formation ceases. The pink to coral flowers reach a diameter of up to 10.5 cm. They stand horizontally on the plant and are scentless, as is usual with hummingbird-pollinated passionflowers. Medium-sized plants in a 20-cm pot may well produce some flowers. However, *P.* 'Coral Glow' is a vigorous vine and will do better when kept in a 30- to 40-cm container outdoors and allowed to reach 2–3 m high. Under ideal conditions planted-out specimens can develop into giants and may even be developed as flowering hedges.

Passiflora 'Coral Glow' tolerates temperatures down to the freezing point, and planted-out specimens will even withstand a short spell of slight frost. Accordingly, container plants can remain outdoors until the first freeze. In well-lighted winter quarters, temperatures of at least 5°C should be guaranteed.

Two other *Tacsonia* hybrids, *Passiflora* 'Coral Sea' and *P.* 'Golden Glow', have come into circulation, and both are easily confused with *P.* 'Coral Glow'. In comparing these 3 hybrids it has turned out that under the climatic conditions prevailing in central Europe, *P.* 'Coral Sea' is quite reluctant to flower.

Vine: pubescent; stem: subangular, striate; stipules: reniform, 1–1.7 × 0.4–0.7 cm, serrate; petioles: 2–4.5 cm long, with 4–8 glands often in upper 1/2; leaves: 3-lobed, 4.5–9.5 × 7–11 cm, serrate; peduncles: 4.5–7.5 cm, solitary; bracts: ovate, 2.5–3.5 × 1.5–2 cm, entire, connate; floral tube: 3–4 cm long, greenish; flowers: pink to coral, 9.5–10.5 cm diam.; sepals: greenish pink outside, pink to coral inside, oblong to oblong-ovate, 4–4.5 × 1.8–2 cm, slightly keeled, keel terminating in an awn 0.4–0.5 cm long; petals: pink to coral, oblong, 4.3–4.7 × 1.6–1.8 cm; corona: 2 series, outer series 0.4–0.6 cm long, erect, bluish violet, inner series a denticulate white band at apex of floral tube; ovary: oblong-ovoid, pubescent; fruit: not seen.

Passiflora 'Corry Rooymans'

named after the breeder's mother
(*P.* 'Amethyst' × *P. subpeltata*) × *P. alata*
Breeder: Henk Wouters (2000)

Passiflora 'Corry Rooymans' (Plate 305) is an exceptional cross of recent origin. Most free-flowering passionflowers, whether botanical species or hybrids, produce flowers with white sepals and petals in combination with a purple, violet, blue, or white corona that can be zoned or banded. Therefore, plants that produce flowers with other colors, above all those with pink and reddish tones, enjoy an increasing demand. *Passiflora* 'Corry Rooymans' is such a hybrid, constituting a pleasing contrast to many other passionflowers. Its pendent to slightly nodding flowers reach a diameter of 7–10 cm and effuse a pleasant scent, though the odor is not as intense as that of its parent plant *P. alata*. The deep pink-purple sepals and petals reflex immediately upon opening so that only the corona, banded purple-violet and white, is visible when the flower is looked at from above.

This easy-to-cultivate, free-flowering hybrid is particularly recommended for cultivation in a container outdoors, though of course it can also be cultivated successfully in a greenhouse or conservatory. Its location should be shaded, with good ventilation, since this plant does not do well in long

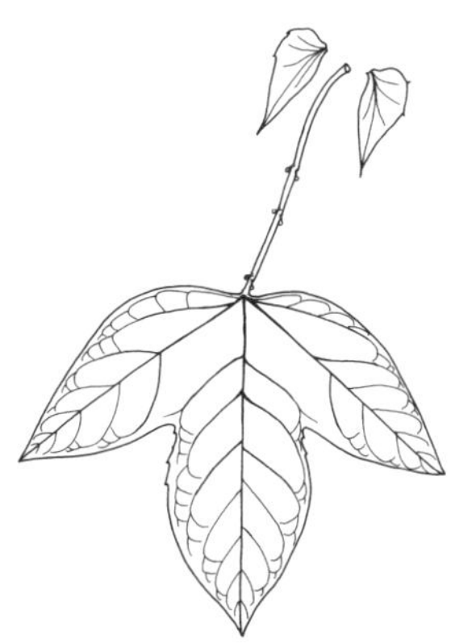

Passiflora 'Corry Rooymans'

spells of hot weather, during which it may produce less flowers.

Passiflora 'Corry Rooymans' tolerates temperatures down to 5°C and even a little lower, though temperatures above 10°C suit it better.

Vine: glabrous; stem: terete to subangular, green with purple, striate; stipules: subreniform, 2–3.5 × 0.8–1.5 cm, serrate; petioles: 4–7 cm long, with 4–6 often paired and slightly elongated glands, 0.1–0.2 cm long; leaves: 3-lobed, 7–12.5 × 7.5–14.5 cm, occasionally purplish violet beneath, entire; peduncles: ca. 4–6 cm, solitary; bracts: ovate to ovate-oblong, 1.7–2.8 × 1.1–1.8 cm, serrate, free; flowers: pinkish purple, purplish violet, and white, 9–10 cm diam.; sepals: green outside, pinkish purple inside, oblong, 4–5 × 1.5–2 cm, slightly keeled outside, keel terminating in an awn 0.4–0.5 cm long; petals: pinkish purple, oblong, 4.3–5.3 × 1.1–1.5 cm; corona: 5 or 6 series, outer 2 series 3–4 cm long, white at base, banded purplish violet and white in lower 1/3, violet and white above, whitish near apex, slightly wavy in upper 1/4, inner series 0.3–1.2 cm long, purplish; ovary: oblong-ovoid, glabrous; fruit: oblong-ovate, ca. 5–6 cm long.

Passiflora 'Crimson Tears'

(((*P. kermesina* × *P. caerulea*) × *P. racemosa*) × *P. gritensis*) × *P. amethystina*
Breeder: Roland Fischer (2002)

To mark the anniversary of the September 11, 2001, terrorist attacks, Roland Fischer named this hybrid *Passiflora* 'Crimson Tears' (see Plate 306). It is dedicated to the victims.

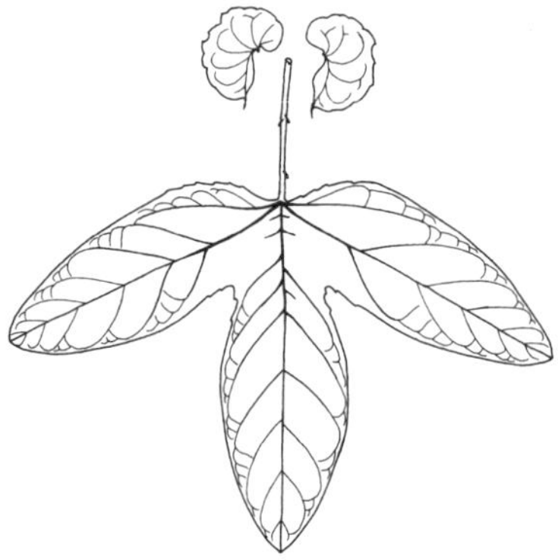

Passiflora 'Crimson Tears'

This is a most beautiful and desirable hybrid with gaudy flowers. Its intense and brilliant color tone, hitherto unknown among passionflowers, is difficult to describe. One might call it crimson-purple. The marked corona of *Passiflora* 'Crimson Tears' is also remarkable. While reddish flowers often have a rather short corona, there are significantly longer coronal series here, slightly reminiscent in structure of *P. gritensis*, one of the parent plants. The complexity of hybrids from a new era becomes evident when considering the parentage of *P.* 'Crimson Tears', which involves 4 plants, one of them a hybrid itself. However, such complex hybrids as this, containing the genetic make-up of several plants, have not yet been made tetraploid.

Because of the involvement of *Passiflora amethystina* (the long-peduncle form) and *P. gritensis*, it is a surprise that this is not a short-day plant. *Passiflora* 'Crimson Tears' reliably sets flower buds during summer months, both outdoors and under glass. The flowers, reaching a considerable size at a diameter of 10–12 cm, appear on long peduncles. At temperatures above 7°C, this hybrid will survive winter without problems.

Vine: glabrous; stem: terete, striate, reddish green; stipules: oblong-ovate, 3.5–4.5 × 1.9–2.3 cm, with reddish nerves, glandular-serrate toward base; petioles: 2.2–4 cm long, reddish, with 5–7 mostly paired, elongated glands, ca. 0.3 cm long; leaves: 3-lobed, 4.5–10(–12.5) × 6.5–15(–17.5) cm, subcoriaceous, with reddish nerves, subentire to slightly serrate toward base; peduncles: 9.5–12.5 cm, solitary; bracts: purplish, ovate, 1.3–2.1 × 0.9–1.3 cm, free; flowers: crimson-purplish, 10–12 cm diam.; sepals: greenish purple outside, crimson-purplish inside, oblong to oblong-lanceolate, 5.2–5.8 × 1–1.3 cm, keeled, keel terminating in an awn 0.8–1 cm long; petals: crimson-purplish, oblong to oblong-lanceolate, 4.4–5 × 0.9–1 cm; corona: 3 or 4 series, outer 2 series 2.6–3.4 cm long, reddish purple at base, banded purplish violet and white in upper 1/2, inner series 0.9–1.2 cm long, reddish purple; ovary: oblong-ovoid, pubescent; fruit: oblong-ovoid, 5–6 × 2.5–3.7 cm, dark green, pubescent.

Passiflora ×*decaisneana* Planch.

Fl. des Serres 8: 267 (1853)
P. alata × *P. quadrangularis*
Synonyms: *P.* ×*bonapartae* Naudin, *P.* ×*innesii* Mast.
Breeder: Gontier

Passiflora ×*decaisneana* (Plate 307) is among the best-known hybrids in Europe and overseas. It resulted from crossbreeding *P. alata* with *P. quadrangularis* in the middle of the 19th century. It has been represented for many years in botanical gardens and enthusiasts' collections. Unfortunately, it was confused with one of its parent plants for many years. In 1996 John Vanderplank pointed out in *Passion Flowers* that almost all plants cultivated in Europe under the name *P. quadrangularis* were actually *P.* ×*decaisneana*. Information provided by the technical literature and by enthusiasts' organizations demonstrated that the differences between the 2 are well known, so that they should no longer be confused.

Passiflora ×*decaisneana* is one of the most gorgeous older hybrids. It proves to be very free-flowering when cultivated in a greenhouse, and when it is planted directly in the ground it rapidly develops into an enormous plant that can hardly be restrained. Flowering often begins in late summer or early fall. Kept in a spacious container of at least 30 cm diam., *P.* ×*decaisneana* can also spend the summer in a sunny outdoor location. Sometimes enthusiasts complain that their newly purchased young plants do not flower in their first year. After being potted, the plants develop a mass of foliage, and only after they have completely permeated the rooting medium and reached an adequate size will they start to flower.

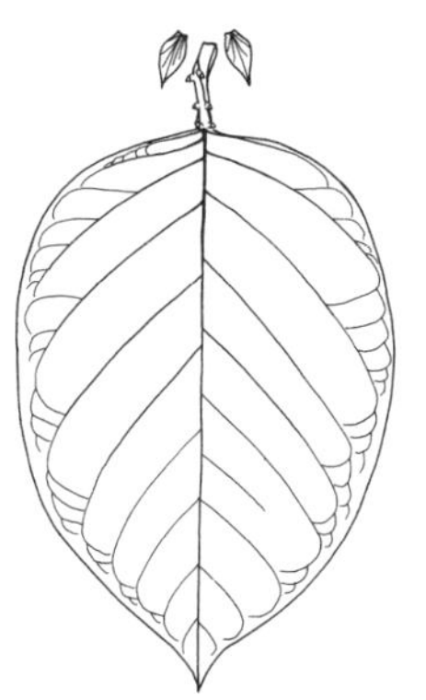

Passiflora ×*decaisneana*

The fragrant flowers of this plant reach a considerable diameter, up to 12.5 cm. The corona is particularly attractive, banded purplish violet and white, with the outermost series reaching a diameter of up to 7 cm. Rich reddish sepals and petals contrast nicely in the background.

When *Passiflora* ×*decaisneana* is cross-pollinated, such as with pollen from *P. caerulea*, it produces large edible fruits that are quite sweet and juicy.

Vine: large, vigorous, glabrous; stem: 4-angular, stout, winged; stipules: oblong-ovate, 1.4–1.9 × 0.5–0.9 cm, subentire to slightly serrate; petioles: 3.6–6 cm long, 3-angular, stout with 4 or occasionally 6 paired glands, 0.2–0.3 cm long and wide; leaves: unlobed, ovate, 15–20(–25) × 9–14(–16.5) cm, entire; peduncles: 1.5–3.5 cm, solitary, stout; bracts: ovate, 2–2.6 × 1.2–1.6 cm, free, serrate toward base; flowers: red, purple to purplish violet and white, 10–12.5 cm diam.; sepals: green outside, red inside, oblong, 4.5–5.3 × 1.8–2 cm, aristate, awn 0.3–0.4 cm long; petals: red, subequal to sepals; corona: 5 or 6 series, outer 2 series 5–7 cm long, banded purplish and white in lower 1/2, banded purplish violet and white above, wavy in upper 1/2, inner series 0.1–0.4 cm long, purple with few whitish bands; ovary: ovoid, glabrous; fruit: ovoid to oblong-ovoid, 15–17.5 × 7.5–9.5 cm, yellow-orange, edible.

Passiflora 'Elizabeth'

named after the breeder's mother
P. phoenicea × *P. incarnata*
Breeder: Patrick J. Worley

By crossbreeding *Passiflora incarnata* and *P. phoenicea*, Patrick Worley succeeded in raising a magnificent new hybrid, which he called *P.* 'Elizabeth' (Plate 308). As a rule its mauve flowers stand horizontally and reach a considerable size of 11–12.5 cm diam., effusing a heavy scent that is difficult to describe. The flowers are dominated by a large undulate corona that is deep purple at its base and whitish at its tips.

'Elizabeth' is included among the free-flowering, recommendable passionflowers that can be cultivated on a balcony or patio during summer where flowers are produced until late fall. Even young plants with shoots not much longer than 30 cm can set flower buds. However, 'Elizabeth' is more difficult to handle than other plants. When the shoots are wound around a trellis or a similar support they are easily damaged and the ends may break off. Large solitary plants do not look so compact because the main shoots cannot be bent down and twisted in again.

Unfortunately, enthusiasts aren't alone in their esteem for 'Elizabeth': this hybrid is a favorite host for spider mites and is often infested by them. Plants should be scrutinized regularly, and necessary steps should be taken to control any infestation.

During winter, 'Elizabeth' is satisfied with temperatures around 5°C, although lower temperatures can be tolerated.

Vine: large, glabrous; stem: terete, striate, stout; stipules: oblong-lanceolate, 0.9–1.3 × 0.3–0.4 cm, serrate; petioles: 3–6 cm long, with 2 paired, large glands below apex, stout; leaves: 3-lobed, occasionally unlobed, 8–15.5 × 9–20 cm, serrate; peduncles: 6–7.5 cm, solitary, stout; bracts: elliptical to ovate, 1–1.5 × 0.9–1.1 cm, with up to 4 green glands at base, serrate, free; flowers: pretty mauve and deep purple, 11–12.5 cm diam.; sepals: greenish outside, densely mottled with mauve speckles inside, oblong-ovate, 4–5 × 1.8–2.2 cm, keeled, keel terminating in an awn 0.5–0.8 cm long; petals: densely mottled with mauve speckles, oblong, 4.5–5.3 × 1.7–1.9 cm; corona: 5 or 6 series, outer 2 series 4.2–5 cm long, deep purple at lower 1/3, mauve above becoming increasingly whitish toward apex, wavy in upper 1/3, inner series 0.3–0.5 cm long, deep purple; ovary: ovoid, pubescent; fruit: ovate, greenish yellow, edible.

Passiflora ×*exoniensis* Mast.

Gard. Chron. 30: 1653 (1872)
P. antioquiensis × *P. tripartita* var. *mollissima*
Breeder: Unknown

Passiflora ×*exoniensis* (Plate 309) is possibly the best-known hybrid from 2 species of *Tacsonia*. It was presented to the public as early as 1872 under its former name, *Tacsonia exoniensis*. Unfortunately, in many cases it is cultivated under the name *P. antioquiensis*, although there are sufficient distinguishing criteria such as the length of the peduncle and floral tube, and the number of petiole glands. Because the real *P. antioquiensis* has been cultivated by an increasing number of enthusiasts during the last years, confusion with *P.* ×*exoniensis* has subsided.

In contrast with many species of *Tacsonia*, which are rather reluctant to flower under cultivation, *Passiflora* ×*exoniensis* is usually quite reliable in producing blossoms. When kept outdoors, it exhibits many pendent flowers in late summer and fall, and when provided optimal conditions in a cold greenhouse or conservatory, it produces flowers even up to winter. Only during longer spells of hot weather in midsummer does it allow itself a rest. Purplish red flowers reach up to 12.5 cm diam. However, the sepals and petals look slightly creased or crumpled.

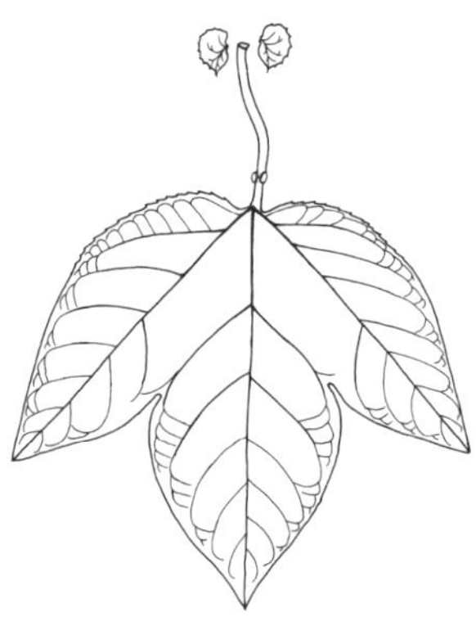

Passiflora 'Elizabeth'

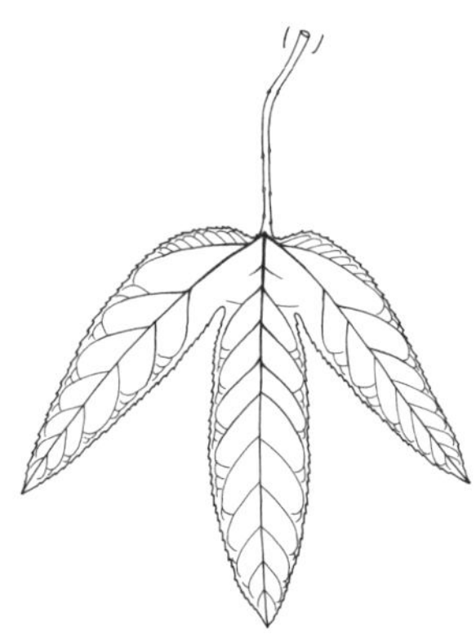

Passiflora ×*exoniensis*

In contrast to many other tacsonias, *Passiflora* ×*exoniensis* does not have to produce shoots of several meters in length in order to flower. Even in a 20- to 25-cm pot it can set its first flower buds. Batches with numerous flowers are reserved for considerably larger plants. After cross-pollination with other tacsonias, especially *P. tarminiana* and *P. tripartita*, *P.* ×*exoniensis* will produce edible fruit. However, pollination must be achieved by hand, as such flowers are normally pollinated by hummingbirds with a long bill.

In summer *Passiflora* ×*exoniensis* will flourish when temperatures are 20°–25°C with nightly drops down to 10°C. In winter it is satisfied with temperatures around 5°C, although temperatures down to –4°C are tolerated for short periods.

Vine: medium to large, pubescent; stem: subterete, striate; stipules: linear-lanceolate, 0.5–0.8 cm long; petioles: 1.5–3.5 cm long, with many subsessile glands scattered along petiole; leaves: deeply 3-lobed, 8.5–17.5 × 9.5–20 cm, serrate; peduncles: 6–10 cm, solitary; bracts: oblong-ovate, 1–2 × 0.5–1 cm, serrate; floral tube: 6.5–9 cm long, green; flowers: purplish red, 10–12.5 cm diam.; sepals: purplish red, oblong, 4.3–5.6 × 1–1.7 cm, keeled, keel terminating in an awn ca. 0.5 cm long; petals: purplish red, subequal to the sepals; corona: 1 or 2 series, purple band with white teeth, ca. 0.1 cm long; ovary: ovoid, glabrous; fruit: oblong-ovate, 7.5–9 cm long, yellow, edible.

Passiflora 'Incense'

Gard. Chron. Ser. 3 174: 33 (1973)
P. incarnata × *P. cincinnata*
Breeder: Robert J. Knight (1973)

Passiflora 'Incense' (Plate 310) was selected by Robert J. Knight in the course of a series of crossbreeding experiments carried out in Florida that were initiated by the U.S. Department of Agriculture. This hybrid is most impressive for its beautiful, dark violet flowers and for its resistance to frosts down to –8°C. As a popular container plant, it has become established in many enthusiasts' collections. It belongs to the 10–15 passionflowers that are propagated on a large scale for flower shops and garden centers. Its flowers, effusing a strong scent of incense, are 10–12.5 cm diam. and appear from July through September when the plant is cultivated outdoors. When it is kept in a greenhouse, flowers appear as early as May.

Passiflora 'Incense' is frequently confused with *P. cincinnata*, although the 2 can be distinguished by the size and color of their flowers (see *P. cincinnata*). Such confusion can be fatal. While 'Incense' is said to withstand temperatures down to –8°C, the warmth-loving *P. cincinnata* needs at least 10°C, even in winter. This confusion may be what led some to mistakenly regard 'Incense' as vulnerable to frost. Nevertheless, this hybrid is only recommended for year-round outdoor cultivation in regions with a temperate climate. Otherwise it can spend winter in a lightly heated greenhouse or conservatory. Although the aboveground parts die off because of the cold, the plant regenerates from its thick, fleshy roots.

Unfortunately, all clones of *Passiflora* 'Incense' are infected with a latent virus that causes yellow mottling and a deformation of the leaves, at least in cooler weather. The virus can be spread to other passifloras through pruning tools, so many growers thoughtfully avoid this hybrid. Attempts to recreate the hybrid, but virus-free, have been successful. Roland Fischer succeeded in breeding several virus-free tetraploid hybrids, as well as a virus-free diploid hybrid of *P. incarnata* with *P. cincinnata* (see *P.* 'Inspiration'). When conditions improve, the symptoms of a viral infection often disappear as swiftly as they appeared, and the plant continues to grow and flower happily.

All in all, *Passiflora* 'Incense' is an absolutely rec-

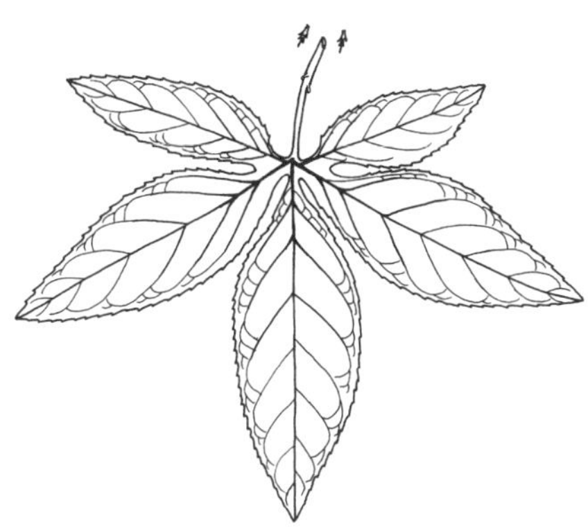

Passiflora 'Incense'

ommendable species, and it may be the most beautiful passionflower that will tolerate such low temperatures.

Problems can arise in propagation. Although the cuttings have often formed callus after 2 weeks, they do not develop roots. Such cuttings do not show any progress and do not look much different after 3–4 months. One must take into account a higher percentage in loss and insert more cuttings. Better results are achieved with root cuttings.

Vine: nearly glabrous; stem: terete, striate; stipules: linear-lanceolate, 0.6–0.9 cm long, soon deciduous; petioles: 4.5–6 cm long, stout, with 2 paired glands near middle; leaves: deeply 5-lobed, 7–16(–20) × 9–17(–22) cm, serrate; peduncles: 7.5–10 cm, solitary; bracts: ovate, 1.5–2.2 × 1–1.5 cm, glandular-serrate toward base, free; flowers: dark violet to purplish violet, (9–)10–12.5 cm diam.; sepals: green outside, purplish violet inside, oblong, 4–5 × 1.8–2 cm, slightly keeled, keel terminating in an awn 0.5–0.8 cm long; petals: purplish violet, oblong, 4–5 × 1.5–1.7 cm; corona: several series, outer 2 series 4–4.8 cm long, banded dark violet and white becoming paler toward apex, wavy in upper 1/3, inner series 0.3–0.9 cm long, purplish violet and white; ovary: ovoid, glabrous; fruit: ovoid, ca. 5 × 4 cm, green, mottled with white speckles, edible.

Passiflora 'Inspiration'

P. incarnata × *P. cincinnata*
Breeder: Roland Fischer (1999)

During the last few years, Roland Fischer has succeeded in breeding several new tetraploid passionflower hybrids that are characterized by particularly large flowers and a vigorous and healthy growth. Among these are *Passiflora* 'Inspiration' (Plate 311), possibly one of the most beautiful hybrids, and its sister plant, *P.* 'Temptation', arising from the same crossing. 'Inspiration' produces fragrant, deep blue-violet flowers that can reach a diameter of 12.5 cm. The color of the flower is so intense and brilliant that a photo can hardly do it justice.

Another positive characteristic of *Passiflora* 'Inspiration' is that with the help of its pollen, the fruits of *P.* 'Byron Beauty' are not hollow as usual but contain a tasty aril.

Passiflora 'Inspiration' is easy to cultivate. It thrives under glass as well as in a protected location in the open during the outdoor season. It does not show the gigantism of other large-flowered species, such as *P.* 'Byron Beauty', and can be handled without problems in a 20- to 30-cm pot. It sets flowers even on shoots about 1 m long.

In contrast to the well-known *Passiflora* 'Incense', also a result of a cross between *P. incarnata* and *P. cincinnata*, *P.* 'Inspiration' is virus-free. However, the breeder is not sure whether it will withstand slight frost. During overwintering it does best at temperatures above 10°C.

Propagation of *Passiflora* 'Inspiration' is a problem. Cuttings often stand for weeks without developing roots, despite developing a thick callus. Root cuttings can achieve better results, but with this method one runs the risk of selecting diploid clones so that the positive tetraploid qualities are lost. Hence this form of propagation is not advisable. Sucker growth should also be removed.

Passiflora 'Blue Velvet', also raised by Roland Fischer, is a diploid hybrid that resulted from the same cross. It is almost as beautiful as *P.* 'Inspiration' and is virus-free, but the flowers are somewhat smaller at a diameter of about 8 cm. However, it can be propagated more reliably by cuttings.

Vine: tetraploid, glabrous; stem: subangular to terete, striate; stipules: lanceolate, 0.8–1 × 0.1 cm, glandular-serrate; petioles: 2.5–5.5 cm long, with 2 paired glands near middle; leaves: deeply 5-lobed, 5–14 × 8–17.5 cm, serrate; peduncles: 5–6.7 cm, solitary; bracts: broadly ovate, 1.5–1.8 × 1.5–2.2 cm, slightly serrate with (1–)2 green glands at base, free;

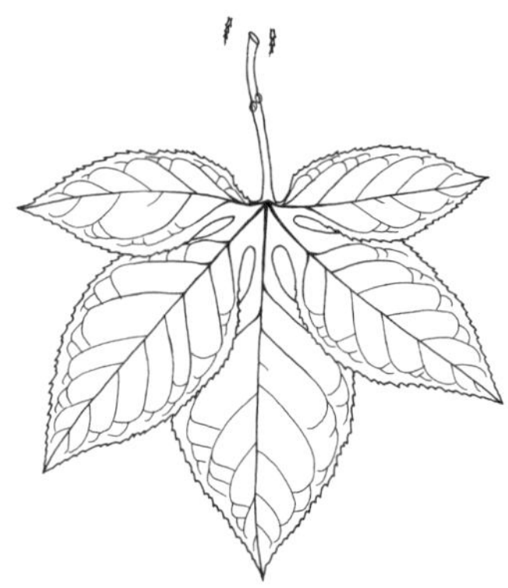

Passiflora 'Inspiration'

flowers: deep violet to bluish violet, 10–12.5 cm diam.; sepals: green outside, deep violet inside, oblong, 3.7–5 × 1.7–2.2 cm, slightly keeled, keel terminating in an awn 0.5–0.7 cm long; petals: deep violet, paler toward base and beneath, oblong, 4.2–5.5 × 1.4–1.6 cm; corona: several series, outer 2 series 3.7–4.5 cm long, violet, banded white at and below middle, wavy in upper 1/3, inner series 0.3–0.9 cm long, violet; ovary: ovoid, glabrous; fruit: ovoid to subglobose, 4–5 × 3–4.2 cm, green, mottled with white speckles.

Passiflora 'Jara'

named after the breeder's daughter
P. caerulea × *P.* 'Purple Haze'
Synonym: *P.* 'Purple Rain'
Breeder: Roland Fischer (1998)

In 1998 Roland Fischer presented *Passiflora* 'Jara' (Plate 312), the first successful result of his tetraploidization experiments. This hybrid is characterized by particularly large flowers, fleshy coronal filaments, deep green leaves, and a healthy growth. It has elongated petiolar glands, a typical sign of a successful tetraploidization.

The flowers of the easy-to-cultivate *Passiflora* 'Jara' reach a diameter of 10–12 cm. Flowering starts in spring and lasts throughout the summer. Even in fall some flowers are set. One special feature is that the coloration of the sepals and petals depends on the temperature. While the flowers of greenhouse plants are often a light purple, the sepals and petals remain white on plants kept out in the open. *Passiflora* 'Jara' may do better under glass than outdoors. Although young plants in a 10-cm pot can set the first flowers on short shoots, *P.* 'Jara' is not a good choice for cultivation on a windowsill because it becomes too unwieldy after some time and the flowers remain rather small. Overall, this hybrid needs good conditions and a spacious container to display all its splendor.

Passiflora 'Jara'

Passiflora 'Jara' belongs to the few hybrids of recent origin that are propagated on a large scale in Europe and that are regularly offered by flower shops and garden centers. Unfortunately, the name *P.* 'Purple Rain' has gained prominence and must be understood as a synonym. Because of the parentage of *P. caerulea*, this hybrid has turned out to be relatively frost-resistant. Plants that are several years old may well tolerate some slight freezes without being damaged, but temperatures above 5°C are recommended in winter. *Passiflora* 'Jara' should be overwintered in a slightly heated greenhouse or conservatory.

Despite its tetraploid chromosome number, this hybrid is not suited to pollinate *Passiflora* 'Byron Beauty'.

Vine: tetraploid, glabrous; stem: terete, striate, green often with purple; stipules: subreniform, 1.8–3.7 × 0.7–1.5 cm, serrate; petioles: 1.7–2.6 cm long, with 2–4 elongated glands, 0.3–0.5 cm long; leaves: deeply 3- or 5-lobed, 4.4–9.5 × 7.3–12 cm, entire; peduncles: 4.7–6.5 cm, solitary; bracts: broadly ovate, 1.6–2.5 × 1.8–3.2 cm, entire, free; flowers: pale purplish violet to white, deep violet, 10–12 cm diam.; sepals: green outside, pale purplish violet to white inside, oblong, 3.7–4.8 × 1.7–2.2 cm, slightly keeled, keel terminating in an awn 0.6–0.8 cm long; petals: greenish white outside, pale purplish violet to white inside, oblong, 4.3–5.3 × 1.6–2 cm; corona: 5 series, outer 2 series 2.9–3.5 cm long, deep purple at base followed by a broad white band, deep violet above middle with few paler bands, inner series 0.4–0.9 cm long, deep purple; ovary: ovoid, glabrous; fruit: not seen.

Passiflora 'Jutta'

named after Jutta Jedzig, German enthusiast
P. umbilicata × *P. tucumanensis*
Breeder: Henk Wouters (2001)

This new creation enriches the available range of hybrids in several respects. Both parent plants (the paternal plant was formerly known as *Passiflora naviculata*) are favorite crossing partners in many recent hybrids, but until 2001 they had never been crossed with one another. Henk Wouters succeeded in doing so and presented an impressing result: *P.* 'Jutta' (Plate 313). Its flowers reach the full 7–8 cm diam. only when they have just opened because the sepals and petals immediately reflex. These are alternately white and blue-violet, which is quite unusual in passionflowers and gives a fascinating impression in combination with the dominating corona, which is banded bluish violet.

Passiflora 'Jutta' owes its greater cold resistance to its parent plants. The breeder states that it will tolerate temperatures down to –10°C. This offers a wide range of options. It is superbly suited as a container plant that can withstand slight spells of frost without problems and can remain outdoors until very late fall. In its winter quarters, temperatures may drop below the freezing point on some nights, though temperatures above 7°C are more suitable. In climatically favored regions, this plant can be kept outdoors throughout the year.

In contrast to many other hybrids, *Passiflora* 'Jutta' produces mature pollen and sets fruit when it is cross-pollinated.

Vine: glabrous; stem: subterete, striate; stipules: subreniform, 1.5–2.5 × 0.9–1.5 cm, serrate; petioles: 2.5–6 cm long, without or with 2 elongated glands ca. 1 cm below leaf base, glands 0.2 cm long; leaves: 3-lobed, 6–12 × 7.5–15 cm, entire; peduncles: 4–6 cm, solitary, occasionally in pairs; bracts: ovate, 1.2–1.6 × 0.8–1 cm, purplish violet, slightly serrate, free; floral tube: 1–1.2 cm long, greenish outside; flowers: bluefish violet and white, 7–8 cm diam.; sepals: greenish white outside, white inside, oblong-lanceolate, 3–3.6 × 0.7–1 cm, keeled, keel terminating in an awn ca. 0.5 cm long; petals: pale violet to blue-violet, oblong-lanceolate, 2.7–3.1 × 0.5–0.7 cm; corona: 3 series, outer series 2–3 cm long, pale violet at base, banded blue-violet and white for the next 1/3 of their length, becoming violet, white at very apex, inner series ca. 0.5 cm long, blue-violet, white at apex; ovary: ovoid, glabrous; fruit: subglobose, 4.5–5 × 4–4.5 cm, green becoming pale yellow when ripe.

Passiflora 'Lady Margaret'

named after Margaret Skimina
P. coccinea × *P. incarnata*
Breeder: Timothy Skimina (1991)

Species with large red flowers are among the most sought-after passionflowers. Unfortunately, however, most red-flowering passionflowers are difficult to care for and will only flourish and flower abundantly in warm climates or within a greenhouse. *Passiflora* 'Lady Margaret' (Plate 314) is one of the few red-flowering hybrids that is comparatively undemanding but nevertheless free-flowering. It doesn't have to be kept in a greenhouse, and in temperate climates it can be cultivated successfully in a container. Although the flowers are not the brilliant blood-red of *P. coccinea,* they nevertheless stand out dramatically, a deep purplish red, with a conspicuous marked corona not found in other red-flowering, hummingbird-pollinated passifloras. The genes of *P. incarnata* account for this distinctiveness and for a certain degree of cold resistance. During overwintering *P.* 'Lady Margaret'

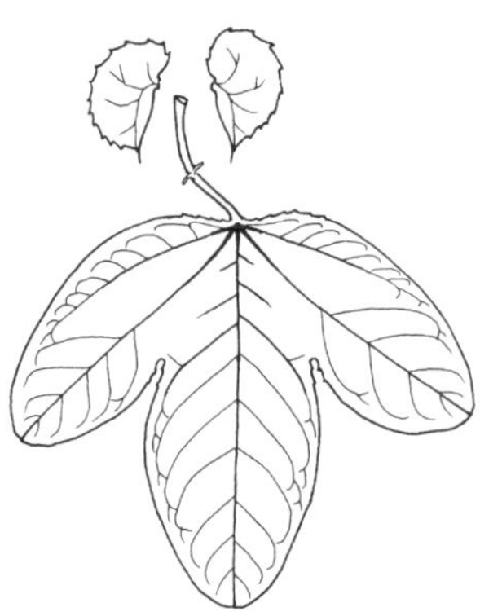

Passiflora 'Jutta'

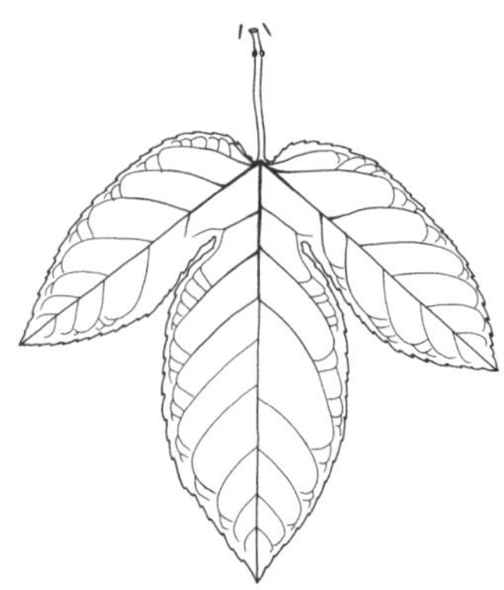

Passiflora 'Lady Margaret'

tolerates temperatures down to the freezing point for a short period, although temperatures above 7°C are preferred.

Passiflora 'Red Inca', raised by Cor Laurens, is another hybrid that resulted from a cross between *P. coccinea* and *P. incarnata*. Though very similar to *P.* 'Lady Margaret' with regard to flower size and color, it is not so free-flowering and has a more irregular corona.

Vine: pubescent; stem: terete, striate; stipules: lanceolate to setaceous, 0.3–0.5 cm long, entire; petioles: 1.7–4 cm long, with 2 paired glands in lower 1/3; leaves: 3-lobed, 6.5–15 × 7.5–16 cm, serrate; peduncles: 5–8 cm, solitary; bracts: ovate to ellipsoid, pale green, 1.1–1.6 × 0.7–1 cm, glandular-serrate, free; flowers: deep purplish red, 7–9 cm diam.; sepals: green outside, deep purplish red inside, oblong, 3.3–4 × 1.1–1.5 cm, keeled, keel terminating in an awn 0.3 cm long; petals: deep purplish red, oblong, 3.6–4.4 × 1.3–1.6 cm; corona: 6 series, outer 2 series 2.7–3.3 cm long, deep purple with some paler spots, whitish at base, inner series 0.5–0.9 cm long, white; ovary: ovoid, pubescent; fruit: ovate, 4–5 × 2–3 cm, green.

Passiflora 'Livie'

Livie is "Olivia" in Czech
P. cinnabarina × *P. herbertiana*
Breeder: Leopold Ševčík (1993)

Passiflora 'Livie' (Plate 315) is the attractive result of a cross between 2 Australian species. No other hybrid has such brilliant orange sepals and petals in combination with a yellow corona. However, the flower does not appear so full, as the short corona stands upright and the sepals and petals are very narrow. The color of the blossom, structure of the corona, and long androgynophore signify that the parent plants are bird-flowered (though not hummingbird-flowered). Consequently, the flowers of *P.* 'Livie' do not give off a scent.

Passiflora 'Livie'

Since *Passiflora* 'Livie' was created, its parent species have been used in additional crossbreeding experiments. Quite recently, Henk Wouters succeeded in raising the red-flowering *P.* 'Red Infinity'. In principle such crosses, including *P.* 'Livie' and *P.* 'Red Infinity', are free-flowering and not finicky. However, propagation by cuttings is problematic. Although the cuttings develop a thick callus within 2 weeks, months may pass before roots develop, if they develop at all. At any rate, *P.* 'Livie' is a virus-free, free-flowering, recommendable hybrid. It prefers a protected, sunny location either in the open during the summer season or under glass. It should be overwintered at temperatures above 7°C.

Vine: glabrous; stem: terete to subangular, striate; stipules: linear-lanceolate, 0.4–0.6 cm long, entire; petioles: (1.7–)4–10 cm, with (1–)2 paired glands at very apex; leaves: 3-lobed, 4.5–15 × 4.2–14 cm, entire; peduncles: 2.5–3.5 cm, solitary, occasionally in pairs; bracts: linear-lanceolate, 0.5–0.6 cm long, scattered in upper 1/2, free; flowers: orange to orange-red and yellow, 7–9 cm diam.; sepals: orange with green midrib outside, orange to orange-red inside, lanceolate, 3.4–4.2 × 0.6–0.8 cm, keeled, without awn; petals: orange to orange-red, lanceolate, 2.2–2.6 × 0.3–0.4 cm; corona: 1 series, 1–1.3 cm long, whitish green at base becoming increasingly yellow above, erect; ovary: ovoid, glabrous; fruit: not seen.

Passiflora 'Marijke'

P. ×*decaisneana* × *P. caerulea*
Breeder: M. G. van der Steeg

Passiflora 'Marijke' (Plate 316) is among the most rewarding hybrids created in the recent past. It is very easy to cultivate, thriving either in a greenhouse or outdoors during summer months. Its flowers are among the most beautiful of the entire genus. The vivid reddish to magenta sepals and petals form the background for a large corona banded purple-violet to white. The flowers, effusing a pleasant smell reminiscent of lily of the valley and lilac, reach a diameter of 9.5–12 cm. The majority of the flowers are produced during the 2nd half of summer. Large plants may open dozens of flowers within a fortnight before taking a rest. It is advisable to grow *P.* 'Marijke' in a spacious container of at least 25 cm diam. so that it can reach the necessary size to flower abundantly.

Passiflora 'Marijke' demonstrates how variable the results of the same cross can be. M. G. van der Steeg selected further hybrids that resulted from a cross between *P.* ×*decaisneana* and *P. caerulea*: *P.* 'Eric', *P.* 'Evelyne', *P.* 'Hermine', and *P.* 'Miranda'. All of these hybrids produce flowers with pink or whitish sepals and petals, somewhat reminiscent of *P.* ×*belotii*; only *P.* 'Marijke' features the intensely reddish to magenta sepals and petals that make it so attractive.

In view of the fact that red-flowering passionflowers are generally difficult to care for during winter, *Passiflora* 'Marijke' has turned out to be surprisingly robust. It tolerates temperatures down to 5°C, and even a few colder nights, without problems. However, it prefers temperatures above 7°C.

Vine: glabrous; stem: quadrangular, green, often becoming reddish green; stipules: subreniform, 1.4–1.7 × 0.5–0.7 cm, slightly serrate toward base; petioles: 2.5–4 cm long, with 4 paired glands below and above middle; leaves: 3-lobed, (6–)8.5–14.5 × (8–)12.5–19 cm, entire; peduncles: 2–3 cm, solitary; bracts: nearly deltoid, 2–2.5 × 1.9–2.3 cm, serrate, free; flowers: reddish to magenta, purple and white, 9.5–12 cm diam.; sepals: green outside, reddish to magenta inside, oblong, 4–4.5 × 1.6–2.3 cm, aristate, awn 0.9–1.4 cm long; petals: reddish to magenta, oblong, 4.2–4.6 × 1.6–2 cm; corona: 5 or 6 series, outer 2 series 3–3.6 cm long, dark purple at base, banded purple and white at middle and violet in upper 1/3, inner series 0.2–0.5 cm long, dark purple; ovary: ovoid, glabrous; fruit: not seen.

Passiflora 'Minai'

named after the breeder's cats
P. aff. *oerstedii* × *P. menispermifolia*
Breeder: Irmi Reiber (2001)

Passiflora 'Minai' (Plate 317) stands out for its deep violet flowers, up to 7.5 cm diam. Like *P. menispermifolia*, *P.* 'Minai' flowers in batches. On a specimen with shoots of several meters in length, up to 60 fragrant flowers can be opened at the same time—a gorgeous spectacle. In comparison with *P. menispermifolia*, *P.* 'Minai' produces flowers that are more intense in color. Although it is more robust than its parent, it is best cultivated in a warm greenhouse. Under glass the main flowering times are early summer and early fall. In a sunny, protected location it can also be kept outdoors during summer, although it may not set as many flowers as in a greenhouse. It is rather warmth-loving and should be overwintered at temperatures above 12°C.

Passiflora 'Mario', another clone from the same cross, has similarly attractive flowers. Irmi Reiber made a reciprocal cross with *P.* aff. *oerstedii* (originally identified as *P. porphyretica*) as a paternal plant. It turned out that in this hybrid the flower buds abort. The influence of the paternal plant is very conspicuous with regard to the leaves. The

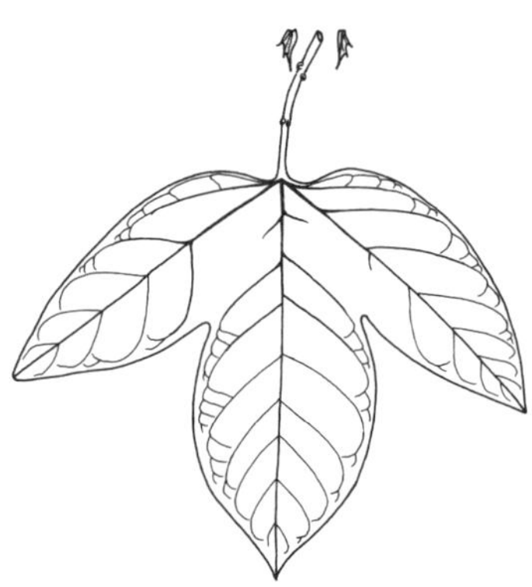

Passiflora 'Marijke'

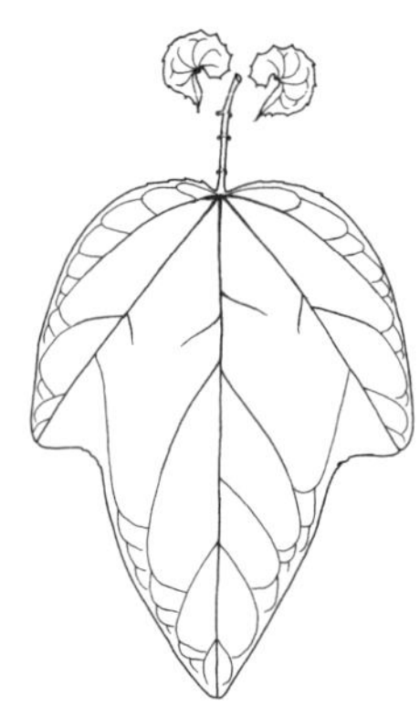

Passiflora 'Minai'

leaves of *P.* 'Minai' look very much like those of *P. menispermifolia*. In the reciprocal cross they are similar to *P.* aff. *oerstedii* (as *P. porphyretica*). It should be noted that the plants represented in many collections under the name *P. porphyretica* are actually a Mexican clone of *P. oerstedii* or a close relative of it (see *P. oerstedii*).

Vine: densely pubescent; stem: terete to subangular, striate; stipules: reniform, 1.7–2.3 × 1–1.4 cm, serrate; petioles: 2.5–4 cm, with 6–8 paired, elongated glands, ca. 0.2 × 0.1 cm; leaves: unlobed or slightly 3-lobed, unlobed leaves cordate, 8–17 × 6.5–12.5 cm, entire; peduncles: 2–3 cm, solitary; bracts: ovate, 2.1–2.5 × 1–1.4 cm, serrate, free; flowers: violet and purple, 6.5–7.5 cm diam.; sepals: green outside, pale violet to violet inside, oblong, 2.5–3 × 0.8–1.1 cm, keeled, keel terminating in an awn ca. 0.3 cm long; petals: violet, oblong, 2.7–3.2 × 0.7–0.9 cm; corona: several series, outer series 1.8–2.4 cm long, purple in lower 1/2, white near middle, violet above, whitish at apex, 2nd series up to 1.5 cm long, inner series 0.2–0.6 cm long, purple, innermost with white apex; ovary: ovoid, glabrous; fruit: not seen.

Passiflora ×*piresea*

named in honor of Joaquina Pires-O'Brien
P. quadrifaria × *P. vitifolia*
Synonym: *P.* ×*piresii*
Breeder: John Vanderplank (1996)

With *Passiflora* ×*piresea* (Plate 318), another large-flowered red passionflower is put at the enthusiast's disposal. Although its color does not equal that of its blood-red parent, *P. vitifolia*, and the contrasting white corona is missing, a direct comparison shows that *P.* ×*piresea* is easier to grow. In Europe *P.* ×*piresea* has become an important factor in plant production for flower shops and garden centers and has superseded *P. vitifolia* to a certain degree. This is due to the fact that *P.* ×*piresea* grows faster and with fewer losses into a marketable flowering plant. Moreover, it has fewer problems during overwintering, although overwintering is the most critical stage of its cultivation. In a greenhouse or conservatory this plant can be overwintered successfully when guaranteed temperatures above 10°C and a high atmospheric humidity. However, under insufficient conditions it soon suffers. It may shed its leaves (a bad sign) and stop growing. When it is also kept too wet, it dies.

The bright orange-red flowers of *Passiflora* ×*piresea* reach 10–12 cm diam. After the flower has opened, the sepals and petals reflex a little, but not as much as those of *P. vitifolia*. The flower buds are covered by large reddish bracts that remain on the plant when the flower has withered away, giving additional color spots. *Passiflora* ×*piresea* will display its full splendor only when grown under glass. It also thrives as a container plant outdoors but will produce fewer flowers there. Even young plants in 12-cm pots can set flowers on relatively short shoots. A 20-cm pot is large enough for a sizeable flowering specimen. Unfortunately, the flowers of this hybrid close as soon as early afternoon; only in cool weather do they remain open longer.

Vine: pubescent; stem: terete, brownish; stipules: linear-lanceolate, 0.7–0.9 × 0.1 cm, glandular-serrate; petioles: 2–5 cm, stout, with 2 paired glands at base; leaves: unlobed, oblong-ovate, 8–16 × 4.3–9 cm, serrate to dentate; peduncles: 5.5–6.7 cm, solitary; bracts: oblong to oblong-ovate, 3.6–4.5 × 1.3–3 cm, reddish, glandular-serrate, free; floral tube: 0.7–1 cm long, reddish green outside; flowers: red, 10–12 cm diam.; sepals: reddish green with green midrib outside, red inside, oblong to oblong-lanceolate, 4.9–5.2 × 1.2–1.4 cm, keeled, keel terminating in an awn 0.6–1 cm long; petals: red, oblong to oblong-lanceolate, 4.6–5.1 × 0.9–1.1 cm; corona: 3 series, outer series 1.5–1.8 cm long, red, erect, inner series 0.9–1.1 cm long; ovary: ovoid, pubescent; fruit: not seen.

Passiflora ×*piresea*

Passiflora 'Pura Vida'

pura vida: good life
P. 'Amethyst' × *P. racemosa*
Breeder: Silvan Kamstra (1999)

When *Passiflora* 'Pura Vida' (Plate 319) became available in the late 1990s, it was one of the first really free-flowering hybrids to offer a welcome contrast to the prevailing purplish hybrids. Consequently, it has become increasingly popular, although its flowers are not the brilliant red preferred by most enthusiasts. A certain touch of blue-violet remains, giving the flowers a color that ranges somewhere between purple and reddish violet. The dark purple corona makes the blossoms look somewhat heavy. Flowers are 8–10 cm diam.

Passiflora 'Pura Vida' tends to open numerous flowers in batches, followed by some weeks in which it sets new flower buds. When this hybrid is cultivated in a container outdoors during summer months, the main flowering period in central Europe is midsummer and late summer. Under glass, the first flowering period begins in spring. Even small plants in a 12-cm pot will reliably set flowers, though *P.* 'Pura Vida' should be cultivated in a pot of 20 or 30 cm in order to produce an abundance of flowers. During overwintering it is satisfied with temperatures around 7°C, although 10°C is better.

The naming of different clones from the same cross seems to be somewhat unfortunate. The best-known and most widely distributed clone may be *Passiflora* 'Pura Vida2', characterized by deep purple flowers. *Passiflora* 'Pura Vida1' produces smaller flowers with a higher content of red, while *P.* 'Pura Vida7' also has reddish flowers but a significantly paler corona. However, these numberings often get mixed up.

Vine: pubescent; stem: terete, striate; stipules: reniform, 0.8–2 × 0.3–1 cm, entire; petioles: 2.5–5 cm, with 2–4 paired, subsessile glands normally above and below middle; leaves: 3-lobed, 4.5–10 × 7.4–15 cm, entire; peduncles: 5–6.5 cm, solitary, occasionally in pairs; bracts: ovate, 0.8–1.4 × 0.4–0.9 cm, greenish purple, free; floral tube: 0.6–0.9 cm long, purple to reddish violet; flowers: purple to reddish violet, 8–10 cm diam.; sepals: purple to reddish violet, oblong to oblong-lanceolate, 3.7–4.5 × 1.1–1.5 cm, keeled, keel terminating in an awn 0.3–0.6 cm long; petals: purple to reddish violet, oblong to oblong-lanceolate, 3.3–4 × 1–1.3 cm; corona: 4 series, outer 2 series 1.3–1.7 cm long, deep purple, whitish at apex, inner series 0.7–1 cm long, deep purple; ovary: oblong-ovoid, glabrous; fruit: oblong-ovoid, 5–6.5 × 2–3 cm, green becoming greenish yellow when ripe.

Passiflora 'Purple Haze'

P. caerulea × *P. amethystina*
Breeder: Cor Laurens (1994)

With the help of this cross, Cor Laurens originally wanted to verify the parentage of *Passiflora* 'Amethyst'. Although it turned out that *P. amethystina* could by no means be a parent plant, in the process of his work Laurens obtained the very beautiful *P.* 'Purple Haze' (Plate 320). Around the same time, new hybrids of this parentage were raised by John Vanderplank and Leopold Ševčík, so that something of a new era of crossbreeding was started in Europe. Since then, many breeders have special-

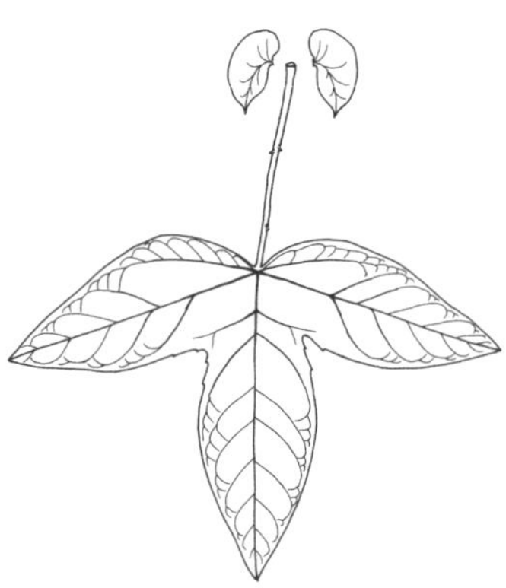

Passiflora 'Pura Vida'

Passiflora 'Purple Haze'

ized in developing new cultivars from these crosses. Laurens himself has raised other beautiful hybrids, such as *P.* 'Maxima', also resulting from a cross between *P. caerulea* and *P. amethystina.*

The paternal plant, *Passiflora amethystina,* known as a short-day plant, contributed its long peduncle to *P.* 'Purple Haze' but not its tendency to flower only when days are getting shorter. From *P. caerulea* it inherited an enormous willingness to flower and a resistance to cold. 'Purple Haze' is excellently suited for cultivation in a container. Because it is so easy to satisfy, it has become firmly established in many collections. The attractive, deep purple and lavender corona dominates the flower, which is 7.5–8.5 cm diam. A wealth of flowers is produced from June until late fall, sometimes dozens appearing in batches before the plant takes a rest.

In a 25- to 30-cm pot, *Passiflora* 'Purple Haze' develops into a sizeable specimen. It should be overwintered at temperatures above 7°C. Older plants will tolerate a short spell of frost, although the tips of the shoots will die off.

Vine: nearly glabrous; stem: terete, striate; stipules: subreniform, 1.5–2.5 × 0.8–1.4 cm, entire; petioles: 3–5 cm, stout, with 4–6 mostly paired, elongated glands, ca. 0.2 cm long; leaves: 3-lobed or occasionally 5-lobed, 5–16 × 8–19 cm, entire; peduncles: 5–10 cm, solitary; bracts: ovate, 1.5–2 × 1–1.3 cm, entire, free; flowers: deep purple, deep violet, and whitish mauve, 7.5–8.5 cm diam.; sepals: green outside, whitish mauve inside, oblong, 2.8–3.3 × 1.2–1.4 cm, keeled, keel terminating in an awn 0.8–1.1 cm long; petals: whitish mauve, oblong, 3–3.5 × 1.4–1.7 cm; corona: at least 4 series, outer series 2.4–2.8 cm long, deep purple in basal 1/3, white near middle, violet above, 2nd series normally slightly shorter than first series, inner series 0.3–0.5 cm long, deep purple; ovary: ovoid, pubescent; fruit: elliptical to oblong-ovoid, 4.5–6 × 2–3 cm, green.

Passiflora 'Roedie'

named after Roelof and Dienie Gommer
P. cincinnata × *P.* 'Pura Vida' (?)
Breeder: Hendrik Jan Gommer (1999)

Passiflora 'Roedie' (Plate 321) is another attractive hybrid. It was raised by Dutch breeder Hendrik Jan Gommer, who runs an excellent nursery in the Netherlands specializing in passionflowers and other unusual plants. Since a cross between *P. cincinnata* and *P.* 'Pura Vida' was supposed to bring about a result other than *P.* 'Roedie', the parentage of this hybrid is often questioned.

Passiflora 'Roedie' is a recommendable hybrid reminiscent of the well-known *P.* 'Byron Beauty', although the color of its flowers is more intense. A closer look at the flower reveals that the basically white sepals and petals are densely mottled with violet speckles. The enormous, undulate corona is particularly attractive. It is intensely violet to purplish violet with only a few white bands. The fragrant flowers reach a diameter of 9–11.5 cm. Within a few years, *P.* 'Roedie' has become a favorite in many collections.

It is best to grow *Passiflora* 'Roedie' as a container plant in a protected location or under glass. The main flowering time is in the 2nd half of summer and first half of fall. Plants grown in the form of a 90-cm-high pyramid may set flowers, and even young plants can set a few flowers, but *P.* 'Roedie' can display all its splendor only as a 2-m-high specimen in a spacious 30- or 40-cm pot. It can tolerate a minimum temperature of around 5°C, although 10°C is advisable in winter.

Passiflora 'Marianne Timmer', named after the popular Dutch speed skater, is another clone from the same cross, also selected by Gommer. In comparison with *P.* 'Roedie', its flowers are paler and the leaves are not as deep green.

Vine: glabrous; stem: terete, striate; stipules: linear-lanceolate, 0.7–1 cm long; petioles: 2.4–4

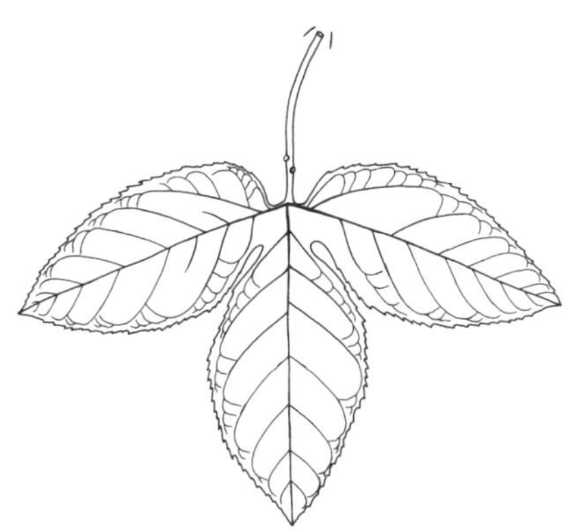

Passiflora 'Roedie'

cm, stout, with 2 paired glands in upper 1/2; leaves: deeply 3-lobed, occasionally with an additional small lobe, 7–12.5 × 9.5–17 cm, serrate; peduncles: 5–6.5 cm, solitary; bracts: broadly ovate, 2.2–2.7 × 1.8–2.3 cm, serrate with 2–4 glands toward base, free; flowers: violet, 9–11.5 cm diam.; sepals: green outside, white, densely mottled with purplish violet speckles inside, oblong, 3.5–4.3 × 1.5–2 cm, slightly keeled, keel terminating in an awn 0.9–1.2 cm long; petals: white, densely mottled with purplish violet speckles, oblong, 3.9–4.6 × 1.2–1.6 cm; corona: several series, outer 2 series 3.7–4.2 cm long, deep purplish violet at base, banded deep purplish violet and white at middle, becoming increasingly violet toward apex, wavy in upper 1/3, inner series 0.4–1.2 cm long, deep purplish violet; ovary: ovoid to globose, glabrous; fruit: not seen.

Passiflora 'Sandra'

P. incarnata × *P.* aff. *oerstedii*
Breeder: Leopold Ševčík (1996)

This recommendable hybrid (Plate 322) stands out pleasantly from the rest. Although its flowers are modest in size at 5–7 cm diam., they impress by their play of color. Depending on the time of year and the temperature, the purplish violet of the flowers varies in intensity, forming a contrasting background for the dominating white corona.

Like many other recent hybrids with a woman's name, *Passiflora* 'Sandra' was created by Leopold Ševčík from the Czech Republic, one of the most successful breeders of the last decades. Many of the fine hybrids raised in the 1990s, among them such beautiful cultivars as 'Aida', 'Colet', 'Libuse', 'Teta', and 'Violetta', are the results of Ševčík's endeavors. Regrettably, due to economic reasons (exploding costs for electric power and fuel, and so on) he had to close down his nursery and is no longer actively breeding.

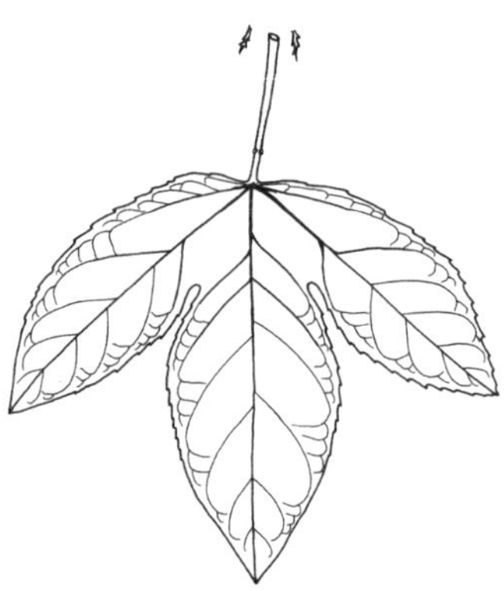

Passiflora 'Sandra'

Unfortunately, like many hybrids created by Ševčík, *Passiflora* 'Sandra' is infected with a latent virus that results in discolored, deformed leaves. Like all virus-infected passionflowers, the plant looks healthy and flowers willingly during summer months, but symptoms recur as soon as conditions become cool.

Passiflora 'Sandra' prefers a warm, protected location under glass. It does not grow as vigorously as many other hybrids and may well be cultivated in a 20-cm pot. As it is easy to tame, it can even be kept on a windowsill. In winter it prefers temperatures above 7°C, although lower temperatures are tolerated for a short while—an obvious inheritance from *P. incarnata*. The paternal plant was originally misidentified as *Passiflora porphyretica*.

Vine: glabrous; stem: terete, striate; stipules: lanceolate, 0.5–0.7 × 0.1 cm, glandular-serrate; petioles: 1.3–3.1 cm, with 2 paired glands in upper 1/3; leaves: 3-lobed, 5.5–10 × 7–14 cm, subcoriaceous, serrate; peduncles: 1.2–2.4 cm, solitary, occasionally in pairs; bracts: ovate, 0.6–0.8 × 0.4–0.5 cm, serrate, free; flowers: purplish violet and white, 5.7–7.2 cm diam.; sepals: green outside, pale purplish violet inside, oblong, 2.6–3.3 × 0.9–1.2 cm, slightly keeled, keel terminating in an awn 0.2–0.3 cm long; petals: whitish outside, purplish violet inside, oblong, 2.7–3.3 × 0.7–0.9 cm; corona: 4 or 5 series, outer 2 series 2–2.4 cm long, white in basal 1/2, pale bluish violet in upper 1/2, slightly wavy near apex, inner series purplish violet, 0.2–0.3 cm long; ovary: oblong-ovoid, glabrous; fruit: not seen.

Passiflora 'Stradivarius'

named after Antonio Stradivari
P. murucuja × *P. cupraea*
Breeder: Maurizio Vecchia (2000)

Passiflora 'Stradivarius' (Plate 323), resulting from a cross between 2 species from subgenus *Decaloba*, stands its ground against large-flowered competitors because of its willingness to flower and the intense purple-red of its blossoms. Starting in late summer its flowers, up to 6 cm diam., appear by the dozen, often in batches. Except for some short periods of rest, it flowers until October. The vivid orange coronal filaments, running parallel to the androgynophore, reach a length of about 0.7 cm. The purple-red color of the flower, the short upright corona, and the elongated androgynophore indicate that *P.* 'Stradivarius' is a cross between 2 hummingbird-flowered species. Consequently, it has no scent. The marked difference in length between the sepals and petals is also remarkable. Pollen is missing.

Passiflora 'Stradivarius' is named after the famous violinmaker who lived in Cremona from 1644 until 1737. Maurizio Vecchia dedicated the hybrid to his hometown. The leaves of this plant are a little reminiscent of the silhouette typical of ancient string instruments. The flower color is similar to alizarine, the reddish extract of *Rubia tinctura* used to paint violins.

Because of its daintiness, *Passiflora* 'Stradivarius' can be grown successfully in a small pot of about 12 cm diam., which makes it suitable for cultivation on a windowsill. It will also flourish in a larger container that can be taken outdoors in May. The minimum temperature is around 10°C.

Vine: glabrous; stem: terete to subangular, slender; stipules: linear, 0.1–0.2 cm long, soon deciduous; petioles: 2–3 cm, without glands; leaves: 3-lobed in upper 1/3, 3–4 × 2.7–3.6 cm, subcoriaceous, bearing laminar nectaries, entire; peduncles: 2.5–3.5 cm, solitary; bracts: minute or absent; floral tube: 0.5–0.6 cm long, greenish red outside; flowers: deep purplish red and orange-red, 5–6 cm diam.; sepals: reddish to reddish green outside, deep purplish red inside, linear-lanceolate, 2.6–3 × 0.4–0.5 cm; petals: deep purplish red, linear-lanceolate, 1.6–1.8 × 0.3 cm; corona: 1 series, ca. 0.7 cm long, orange-red, erect, filaments partly connate; ovary: ovoid, green, glabrous; fruit: not seen.

Passiflora 'Sunburst'

P. gilbertiana × *P. jorullensis*
Breeder: Patrick J. Worley (1983)

By crossing *Passiflora gilbertiana* with *P. jorullensis*, Patrick Worley succeeded in raising this charming hybrid (Plate 324), which was first offered for sale in 1983. Since then it has spread very rapidly and is represented in collections far more often than its parent plants.

Passiflora 'Sunburst' was named for its unusually intense yellow-orange flowers, a rare color in *Passiflora*. It is very free-flowering, many blossoms of around 2.5 cm diam. appearing every day. There may be 1, 2, or up to 6 of them at a node. Unfortunately, the flowers exude an unpleasant odor reminiscent of mothballs or bad breath, a characteristic inherited from both parents but grown worse in the child. If possible, position the vigorous *P.* 'Sunburst' away from any seating areas, or at least place it downwind. The flowers can stay open for up to 3 days while the intensity of color and odor decreases.

The dark leaves, mottled greenish cream along the main veins on the upper side, form a beautifully contrasting background for the brilliant flowers. Moreover, there are 2 rows of conspicuous gland spots on the leaves, and the undersides have a violet shimmer. This is among the few hybrids sometimes grown just for its unusually beautiful foliage.

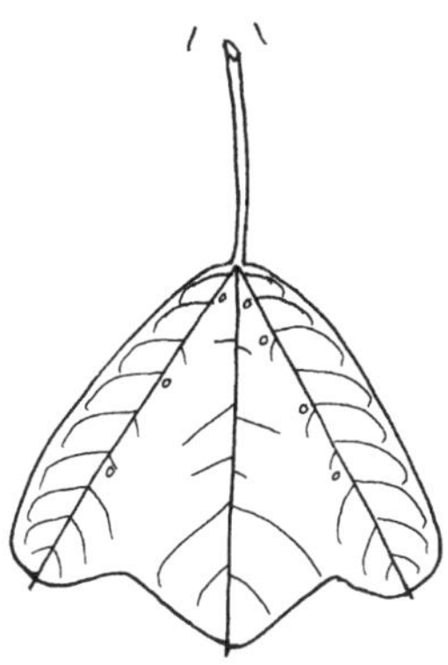

Passiflora 'Stradivarius'

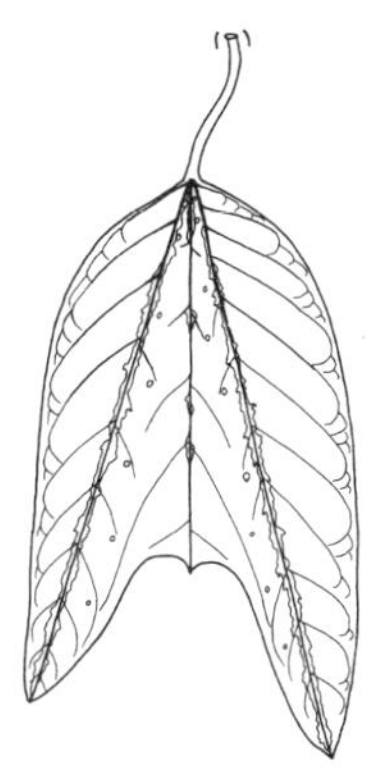

Passiflora 'Sunburst'

Passiflora 'Sunburst' is easy to cultivate, vigorous, and can be successfully grown in a container. Provided a sunny location, it develops into an imposing plant. It can be overwintered without problems in a conservatory, greenhouse, or even on a windowsill when temperatures of at least 10°C are guaranteed.

Vine: glabrous; stem: subangular, striate, more or less slender; stipules: linear, 0.3–0.5 cm long, purplish, soon deciduous; petioles: 2–4 cm, without glands; leaves: 2- or 3-lobed at apex, 5.5–15 × 3–7.5 cm, variegated along the main veins above, bearing yellowish laminar nectaries, purplish green beneath, entire; peduncles: 2–3.5 cm, generally in pairs, often dichotomous, up to 6 flowers per node; bracts: minute or absent; flowers: greenish yellow, orange, and yellow, 2.2–2.7 cm diam.; sepals: green outside, greenish yellow inside, linear-lanceolate, 1–1.4 × 0.2–0.3 cm; petals: greenish yellow, linear-lanceolate, 0.2–0.4 × 0.1 cm; corona: 2 series, outer series ca. 1 cm long, orange becoming increasingly yellow toward apex, inner series up to 0.2 cm long, purple, few filaments; ovary: subglobose to ovoid, green, glabrous; fruit: subglobose, 1–1.5 × 1 cm, black to blackish violet.

Passiflora ×*violacea* Loiseleur-Des.

Herb. gén. Ama. 7: t. 499 (1824)
violaceus: violet
P. racemosa × *P. caerulea*
Synonyms: *P.* ×*caeruleo-racemosa*, *P.* 'Eynsford Gem', *P.* 'Lilac Lady', *P.* ×*tresederi*, *P.* 'Victoria'
Breeder: Thomas Milne (1819)

The recommendable *Passiflora* ×*violacea* was raised by Thomas Milne of Britain in 1819 and described by Joseph Sabine 3 years later. Until a few years ago this hybrid was known under its hybrid formula, *P.* ×*caeruleo-racemosa*. We owe the correct name to Emil Kugler and Wolfgang Wetschnig, who investigated the matter and published their findings in an excellent article in 1991. The crossbreeding of *P. caerulea* and *P. racemosa* brought about various hybrids, so that several forms are distinguished. While usually solitary flowers prevail, *P.* ×*violacea* 'Eynsford Gem' produces racemes. It also stands out for its bushy growth and somewhat sickly appearance.

The color intensity of the sepals and petals varies, depending on the form, from pale purple through old pink to rose-pink (see *Passiflora* ×*violacea* 'Atropurpurea', Plate 325). The corona is mainly dark purple up to the middle and white toward the tip. Occasionally another purple band can be found at the tip. *Passiflora* ×*atropurpurea* and *P.* ×*atrorubens* have particularly large and beautiful flowers. With *P.* ×*violacea* there can be 3- or 5-lobed leaves on the same plant, although 3-lobed forms are far more common.

Passiflora ×*violacea* is superbly suited for cultivation in a container but will also produce numerous flowers up to 10.5 cm diam. when grown on a windowsill. Although it is said to withstand temperatures down to 0°C for a short while, overwintering temperatures should be above 7°C.

Vine: glabrous; stem: terete, striate; stipules: subreniform, 2–3.5 × 1.2–1.8 cm, subentire to slightly serrate; petioles: 4–6.5 cm, with 2–6 paired glands; leaves: 3-lobed or occasionally 5-lobed, 7.5–16 × 9.5–19 cm, entire; peduncles: 2.5–4.5 cm, solitary, occasionally in small fascicles; bracts: broadly ovate, 2–3 × 1.4–2.3 cm, subentire, purplish green, free; flowers: purplish, 8–10.5 cm diam.; sepals: brownish purple to purple outside, more or less deep purple to pinkish red inside, oblong, 3.7–5 × 1.4–2 cm, keeled, keel terminating in an awn 0.4–0.7 cm long; petals: purple to pinkish red, oblong, 3.5–4.7 × 1.2–1.7 cm; corona: 4 or 5 series, outer 2 series 1.8–3 cm long, deep purplish in lower 1/2, white in upper 1/2, purplish at apex, inner series 0.3–0.9 cm long, deep purple; ovary: oblong-ovoid, glabrous; fruit: oblong-ovate, 5–7 × 2.2–3 cm, green.

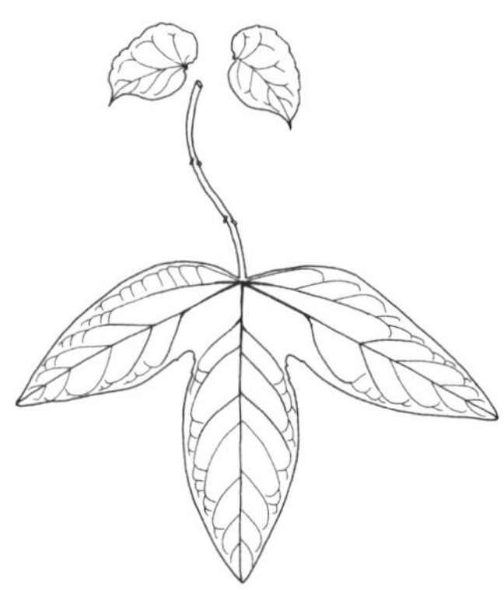

Passiflora ×violacea

19

Heliconius Butterflies and *Passiflora*

Ronald Boender

THE BEAUTIFUL NEOTROPICAL BUTTERFLIES found in the family Heliconidae, called longwings or maracuja butterflies, are always associated with passionvines, since the butterfly larvae in this family feed on nothing but leaves of *Passiflora*. Heliconidae butterflies encompass a number of genera in this family, but the most familiar is the genus *Heliconius*. The butterflies of this genus are the most colorful of the family Heliconidae and are long-lived, slow flyers that can be kept successfully in small enclosures. Many *Heliconius* species will only lay eggs on one species of *Passiflora*, and consequently their existence depends on that species of plant: the demise of the plant would mean the demise of the butterfly. Other *Heliconius* butterflies exhibit more general feeding habits for their larvae and lay eggs on quite a few species of *Passiflora*.

In addition to *Heliconius*, the variegated fritillary, *Euptoieta claudia* (Nymphalidae), and the day-flying moths in the tribe Josiini (Dioptidae) feed on passionvines (Miller 1996). Also included among *Passiflora* feeders are 2 Indo-Australian genera, *Vindula* and *Cethosia* (Corbet and Pendlebury 1992), as well as colorful flea beetles and flag-legged bugs.

Egg Laying

Most species oviposit only one egg at a time (*Heliconius melpomene, H. cydno, H. ethilla*), while others lay all their eggs at once in beautiful circular patterns (*H. doris, Dione juno*). Some species communally oviposit all at once on a new shoot for a mass of eggs (*H. sara*), while others lay small clusters on new growth (*H. charitonia, H. wallacei*) or singly lay in neat rows on new meristems (*H. sapho, H. eleuchia, H. hewitsoni* [Plate 326]). Eggs take 3–9 days to hatch. (See Plates 327 and 328.)

Larva or Caterpillar Stage

Most caterpillars start their life by eating the shell of their egg for nutrients, but a number of species abandon the shell once they have eaten their way out, as with *Heliconius sapho*. Most of the species that lay single eggs are very cannibalistic, as with *H. erato*, and will quickly eliminate any competition. On the other end of the scale are gregarious feeders like *H. sara* that move communally as larvae, lining up neatly to consume the plant together. (See Plates 329–331.)

During its 2- to 3-week larval stage, the larva molts, shedding its skin 4 times for a total of 5 "instars," or larval stages. At the completion of the larval stage it searches for a place to spin a silk pad or "cremaster." It will then attach itself to it with a series of cremastral hooks on the end of the body and slowly release its feet to hang freely. The first metamorphosis takes place as the larva becomes a pupa or chrysalis. This miraculous change occurs about 24–48 hours after the larva suspends itself in this position. The chrysalides of *Heliconius* butterflies are usually tan or white. They are covered with many spines and have some metallic markings.

Pupa or Chrysalis

It is difficult in nature to find the chrysalis, as it is cryptic or camouflaged (see Plates 332 and 333). It usually takes 8–12 days to mature before emergence occurs, and during this period many changes take place inside it. The body of the larva must transform from an ugly eating machine into a beautiful butterfly. When time is near, the skin of the chrysalis becomes clear and the coloration of the butterfly begins to show through. The 2nd stage of metamorphosis takes place as the butterfly emerges as an adult. Pulling its swollen body from the chrysalis, it begins to pump fluid into its wings to enlarge and straighten them. The butterfly is vulnerable at this stage, so it usually emerges early in the morning, pumping its wings quickly to size and releasing an enzyme to harden them. The process is complete in only a matter of minutes, and the butterfly is ready to fly within the hour.

Courtship and Copulation

Many heliconid males, including *Heliconius doris melpomene*, *H. cydno*, and *H. ethilla*, have pheromone patches on their hind wings and are able to hover over the female showering her with aphrodisiacal pheromone particles. They accomplish this by scraping the forewing over the hind wing while hovering, thereby releasing the particles. If the female is receptive, she will allow copulation. If not, she will expose a gland on her abdomen and flutter her wings to push an odor up to the male, repelling him. Some call this an antiaphrodisiac.

In some species, such as *Heliconius doris*, *H. erato*, *H. charitonia*, *H. hecalesia*, *H. sapho*, *H. eleuchia*, and *H. hewitsoni*, the male butterfly will find and hang on to the female chrysalis for days. Shortly before she ecloses (emerges), he will penetrate the skin and insert up to 3 abdominal sections; then, upon emergence, he will mate with her (see Plate 334) (Deinert et al. 1994). Oftentimes a number of males will struggle for the right position on the chrysalis, attaching themselves one on top of another—a strange behavior that can be life-threatening to the female if it becomes too excessive. If the males remain this way for days, some will be unable extricate themselves from the rest to feed and will find themselves in danger of starvation or desiccation (Lawrence E. Gilbert, personal communication).

Life Spans

The heliconids are considered to be a tribe of the family Nymphalidae and are divided into 9 genera: *Dione*, *Philaethria*, *Eueides*, *Agraulis*, *Dryas*, *Dryadula*, *Neruda*, *Podotricha*, and *Heliconius* (Holzinger and Holzinger 1994).

Members of the first 8 genera are like most other butterflies in that their diet consists of drinking nectar from a wide source of flowers. With this source of sugar energy the typical butterfly has a life span of about a week. In contrast, *Heliconius* butterflies are able to extract nutrients from pollen as well as feed on nectar. After pollen is collected on the proboscis, the butterfly excretes a fluid to suspend the pollen into a wet mass, which is held for several hours (see Plate 335). Apparently the pollen is stimulated to release amino acids and proteins, as if germinating. The butterfly simply sucks up the slurry without ever digesting or breaking the pollen grains. This ability to collect and exploit pollen is unique to this genus and gives these butterflies the ability to live for months. William Beebe, a researcher in tropical butterflies, had a pet *H. charitonia* named Higgins that lived for more than 5 months, and Lawrence E. Gilbert (1971) marked a wild *H. ethilla* that lived for 6 months.

Source of Pollen

While small amounts of pollen are found in most flowers, especially *Lantana* on the forest edges, the most often visited flowers are in the family Cucurbitaceae, particularly the genera *Psiguria* and *Gurania* (see Plates 336–338). *Heliconius* butterflies have a mutualistic relationship with some species of these genera. The male flowers supply pollen and nectar for the butterfly, and consequently the butterfly performs the pollination for the plant. These canopy vines are very interesting in that they produce huge amounts of nectar, so much so that in the morning one can see a large drop of nectar protruding from each corolla. Almost inexhaustible, the flowers refill themselves with nectar constantly throughout the day. Unlike most flowers, however, these cucurbits present pollen before nectar flow, and the pollen is collected efficiently with the help of long papillae on the tip of the butterfly's proboscis.

The plants produce flowers in an inflorescence on a long peduncle at the rate of about one a day and can continue producing flowers in the male form for up to about a year (Gilbert 1975). These plants are also unusual in that they start out in the male form, producing pollen from male flowers until quite mature, and then are able to switch sex and begin producing female flowers, which in turn are pollinated by the *Heliconius* and produce fruit. Some time later, the plant switches back to the male form and the process begins again.

The plants can be identified by the flowers. *Psiguria* species produce a new flower every day, each with a green calyx and petals that are usually red, pink, or orange. *Gurania* flowers have a calyx of orange or yellow with yellow petals. Some *Gurania* species have flowers that are rather spectacular in shape and color.

For *Heliconius* in greenhouses, plants such as *Pentas*, *Stachytarpheta*, *Jatropha*, *Palicourea*, and *Hamelia* may be useful since all are good sources of both nectar and pollen.

Color

Heliconius species are probably the slowest-flying butterflies, and many believe them to be the most beautiful. While they may not be as stunning as a large blue morpho within the forest, their slow wing beats and bold color patterns make them highly conspicuous. In spite of these showy traits, they are able to survive due to aposematic coloring, which warns birds of their poisonous nature. Like their *Passiflora* hostplants, they contain cyanogenic compounds. The butterfly cyanogens were once thought to be derived from these plants (Brower and Brower 1964), but it is now known that most *Heliconius* species manufacture their own cyanogens de novo from amino acids. *Heliconius* have a good amino acid supply from pollen, and this might be why they are more distasteful than other passionvine butterflies.

In the neotropics many *Heliconius* species have large numbers of subspecies, with each subspecies inhabiting a different valley or other isolated region. Where races come into contact and hybridize, one might find many other forms and colors. For example, Butterfly World (in Florida) has farmed over a million *Heliconius* and artificially mixed sub-

Heliconius butterflies and their corresponding _Passiflora_ hostplants.

BUTTERFLY	HOSTPLANT
H. atthis	*P. resticulata*
H. charitonia	*P. adenopoda, P. boenderi, P. biflora, P.* 'Incense', *P. lobata, P. morifolia, P. suberosa, P. trifasciata, P. tricuspis*
H. clysonymus	*P. apetala*
H. cydno	Most species of *Passiflora*
H. doris	*P. ambigua, P. laurifolia, P. serratodigitata*
H. eleuchia	*P. macrophylla, P. tica*
H. erato	*P. biflora, P. discophora, P. rubra, P. talamancensis, P. vespertilio*
H. ethilla	Most species of *Passiflora*
H. hecale	*P. oerstedii, P. platyloba, P. vitifolia*
H. hecalesia	*P. arbelaezii, P. discophora, P. lancearia, P. obovata*
H. hewitsoni	*P. pittieri*
H. ismenius	*P. alata, P. ambigua, P. platyloba, P. quadrangularis*
H. melpomene	*P. caerulea, P. menispermifolia, P. oerstedii,* and many others
H. numata	Most species of *Passiflora*
H. sapho	*P. pittieri*
H. sara	*P. auriculata, P. mucronata*

species for many years (see Plate 339). The resulting hybridization has resulted in some spectacular forms similar to those found in natural hybrid zones.

Traplines and Roosting

In the wild, *Heliconius* butterflies find psigurias and guranias while traveling over a circular path that can span hundreds of meters. The same butterflies can be found in the same area, going from plant to plant feeding on the flowers during the day, and then returning to the same roosting area every night (Murawski and Gilbert 1986). As darkness begins to fall, they cluster together and roost for the night in large numbers on fragile plants or hanging roots, usually over an open area, reducing the chance that they will be disturbed. These roosts, which often involve more than one species of *Heliconius*, are abandoned in the early morning as each butterfly leaves to seek flowers along its particular trapline route. See Plates 340 and 343 for a survey of various *Heliconius* species.

Greenhouse Beauty

Heliconius butterflies, despite their small size, have one of the largest heads of all butterflies, better enabling them to perform their tasks. They also have the ability to fly, live, and breed successfully in enclosures as small as 2 × 2 × 3 m, making them perfect for captive study (see Plates 341 and 342). This is probably why they have become the most studied, most farmed butterfly in history, and it is certainly why they make such perfect additions to a greenhouse. Many botanists who cultivate *Passiflora* are becoming aware of these beautiful creatures and have begun adding them to their greenhouses. Some countries have laws governing importation and containment of live exotic insects, so one should check regulations before beginning one's own Butterfly World.

Artificial Nectar and Feeders

Many times due to latitude, climate, or greenhouse conditions it is difficult to keep flowers in bloom that are necessary to keep *Heliconius* butterflies alive. Feeding them only sugar water will result in one-week life spans. One easy way to sustain *Heliconius*, while eliminating flowers completely if desired, has been developed at Butterfly World in the form of a commercially available product called Nektar Plus. Although Nektar Plus was developed for hummingbirds, it combines sugar and amino acids and supplies everything these butterflies need for a normal life span. It arrives packed in a powdered form that must be mixed with water. Unfortunately, bacteria growing on the rich mixture of the amino acids break the solution down quickly, and it becomes rancid in about 4–12 hours, depending on ambient temperature, making this a slightly labor-intensive way to feed butterflies.

When dispensing Nektar Plus, small acrylic platforms developed by Butterfly World are used. Each platform is suspended to hang horizontally. Prior to this, 2 holes are drilled into each one, and the holes are circled with affixed plastic red petals made from vinyl sign material. Two vials are then taken (one can use an Aquapick, used in the floral trade, by cutting it to about 2.5 cm and flanging it with an old iron) and dropped into the holes to form an artificial corolla. The butterflies quickly learn how to use this method and feed without difficulty.

Conversion Tables

MILLIMETERS (mm)	INCHES	FEET
0.5	0.02	
1	0.04	
2	0.08	
3	0.12	
4	0.16	
5	0.20	
6	0.24	
7	0.28	
8	0.32	
9	0.36	
10	0.4	
15	0.6	
20	0.8	
25	1.0	
1000		3.3
2000		6.6
3000		10.0
4000		13.3
5000		16.6
10,000		33.3

CENTIMETERS (cm)	INCHES
0.5	0.2
1	0.4
2	0.8
3	1.2
4	1.6
5	2.0
6	2.4
7	2.8
8	3.2
9	3.6
10	4.0
15	6.0
20	8.0
25	10.0
30	12.0
40	16.0
50	20.0
100	40.0

GRAMS (g)	OUNCES
1	0.035
2	0.07
10	0.35
35	1.2
50	1.75
100	3.5

METERS (m)	FEET
1	3
2	6
3	9
4	12
5	15
6	18
7	21
8	24
9	27
10	30
15	45
20	60
30	90
40	120
50	150
75	225
100	300
500	1500
1000	3000
2000	6000
2500	7500
3000	9000
3500	10,500
4000	12,000
4500	13,500

KILOGRAMS(kg)	POUNDS
1	2.2
1.5	3.3
2	4.4
3	6.6
4	8.8
5	11
10	22

KILOMETERS (km)	MILES
50	31
100	62
250	155
1000	620
1500	930

HECTARES	ACRES
1	2.5
5	12.5
10	25

°CELSIUS	°FAHRENHEIT
–30	–22
–15	5
–12	10
–8	18
–5	23
0	32
2	36
5	41
7	45
10	50
12	54
15	59
20	68
25	77
30	86
35	95
40	104

Plant Societies, Suppliers, and Web Sites

PLANT SOCIETIES

Interessengemeinschaft Passionsblumen
Kontaktadresse Redaktion
Emil Kugler
Treustr. 92/6/8
1200 Wien
Austria

Interessengemeinschaft Passionsblumen
Kontaktadresse Vertrieb
Heinz Opitz
Roeberstr. 7
42117 Wuppertal
Germany

Passiflora Society International
c/o Butterfly World
Tradewinds Park
3600 W. Sample Road
Coconut Creek, Florida 33073
United States
http://www.passiflora.org/

SUPPLIERS

The following list includes American and European suppliers of Passiflora *plants,* Passiflora *seeds, and Nektar Plus, a commercial product useful for feeding* Heliconius *butterflies.*

Plants (United States)

Butterfly World
Tradewinds Park
3600 W. Sample Road
Coconut Creek, Florida 33073
http://www.butterflyworld.com/
Walk-ins only; no shipping.

Glasshouse Works
Church Street, P.O. Box 97
Stewart, Ohio 45778
http://www.glasshouseworks.com/

Grassy Knoll Exotic Plants
422 J Street
Washougal, Washington 98671
http://www.gkexoticplants.com/

Kartuz Greenhouses
1408 Sunset Drive
Vista, California 92085
http://www.kartuz.com/

Logee's Greenhouses
141 North Street
Danielson, Connecticut 06239
http://www.logees.com/store/

Ransom Gardens
P.O. Box 941625
Houston, Texas 77094
https://www.ransomgardens.com/

Plants (Europe)

Blumen & Passiflora
Bettina and Torsten Ulmer
Hevener Str. 18
58455 Witten
Germany
http://www.blumen-passiflora.de/

Gärtnerei Baum
Strohgäustr. 51
71229 Leonberg
Germany

Kwekerij Gommer
Vossebelt 41a
7751 SW Dalen
Netherlands
http://www.kwekerijgommer.com/

National Collection of Passiflora
John Vanderplank
Lampley Road
Kingston Seymour
Clevedon
North Somerset BS21 6XS
United Kingdom
http://www.passiflora-uk.co.uk/

Plantencentrum Exotica
Guido Van Herck
Nijlensesteenweg 268
2270 Herenthout
Belgium

Toms Treibhaus
Thomas Schneider
Schönarts 6
97776 Eußenheim
Germany
http://www.toms-treibhaus.de/

Seeds

Direct Seed Co.
Mark Cooper
3303 Harbor Boulevard, Suite E-7
Costa Mesa, California 92626
http://members.aol.com/pasiflora1/

Sementes
Margrit Reiner
Am Ringenstall 5
72469 Meßstetten
Germany
http://www.sementes.de/

Sunshine-Seeds
Renate Krebs
Warendorfer Str. 58
59227 Ahlen
Germany
http://www.sunshine-seeds.de/

Tropical Seeds
Barbadine B. P.
100 Le Tampon cedex
97832 Ile de la Réunion
France
http://www.barbadine.com/

Nektar Plus

Günther Enderle
27 W. Tarpon Avenue
Tarpon Springs, Florida
phone: 813-938-1544

Günther Enderle
Kieselbronnerstr. 28
D-75177 Pforzheim
Germany
phone: 07231-9546-0

USEFUL WEB SITES

Cor Laurens's *Passiflora* site. Nationale Collectie Passiflora. Netherlands.
http://www.passiebloem.nl/

Ecology, systematics, and evolution of passionvines (Passifloraceae). Lawrence E. Gilbert Laboratory. United States.
http://uts.cc.utexas.edu/~gilbert/research/passionvines/index.html

Gerrit Graat's *Passiflora* site. Netherlands.
http://www.passiflorasite.100.nl/

Henk Wouters's *Passiflora* site. Netherlands.
http://home.wanadoo.nl/henkwouters/

Maurizio Vecchia's *Passiflora* site. Passiflora! Italy. http://www.passiflora.it/index.htm

Mauro Peixoto's *Passiflora* site, with photos of Brazilian species. Brazil.
http://mpeixoto.sites.uol.com.br/passif/fotopassiflora.html

Missouri Botanical Garden research page. Very informative Web page related to many current research projects at the garden. Plant Image Index includes photos of passionflowers.
http://www.mobot.org/MOBOT/research/

Myles Stewart Irvine's *Passiflora* site. United Kingdom.
http://www.passionflow.co.uk/

Passiflora enthusiast discussion list.
http://listserv.surfnet.nl/archives/passiflora-l.html

Passiflora Images. Online group providing a way for people to share images via e-mail.
http://groups.yahoo.com/group/Passiflora_Images/

Glossary

Abaxial. Away from the stem of the plant.
Adaxial. Toward the stem of the plant.
Adnate. Attached to something else.
Anatomy. Inner structure.
Androecium. A collective term for all the stamens in a single flower.
Androgynophore. A column bearing both the ovary and the anthers of a flower.
Annular. Ringlike.
Anther. A pollen sac.
Aril. One of the individual juice sacs, or pulp, that surround each seed.
Auriculate. Ear-shaped.
Bullate. Puckered, bumpy.
Callus. A swollen, pale mass of undifferentiated tissue that develops over an injured or cut plant surface.
Capillary. Hairlike.
Capitellate. Having a small head.
Carinate. Keeled.
Cernuous. Nodding.
Compound. So deeply lobed as to be composed of distinct leaflets.
Concentric rings. Normal secondary growth.
Connate. Fused to a structure of the same kind.
Cordate. Heart-shaped.
Coriaceous. Leathery.
Cotyledon. A leaf of the embryo of a seed.
Cuneate. Wedge-shaped.
Cyanogenic. Cyanide-producing.
Dehiscent. Splitting open.
Dentate. Having teeth.
Denticulate. Having tiny teeth.
Dimidiate-ovate. Shaped like half of an egg that is lengthwise cut into 2 parts.
Diploid. Having 2 sets of chromosomes.
Epidermis. The outermost layer of cells, as of leaves, stems, and roots.
Extrafloral. Outside the flower.
Falcate. Slender and curved like a sickle.
Filiform. Threadlike.
Gamete. A haploid reproductive cell.
Genetically. Referring to heredity and mutation, with respect to genesis or origin.
Genome. The total genetic information contained within a cell.
Genotype. The genetic constitution of an organism, as contrasted with the phenotype.
Glabrescent. Becoming smooth with age.
Glabrous. Smooth.
Gynoecium. The aggregate of carpels in the flower of a seed plant.
Haploid. Having only one set of chromosomes.
Hispid. Having straight, stiff hairs.
Homologous. Of the same essential nature and of common descent. Homologous chromosomes associate in pairs in the first stage of meiosis.
Lamina. Leaf blade.
Lanate. Woolly.
Lanceolate. Lance-shaped.
Meiosis. Two successive nuclear divisions in which the chromosome number is reduced from diploid ($2n$) to haploid (n) and segregation of the genes occurs.

Morphology. Outer form. The study of form and its development.
Node. Leaf joint or bud.
Obliquely ovate. Egg-shaped at an angle.
Ocellae. Eyespots.
Operculum. Innermost coronal row, often a membrane.
Ovate-amplexicaule. Egg-shaped and clasping around the stem.
Ovate-auriculate. Shaped somewhere between an egg and an ear.
Perianth. Petals and sepals.
Pericarp. The wall of a ripened ovary.
Petiole. Stalk.
Phenotype. The physical appearance of an organism, as contrasted with the genotype.
Pilose. Having long, mostly straight trichomes.
Plicate. Pleated or folded.
Pollen. Collective term for pollen grains, microspores containing a male gamete.
Pubescent. Hairy.
Radicle. Embryonic root.
Reniform. Kidney-shaped.
Reticulate. Having a network of pits or dots.
Rhizomes. Specialized underground stems that act as storage organs and produce aerial shoots at their apex and along their lengths.
Rugulose. Finely wrinkled.
Self-incompatible. Self-sterile or clone-sterile, incapable of effective self-pollination; said of a plant that does not fruit with its own pollen, making cross-pollination necessary.
Semiovate. Partly egg-shaped.
Semireniform. Shaped like half a kidney.
Serrate. Saw-toothed.
Serrulate. Finely saw-toothed.
Sessile. Stalkless.
Setaceous. Bristlelike.
Stipules. Paired leafy or bristlelike appendages at base of petiole.
Striate. Lined.
Subreniform. Somewhat kidney-shaped.
Sulcate. Furrowed.
Terete. Round.
Trichome. Plant hair.
Truncate. Appearing cut off.
Vegetatively. Asexually.
Xylem. Vascular tissue.
Zygomorphic. Having lateral symmetry.

Bibliography

Acosta, J. de. 1590. *Historia Natural y Moral de las Indias* (The Natural and Moral History of the Indies). Seville: J. de León. Trans. E. Grimston, 1604. Reprint, ed. C. R. Markham. London: Hakluyt Society, 1880.

Adams, C. D. 1967. Additions to the flora of Jamaica. Passifloraceae. *Bulletin of the Institute of Jamaica, Science Series* 16: 27.

——. 1972. Passifloraceae. In *Flowering Plants of Jamaica*. Jamaica: University of the West Indies. 492–496.

Akamine, E. K., and G. Girolami. 1959. Pollination and fruit set in the yellow passion fruit. Hawaii Agricultural Experiment Station, University of Hawaii, Technical Bulletin 39: 1–44.

Aldinus, T. [Castelli, P.]. 1620. *Vera, e natural effigie della pianta indiana, chiamata maraco, granadilla, e fior della passione D-N-S*. Venice.

——. 1625. *Exactissimo descriptio rariorum quarundam plantarum quae continentur Romae in horto farnesiano*. Rome: J. Mascardi.

Ambühl, H. 1966. *Anatomische und Chemische Untersuchungen an* Passiflora coerulea *L. und* Passiflora incarnata *L.* Thesis, No. 3830, ETH Zürich.

Amela García, M. T., B. G. Galati, and A. M. Anton. 2002. Microsporogenesis, microgametogensis and pollen morphology of *Passiflora* spp. (Passifloraceae). *Botanical Journal of the Linnean Society* 139: 383–394.

Annonay, H., and C. Feuillet. 1997. *Passiflora cerasina*, a new species of *Passiflora* subgenus *Passiflora* (Passifloraceae) from French Guiana. *Sida* 17 (3): 551–554.

——. 1998. *Passiflora aimae* (Passifloraceae), une espèce nouvelle de Guyane française. *Adansonia* 20 (2): 295–298.

Anonymous. 1609. *Copia del fior, e frvtto chiamato Granadilla, che nasce dell' Indie Occidentali*. Bologna: G. Ardizzoni.

Anonymous. 1614. *Le vray portraict du Maraco des Indes dicte granadille ou fleur de la passion*. Paris: N. Mathonière.

Arjona, E. H., and B. F. Matta. 1991. Postharvest quality of passion fruit as influenced by harvest time and ethylene treatment. *HortScience* 26 (10): 1297–1298.

Arjona, E. H., B. F. Matta, and O. J. Garner. 1991. Growth and composition of passion fruit (*Passiflora edulis*) and maypop (*P. incarnata*). *HortScience* 26 (7): 921–923.

Ayensu, E. S., and W. L. Stern. 1964. Systematic anatomy and ontogeny of the stem in Passifloraceae. *Contributions United States National Herbarium* 34 (3): 45–73.

Barco-Centenera, M. del. 1602. *Argentina y conquista del Río de la Plata*. Lisbon: P. Crasbeeck.

Benson, W. W., K. S. Brown, Jr., and L. E. Gilbert. 1975. Coevolution of plants and herbivores: passion flower butterflies. *Evolution* 29: 659–680.

Bernhard, A. 1999. Flower structure, development, and systematics in Passifloraceae and in *Abatia* (Flacourtiaceae). *International Journal of Plant Sciences* 160 (1): 135–150.

Boender, R., and T. Ulmer. 2001. *Passiflora tina* spec. nov., a new species of *Passiflora*, subgenus *Astrophea* (Passifloraceae) from Ecuador. *Sendtnera* 7: 5–12.

Bosio, G. 1610. *La trionfante e gloriosa croce*. Rome: A. Ciacone.

Brako, L., and L. J. Zarucchi, eds. 1993. Passifloraceae. *Catalogue of the Flowering Plants and Gymnosperms of Peru: Monographs in Systematic Botany from the Missouri Botanical Garden* 45: 876–882.

Brizicky, G. K. 1961. The genera of Turneraceae and Passifloraceae in the southeastern United States. *Journal of the Arnold Arboretum*. 42: 204–218.

Brower, L. P., and J. V. Z. Brower. 1964. Birds, butterflies and plant poison: a study in ecological chemistry. *Zoologica* 49: 137–158.

Buzato, S., and A. L. M. Franco. 1992. *Tetrastylis ovalis*: a second case of bat-pollinated passionflowers (Passifloraceae). *Plant Systematics and Evolution* 181: 261–267.

Caesar, W., and R. Spohn. 1997. Passionsblume. Kulturhistorische Aspekte einer Arzneipflanze. *Deutsche Apotheker-Zeitung* 137 (8): 51–56, 61.

Cavanilles, A. J. 1790. *Monadelphia Classis Dissertationes Decem*. Diss. 10: Decima dissertatio botanica de *Passiflora*. Madrid: Typographia regia. 439–463.

Cervi, A. C. 1994a. *Passiflora hatschbachii*, nueva especie de Minas Gerais (Brasil). *Fontqueria* 40: 45–47.

——. 1994b. Studies in Brazilian Passifloraceae III. A new species of *Passiflora*. *Brittonia* 46 (2): 144–146.

——. 1997. Passifloraceae do Brasil. Estudo do genero *Passiflora* L., subgenero *Passiflora*. *Fontqueria* 45: 1–92.

Cervi, A. C., and V. A. de Pessoa S. 1992. *Passiflora farneyi*, a new species of Passifloraceae, subgenus *Passiflora*, series *Serratifoliae*, for Brazil. *Candollea* 47: 631–634.

Cesi, F. 1651. Phytosophicarum tabularum pars I. In *Rerum medicarum Novae Hispaniae thesaurus* by F. Hernandez. Ed. Nardo A. Recchi. Rome: B. Deversinus and Z. Masottus.

Cheesman, E. E. 1940. *Flora of Trinidad and Tobago* 1 (7): 428–436.

Cieza de León, P. de. 1553. *Parte primera de la Chronica del Peru* (First part of the Chronicle of Peru). Seville: M. de Montesdoca. Trans., ed. C. R. Markham, 1864. London: Hakluyt Society.

Colonna, F. 1592. *Phytobasanos sive plantarum aliquot historia*. Naples: J. J. Carlinus.

Coppens d'Eeckenbrugge, G., V. E. Barney, P. M. Jørgensen, and J. M. MacDougal. 2001. *Passiflora tarminiana*, a new cultivated species of *Passiflora* subgenus *Tacsonia* (Passifloraceae). *Novon* 11: 8–15.

Corbet, A. S., and H. M. Pendlebury. 1992. *The Butterflies of the Malay Peninsula*. Malayan Nature Society.

Correll, D. S., and H. B. Correll. 1982. Passifloraceae. *Flora of the Bahamas*. 990–997.

Cowley, A. 1668. *Poemata latina in quibus continentur sex libri plantarum*. London: T. Roycroft.

Cusset, G. 1965. Les nectaires extra-floraux et la valeur de la feuille de Passifloracées. *Rev. Gén. Bot.* 72: 145–219.

——. 1967. Les Passifloracées Asiatiques. *Adansonia* 2 (7): 371–385.

——. 1968. Les vrilles de Passifloracées. *Bulletin de la Société Botanique de France* 115: 45–61.

——. 1970. Remarques sur des feuilles de dicotylédones. *Boissiera* 16: 1–210.

Darwin, C. 1884. *The Movements and Habits of Climbing Plants*. 2nd ed. New York: D. Appleton.

de Candolle, A. P. 1822. *Mémoires de la Société de Physique et d'Histoire Naturelle de Gèneve* 1: 434–436.

——. 1828. Passifloraceae. *Prodromus Systematis Naturalis Regni Vegetabilis* 3: 321–338. Paris: Treuttel & Würtz.

Deginani, N. 2001. Las especies Argentinas del género *Passiflora* (Passifloraceae). *Darwiniana* 39 (1–2): 43–129.

Deinert, E. I., J. T. Longino, and L. E. Gilbert. 1994. Mate competition in butterflies. *Nature* 370: 23–24.

de Wilde, W. J. J. O. 1971a. A monograph of the genus *Adenia* Forssk. (Passifloraceae). *Mededeelingen van de Landbouwhogeschool* 71 (18): 1–281.

——. 1971b. The systematic position of tribe Paropsieae, in particular the genus *Ancistrothyrsus*, and a key to the genera of Passifloraceae. *Blumea* 19: 99–104.

——. 1972a. The indigenous Old World Passifloras. *Blumea* 20: 227–250.

——. 1972b. Passifloraceae. In *Flora Malesiana*. Ed. C. G. G. J. Van Steenis. Jakarta: Nordhoff-Kolff N. V.

——. 1974. The genera of the tribe Passifloreae (Passifloraceae) with special reference to flower morphology. *Blumea* 22: 37–50.

Diener, H. 1995. Die Passionsblumen. *PTA heute* 9 (4): 306–307.

Donato d'Eremita. 1619. *Vera effigie della Granadiglia, detta fior della Passione*. Naples: Donato d'Eremita.

——. 1622. *Fnera vix norant Christi crvdelia vates, praecisa cvm tellvs protvlit ante notes. Granadiglia ouero Fior della passione*. Naples: Donato d'Eremita.

Dornelas, M. C., F. C. A. Tavares, J. C. de Oliviera, and M. L. C. Vieira. 1995. Plant regeneration from protoplast fusion in *Passiflora* spp. *Plant Cell Reports* 15: 106–110.

Durkee, L. T. 1977. *Structure and Function of Extra-Floral Nectaries of* Passiflora. Ph.D. Dissertation, University of Iowa.

——. 1981. The floral and extrafloral nectaries of *Passiflora*. I. The floral nectary. *American Journal of Botany* 68: 453–462.

——. 1982. The floral and extrafloral nectaries of *Passiflora*. II. The extrafloral nectary. *American Journal of Botany* 69: 1420–1428.

——. 1983. The ultrastructure of floral and extrafloral nectaries. In *The Biology of Nectaries*, ed. B. Bentley and T. Elias. New York: Columbia University Press. 1–29.

Esau, K. 1965. *Plant Anatomy*. 2nd ed. New York: J. Wiley.

Escobar, L. K. 1980. Interrelationships of the edible species of *Passiflora* centering around *Passiflora mollissima* (H. B. K.) Bailey subgenus *Tacsonia*. Ph.D. Dissertation, University of Texas.

——. 1986. New species and varieties of *Passiflora* (Passifloraceae) from the Andes of South America. *Systematic Botany* 11 (1): 88–97.

——. 1987. A taxonomic revision of the varieties of *Passiflora cumbalensis* (Passifloraceae). *Systematic Botany* 12 (2): 238–250.

——. 1988a. Passifloraceae: *Passiflora*, Subgeneros: *Tacsonia, Rathea, Manicata* y *Distephana*. In *Flora de Colombia*, monographía 10. Ed. P. Pinto and G. Lozano. Bogotá: Instituto de Ciéncias Naturales.

——. 1988b. Novedades en *Passiflora* (Passifloraceae) de Colombia. *Mutisia* 71: 1–8.

——. 1989a. Passionfruits. In *Lost Crops of the Incas: Little Known Plants of the Andes with Promise for Worldwide Cultivation*. Ed. Board on Science and Technology for International Development. Washington: National Academy Press. 287–295.

——. 1989b. A new subgenus and five new species in *Passiflora* (Passifloraceae) from South America. *Annals of the Missouri Botanical Garden* 76: 877–885.

——. 1989c. Two new species of *Passiflora* (Passifloraceae) from South America. *Phytologia* 66 (1): 80–82.

——. 1989d. *Passiflora chlorina*, a new species of subgenus *Astrophea* (Passifloraceae) from the cerrado vegetation of Brazil. *Phytologia* 67 (2): 132–133.

——. 1990. Two new species of *Passiflora* (Passifloraceae) from northern South America. *Phytologia* 69 (5): 364–365.

——. 1992. *Passiflora brachyantha* (Passifloraceae), a new species from the Andes of southern Ecuador. *Novon* 2: 198–200.

——. 1994. Two new species and a key to *Passiflora* subg. *Astrophea*. *Systematic Botany* 19 (2): 203–210.

Ferrari, G. B. 1633. *Flora sev de floram cvltvra libri 4*. Rome: S. Pavlinus.

Feuillet, C. 1986. Etudes sur la flore des Guyanes. 22. Deux Passifloraceae nouvelles et quelques especies rares en Guyane française. *Candollea* 41: 173–178.

——. 1989. Diversity and distribution of Guianan Passifloraceae. In *Tropical Forests: Botanical Dynamics, Speciation and Diversity*. Ed. L. B. Holm-Nielsen. 311–315.

——. 1994. Two new species of *Passiflora* (Passifloraceae) from French Guiana. *Novon* 4: 236–241.

——. 1998. *Passiflora amoena* and *P. fuchsiiflora*. *Passiflora* 8: 3.

——. 2002. A new series and three new species of *Passiflora* subgenus *Astrophea* from the Guianas. *Brittonia* 54: 18–29.

Feuillet C., and G. Cremers. 1984. Studies on the flora of the Guianas. 6: Passifloraceae nouvelles ou meconnues de Guyane française. *Proc. Kon. Ned. Akad. Wetensch. C* 87: 377–386.

Feuillet, C., and J. M. MacDougal. 1996. *Passiflora trialata* (Passifloraceae), a new species of granadilla (*Passiflora* subg. *Passiflora*) from French Guiana. *Novon* 6: 351–355.

——. 1997. New infrageneric names in *Passiflora* (Passifloraceae). *Biollania, Edición Especial* 6: 335–340.

——. 2004. A new infrageneric classification of *Passiflora*. *Passiflora* 14 (1).

Fouque, A. 1982. Quelques *Passiflora* de Guyane. *Fruits* 37 (10): 599–608.

Frank, A., and E. E. Kugler. 2001a. Hybriden und Sorten der Passionsblumen. Eine Checklist der Gattung Passiflora. *Passiflorunde*. Sonderheft 2.

——. 2001b. Hybrids and cultivars of passionflowers. A checklist for the genus *Passiflora*. *Passiflorunde*. Special Issue 2.

Frank, A., and R. Gamma. 2002. A search for passionflowers in Brazil: a journey report. *Passiflora* 12 (1): 3–6, 10–16.

Galarda Varassim, I., J. R. Trigo, and M. Sazima. 2001. The role of nectar production, flower pigments and odour in the pollination of four species of *Passiflora* (Passifloraceae) in south-eastern Brazil. *Botanical Journal of the Linnean Society* 136: 139–152.

Gentry, A. H. 1976. Additional Panamanian Passifloraceae. *Annals of the Missouri Botanical Garden* 63: 341–345.

——. 1981. Distributional patterns and an additional species of the *Passiflora vitifolia* complex: Amazonian species diversity due to edaphically differentiated communities. *Plant Systematics and Evolution* 137: 95–105.

Gerlach, G., and T. Ulmer. 2000. Una nueva especie de *Passiflora*, subgénero *Passiflora* (Passifloraceae) del oeste de Colombia. *Caldasia* 22 (2): 231–235.

Gilbert, L. E. 1971. *Influence of Resource Distribution and Abundance on Butterfly Population Structure.* Ph.D. Dissertation, University Microfilms International.

——. 1975. Ecological consequences of a coevolved mutualism between butterflies and plants. In *Coevolution of Animals and Plants.* Ed. L. E. Gilbert and P. R. Raven. Boulder, Colorado: International Congress of Systematics and Evolutionary Biology. Austin: University of Texas Press. 210–240.

——. 1982. The coevolution of a butterfly and a vine. *Scientific American* 247: 110–121.

Gilbert, L. E., and J. M. MacDougal. 2000. *Passiflora microstipula*, a new species of Passifloraceae from southeast Mexico. *Lundellia* 3: 1–5.

Gómez, L. D., and P. Gómez. 1981. A new species of arborescent *Passiflora* (Astrophea) from Costa Rica. *Phytologia* 49 (1): 56–57.

Gottschalk, M. 1991a. Alte und neue Hybriden der Passionsblumen. *Gartenpraxis* 1: 46–51.

——. 1991b. Pflegenswerte Passionsblumen. *Gartenpraxis* 7: 52–59.

Green, P. S. 1972. *Passiflora* in Australasia and the Pacific. *Kew Bulletin* 26 (3): 539–558.

Grüneberg, H. 1994. *Passiflora racemosa. Taspo Gartenbaumagazin* 10: 50.

Guédès, M. 1979. *Morphology of Seed-plants.* Hirschberg: J. Cramer.

Hahn, G., A. Mayer, and H. Soicke. 1983. Bewährte Wirksubstanzen aus Naturstoffen. Passionsblume. notabene medici. *Journal für Ärzte. Sonderdruck.* 13: 937–944.

Harms, H. 1893a. Über die Verwertung des anatomischen Baues für die Umgrenzung und Einteilung der Passifloraceae. *Botanisches Jahrbuch für Systematik, Pflanzengeschichte und Pflanzengeographie* 15: 578–633.

——. 1893b. Passifloraceae. In *Die natürlichen Pflanzenfamilien* 3 (6a). Ed. H. Engler and K. Prantl. Leipzig, Germany: W. Engelmann. 69–82.

——. 1898. Zur Morphologie der Ranken und Blütenstände bei den Passifloraceae. *Botanische Jahrbücher für Systematik* 24: 163–178.

——. 1925. Passifloraceae. In *Die natürlichen Pflanzenfamilien* 21, 2nd ed. Ed. H. Engler and K. Prantl. Leipzig, Germany: W. Engelmann. 470–507.

Hegnauer, R. 1990. Passifloraceae. *Chemotaxonomie der Pflanzen.* Birkhäuser Verlag. 208–215.

Herklots, G. 1976. Passifloraceae. *Flowering tropical climbers.* New York: Science History Publications. 145–158.

Hernández, F. 1651. *Rerum medicarum Novae Hispaniae thesaurus.* Ed. Nardo A. Recchi. Rome: B. Deversinus and Z. Masottus.

Hickey, L. J., and J. A. Wolfe. 1975. The bases of angiosperm phylogeny: vegetative morphology. *Annals of the Missouri Botanical Garden* 62: 538–589.

Holm-Nielsen, L. B., P. M. Jørgensen, and J. E. Lawesson. 1988. Passifloraceae. In *Flora of Ecuador* 31. Ed. G. Harling and L. Andersson. Copenhagen: Nordic Journal of Botany.

Holzinger, H., and R. Holzinger. 1994. *Heliconius* and related genera. *Sciences Nat. Venette.* France.

Howell, C. W. 1976. Edible fruited *Passiflora* adapted to south Florida growing conditions. *Proceedings of the Florida State Horticultural Society* 89: 236–238.

Hultin, E. 1965. Partition coefficients of ether-extractable passionflower alkaloids. *Acta Chemica Scandinavica* 19: 1431–1434.

Hutchinson, J. 1967. Passifloraceae. *The Genera of Flowering Plants.* Vol. 2. Oxford: Clarendon Press. 364–374.

Iwert, E. 1980. *Das Gleichnis von der Passionsblume.* Kriens, Switzerland: E. Iwert.

Janzen, D. H. 1968. Reproductive behavior in the Passifloraceae and some of its pollinators in Central America. *Behaviour* 32: 33–48.

Jørgensen, P. M., J. E. Lawesson, and L. B. Holm-Nielsen. 1984. A guide to collecting passion-flowers. *Annals of the Missouri Botanical Garden* 71: 1172–1174.

Jørgensen, P. M., and J. M. MacDougal. 1997. Three new species of *Passiflora* (Passifloraceae) from Ecuador and notes on *Passiflora viridescens. Novon* 7: 379–386.

Jussieu, A. L. de. 1789. *Genera Plantarum Secundum Ordines Naturales Disposita.* 8 vols. Paris.

——. 1805. Premier mémoire sur quelques nouvelles espéces du genre *Passiflora*, et sur la nécessité d'etablir une famille des Passiflorées. *Annales du Muséum d'Histoire Naturelle* 6: 102–116.

Kay, E. 2001. Observations on the pollination of *Passiflora penduliflora. Biotropica* 33: 709–713.

Killip, E. P. 1938. The American species of Passifloraceae. *Publications of the Field Museum of Natural History, Botanical Series* 19 (1, 2): 1–613.

——. 1939a. *Passiflora pachyantha. Lloydia* 2 (3): 198–199.

——. 1939b. *Passiflora cardonae. Brittonia*: 172.

——. 1947. *Passiflora macrostemma. Publications of the Field Museum of Natural History, Botanical Series* 23: 217–218.

——. 1966. Passifloraceae. In *Flora of Suriname.* Ed. A. Pulle. Reprint. Leiden, Netherlands. 307–327.

Killip, E. P., and J. Cuatrecasas. 1960. Supplemental notes on the American species of Passifloraceae, with descriptions of new species. *Contributions from the United States National Herbarium* 35 (1): 1–23.

Kirchoff, B. 1984. On the relationship between phyllotaxy and vasculature: a synthesis. *Botanical Journal of the Linnean Society* 89: 37–51.

Klock, P. 1996. *Das große Buch der Passionsblumen.* Hamburg: Lagerstroemia.

Klucking, E. P. 1992. Passifloraceae. *Leaf Venation Patterns.* Vol. 6. Berlin: J. Cramer. 222–262.

Knapp, S., and J. Mallet. 1984. Two new species of *Passiflora* (Passifloraceae) from Panama, with comments on their natural history. *Annals of the Missouri Botanical Garden* 71: 1068–1074.

——. 1998. A new species of *Passiflora* (Passifloraceae) from Ecuador with notes on the natural history of its herbivore, *Heliconius* (Lepidoptera: Nymphalidae: Heliconiiti). *Novon* 8: 162–166.

Knight, R. J. 1994. Problems and opportunities in passion fruit culture and development. *Fruit Varieties Journal* 3: 159–162.

Knight, R. J., and H. F. Winters. 1962. Pollination and fruit set of yellow passionfruit in southern Florida. *Proceedings of the Florida State Horticultural Society* 75: 412–418.

Knock, F. 1965. *Passifloras for Your Garden.* Kansas City, Missouri: Diversity Books.

Kuechen, U. B. 1989. Das späte Eindringen der Passionsblume in den "Mundus symbolicus." In *Festschrift für Herbert Kolb.* Ed. K. Matzel and H. G. Roloff. Bern: P. Lang. 361–388.

Kugler, E. E. 1995. Passifloraceen Passionsblumen. Kurze Geschichte der Namensgebung 1/2. *Passiflorunde* 3 (2): 5–12, 3 (3): 6–13.

——. 2002. Passionsblumen in der Dichtung. 1. Fruchtbringende Gesellschaft / Pegnesischer Blumenorden. *Passiflorunde* 10 (2): 2–9.

Kugler, E. E., and W. Wetschnig. 1991. Bibliography and nomenclature of *Passiflora violacea* Loisel. and *P. amethystina* J. C. Mikan (*P. violacea* Vell.). *Linzer Biologische Beiträge* 23 (2): 753–774.

Labillardière, J. J. H. de. 1825. *Sertum Austro-Calédonicum* 78: 29–31.

Lamarck, J. B. Monnet de. 1789. *Encyclopédie Méthodique, Botanique* 3 (1). Paris: Panckoucke.

La Rosa, A. M. 1984. *The Biology and Ecology of* Passiflora mollissima *in Hawaii.* Technical Report 50. Cooperative National Park Resources Studies Unit. Honolulu: University of Hawaii at Manoa.

Lawrence, G. H. M. 1960a. Names of *Passiflora* hybrids. *Baileya* 8: 118–120.

——. 1960b. Identification of the cultivated passionflowers. *Baileya* 8: 121–132.

Lindley, J. 1824. Masters's hybrid Passiflora: *P. alatocaerulea. Botanical Register* 10.

Lindman, C. A. M. 1906. Zur Kenntnis der Corona einiger Passifloren. In *Botaniska Studier Tillägnade F. R. Kjellman.* Ed. R. Sernander, N. Svedelius, and C. O. Norén. Uppsala. 55–79.

Linnaeus, C. [Linné, C.]. 1742. *Genera Plantarum.* Leiden, Netherlands: C. Wishoff.

——. 1745. *Dissertatio Botanica de* Passiflora, *quam cum consensu amplis.* Stockholm: L. Salvius. Written by J. G. Hallmann under the direction of Linnaeus.

——. 1753. *Species Plantarum.* Vol. 2. Stockholm: L. Salvius.

Lutomski, J. 1959. *Biul. Inst. Roslin Lecniczych* 5: 197. Cited in Ambühl 1966.

MacDougal, J. M. 1984. *Revision of Passiflora Section Pseudodysosmia.* Ph.D. Dissertation, Duke University.

——. 1988a. *Passiflora pusilla* (Passifloraceae), a new species from Central America. *Annals of the Missouri Botanical Garden* 75: 392–395.

——. 1988b. *Passiflora eglandulosa*, a new species in section *Cieca* formerly included with *P. trinifolia* Masters. *Annals of the Missouri Botanical Garden* 75: 1658–1662.

——. 1989a. *Passiflora citrina*, a new species in section *Xerogona* (Passifloraceae) from Mesoamerica. *Annals of the Missouri Botanical Garden* 76: 354–356.

——. 1989b. Two new species of *Passiflora* section *Decaloba* (Passifloraceae) from Costa Rica. *Annals of the Missouri Botanical Garden* 76: 608–614.

——. 1989c. *Passiflora mayarum*, a new species related to *P. prolata* in subgenus *Passiflora* (Passifloraceae). *Annals of the Missouri Botanical Garden* 76: 615–618.

——. 1989d. *Passiflora malletii*, a new species in section *Decaloba* (Passifloraceae) from Mesoamerica. *Annals of the Missouri Botanical Garden* 76: 1172–1174.

——. 1989e. The correct name for the hybrid *Passiflora*

alata × *P. caerulea* (Passifloraceae). *Baileya* 23(2): 68–73.

——. 1992. New species of *Passiflora* subgenus *Plectostemma* (Passifloraceae). *Novon* 2: 358–367.

——. 1994. Revision of *Passiflora* subgenus *Decaloba* section *Pseudodysosmia* (Passifloraceae). *Systematic Botany Monographs* 41: 1–146.

——. 1995. *Passiflora sanctae-mariae* in series *Luteae* (Passifloraceae), a new species and series in subgenus *Decaloba*. *Novon* 5: 48–51.

——. 2001. Two new species of passionflower (*Passiflora*, Passifloraceae) from southwestern Mexico. *Novon* 11: 69–75.

MacDougal, J. M., and A. K. Hansen. 2003. A new section of *Passiflora*, subgenus *Decaloba* (Passifloraceae), from Central America, with two new species. *Novon* 13: 459–466.

Mang, C. 1609. *Abriß der Blü vnd Frucht, die in den Niedergängigen Indien wächst*. Augsburg, Germany: D. Custos.

Marcgrav von Liebstad, G., and W. Piso. 1648. *Histora rerum naturalium Brasiliae*. Leiden, Netherlands: F. Hack. Amsterdam: D. Elzevir.

Markham, C. R., trans., ed. 1864. *The Travels of Pedro de Cieza de León*. London: Hakluyt Society.

Martin, F. W., and H. Y. Nakasone. 1970. The edible species of *Passiflora*. *Economic Botany* 24: 333–343.

Masters, M. T. 1871. Contributions to the natural history of the Passifloraceae. *Transactions of the Linnean Society of London* 27: 593–645.

——. 1872. Passifloraceae. In *Flora Brasiliensis* 13 (1). Ed. K. Martius and A. Eichler. 529–628.

——. 1877. A classified synonymic list of all the species of Passifloreae cultivated in European gardens, with references to the works in which they are figured. *Journal of the Royal Horticultural Society* 4: 125–149.

May, P. G., and E. E. Spears. 1988. Andromonoecy and variation in phenotypic gender in *Passiflora incarnata* (Passifloraceae). *American Journal of Botany* 75: 1830–1841.

McGuire, C. M. 1999. *Passiflora incarnata* (Passifloraceae), a new fruit crop. *Economic Botany* 53 (2): 161–176.

Medikus, F. K. 1787. *Über einige künstliche Geschlechter aus der Malven-Familie*. Mannheim, Germany: Neue Hof- und akademische Buchhandlung.

Meerman, J. C. 1993. Finding the unknown. *Passiflora* 3 (2): 1–2.

——. 1996a. Description of *Passiflora pittieri* from Belize. *Passiflora* 6 (1): 3–4.

——. 1996b. Vegetative key to the passionflowers of Belize. *Passiflora* 6 (3): 25–28.

——. 1997. *Passiflora obovata* from Belize. *Passiflora* 7 (2): 14, 17–18.

Mendel, G. 1865. *Experiments in Plant Hybridization*. http://www.netspace.org/MendelWeb/Mendel.plain.html

Miller, J. S. 1996. Phylogeny of the Neotropical moth tribe Josiini [Notodontidae: Dioptinae]: a hidden case of Mullerian mimicry. *Zoological Journal of the Linnean Society* 118: 1–45.

Monardes, N. 1569. *Dos libros, el uno que trata de toda las cosas que traen de nuestras indias occidentales, que sirven al uso de la medicina*. Seville: F. Diaz.

——. 1574. *Historia medicinal de las cosas que se traen de nuestras Indias occidentales que sirven al uso medicina*. 2nd ed. Seville: F. Diaz.

——. 1596. *Ioyfvll newes out of the new-found worlde, wherein are declared the rare and singular vertues of divers herbs*. Trans. John Frampton. London: E. Alldee.

Moore, S. L. 1982. *The Legend of the Passionflower*. Bradenton, Florida: Maracujá.

Morton, J. F. 1987. *Fruits of Warm Climates*. Miami, Florida: J. F. Morton. 320–333.

Murawski, D. A., and L. E. Gilbert. 1986. Pollen flow in *Psiguria warscewiczii*: a comparison of *Heliconius* butterflies and hummingbirds. *Oecologia* 68: 161–167.

Neu, R. 1954. *Arzneimittelforschung* 4: 601.

Otoni, W. C., N. W. Blackhall, F. B. d'Utra Vaz, V. W. Casali, J. B. Power, and M. R. Davey. 1995. Somatic hybridization of the *Passiflora* species, *P. edulis* f. *flavicarpa* Degener, and *P. incarnata* L. *Journal of Experimental Botany* 46 (288): 777–785.

Otoni, W. C., G. Manders, F. B. d'Utra Vaz, N. W. Blackhall, J. B. Power, and M. R. Davey. 1994. Transformation of passionfruit (*Passiflora edulis* f. *flavicarpa* Degener) using *Agrobacterium tumefaciens*. *Plant Cell Reports* 13: 697–702.

Panero, J. L. 1996. *Passiflora linda*, a new species from southern Ecuador. *Brittonia* 48 (2): 192–194.

Parkinson, J. 1629. *Paradisi in sole paradisus terrestris, or A Garden of All Sorts of Pleasant Flowers*. London: H. Lowes and R. Young.

Parlasca, S. 1609. *Il fiore della granadiglia, overo della passione di nostro signore Giesu Christo*. Bologna: B. Cocchi.

Peckoldt, T. 1909. Heil- und Nutzpflanzen Brasiliens Passifloraceae. *Berichte der Deutschen Pharmazeutischen Gesellschaft* 19: 343–363.

Pessoa S., V. A. de, and A. C. Cervi. 1992. *Passiflora farneyi*, a new species of Passifloraceae, subgenus

Passiflora, series *Serratifolia*, for Brazil. *Candollea* 47 (2): 631–634.

Petrelli, E. 1610. Vera narratio fruticis florum et fructuum novissime in Occidentalibus Indijs nascerntium. In *Cultura ingeniorum*, 7th ed. Ed. A. Possevini. Cologne: J. Gymnicus.

Plukenet, L. 1691. *Phytographia*. London: L. Plukenet.

———. 1696. *Almagestum botanicum*. London: L. Plukenet.

Plumier, C. 1693. *Description des plantes de l'Amérique*. Paris: Imprimerie Royale.

Pomié, P. 1996. Les Passiflores. *Hommes & Plantes* 18: 24–28.

Porter-Utley, K. E. 2003. Revision of *Passiflora* Subgenus *Decaloba* Supersection *Cieca* (Passifloraceae). Ph.D. Dissertation, University of Florida, Gainesville.

Presting, D. 1965. Zur Morphologie der Pollenkörner der Passifloraceen. *Pollen and Spores* 7: 193–247.

Proctor, G. R. 1982. *Passiflora calcicola*. Flora of Jamaica. *Journal of the Arnold Arboretum* 63: 270.

Puri, V. 1948. Studies in floral anatomy. V. On the structure and nature of the corona in certain species of the Passifloraceae. *Journal of the Indian Botanical Society* 17: 130–149.

Rabel, D. 1622. *Theatrum Florae*. Paris: N. Mathonière.

Rafinesque, C. S. 1838. Passiflorea. *Flora Telluriana* 4. Philadelphia, Pennsylvania: H. Probasco. 102–104.

Rasciotti, D. 1609. *Copia del fiore et frutto, che nasce nelle Indie occidentali*. Venice.

Rehwald, A., and O. Sticher. 1995. Trace analysis of harman alkaloids in *Passiflora incarnata* by reversed-phase high-performance liquid chromatography. *Phytochemical Analysis* 6: 96–100.

Reichenbach, H. G. L. 1828. Passifloraceae. *Conspectus Regni Vegetabilis*. Leipzig, Germany: Carl Cnobloch. 132.

Roemer, M. J. 1846. Passifloraceae. *Familiarum Naturalium Regni Vegetabilis Synopses Monographicae* 2. Weimar, Germany. 125–207.

Roth, I. 1974. Morphología, anatomía, y desarollo de la hoja pinnada y de las glandulas laminales en *Passiflora* (Passifloraceae). *Acta Botánica Venezuelica* 9: 363–380.

Roussel, H. F. de. 1806. *Flore du Calvados et terrains adjacents, composée suivant le méthode de Jussieu*. Caen: L. J. Poisson.

Sabine, J. 1822. Account of a newly produced hybrid *Passiflora*. *Transactions of the Horticultural Society of London* 4: 258–268.

Sacco, J. C. 1967. Contribuicao ao estudo das Passifloraceae do Brasil. *Passiflora margaritae Sellowia* 19: 59–61.

———. 1973. Uma nova especie de *Passiflora* da Bolivia. *Bradea* 1 (33): 349–352.

———. 1980. Passifloraceae. In *Flora Ilustrada Catarinense*. Ed. R. Reitz. Itajaí, Santa Catarina.

Sandwith, N. Y. 1956. *Passiflora lonchophora*. *Icones Plantarum London* 36 (3): 1–6.

———. 1960. *Passiflora retipetala* Masters and *P. lonchophora* Harms. *Kew Bulletin* 14: 318.

Sazima, M., and I. Sazima. 1978. Bat pollination of the passionflower, *Passiflora mucronata*, in southeastern Brazil. *Biotropica* 10: 100–109.

Schilcher, H. 1967. Qualitätsprüfung von Handelsdrogen. 4. Mitteilung: Herba Passiflorae. *Deutsche Apotheker-Zeitung*. 849–852.

Schwerdtfeger, M. 1997. Three new species of *Passiflora* from Ecuador. *Haussknechtia* 6: 41–46.

Seidemann, J. 1995. Zur Kenntnis von wenig bekannten exotischen Früchten 6. Mitt.: Curuba (*Passiflora mollissima* (H. B. K.) Bailey). *Deutsche Lebensmittel-Rundschau* 91 (6): 185–190.

Shull, G. H. 1948. What is heterosis? *Genetics* 33: 439–446.

Skrabal, J., H. J. Tillich, and M. Weigend. 2001. A revision of the *Passiflora lobbii* group (Passifloraceae) including some new species and subspecies. *Harvard Papers in Botany* 6 (1): 309–338.

Smith, J. 1884. *Works, 1608–1631*. Ed. Edward Arber. Birmingham: E. Arber.

Snow, A. A. 1982. Pollination intensity and potential seed set in *Passiflora vitifolia*. *Oecologia* 55: 231–237.

Snow, A. A., and R. Gross. 1980. Pollination and seed set in passionflowers: *P. vitifolia*. Corcovado Independent Research, Student Report. *Organizational for Tropical Studies Course Book 80-1*. San José, Costa Rica.

Snow, N., and J. M. MacDougal. 1993. New chromosome reports in *Passiflora* (Passifloraceae). *Systematic Botany* 18: 261–273.

Sowerby, J. 1794. Account of the difference of structure in the flowers of 6 species of *Passiflora*. *Transactions of the Linnean Society of London* 2: 19–28.

Spirlet, M-L. 1965. Utilisation taxonomique de grains de pollen de Passifloracées. I. *Pollen and Spores* 7: 249–301.

Sprengel, C. K. 1793. *Das entdeckte Geheimnis der Natur im Bau und der Befrüchtung der Blüten*. Berlin: Fr. Vieweg.

Standley, P. C., and L. O. Williams. 1961. Passifloraceae. Flora of Guatemala. *Fieldiana: Botany* 24 (7): 115–146.

Steyermark, J. A., and Y. O. Huber. 1978. Passifloraceae. *Flora del Avila*. Caracas: Sociedad Venezolana de Ciencias Naturales. 715–724.

Steyermark, J. A., and B. Maguire. 1967. *Passiflora bomareifolia. Memoirs of the New York Botanical Garden* 17: 455–456.

Stoffers, A. L. 1966. Passifloraceae. *Flora of the Netherlands Antilles.* Vol. 2. Uitg. Natuurwet. Studiekr. Suriname Ned. Ant. 42: 56–64.

Stopp, K. 2001. *Botanische Einblattdrucke und Flugschriften vor 1800.* Band 2. Stuttgart: A. Hiersemann.

Strachey, W. 1849. *The historie of travaile into Virginia Britannica (1610–1612).* Ed. R. H. Major. London: Hakluyt Society.

Sweet, R. 1823–1829. *Passiflora ×colvillii.* In *The British Flower Garden,* vol. 2. London: W. Simpkin and R. Marshall.

——. 1825. Description of a beautiful new hybrid species of passionflower. *The News of Literature and Fashion* 3 (6): 94.

Tillett, S. S. 1970. *Passiflora.* In *Flowering Vines of the World* by D. E. Menninger. New York: Hearthside Press. 258–267.

——. 1988. Passionis Passifloris II. Terminologia. *Ernstia* 48: 1–40.

——. 2003. Passifloraceae. In *Flora of the Venezuelan Guayana* 7. Ed. P. E. Berry, K. Yatskievych, and B. K. Holst. St. Louis: Missouri Botanical Garden Press. 625–667.

Triana, J. J., and J. E. Planchon. 1873. Passifloraceae. *Annales des Sciences Naturelles: Botanique* 5 (17): 121–186.

Turre, G. A. 1662. *Catalogus plantarum horti patavini novo incremento locupletior.* 2nd ed. Padua: Trambotti.

Ulmer, B., and T. Ulmer. 1997. *Passionsblumen: Eine faszinierende Gattung.* Witten, Germany: T. and B. Ulmer.

——. 1999. *Faszinierende Pflanzenwelt: Passionsblumen.* Witten, Germany: Formosa.

Ulmer, T. 1995. Schmackhafte Passionsfrüchte. *Gartenpraxis* 9: 54–59.

——. 1996. Vermehrung der Passionsblumen. *Gartenpraxis* 9: 53–55.

——. 1999. *Passiflora formosa* sp. nov., a hitherto misunderstood taxon in *Passiflora* subgenus *Tacsonia* (Passifloraceae) from Colombia. *Edinburgh Journal of Botany* 56 (2): 195–198.

Ulmer, T., and M. Schwerdtfeger. 2000. A new species of *Passiflora* subgenus *Tacsonia* (Passifloraceae) from the Andes of Peru. *Nordic Journal of Botany* 20 (1): 47–49.

Ulmer, T., and B. Ulmer. 1997. *Passionsblumen: Blütenzauber für Jedermann.* Witten, Germany: T. and B. Ulmer.

Uribe, L. U. 1955. Passifloráceas. In *Flora de la Real Expedición Botánica del Nuevo Reino de Granada,* tomo 27. Madrid: Ediciones Cultura Hispánica.

——. 1957. Una nueva e interesante Passiflora de Colombia. *Caldasia* 7 (35): 335–338.

——. 1958. *Passiflora albicans. Caldasia* 8 (37): 127–128.

Vanderplank, J. 1991. *Passion Flowers and Passion Fruit.* London: Cassell. Cambridge, Massachusetts: MIT Press.

——. 1994. A growing passion. *The Garden* 119 (1): 28–33.

——. 1996a. *Passion Flowers.* 2nd ed. Kingston Seymour, Clevedon, United Kingdom: National Collection of Passiflora.

——. 1996b. *Passiflora quadrifaria* (Passifloraceae). *Curtis's Botanical Magazine* 13: 63–69.

——. 1998. *Passiflora cuneata* 'Miguel Molinari' (Passifloraceae). *Bentham-Moxon Trust.* 109–114.

——. 2000. *Passion Flowers.* 3rd ed. Cambrige, Massachusetts: MIT Press.

——. 2003. The International *Passiflora* Register. Passiflora Society International.

Vásquez Ch., R. 1994. *Passiflora nigradenia* (Passifloraceae) especie poco conocida de Bolivia. *Revista de la Sociedad de Estudios Botanicos* 1 (1): 49–55.

——. 1998. Las especies de *Passiflora* subgénero *Granadilla* serie *Laurifoliae* (Passifloraceae) en Bolivia. *Revista de la Sociedad Boliviana de Botanica* 2 (1): 36–45.

Vieira, M. L. C., and M. C. Dornelas. 1996. Regeneration of plants from protoplasts of *Passiflora* species (Passion Fruit). In *Biotechnology in Agriculture and Forestry,* 38, *Plant Protoplasts and Genetic Engineering* VII. Ed. Y. P. S. Bajaj, ed. 108–119.

Vitta, F. A. 1997. *Passiflora loefgrenii* (Passifloraceae), a new species in subgenus *Passiflora* from the Brazilian Atlantic Rainforest. *Novon* 7: 210–212.

Wehtje G., R. B. Reed, and R. R. Dute. 1985. Reproductive biology and herbicidal sensitivity of maypop passionflower (*Passiflora incarnata*). *Weed Science* 33: 284–290.

Williams, K. S., and L. E. Gilbert. 1981. Insects as selective agents on plant vegetative morphology: egg mimicry reduces egg laying by butterflies. *Science* 212: 467–469.

Woodson, R. E., Jr., and R. W. Schery. 1958. Passifloraceae. Flora of Panama. *Annals of the Missouri Botanical Garden* 45: 1–22.

Wörle, M. 1609. *Warhafftes Contrafet eines vberauß seltzamen und Wunderbarlichen Gewächß.* Augsburg, Germany: M. Wörle.

Index